SUBMICRON
INTEGRATED CIRCUITS

SUBMICRON INTEGRATED CIRCUITS

Edited by

R. K. Watts

WILEY

A Wiley-Interscience Publication

JOHN WILEY & SONS

New York · Chichester · Brisbane · Toronto · Singapore

Library of Congress Cataloging in Publication Data:

Submicron integrated circuits/edited by R. K. Watts.
        p.   cm.
    "A Wiley-Interscience publication."
    Includes bibliographies and index.
    ISBN 0-471-63523-5
    1. Integrated circuits—Design and construction. I. Watts, R. K.
(Roderick K.), 1939–
TK7874.I5469 1989
621.381'73—dc19   88-30275
                                                           CIP

Printed in the United States of America

10 9 8 7 6 5 4 3 2 1

Contributors

MASAYUKI ABE,   Fujitsu Laboratories Ltd., Fujitsu Limited, Atsugi 243-01, Japan

L. A. AKERS,   Center for Solid State Electronics Research, Arizona State University, Tempe, AZ 85287-6206

FABIO BELTRAM,   AT&T Bell Laboratories, Murray Hill, NJ 07974

G. H. BERNSTEIN,[†]   Center for Solid State Electronics Research, Arizona State University, Tempe, AZ 85287-6206

J. R. BREWS,   AT&T Bell Laboratories, Murray Hill, NJ 07974

FEDERICO CAPASSO,   AT&T Bell Laboratories, Murray Hill, NJ 07974

ALFRED Y. CHO,   AT&T Bell Laboratories, Murray Hill, NJ 07974

D. K. FERRY,   Center for Solid State Electronics Research, Arizona State University, Tempe, AZ 85287-6206

R. O. GRONDIN,   Center for Solid State Electronics Research, Arizona State University, Tempe, AZ 85287-6206

AVID KAMGAR,   AT&T Bell Laboratories, Murray Hill, NJ 07974

JUNJI KOMENO,   Fujitsu Laboratories Ltd., Fujitsu Limited, Atsugi 243-01, Japan

KAZUO KONDO,   Fujitsu Laboratories Ltd., Fujitsu Limited, Atsugi 243-01, Japan

TAKASHI MIMURA,   Fujitsu Laboratories Ltd., Fujitsu Limited, Atsugi 243-01, Japan

K. K. NG,   AT&T Bell Laboratories, Murray Hill, NJ 07974

SUSANTA SEN,[‡]   AT&T Bell Laboratories, Murray Hill, NJ 07974

RAMAUTAR SHARMA,   AT&T Bell Laboratories, Murray Hill, NJ 07974

[†]Present address: Department of Electrical and Computer Engineering, University of Notre Dame, Notre Dame, IN 46556.

[‡]On leave from the Institute of Radio Physics and Electronics, University of Calcutta, Calcutta 700 009, India.

MICHAEL SHUR,   Department of Electrical Engineering, University of Minnesota, Minneapolis, MN 55455

DENNY D. TANG,   IBM Thomas J. Watson Research Center, Yorktown Heights, NY 10598

R.K. WATTS,   AT&T Bell Laboratories, Murray Hill, NJ 07974

Preface

Three decades after the invention of the integrated circuit the technology of integrated circuits continues to advance rapidly. One of the main trends has been the scaling down of integrated circuit features to ever smaller size. This book treats aspects of the scaling process. It is intended for engineers, scientists, and managers in the field of integrated circuits. Graduate students and faculty will also find it worthwhile, although it is not a text book.

The first two chapters deal with silicon MOSFETs and bipolar transistors. Chapter 3 treats the variety of GaAs and mixed crystal III–V devices. Chapter 4 covers circuits made with III–V high electron mobility or HEMT transistors. Chapter 5 deals with resonant tunneling devices, a relatively new field. One of the most important aspects of circuit design is the layout of the interconnecting wiring. Chapter 6 treats the modeling of interconnect. Chapters 7, 9, and 10 discuss circuit architecture, Chapters 9 and 10 being dedicated to neural networks. Chapter 8 treats devices which may represent the next generation of integrated circuits. Chapters 11 and 12 deal with two important parts of wafer processing — rapid thermal processing and lithography. For smaller dimensions thermal cycles must be restricted to control diffusion. Lithographic equipment represents the largest fraction of the cost of a modern integrated circuit factory.

R. K. WATTS

Contents

Introduction   1

R. K. Watts

    References   8

1 The Submicrometer Silicon MOSFET   9

J. R. Brews, K. K. Ng, and R. K. Watts

  1.1   Introduction   9
  1.2   Overview of the Environment   10
  1.3   Scaling   15
  1.4   HMOS Design   19
  1.5   Parasitic Resistance of Source and Drain   37
  1.6   Transport   42
  1.7   Hot-Carrier Degradation   51
  1.8   Device Designs   62
  1.9   Closing Remarks   75
       References   76

2 Scaling the Silicon Bipolar Transistor   87

Denny D. Tang

  2.1   Introduction   87
  2.2   Bipolar Transistor Design and Scaling   88
  2.3   Scaling Properties and Limits   99
  2.4   Transistor Structures   104
  2.5   Bipolar Circuits   107
  2.6   Future Possibilities   113
       References   116

3 Submicron GaAs, AlGaAs/GaAs, and AlGaAs/InGaAs Transistors   122

Michael Shur

  3.1   Introduction   122

3.2   Scaling Considerations for Field-Effect Transistors   126
3.3   Current–Voltage Characteristics of Submicron GaAs and
       AlGaAs/GaAs Field-Effect Transistors   131
3.4   Short-Channel Effects in Submicron GaAs and AlGaAs Field-Effect
       Transistors   146
3.5   Vertical Device Structures   153
3.6   Conclusion   168
       References   169

4 Ultrahigh-Speed HEMT LSI Circuits   **176**
       Masayuki Abe, Takashi Mimura, Kazuo Kondo, and Junji Komeno

4.1   Introduction   176
4.2   Performance Advantages of HEMT Approaches   177
4.3   HEMT Technology for VLSI   184
4.4   HEMT LSI Circuit Implementations   190
4.5   Summary   198
       References   201

5 Resonant Tunneling Devices and Their Applications   **204**
       Federico Capasso, Susanta Sen, Fabio Beltram, and Alfred Y. Cho

5.1   Introduction   204
5.2   Resonant Tunneling Diodes   205
5.3   Resonant Tunneling Bipolar Transistor (RTBT)   228
5.4   Resonant Tunneling Unipolar Transistors   243
       References   264

6 Electrical Modeling of Interconnections   **269**
       J. R. Brews

6.1   Introduction   269
6.2   *RC* Line Models   276
6.3   *RLC* Line Models   283
6.4   *RLCG* Line Models   294
6.5   Finding Interconnection Parameters   297
6.6   Pulse Propagation   307
6.7   Crosstalk   312
6.8   Closing Remarks   321
       References   322

7 Impact of VLSI Technology Scaling on Computer Architectures   **332**

    Ramautar Sharma

    7.1   Introduction   332
    7.2   Historical Perspective   334
    7.3   Classification of Computer Architectures   339
    7.4   CPU Requirements and VLSI Technology   341
    7.5   Scaling Upsets Architectural Balance   351
    7.6   GaAs Digital ICs for High-Performance Computers   353
    7.7   Applications of VLSI Technology   353
    7.8   Summary   355
         References   357

8 Lateral Surface Superlattices   **360**

    D. K. Ferry and G. H. Bernstein

    8.1   Lateral Superlattices   361
    8.2   GaAs Structures   364
    8.3   Transport Theory   369
    8.4   Conclusions   374
         References   375

9 Two-Dimensional Automata in VLSI   **377**

    D. K. Ferry, R. O. Grondin, and L. A. Akers

    9.1   Introduction   377
    9.2   Massively Interconnected Systems   387
    9.3   Chip Architectures   397
    9.4   The Future   409
         References   410

10 VLSI Electronic Neural Networks   **413**

    Ramautar Sharma

    10.1   Introduction   413
    10.2   Basic Neural Models   414
    10.3   Neural Networks   416
    10.4   Neural Network Implementations   421
    10.5   Some Applications   431
    10.6   Summary   432
         References   432

11 Rapid Thermal Processing of Silicon   **434**

    Avid Kamgar

  11.1   Introduction   434

  11.2   Advantages of Rapid Thermal Processing   435

  11.3   Fundamentals of Rapid Thermal Processing   435

  11.4   Rapid Thermal Processing Equipment   439

  11.5   Temperature Measurements   440

  11.6   Stress Due to Rapid Thermal Anneal   441

  11.7   Gate Dielectrics   442

  11.8   Post-Implantation Anneal   445

  11.9   Junction Formation   452

  11.10 Polycrystalline Silicon Anneal   455

  11.11 Silicide Formation   460

  11.12 Glass Reflow   463

  11.13 *In Situ* Processing   463

  11.14 Conclusions   464

        References   465

12 Lithography   **470**

    R.K. Watts

  12.1   Introduction   470

  12.2   Optical Lithography   470

  12.3   Electron Lithography   482

  12.4   X-Ray Lithography   488

  12.5   Ion Lithography   494

  12.6   Metrology   499

  12.7   Conclusion   503

        References   504

Index   **509**

SUBMICRON
INTEGRATED CIRCUITS

Introduction

R. K. WATTS

AT&T Bell Laboratories
Murray Hill, New Jersey

The era of modern microelectronics began in 1958 with the invention of the integrated circuit by J. S. Kilby of Texas Instruments.[1] His first chip is shown in Fig. I.1 For comparison, Fig. I.2 shows a modern microprocessor chip.[2] After the announcement of the integrated circuit in 1959, three objections were immediately raised. (1) The use of materials was not optimum. For example, better resistors could be made with nichrome wire than with silicon. (2) Such circuits could not be produced because the yield would be too low. (3) Designs would be expensive and difficult to change. These objections all contained elements of truth, but they reflected an unwarranted pessimism which kept several large corporations out of the new field for many years.

Integrated circuits moved rapidly into production because the manufacturing technology was similar to that already in place for making discrete devices. In 1961, Fairchild announced a family of digital circuits, and Texas Instruments delivered to the U.S. Air Force a small computer with several hundred bits of semiconductor memory. NASA and Air Force contracts helped the young technology move along a learning curve. Direct experience contradicted objection number two.

During the late 1960s, sales of integrated circuits grew rapidly, and they found application in data processing, communications, and industrial products. In the late 1960s and early 1970s, many new companies entered the field. MOS circuits began rising rapidly in production, and many U.S. companies began foreign assembly of products.

In Japan, integrated circuit production began in 1963. Without the military market, Japanese companies relied on consumer products, the first being the electronic calculator, which emerged in 1964. That year, further stimulus was provided by two events. IBM introduced the System 360 computer, and Texas Instruments (TI) sought a wholly owned Japanese subsidiary. The next year, TI applied for a Japanese patent for integrated circuit technology. This action was known in Japan as the "TI Shock." Part of the Japanese reaction to these events was the recommendation of a national computer strategy and the funding of a cooperative high-performance computer project by the Ministry of International Trade and Industry

Fig. I.1   The first integrated circuit. The chip contains one transistor, three resistors, and one capacitor. (Figure provided by J. S. Kilby.)

(MITI). Japanese manufacturers emphasized high-volume production of MOS chips for calculators, capturing 50% of the Japanese market for LSI calculators by 1973. In 1976, the cooperative VLSI Technology Research Association was formed in Japan to expedite development of manufacturing methods and basic technology. By 1980, this project had produced more than 1000 patents. This was the most successful of a number of such joint projects in Japan. In the United States, two cooperative efforts appeared. In 1984, the Semiconductor Research Corporation was established to disperse industry funds to universities for research in semiconductors. And in 1987, Sematech was founded — an industry consortium whose purpose is to develop improved chip manufacturing techniques.

In the late 1970s, the industry and competition became more global. Figures I.3 and I.4 illustrate the trends of market share for the major geographical areas and for four of the largest manufacturers. The United States has steadily lost market share to Japan. The European share has remained nearly constant, and the share held by the rest of the world (not shown) is small but rapidly growing.

In the early 1980s, the complexity of circuits continued to increase. Thirty-two bit microprocessors appeared, and dynamic random access memories (DRAMs) grew in size. Peak production of 16 kbit DRAMs occurred in 1982, and the ratio of U.S. production to Japanese production at the peak was 2. For the peak of 64 kbit DRAM production in 1984, this ratio had dropped to 0.7, and later in

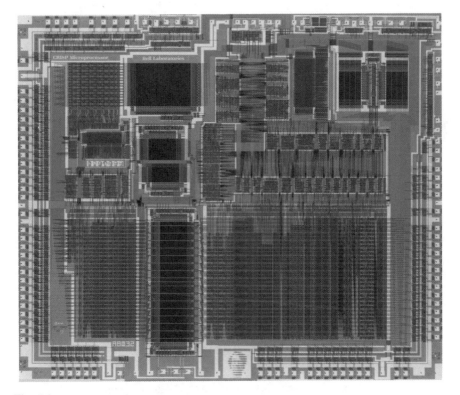

Fig. I.2 A 32 bit microprocessor from AT&T. Die size is 1.04×1.22 cm. The CMOS chip contains 1.7×10^5 transistors. (Figure provided by K. J. O'Connor.)

256 kbit DRAMs Japanese dominance increased dramatically. Giving impetus to this trend was the base laid for efficient manufacturing: from 1978 to 1982, total capital spending by the nine largest Japanese producers had increased from 0.2 to 0.9 billion dollars. The government had become convinced that the pervasive influence of microelectronics and computers would have large beneficial effects throughout most of the economy. In 1979, the industrial plan drafted by MITI specified the industries that would benefit most: telecommunications, automobiles, machine tools, robotics, ocean development, aerospace, nuclear engineering, and bioengineering.

If the integrated circuit industry is well known for rapid progress, it is also notorious for turbulence. The ascending curve of sales growth in Fig. I.5 is compared with a similar curve for a mature business, the U.S. automobile industry. The turbulence masked by the first curve is revealed in the other curve plotted in Fig. I.5, the "book-to-bill ratio," the ratio of orders booked to billings or sales. During an industry boom, shortages develop. Users place multiple orders for the same part when only a single part is needed. Manufacturers increase capital spending and add manufacturing capacity. The peak in demand passes. Orders are can-

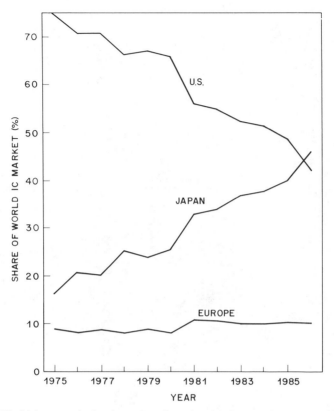

Fig. I.3 World integrated circuit market share for the three major geographical areas. (Data from Dataquest, Inc.)

celed. Inventories are reduced. Manufacturers find that they have excess capacity. Large peaks in production with accompanying gyrations of this type occurred in 1974 and 1984. Fluctuating inventories of the purchasers of integrated circuits increase the instability. In 1984, about one-third of the growth in sales went into increased inventories.

The integrated circuit industry is characterized by speed in product development. Other industries also have learning curves, reduce production costs, and offer enhanced products at lower real prices. However, these actions take place with greater speed in this industry.

In the late 1980s, integrated circuit products have split into two main types: expensive chips designed for a particular application and cheap commodity chips manufactured on a larger scale. For the "application specific" chips, we may in the future see more circuit design by end users who have the chips fabricated by specialized "foundries," or companies that process wafers but do not design circuits. Commodity chips continue to evolve also.[3] Because of their various different applications, there is growing demand for a wider variety of on-chip functions

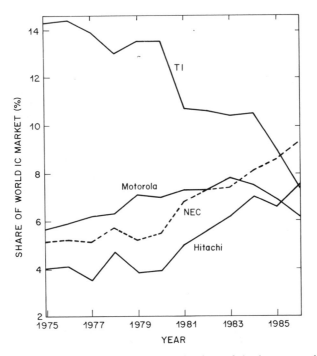

Fig. I.4 World integrated circuit market share for four of the largest producers. (Data from Dataquest, Inc.)

in memories. Large-scale production technology comprising high-throughput equipment, computer-controlled logistics, and robotics becomes of increasing importance.

The growth in circuit density shown in Fig. I.6 was made possible by shrinking sizes of features of circuits. This trend is shown in Fig. I.7. As incremental progress becomes more difficult, the rate of progress usually decreases. This is indicated by the departure from the historical trend, represented by the shaded area. A lower rate of reduction may be brought about by the increasingly difficult fabrication technology. The trends of Figs. I.6 and I.7 have led to a reduction in unit cost, as the cost of processing a wafer has increased more slowly than chip component density. Device performance has also increased. This cost reduction and competition have brought lower selling prices. This, with the increased performance, has led to a lower price per electronic function with higher performance levels.

Integrated circuits may also be divided into two categories by material: silicon, of which the vast majority is made, and gallium arsenide. Gallium arsenide integrated circuit technology grew out of work on GaAs microwave transistors designed for military applications, largely in radar systems. Many small companies, or "start-ups," have appeared to produce GaAs circuits. One of these, Vitesse Electronics, announced in 1986 the first GaAs large-scale integration (LSI) available commercially. The growing array of uses for high-speed GaAs parts will in-

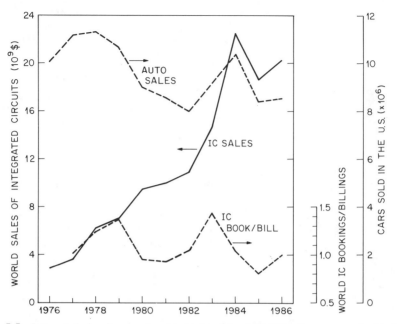

Fig. I.5 Integrated circuit sales (worldwide) and book-to-bill ratio compared with U.S. auto sales.

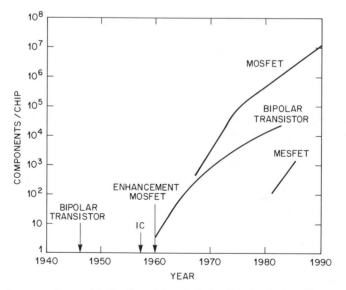

Fig. I.6 Components per chip for three integrated circuit technologies. The curve labeled "MESFET" pertains to GaAs MESFET circuits. (Figure provided by S. M. Sze.)

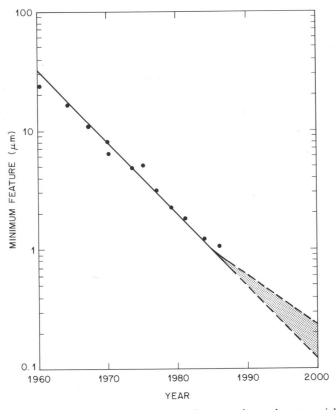

Fig. I.7 Trend of size of the smallest feature on the most advanced commercial integrated circuits. (Figure provided by S. M. Sze.)

clude applications in real-time signal processing, computer register files for fast machines, cache memory, and high-performance graphics systems. In addition to speed, radiation hardness is another advantage of GaAs circuits, important in some applications.

Still another way to divide circuits into two categories is by speed of operation. The very high speed circuits — above 1 Gbit/s or 1 GHz — are largely GaAs, although silicon bipolar circuits and silicon NMOS circuits occur in this category also. The level of integration of the highest-speed circuits is generally low and they are not commercially available, representing advanced development efforts. Dimensions of circuit components are deep within the submicron range.

Figures I.6 and I.7 imply that the benefits provided by integrated circuit technology are closely tied to circuit scaling or reduction of circuit component size. Continuation of this historical trend requires advances in many areas — most evolutionary and some revolutionary. On following pages, the issues involved with the scaling of technologies of the present day are examined as well as some candidates for replacement of present technologies when their scaling comes to an end.

REFERENCES

1. J. S. Kilby, "Invention of the Integrated Circuit," *IEEE Trans. Electron Devices,* **ED-23,** 648 (1976).
2. A. Berenbaum, B. W. Colbry, D. R. Ditzel, R. D. Freeman, and K. J. O'Connor, "A Pipelined 32b Microprocessor with 13 kb of Cache Memory," *Int. Solid State Circuit Conf., Dig. Tech. Pap.,* p. 34 (1987).
3. S. Asai, "Semiconductor Memory Trends," *Proc. IEEE,* **74,** 1623 (1986).

1 The Submicrometer Silicon MOSFET

J. R. BREWS, K. K. NG, and R. K. WATTS

AT&T Bell Laboratories
Murray Hill, New Jersey

The sheer volume of answers can often stifle insight . . . The purpose of computing is insight, not numbers.

— Hamming[1]

Numerical analysis is sometimes difficult to interpret unless one has a basic idea of the physics involved.

— Cham, Oh, Chin, and Moll[2]

A simple analytical model should provide additional physical insight into the optimization of these devices.

— Lee, Mayaram, and Hu[3]

1.1 INTRODUCTION

MOSFET design used to be simpler. Devices once within the boundaries of long-channel behavior now are near or even outside these boundaries. Models are becoming more quantitative and complex as devices shrink in the pursuit of better system performance, greater reliability, and lower cost.

With MOSFET design crossing the long-channel frontier, parameters once secondary now are crucial. Examples are source/drain doping profiles, channel doping profiles, contact construction, and device geometry around the gate periphery. Moreover, this detail is not controlled precisely enough to allow neglect of fabricational variations. As a result, fabricational variations input a noise level into device design that carries through to the circuit level. Conversely, circuit requirements work backward through device simulation to become fabricational constraints. Thus, device design serves as a complex interpreter, translating back and forth between circuit and fabricational constraints during the search for an optimal structure.

This chapter outlines how miniaturized MOSFET design is developing. An optimal design is found following a complex combination of empiricism, simulation,

9

and iteration. No best path through this labyrinth has been pointed out, but it helps to have some overview of the design environment, despite a lack of unanimity. An overview is provided in the next section, followed by discussion of miniaturization methods and some specific trade-offs. Space restrictions force omission of basic MOSFET behavior, found in a number of review articles and books.[4-7]

Simple models are used. As the beginning quotations indicate, these models are not quantitative, but they do bring out key features in miniaturization and relate them. With a clear overall structure, a strategy can be planned: minimizing iterations, selecting appropriate computer modeling, and introducing monitoring and measurement sensibly. Then quantitative computer calculation can sharpen these broad considerations.

1.2 OVERVIEW OF THE ENVIRONMENT

The VLSI design environment is complex. It incorporates feedback, iteration, and appeals back and forth from measurement and adjustment of parameters to theoretical simulation. One description of this environment is a Warnier–Orr diagram, as described by Rood,[8] for example. A more common description is the flow chart, and many other diagraming techniques could be used (see, e.g., Steward[9]). However, if the reader tries to construct a Warnier–Orr diagram for the VLSI design environment, he will find this exercise forces more thought about order and hierarchy than does a flow chart. The treatment of feedback and iteration in a Warnier–Orr diagram is more disciplined, and reflects what information is passed between stages, not just the order of events.

Figure 1.1 illustrates a Warnier–Orr diagram. The diagram consists of branch headings which can be subdivided to create new levels with more branches. Each branch that is subdivided may have a "loop until . . ." criterion that specifies an it-

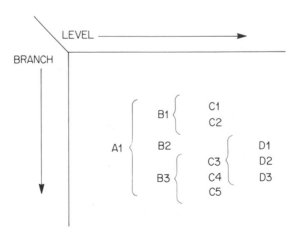

Fig. 1.1 A Warnier–Orr diagram, showing lettered levels and numbered branches.

eration of its subdivisions until some condition is satisfied, like a "do while . . ." loop in programming.

In Fig. 1.1, each level is labeled by a letter and each branch is numbered. The branch heading A1 is read first. Because this branch is subdivided, one moves to the next level B and reads the branch heading B1. Because B1 is divided, one proceeds to level C and reads the subdivisions of B1, namely C1 and C2. As there are no further subdivisions in this brace, one returns to level B and reads branches B2 and B3.

The reading of Fig. 1.1 is completed as follows. Because B3 is subdivided, one goes to level C and reads branch C3. As C3 is subdivided, one goes to level D and reads D1, D2, D3. As there are no more subdivisions, one returns to level C and reads C4, C5. Finally, one returns to level B, and as B3 is the last branch on this level, one continues back to level A, ending the reading.

Figure 1.2 is a possible Warnier–Orr breakdown of the IC manufacturing environment. It begins with a specification, which includes cost and performance goals. The remaining steps, design, fabricate, and final test, are self-explanatory. The "IC manufacture" loop is repeated until testing shows that the specifications have been met. Although all the steps can be subdivided, only IC design is tentatively broken down into further steps. As indicated by the instruction "loop until estimated yield is satisfactory," these steps are repeated until this condition is met.

Figure 1.3 is an expansion of the fourth step in the IC design breakdown, namely, the step: "find device behavior." Here, "device" could mean a single MOSFET, a number of different MOSFETs used in the circuits under design, or some simple structures that could test isolation or latch-up, or some simple building block such as an inverter. The first step in Fig. 1.3 is to decide how the origi-

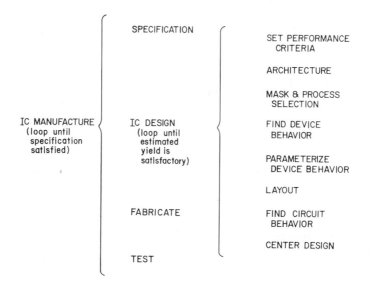

Fig. 1.2 A Warnier–Orr diagram for integrated circuit manufacture.

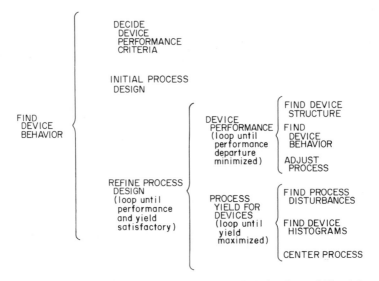

Fig. 1.3 Expansion of the fourth step in the IC design loop of Fig. 1.2.

nal IC performance criteria are reflected at the device level as device criteria. Examples could be translating clock rates into off-currents or voltage levels into hot-electron reliability constraints. The initial process design follows, uses simple models, and depends very strongly on the criteria of step one. Its objective is a first estimate of device dimensions, implant parameters, and times and temperatures of anneals and oxidations.

Once the initial process parameters are decided, one enters the loop to refine process design. Initial process parameters are fed into a process simulator, and realistic computer analysis is used extensively to refine the initial guess. The elements of a workstation for such a process synthesis were described by Strojwas and Director.[10] Then the process is centered for best yield. Figure 1.3 assumes yield and performance can be optimized by optimizing each in succession, as done by Styblinski and Opalski.[11] Within the breakdown of this loop, the step "device performance" is expanded in Fig. 1.4.

In Fig. 1.4, the first step determines the device structure, as expressed by doping profiles, surface topology, oxide thickness, and so forth. The determination of this microscopic structure is based on process models for all the major processes, such as oxidation, implantation, annealing, and so forth. Because of a lack of basic understanding, these models contain parameters that are adjusted empirically to agree with measurements made on test structures. These appeals to empiricism have not been included in the diagram, and would appear if a further breakdown level were added.

The second step in this loop is the determination of device behavior, not only electrical behavior (ac, dc, or large signal), but also long-term reliability (hot-electron trapping) and process sensitivities. Device behavior is found from two- or

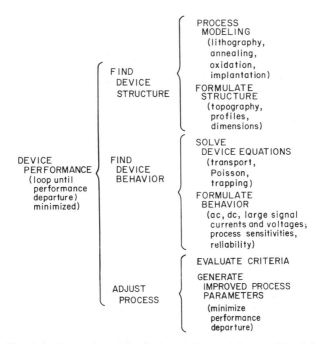

Fig. 1.4 Expansion of the device performance loop in Fig. 1.3.

three-dimensional numerical solutions of the partial differential equations governing transport, electrical potential, and so forth. Process sensitivities can be generated at this stage using an ac analysis (see Yang et al.[12]).

The final step in Fig. 1.4 is optimization, in which the device performance criteria are compared with the simulated results, and the process is changed to minimize any discrepancies.

Figure 1.5 is a flow chart of a simplified version of Fig. 1.4, as proposed by Tazawa et al.[13] In their procedure, the criterion for device performance is the departure of the simulated $I-V$ curves from a set of target $I-V$ curves. These authors also used this simulation loop to find process sensitivities by varying the process parameters within reasonable windows of variation (designated "upper and lower bounds for optimization parameters"). The understanding incorporated in this design scheme leads to a diagnostic tool incorporated in the manufacturing line. Their diagnostic approach, shown in Fig. 1.6, identifies common failures.

More elaborate diagnostics were discussed by Director,[14] and process characterization (i.e., the identification, statistical modeling, and monitoring of process variations) was programmed by Spanos and Director,[15] and related to test structures by Chen and Strojwas.[16] An example of the interaction between process optimization and original specifications was presented by Styblinski and Opalski.[11] Various alternative proposals to cut computational overhead by treating statistical variations at the level of device parameters, rather than at the process level, were presented by Yu et al.,[17] Aoki et al.,[18] and Yang et al.[12]

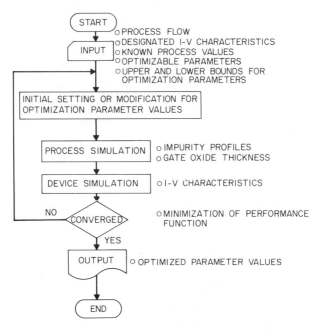

Fig. 1.5 A flow chart for a simplified version of Fig. 1.4. (After Tazawa et al.[13] Reprinted with permission, copyright 1987, Pergamon Journals, Ltd.)

As indicated by the quotations beginning the chapter, we are interested in the role of qualitative models in downsizing. From the overview of the design environment, there are several roles for simple models.

First, as shown in Fig. 1.3, performance criteria and the initial process design are guided by simple models — that is, simple models identify primary factors to be monitored throughout the design and provide a starting point for the detailed computer design that refines the process.

Second, the qualitative understanding gained with simple models guides the subsequent computer calculations: what is calculated and what it means. If models identify the correct macroscopic parameters to exchange between design levels, they also give us a grip on an effective hierarchy for the Warnier–Orr diagrams. That is, another role for simple models is to render the design more transparent and more intuitive and to organize it better.

Third, if simple models are sufficiently accurate, they lessen computation and iteration. At a minimum, these models should locate a viable region in the design space. More ideally, nonlinearities of the design might be soluble at this level.

Device downsizing is not controlled today simply by device physics, because there is a coupling with many aspects of manufacturing and with the circuit in which the device must function. Nonetheless, we begin with traditional approaches to downsizing based on scaling. This discussion shows these approaches must be improved, opening the way to a wider variety of trade-offs in MOSFET design and to an appreciation of how the design environment plays a part.

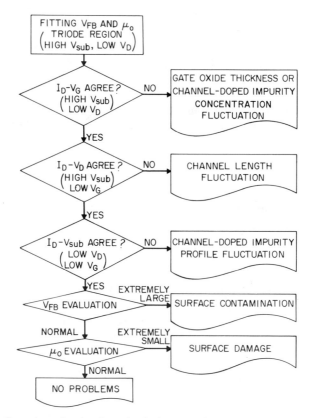

Fig. 1.6 A flow chart showing how the design procedure can be applied to a process line monitor and diagnostic tool. (After Tazawa et al.[13] Reprinted with permission, copyright 1987, Pergamon Journals, Ltd.)

1.3 SCALING

Our main goal is to explore device downsizing, and we begin with electrostatic scaling as introduced by Baccarani et al.[19] This approach extends the earlier constant field scaling of Dennard et al.,[20] but still is based on an assumed electrostatic similarity between a good large device and a good small device. That is, Poisson's equation is scaled without regard for coupling to current.

Next, we discuss the subthreshold scaling of Brews et al.[21] This approach stresses good turn-off behavior and avoidance of punchthrough. As it happens, subthreshold behavior does not scale under electrostatic scaling, so the two approaches are different.

1.3.1 Electrostatic Scaling

Baccarani et al.[19] noted that Poisson's equation for two different device structures is unaltered provided a potential scaling factor κ and a length scaling factor λ are introduced such that the potential ϕ satisfies the following scaling rule:

$$\phi \rightarrow \phi' = \frac{\phi}{\kappa}. \tag{1}$$

In addition, the boundary conditions expressed in terms of the potentials on the various electrodes also are scaled down by κ. The distances, (x, y, z), satisfy the rule

$$(x, y, z) \rightarrow (x', y', z') = \frac{(x, y, z)}{\lambda}, \tag{2}$$

and the charge densities, $q(n, p, N_A, N_D)$, satisfy

$$q(n, p, N_A, N_D) \rightarrow q(n', p', N_A', N_D') = q(n, p, N_A, N_D)\left(\frac{\lambda^2}{\kappa}\right). \tag{3}$$

Naturally, the mobile carrier densities cannot really scale in this way, because they are determined by exponential functions of the potential. However, in the depletion regions of the MOSFET, the charge scaling requirement can be satisfied by adjustment of the dopant distribution, and the depletion layer dimensions then scale approximately according to the lengths. For example, for uniform doping in one dimension, the depletion width w, in terms of the Debye length L_D, and potential satisfies

$$w = \sqrt{2}L_D(\beta\phi - 1)^{1/2}, \tag{4}$$

$$L_D \rightarrow L_D' = \left(\frac{\kappa^{1/2}}{\lambda}\right)L_D, \tag{5}$$

$$w \rightarrow w' = \sqrt{2}\left(\frac{\kappa^{1/2}}{\lambda}\right)L_D\left(\frac{\beta\phi}{\kappa} - 1\right)^{1/2} \sim \frac{w}{\lambda}, \tag{6}$$

with $\beta = q/(kT)$, q = electronic charge, k = Boltzmann constant, and T = absolute temperature. The approximation in (6) assumes $\phi/\kappa \gg kT/q \sim 25$ mV, which usually is valid.

Although the depletion layer width under the channel approximately scales, the junction depletion widths scale only under much more restrictive conditions. The depletion width of the drain junction in a one-dimensional approximation for uniform doping (assuming the drain doping to be much higher than the substrate doping) is

$$w_D = \sqrt{2}L_D[\beta(V_{BI} + V_{BS} + V_{DS}) - 1]^{1/2}, \tag{7}$$

where V_{BI} is the built-in voltage dictated by the difference in Fermi levels of the substrate and drain, V_{BS} the substrate-to-source reverse bias, and V_{DS} the drain-to-source voltage. The built-in voltage does not scale, as the Fermi level difference is a logarithm of the doping ratio. As a result, a scaling of w_D results only under the condition

$$V_{DS} + V_{BS} \gg V_{BI} \sim 0.8 \text{ V}. \tag{8}$$

The condition (8) is hard to satisfy when voltages are reduced so that, more than its larger original, an electrostatically scaled device is subject to two-dimensional problems such as punchthrough and barrier lowering. However, in the linear region of MOSFET operation, the current–voltage characteristics do scale. Some of the scaling properties of various quantities are shown in Table 1.1. The original constant field scaling of Dennard et at.[20] is obtained by setting $\kappa = \lambda$.

Electrostatic scaling has the flexibility of choosing voltages to scale differently than dimensions. This flexibility is important because voltages usually cannot be scaled down by the same factor as channel length, for example. Multiple scaling factors are needed, according to Chatterjee et al.,[22] because: (a) fabricational variations do not scale with dimensions, leading to noise margin problems if voltages are scaled with dimensions; and (b) current drive capability cannot be scaled because the fringing capacitance of interconnect does not scale, so for acceptable circuit speed, the drive current must be kept larger than scaling suggests.

In addition, as pointed out by Dennard et al.,[20] subthreshold turn-off does not scale, because it is dominated by the exponential dependence on potential of the channel carrier density. According to Pfiester et al.,[23] it is this turn-off behavior that sets the performance limits in ULSI CMOS applications. It also is a limiting factor in dynamic circuits, as better turn-off allows longer refresh times. This limitation is described, for example, by Wordeman et al.[24]

1.3.2 Subthreshold Scaling

Another approach to scaling was proposed by Brews et al.[21] Perhaps scaling is a misnomer here, since miniaturization based on subthreshold behavior is meant.

TABLE 1.1 Effects of the Two Types of Scaling Proposed by Dennard and Co-workers.[a,b]

Parameter	κ Scaling	κ, λ Scaling
Dimension	κ^{-1}	λ^{-1}
Voltage	κ^{-1}	κ^{-1}
Current	κ^{-1}	λ/κ^2
Dopant concentration	κ	λ^2/κ
Power density	1	λ^3/κ^3
Circuit power	κ^{-2}	λ/κ^3
Gate delay	κ^{-1}	κ/λ^2
Power · delay product	κ^{-3}	$1/(\lambda\kappa^2)$
Line current density	κ	λ^3/κ^2

[a]Refs. 19 and 20.
[b]$\kappa, \lambda > 1$.

These authors identified a bound upon a single combination of parameters. If this bound is satisfied, the resulting device has long-channel subthreshold behavior. Unlike electrostatic scaling, one does not have to start with an optimal large device and then scale it down. Rather, any device that fits the criterion is suitable. Based on fitting experimental and computer-generated subthreshold characteristics, their empirical condition is that the channel length of an acceptable device must be larger than L_{ch}^{min}, where

$$L_{ch}^{min} = A[r_j t_{ox}(w_S + w_D)^2]^{1/3} . \qquad (9)$$

Here, L_{ch}^{min} is the minimum channel length for which long-channel subthreshold behavior obtains, r_j is the junction depth, w_S and w_D are the source and drain depletion widths in a one-dimensional abrupt junction approximation respectively, and t_{ox} is the oxide thickness. If the depletion widths and junction depth are in microns and the oxide thickness in angstroms, then the proportionality factor $A = 0.41 \text{ Å}^{-1/3}$, and L_{ch}^{min} is in microns. The doping dependence of (9) is contained in the depletion widths, inasmuch as [see (7)] they are proportional to the Debye length, and $L_D \propto N^{-1/2}$.

To arrive at (9), it was noted that in the subthreshold region for a long-channel device the current is independent of drain-to-source voltage once this voltage exceeds a few kT/q. The criterion used for long-channel behavior was that no more than a 10% departure in current should occur for a 0.5 V change in drain-to-source voltage. The experimental and computer-generated points then were fitted to the expression (9).

Figure 1.7 shows experimental and computer-generated points clustered near the line described by (9). Perhaps a 10% criterion is stricter than necessary. Depending on the application, a larger departure could be allowed without serious consequences.

Subthreshold scaling based on (9), like electrostatic scaling, is more flexible than constant field scaling. It allows voltages to be held fixed or only modestly re-

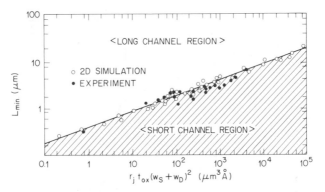

Fig. 1.7 A comparison of simulation and experiment with the long-channel boundary description of Eq. (9). (After Brews et al.[21] Reprinted with permission, copyright 1980, IEEE.)

duced provided the other factors are changed to compensate. Subthreshold scaling is most useful when good turn-off behavior must be maintained, for example, to control residual current at zero gate bias (CMOS) or to contain refresh cycling (dynamic circuits).

A major failing of both types of scaling is that they provide no insight into miniaturization. With electrostatic scaling, all wisdom is embodied in the large device chosen to be scaled: one's hope is simply that scaling disturbs operation as little as possible. With subthreshold scaling, all insight is buried in empiricism: we do not know why Eq. (9) works. Moreover, neither scaling method incorporates all the necessary trade-offs, so the reduced device still needs adjustments. Obviously, trade-offs optimal at 1 μm may not be optimal at 0.25 μm, but these scaling procedures do not reexamine these trade-offs. In particular, neither method treats transport-related problems such as velocity saturation, nor hot-electron trapping, nor circuit-related restrictions such as noise margin or off-current requirements, nor fabricational variations. Thus, we are led to consider in more detail the optimization of the MOSFET structure. First, some basic design constraints are presented. Then an initial design strategy based on the off-current parameter is outlined.

1.4   HMOS DESIGN

The most commonly used MOSFET structure is shown in Fig. 1.8. One name for this structure is HMOS, for high-performance MOS. Methods for making HMOS structures have changed with shrinking design rules, but the basic concepts have proved robust. They have been described by Shannon et al.,[25] Pashley et al.,[26] and Dennard et al.[20] The structure is meant to take full advantage of ion implantation by use of (a) threshold control implants, (b) punchthrough control implants, (c) source

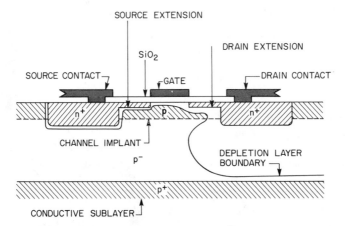

Fig. 1.8   A typical HMOS structure. (After Shannon et al.[25] Reprinted from ELECTRONICS, February 3, 1969 issue. Copyright 1969, McGraw-Hill, Inc. All rights reserved.)

and drain implants for shallow junctions or junction extensions to control two-dimensional field effects and gate-junction overlap capacitances, and (d) a self-aligned gate to minimize overlap capacitances. Additional features are (e) a lightly doped substrate or epitaxial layer to minimize junction capacitance and back-bias sensitivity of threshold and, (f) a low-resistivity sublayer to reduce IR drops, delay, latch-up, and noise.

The description of an HMOS structure can be summarized in terms of the following parameters: (a) gate material, (b) oxide thickness, (c) source/drain doping profile parameters, (d) surface doping profile parameters, and (e) lightly doped substrate doping N_{sub}. To fabricate the structure, this description must be translated into mask descriptions, process sequences, implant doses and energies, and annealing and oxidation times and temperatures. This translation can become involved at the stage of computer design of the processing, but at the simple level that concerns us here, it does not require much discussion.

What is of greater interest is to relate the HMOS parameters to various criteria imposed by device and circuit operation. These criteria are set to avoid a number of maladies, including (a) barrier lowering or surface punchthrough, (b) bulk punchthrough, (c) parasitic capacitance, (d) back-bias sensitivity of threshold, (e) hot carrier effects (trapping), (f) residual (zero gate bias) current, and (g) inadequate noise margin. A number of authors have selected from these maladies those they consider of greatest importance and have suggested various rules of thumb or formulas to relate the HMOS parameters in a way that limits the impact of these problems. These prescriptions now are discussed.

1.4.1 NMOS Design

NMOS (n-channel MOS) enhancement–depletion logic design in the submicron region has been discussed by several authors. We mention particularly the work of Klaassen,[27] Wordeman et al.,[24] Baccarani et al.,[19] and Shichijo.[28] Although details vary, the basic criteria employed are as follows:

1. *A Channel Doping to Substrate Doping Relation*

$$N_{sub} > \frac{N_{ch}}{10}.$$ (10)

Here, N_{sub} is the uniform bulk substrate doping/cm^3 and N_{ch} is the average channel doping/cm^3. This condition is a rule of thumb for prevention of bulk punchthrough. This condition does not depend on junction depth or channel length, suggesting that in practice punchthrough dependence on these variables is weak when barrier lowering is satisfactory.

2. *A Supply Voltage to Threshold Voltage Ratio*

$$V_T \approx \frac{V_{DD}}{4} \quad \text{or} \quad \frac{V_{DD}}{5}.$$ (11)

Here, V_{DD} is the supply voltage and V_T the threshold voltage. This condition is viewed variously as a symmetry condition on an inverter characteristic,[27] as a noise margin requirement,[27] or as a compromise between drive current (which is higher if V_T is lower as it is proportional to $(V_G - V_T)$) and turn-off (which is better if V_T is larger).[23,29]

3. *A Channel Profile Requirement.* Klaasen[27] suggests that the channel implant should just deplete at threshold. This condition ensures low back-bias sensitivity of threshold, because the depletion edge always resides in the lightly doped substrate. Baccarani et al.[19] suggest that the channel profile should be tailored to produce a depletion width as small as is consistent with acceptable back-bias sensitivity of threshold. That is, minimization of barrier lowering (small *w*) must be included in the channel profile design.

4. *A Junction Depth Requirement.* Klaassen[27] suggests that the junction depth should be deeper than the channel implant region to reduce parasitic capacitance between junctions and substrate. Junction depths can be even deeper provided bulk punchthrough is avoided and barrier lowering is not severe. Chatterjee et al.[22] and Shichijo[28] drew attention to the parasitic resistance introduced when junctions are very shallow, a subject discussed in detail later. Baccarani et al.[19] appear to choose junction depths as large as possible, consistent with keeping barrier lowering low.

5. *A Noise Margin Requirement.* The limiting amount of series resistance and barrier lowering allowable is set by noise margin requirements, according to Baccarani et al.[19]

6. *An Oxide Thickness Requirement.* Short-channel effects are minimized if the oxide is thin [e.g., see Eq. (9)], so this parameter is usually taken to be as thin as possible, within reasonable yield. Of course, voltage must also be kept commensurate to avoid corner breakdown or oxide breakdown.

7. *A Maximum Field Limitation.* This constraint is imposed to avoid hot-electron degradation. The maximum field is not a direct indicator of such degradation, as discussed in a later section.

For explanations of various terms such as "punchthrough," "barrier lowering," and "parasitic capacitance," the reader is referred to the reviews mentioned earlier.[4-7]

Approximate formulas for the above criteria were provided by Klaassen.[27] A somewhat different scheme was suggested by Baccarani et al.[19] Both groups suggest that their simple analysis schemes are adequate to predict where successful designs can be found, and that computer analysis introduces no unexpected features into the design, but simply makes the design quantitative, particularly as regards barrier lowering and punchthrough. At this point, we defer further discussion of these constraints on NMOS to consider those for CMOS.

1.4.2 CMOS Design

Here we discuss the work on CMOS (complementary MOS) design as presented by Cham et al.[2] and by Pfiester et al.[23] The main difference here from the NMOS design problem is an increased emphasis on subthreshold behavior. Thus, in addition to the NMOS conditions 1–7 just mentioned, we add:

8. *A Leakage Current Criterion.* This criterion is intended to keep the off-current low at zero gate bias. The criterion proposed by Pfiester et al.[23] is

$$V_T \geq S \, \log(R) \, , \tag{12}$$

where S is the subthreshold slope (the gate swing needed to reduce the gate current a decade in subthreshold) and R the ratio of "on" to "off" current. In our later discussion, (12) is not used. Instead, off-current level is used as a design parameter.

The emphasis on subthreshold leakage reflects interest in a low standby power consumption per device, necessary to keep heating under control for VLSI circuits (Ref. 2, p. 134). Another reason for such a criterion is dynamic CMOS where it is required that some nodal capacitances must not discharge during a clock cycle despite undesirable leakage to ground via a transistor that nominally is off at zero gate bias. Depending on the application, meeting these leakage current requirements is not automatic, and it places constraints on the design that are more severe for submicron devices than for larger devices because of larger short-channel effects.

For NMOS, the channel profile is chosen, according to point 3, to balance (a) back-bias sensitivity of threshold and (b) barrier lowering. For many typical CMOS circuits, as shown in Fig. 1.9,[30] the back-bias sensitivity requirement is not important because the source and substrate of some transistors like C1 are tied to V_{DD}, while those of others like C2 are tied to ground. Hence, no source-to-substrate bias variation occurs. The same is true of the output inverter in Fig. 1.9. For such cases, the requirement on channel profile might seem to reduce to only a

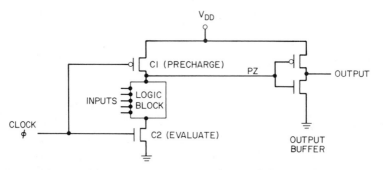

Fig. 1.9 A typical domino CMOS logic gate. For a discussion of its operation, see Weste and Eshraghian.[30]

requirement on barrier lowering in the CMOS case. However, the requirement on leakage for CMOS, point 8, places a constraint on the channel profile, so a balance of a different kind is required.

Although the various criteria mentioned seem reasonable, how they are interconnected, whether they all can be satisfied, and what they mean in terms of implant parameters, etc., is not clear. To clarify these criteria, we now discuss a tentative intitial design that uses off-current as a basic parameter for the n-channel device in a symmetric CMOS technology.

1.4.3   Symmetric CMOS Design

A symmetric CMOS technology means an approach like that discussed by Hillenius et al.[31] and by Hillenius and Lynch,[32] in which the n-channel and p-channel devices are as similar as possible. In particular, p^+ polysilicon/silicide gates are used for the p-channel devices and n^+ polysilicon/silicide gates for the n-channel devices. As discussed by Noguchi et al.[33] and by Hillenius and Lynch,[32] the use of common $MoSi_2$ gates also allows a symmetric technology. Such a gate allows very low off-currents to be achieved, but usually requires a compensating implant to lower threshold, which can result in barrier lowering problems. Hu and Bruce[34] also found experimentally and by computer simulation that buried channel devices are more subject than surface channel devices to barrier lowering and bulk punchthrough. In addition, buried channel devices are subject to more process variability stemming from control of the compensating implant. Thus, for minimally sized devices, the polysilicon-gate surface channel devices seem the best choice and are taken here for illustration. Another advantage of symmetric polysilicon gates is that the oxide fields over the drains are kept to a minimum, being either p^+ gates of p^+ drains, or n^+ gates over n^+ drains. Pfiester et al.[23] and Sun et al.[35] also recommend this choice.

A design based on off-current as a major parameter now will be presented. Ordinarily, off-current does not play a dominant role, but as devices shrink such a strategy assumes greater significance. Off-current is left as an undetermined parameter, so the following discussion of trade-offs can be applied no matter what off-current actually is required for a particular application.

Most simple approaches are based on modifications of long-channel theory. Moreover, departures from long-channel behavior generally are small. Therefore, maximum use is made of long-channel theory, and short-channel effects are taken as a correction. We use long-channel formulas for threshold voltage, off-current, and depletion layer width.

Although we design an optimal long-channel device, the channel length ultimately is to be shrunk until short-channel effects become excessive. Therefore, escalating a suggestion of Baccarani et al.,[19] we include a peculiar condition that is a zero-order protection of the device from barrier lowering: the condition that the depletion layer width be a minimum. This condition helps to isolate the channel from the drain and to maintain a long-channel behavior down to as short a channel length as possible. In effect, we trade a poor long-channel subthreshold slope for immunity from short-channel barrier-lowering effects on subthreshold slope. This condition has no place in long-channel design, but anticipates later channel length shrinkage. As a final step, hot-electron effects should be estimated, possibly leading to modifications of the design for example via drain profiling (e.g., lightly doped drain structures). An outline of the overall procedure is shown in Fig. 1.10.

1.4.3.1 Long-Channel Design. We first make a long-channel design and then introduce barrier lowering. A simple correction to the long-channel design is generated to obtain the initial *n*-channel CMOS device design.

Select Oxide Thickness. The thinner the gate oxide, the smaller two-dimensional effects become. Thus, a thin oxide is best. However, usually yield suffers when the oxide is made too thin. Also, voltage must be proportionately reduced to avoid corner or edge breakdown. For a recent discussion, see Feng et al.[36] and Chen et al.[37] Thus, we leave oxide thickness as a basic parameter of the design, to be set later by process capabilities.

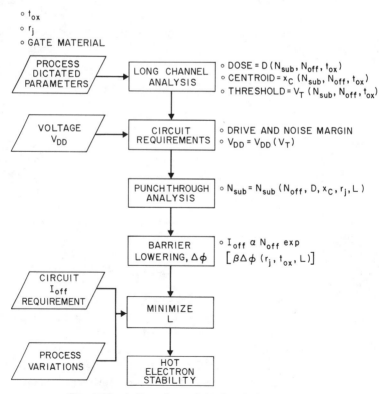

Fig. 1.10 A flow chart of the I_{OFF} design strategy.

Select Junction Depth. The shallower the junctions, the smaller two-dimensional effects become. However, series resistance and fabricational difficulties increase. Hence, like oxide thickness, we leave junction depth as a basic parameter of the design, to be set later by process capabilities.

Choose Gate Material. For the *n*-channel device, we chose an n^+ polysilicon gate, for the reasons already given.

Design Channel Profile. The channel profile is constrained by two factors: the off-current at zero gate bias, and the built-in potential at zero gate bias due to the difference in Fermi levels of the gate and substrate. The off-current is a parameter in the model. Ultimately, the off-current is related to channel length, and setting the off-current to the circuit specification determines the minimum channel length compatible with the circuit off-current requirement.

The design of the channel profile can be expressed in terms of the exposed dose D and the centroid of the exposed dose x_C. "Exposed" means the portion of the channel implants contained within the depletion layer under the channel. Thus, any implanted ions in the oxide or in the bulk beyond the depletion layer edge are irrelevant. The centroid is defined by (see Refs. 7 and 38)

$$
x_C = \frac{\int_0^w dx\, x[N(x)/N_{\text{sub}} - 1]}{\int_0^w dx\, [N(x)/N_{\text{sub}} - 1]}, \tag{13}
$$

where x measures depth from the interface, w is the depletion layer width, and $N(x)$ is the actual channel profile, so that it is the departure of $N(x)$ from the substrate doping N_{sub} within the depletion layer that contributes to x_C. We now will relate D and x_C to the off-current parameter. The off-current I_{OFF} is given in terms of the off-carrier density per unit area N_{off}, as

$$
I_{\text{OFF}} = \mu \left(\frac{W}{L}\right) \left(\frac{kT}{q}\right) q N_{\text{off}}, \tag{14}
$$

where W = device width, L = device length, and μ = channel mobility. The off-carrier density/cm^2 N_{off} is given in the subthreshold region by

$$
N_{\text{off}} = N_0 \frac{\exp(\beta \phi_s)}{Q}, \tag{15}
$$

where $\beta = q/kT$, and N_0 is given by

$$
N_0 = \left(\frac{n_i}{N_{\text{sub}}}\right)(n_i L_D) \tag{16}
$$

with n_i the intrinsic carrier density (taken here as $n_i \approx 10^{10}$ carriers/cm³ at $T = 290$ K). Here, Q is a dimensionless surface charge given by

$$Q = \frac{D}{N_{sub}L_D} + \frac{N_{off}}{N_{sub}L_D} + \frac{w}{L_D}.$$ (17)

The terms in (17) are contributed by the charge from the implant, the channel carriers, and from the depletion layer of width w, respectively. To find D and x_C, it is convenient to find Q first, using two constraints on the surface band bending ϕ_s. Using N_{off} from (15), the first constraint is

$$\beta\phi_s = \ln\left(\frac{QN_{off}}{N_0}\right).$$ (18)

The other constraint on ϕ_s, that of the gate potential at zero gate bias ϕ_g, can be understood from Fig. 1.11, which shows ϕ_s satisfies

$$C_{ox}(\phi_g - \phi_s) = qD + qN_{sub}w + qN_{off}$$

$$= qN_{sub}L_D Q,$$ (19)

which is Gauss's law, where C_{ox} is the oxide capacitance per unit area. Rearranging this expression, the second constraint on ϕ_s, additional to (18), is

$$\beta\phi_s = \beta\phi_g - \left(\frac{\kappa_s}{\kappa_{ox}}\right)\left(\frac{t_{ox}}{L_D}\right)Q,$$ (20)

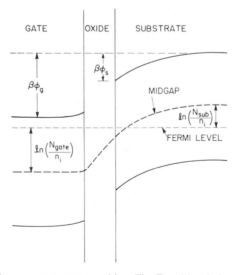

Fig. 1.11 An MOS structure at zero gate bias. The Fermi levels in gate and substrate line up, necessitating a gate potential at zero volts.

where κ_s and κ_{ox} are the dielectric constants of silicon and SiO_2 (11.7 and 3.9), respectively. For our symmetric technology, the gate potential at zero gate bias (see Fig. 1.11) is

$$\beta\phi_g = u_G + \ln\left(\frac{N_{sub}}{n_i}\right). \tag{21}$$

Here, u_G is the contribution of the gate material, easily evaluated by assuming that the gate is so heavily doped that the Fermi level in the gate resides at the conduction band edge, resulting in the approximation

$$u_G \approx \frac{\beta E_G}{2q} \approx 22.4, \tag{22}$$

where E_G is the bandgap energy of silicon. Combining (18), (20), and (21), an equation determining Q can be found, namely,

$$u_G - \left(\frac{\kappa_s}{\kappa_{ox}}\right)\left(\frac{t_{ox}}{L_D}\right)Q = \ln\left(Q\frac{N_{off}}{n_i L_D}\right), \tag{23}$$

which can be solved by iteration.

Equation (23) can be arranged more conveniently in terms of midgap values as follows. At midgap, the surface band bending is $\beta\phi_s = \beta\phi_{mg} = \ln(N_{sub}/n_i)$, and from (20) and (21), if midgap occurs at zero bias, then $Q = Q_{mg}$, given by

$$Q_{mg} = \left(\frac{\kappa_{ox} L_D}{\kappa_s t_{ox}}\right)u_G. \tag{24}$$

For (24) and (23) to agree, the corresponding value of N_{off} that applies is N_{mg}, found by setting the argument of the logarithm to unity,

$$N_{mg} = \frac{n_i L_D}{Q_{mg}}$$

$$= \frac{n_i t_{ox}(\kappa_s/\kappa_{ox})}{u_G}, \tag{25}$$

independent of N_{sub}. The result (25) also is obtained from (15) and (24) as $N_{mg} = (N_0/Q_{mg}) \exp(\beta\phi_{mg})$. That is, N_{mg} is the value of N_{off} when a midgap condition arises at zero gate bias.

In terms of N_{mg} and Q_{mg}, (23) can be rewritten as

$$\frac{Q}{Q_{mg}} = 1 - \left(\frac{1}{u_G}\right)\ln\left[\left(\frac{Q}{Q_{mg}}\right)\left(\frac{N_{off}}{N_{mg}}\right)\right], \tag{26}$$

which expresses Q/Q_{mg} as a function of N_{off}/N_{mg} independent of t_{ox} and N_{sub}.

Figure 1.12 shows Q/Q_{mg} from (26) as a function of N_{off}/N_{mg}. The behavior of Q can be understood as follows. A larger N_{off} means a larger band bending in the silicon. However, the gate potential is fixed at ϕ_g, so a larger drop in the silicon necessarily means a smaller field in the oxide, that is, a smaller Q.

Once Q is determined, so is the surface band bending, ϕ_s [see (20)], which allows the dose and centroid of the channel profile to be determined. To find the centroid, x_C, we use the result for the depletion width in the presence of nonuniform doping, namely (see Refs. 7 and 38),

$$\frac{1}{2}\left(\frac{w}{L_D}\right)^2 = \beta\phi_s - \left(\frac{D}{N_{sub}L_D}\right)\left(\frac{x_C}{L_D}\right) - 1 . \tag{27}$$

Now, the centroid must lie within the depletion region, so that

$$\frac{w}{L_D} \geq \frac{x_C}{L_D} . \tag{28}$$

The least value of w (which we interpret as the greatest immunity to barrier lowering) occurs when the condition (28) is an equality, and equality occurs according to (27) when

$$\frac{1}{2}\left(\frac{x_C}{L_D}\right)^2 + \left(\frac{D}{N_{sub}L_D}\right)\left(\frac{x_C}{L_D}\right) - \frac{1}{2}\left(\frac{w_u}{L_D}\right)^2 = 0$$

or

$$\frac{x_C}{L_D} = -\frac{D}{N_{sub}L_D} + \sqrt{\left(\frac{D}{N_{sub}L_D}\right)^2 + \left(\frac{w_u}{L_D}\right)^2} , \tag{29}$$

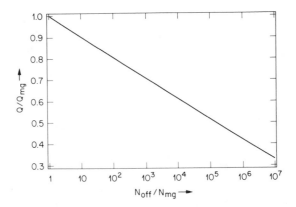

Fig. 1.12   The normalized field parameter Q/Q_{mg} of Eq. (26) as a function of zero-bias carrier density to midgap carrier density ratio N_{off}/N_{mg}.

where the positive square root is chosen because $\lim_{D \to 0} x_C = w_u$ is the depletion width for a uniformly doped substrate with band bending ϕ_s, given by

$$w_u = \sqrt{2} L_D (\beta \phi_s - 1)^{1/2}. \tag{30}$$

The relation (29) can be reexpressed as

$$\frac{x_C}{w_u} = -\frac{D}{N_{sub} w_u} + \left\{ \left(\frac{D}{N_{sub} w_u} \right)^2 + 1 \right\}^{1/2} \tag{31a}$$

or as

$$\frac{D}{N_{sub} w_u} = \frac{1 - (x_C/w_u)^2}{2(x_C/w_u)}. \tag{31b}$$

This relation is independent of N_{off} and simply expresses the relation between D and x_C necessary to force the condition $w = x_C$.

With (31), the profile is determined by either D or x_C. Let us find D as a function of N_{off} using (17) with $w = x_C$. We find from (17) and (31a),

$$Q\left(\frac{L_D}{w_u} \right) = \frac{N_{off}}{N_{sub} w_u} + \left[\left(\frac{D}{N_{sub} w_u} \right)^2 + 1 \right]^{1/2}. \tag{32}$$

By rearrangement of (32) and introduction of N_{mg} from (25),

$$\frac{D}{N_{sub} L_D} = \left\{ \left[Q - \left(\frac{N_{off}}{N_{mg}} \right) \left(\frac{n_i}{N_{sub}} \right) \frac{1}{Q_{mg}} \right]^2 - \left(\frac{w_u}{L_D} \right)^2 \right\}^{1/2}, \tag{33}$$

which determines the dose D for a given N_{off}, inasmuch as Q already is known from (26) or Fig. 1.12 and w_u is known from (30) and (20) as

$$\frac{1}{2} \left(\frac{w_u}{L_D} \right)^2 = \beta \phi_s - 1$$

$$= u_G + \ln \left(\frac{N_{sub}}{n_i} \right) - 1 - \left(\frac{\kappa_s}{\kappa_{ox}} \right) \left(\frac{t_{ox}}{L_D} \right) Q. \tag{34}$$

Generally, the explicit term in N_{off} in (33) is negligible compared to Q, so D and x_C are both given as explicit functions of Q using (33) and (34), as plotted in Fig. 1.13. As this figure shows, at larger values of Q the net charge in the silicon is dominated by the implant dose D, resulting in a straight-line behavior of D vs. Q. As this happens, the centroid and the depletion width approach the interface. At the other extreme, as D approaches zero, the centroid and the depletion width tend to the value for the uniformly doped case with no implant.

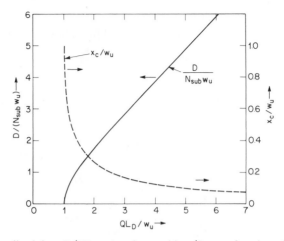

Fig. 1.13 Normalized dose $D/(N_{sub}w_u)$ and centroid x_C/L_D as a function of $Q/(w_u/L_D)$ for the condition that the implant centroid coincides with the depletion layer edge.

In Fig. 1.14, the dose is plotted using (33) as a function of N_{off}/N_{mg}. The dose is normalized to the charge Q using (17), that is,

$$\frac{D}{N_{sub}L_D Q} = \frac{D}{D + N_{off} + N_{sub}w}.$$ (35)

Figure 1.14 shows that for lightly doped substrates the charge Q is dominated by D, and the above ratio has the asymptotic value of unity. As the substrate doping increases, D no longer is dominant, and for a chosen N_{sub} the curve drops to $D = 0$ at a value of N_{off}/N_{mg} that applies for a substrate uniformly doped at N_{sub}.

The curves in Fig. 1.14 show that for a given N_{off}/N_{mg}, a maximum substrate doping exists, beyond which the desired N_{off}/N_{mg} is unobtainable. For thinner oxides (compare Figs. 1.14a and b), this maximum doping is larger. For substrate dopings less than this maximum, a particular dose is necessary to realize the given N_{off}/N_{mg}. This dose becomes independent of the substrate doping for light enough dopings, becoming [using (24) and $L_D^2 = (kT/q)\kappa_s \varepsilon_0/(qN_{sub})$]

$$qD \approx \left(\frac{Q}{Q_{mg}}\right)C_{ox}\left(\frac{kT}{q}\right)u_G,$$ (36)

where the right side depends on t_{ox} and N_{off}/N_{mg}, but not on N_{sub}. This substrate doping independence of D continues to larger N_{sub} (for fixed N_{off}) or to larger N_{off} (for fixed N_{sub}) the thinner the oxide. In most cases of interest, (36) will apply, so that t_{ox} and N_{off}/N_{mg} determine D, and N_{sub} has no effect.

In Fig. 1.15, for 50 Å oxides the unnormalized dose and centroid (depletion width) are plotted vs. N_{off}/N_{mg}. From the small values of x_C, it is clear that a delta function implant is not a realistic approximation. For realistic profiles, the cen-

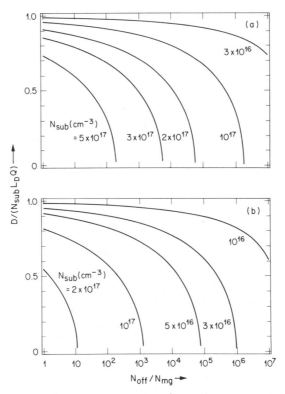

Fig. 1.14 Normalized dose as a function of N_{off}/N_{mg}. The normalized dose is given by Eq. (35): (a) $t_{ox} = 50$ Å; (b) $t_{ox} = 100$ Å.

troid cannot be forced to coincide with the depletion edge as done here, because the profile always has a nonuniformity of nonzero width. However, generalization of this analysis does not affect the observation that the dose and centroid of the exposed portion of the surface profile are determined once the off-carrier density and the gate material are specified, provided the condition of minimal depletion width is invoked.

With D and x_C, the channel profile is known in terms of the basic design parameters. Now the threshold voltage and substrate doping remain to be settled.

Select Threshold Voltage. This selection is based on drive considerations and on noise margin at the specified supply voltage. One could use the rule of thumb (11), or a more sophisticated analysis. We choose here to use (11).

Select a Substrate Doping. The main consideration here is avoidance of bulk punchthrough. Parasitic capacitance and back-bias sensitivity of threshold also can be factors. For discussion, (10) will be adopted, although more sophisticated punchthrough models, such as that of Skotnicki et al.,[39] also could be used.

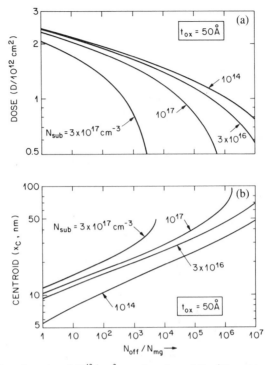

Fig. 1.15 (a) Dose in units of $10^{12}/cm^2$ as a function of N_{off}/N_{mg} with substrate doping as parameter. (b) Centroid or depletion layer width $x_C = w$ in nanometers vs. N_{off}/N_{mg}. In both figures, the oxide thickness is $t_{ox} = 50$ Å.

Our design scenario assumes the supply voltage as given, determining the target threshold voltage via (11). This target value of threshold V_T can be achieved by various combinations of substrate doping and implants. But N_{off}/N_{mg} and the substrate doping determine the implant parameters using (33) and (34), or Fig. 1.15. Consequently, V_T is determined by N_{off}/N_{mg} and N_{sub}. To find this dependence, the threshold voltage corresponding to the chosen D and x_C must be found.

The threshold *shift* from the uniformly doped case can be found following Ref. 38. To this shift must be added the threshold for the uniformly doped case. Sometimes this uniform threshold is taken as the gate bias corresponding to a bandbending of $2(kT/q)\ln(N_{sub}/n_i)$, but a more accurate estimate is obtained by extrapolation to zero carrier density of the asymptote of a carrier density vs. gate bias plot in strong inversion. Such extrapolated thresholds are shown in Fig. 1.16 as a function of oxide thickness for several choices of uniform substrate doping.

When these uniform thresholds are combined with the threshold shifts due to the dose D at depth x_C, the final thresholds are obtained, as shown in Fig. 1.17a as a function of N_{off}/N_{mg} for an oxide thickness of 50 Å, and in Fig. 1.17b for 100 Å. The curves terminate at a value of N_{off}/N_{mg} large enough that D from (33) is zero, that is, $V_T = V_T$ (substrate). To reduce V_T beyond this point requires

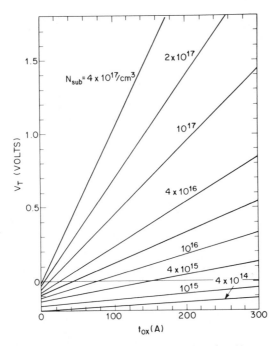

Fig. 1.16 Threshold voltage for substrates uniformly doped at N_{sub} as a function of oxide thickness t_{ox}. These thresholds are found by extrapolation of straight-line asymptotes to carrier density vs. gate bias curves in strong inversion down to zero carrier density. An n^+ polysilicon gate is assumed, using Eq. (21) for the gate potential at zero bias.

$D < 0$, that is, a compensating implant. If a compensating implant is large enough, buried channel behavior results with degradation of the short-channel properties of the device.[32-34] Hence, we do not pursue the case $D < 0$.

The curves in Fig. 1.17 allow a discussion of the trade-offs involved in selecting a combination of V_T and N_{off}/N_{mg}. Consider the case where a 1 V supply is desired. Then threshold [using (11)] is about 0.25 V. Figure 1.17 then shows that such a design is possible for a range of substrate dopings (those corresponding to curves that cross the dashed 0.25 V threshold line in Fig. 1.17) with lower N_{off} values corresponding to lower substrate dopings. However, when the substrate doping drops, the depletion width increases, so that containment of both bulk punchthrough and barrier lowering is expected to result in longer channel lengths.

If V_T is given, a larger upper bound on N_{off} permits larger N_{sub} and, hence, shorter minimum channel lengths. Similarly, Fig. 1.17 shows that for a fixed upper bound on N_{off}, a larger V_T implies a larger N_{sub}, again implying a shorter minimum channel length (at the cost of decreased drive and noise margins). As for oxide thickness, Figs. 1.17a and b show that thinning the oxide has little effect at lighter substrate dopings where D dominates the silicon charge density and (36) applies. But as substrate doping increases and the contribution of the uniform dop-

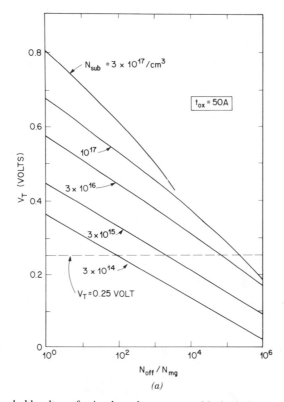

Fig. 1.17 Threshold voltage for implanted structures with the indicated substrate doping, as a function of $N_{\text{off}}/N_{\text{mg}}$. The dose and centroid of the implants are chosen for minimum depletion width as described by Eqs. (29) and (33): (a) $t_{\text{ox}} = 50$ Å; (b) $t_{\text{ox}} = 100$ Å.

ant begins to compete with D (corresponding to the drop of the curves in Fig. 1.14), and increase in t_{ox} forces a shift to lighter substrate doping and, hence, to longer minimum channel lengths.

Consider the effects of a fabricational variation in threshold. Figure 1.17 shows, for example, that a variation of threshold of approximately 0.06 V is sufficient to make N_{off} vary by one order of magnitude. Consequently, to design a circuit with many devices, a conservative design will require a choice of N_{off} somewhat lower than absolutely necessary, just to be sure that the devices with the lowest thresholds have sufficiently low off-current. This in turn implies a lower N_{sub} (for given V_T) and, hence, longer minimum sized devices. An additional complication is that the amount of threshold variation is a function of channel length. Hence, some iteration of the design is necessary to trade threshold variations against channel length.

For the choice of a 0.25 V threshold, we can plot the variation of the doping ratio $N_{\text{sub}}/N_{\text{ch}}$ as a function of $N_{\text{off}}/N_{\text{mg}}$. This doping ratio can be compared with the punchthrough criterion (10), namely, that this ratio should not be less than 0.1.

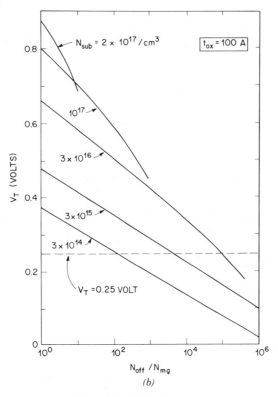

Fig. 1.17   (Continued)

The result is shown in Fig. 1.18. To estimate the doping of the channel region, we have used $N_{ch} \sim D/w_i$, where w_i is the depletion width of the implanted structure from (27). The curves in Fig. 1.18 show that bulk punchthrough imposes a lower bound on substrate doping. For example, for a 50 Å oxide, N_{sub} must be greater than $5 \times 10^{15}/cm^3$. This lower bound also sets a lower bound on $N_{off}/N_{mg} \gtrsim 5 \times 10^3$ and on the parasitic source- and drain-to-substrate capacitances because these capacitances decrease with substrate doping as the junction depletion widths expand.

Although (10) is a crude punchthrough criterion, it is clear that any punch-through model will set a lower bound on N_{sub}, whereas specification of N_{off} sets an upper bound. Thus, there is a limited range of substrate dopings available, which becomes further restricted by barrier lowering as the channel length is reduced, as will be shown in the next section.

At this point, the long-channel design is complete. The next step is to shrink the channel length while monitoring punchthrough and N_{off} to see how small the channel length can be made. To monitor short-channel effects, a two-dimensional barrier lowering model is necessary, and this model is discussed next.

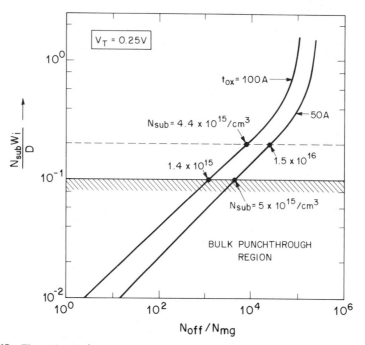

Fig. 1.18   The ratio N_{sub}/N_{ch} as a function of N_{off}/N_{mg} for a threshold of $V_T = 0.25$ V. The bulk punchthrough region is marked, according to Eq. (10), and also a more conservative condition, $N_{sub}/N_{ch} = 0.2$, is shown. Substrate doping at the intersections of these curves with the two punchthrough limits are indicated.

1.4.3.2   Barrier-Lowering Model.

A number of barrier-lowering models exist in the literature. Many of them are discussed by Jain and Balk.[40] The real challenge to these models is to provide a reasonable barrier-lowering estimate for a variety of device structures. At present, these models do not appear to be very accurate, and do not include nonuniform surface doping profiles, with the exception of the models of Skotnicki et al.[41] and of Kendall and Boothroyd.[42] Marash and Dutton[43] have discussed a general approach based on that of Kendall and Boothroyd.[42]

Regardless of the model, a predicted lowering of the barrier by an amount $\Delta\phi$, say, has the effect of increasing N_{off} by a factor $\exp(\beta\,\Delta\phi)$. Thus, as a first correction to the long-channel design, we must choose the initial long-channel N_{off} parameter to be small enough that multiplication of N_{off} by the barrier-lowering factor results in the target value of N_{off}. Referring to Fig. 1.17, this anticipatory reduction in N_{off} means a choice of substrate doping lower than a purely long-channel design would indicate. Then, referring to Fig. 1.18, we must check that our punchthrough requirement still can be met. As channel length is reduced, the barrier lowering increases, and for a sufficiently small channel length it becomes impossible to satisfy the bulk punchthrough restriction. Then one either settles for this minimum channel length or increases N_{off} by changing the circuit design.

Another consideration as channel length is reduced is the sensitivity of threshold to channel length variations. As channel length is reduced, variations in channel length cause greater variations in threshold. As already remarked, the effect of these variations is to require a more conservative setting of N_{off} so that low threshold devices will still have acceptable off-currents. As channel length is reduced, at some point this lowered N_{off} estimate will encounter the same punchthrough limit already discussed.

There are, therefore, two conditions, either of which may decide the minimum channel length: either direct increases in N_{off} due to barrier lowering or threshold variations that set a lower bound on N_{off}. If channel length is under very tight control, then threshold variation from length variation remains small, and only threshold variation due to drain voltage variations will matter. However, it is more common that channel length variations dominate the choice of minimum channel length.

At this point, the overall design strategy as shown initially in Fig. 1.10 is clear, although no specific barrier-lowering model has been chosen. Refinement of the long-channel design to include profiles like Gaussian or Pearson distributions affords no change in strategy and results in minor numerical complications extremely simple compared with full-blown device analysis. Despite the basic implications of any such design strategy, there has been remarkably little study of strategies in the literature and it is hoped that this discussion will stimulate further activity.

1.4.3.3   *Other Effects.* Up to this point, emphasis has been on design of the channel profile, based on adequate off-current combined with prevention of barrier lowering and punchthrough in the bulk. However, as devices shrink, two other factors become increasingly important: parasitic elements and hot-electron effects. In the following sections, we discuss series resistance and hot-electron degradation as two areas receiving much attention at the moment.

1.5   PARASITIC RESISTANCE OF SOURCE AND DRAIN

As MOSFETs are scaled into the submicron regime, intrinsic source and drain series resistance is becoming more and more important. Because both the g_m and the current of a MOSFET increase with miniaturization, the IR drop across the series resistance becomes a non-negligible fraction of the applied bias. In addition, the parasitic resistance itself is not fully scalable due to fundamental limitations such as doping solid solubility. In practice, the parasitic resistance actually may increase upon scaling to smaller dimensions. The resistance components associated with a MOSFET structure are shown in Fig. 1.19. These include the contact resistance (R_{co}), the diffusion sheet resistance (R_{sh}), the spreading resistance (R_{sp}), and the accumulation layer resistance (R_{ac}), which now will be discussed.[44]

The resistance R_{co} is defined as the resistance between the contact metal and the Si underneath the leading edge of the contact. This component is well character-

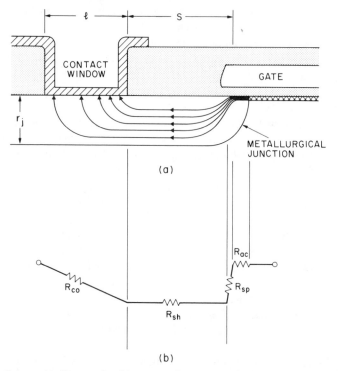

Fig. 1.19 Schematic diagram for (*a*) current flow pattern in the source/drain region and (*b*) the associated resistance components. (After Ref. 44. Reprinted with permission, copyright 1987, IEEE.)

ized by a one-dimensional R–G transmission line model[45,46] which predicts that

$$R_{co} = \frac{\sqrt{\rho_\square \rho_c}}{W} \coth\left(\ell \sqrt{\frac{\rho_\square}{\rho_c}}\right), \tag{37}$$

where $\rho_\square$ is the Si diffusion sheet resistance per square, ρ_c is the interfacial specific contact resistivity between Si and foreign material, ℓ and W are the length and width of the contact, respectively. Equation (37) includes the current crowding effect that increases the value from the familiar simplified expression $\rho_c/$area. Equation (37) has two extremes for most of the values. Within 10% accuracy,

$$R_{co} \approx \frac{\rho_c}{W\ell}, \qquad \ell \lesssim 0.6\sqrt{\frac{\rho_c}{\rho_\square}}, \tag{38}$$

and

$$R_{co} \approx \frac{\sqrt{\rho_\square \rho_c}}{W}, \qquad \ell \gtrsim 1.5\sqrt{\frac{\rho_c}{\rho_\square}}. \tag{39}$$

The former case represents negligible current crowding, whereas the latter case represents total domination by current crowding (R_{co} is independent of ℓ).

For self-aligned silicide contacts, the dominant contact resistance is between silicide and Si, and relevant parameters should be used.[47] The finite sheet resistance of the silicide can be neglected, as it is usually much smaller than the Si $\rho_\square$, and the silicide can be treated as a metal contact. It is a common mistake to think that since silicide is in parallel with the Si diffusion, it reduces the sheet resistance $\rho_\square$. More correctly, the advantage of silicide is that it is a self-aligned process which results in an area for the contact to Si almost as large as the area of the source and drain regions themselves. This contact area is much larger than that available using a process that must allow room for realignment tolerances.

The resistance R_{sh}, as shown in Fig. 1.19, is given simply by

$$R_{sh} = \frac{\rho_\square S}{W} \tag{40}$$

and

$$\rho_\square = \frac{\rho}{r_j}, \tag{41}$$

where ρ is the resistivity of the heavily doped region and S the spacing between the contact edge and the channel. For self-aligned silicides, S in Eq. (40) is very small. Equation (41) assumes that the maximum doping level in the source and drain is independent of junction depth r_j. Some authors claim that in order to obtain shallow junctions the maximum doping level has to be reduced. They propose that[28,47]

$$\rho_\square \propto \left(\frac{1}{r_j}\right)^n, \tag{42}$$

where n can be as high as five.

The details concerning resistances R_{sp} and R_{ac} are the main subjects of Ref. 48. A very important parameter is the slope of the doping profile near the metallurgical junction at the Si surface [the horizontal profile is approximated by $\exp(Kx)$ where x is the horizontal distance from the junction]. At large K, the profile approaches a box profile which minimizes both R_{ac} and R_{sp}. In general, a steep junction profile is desirable for series resistance reduction, but it does increase electric fields, thus increasing substrate current and lowering breakdown voltages.

Qualitatively, the resistance R_{sp} arises from the radial pattern of current spreading from the MOSFET channel, which has a thickness on the order of 50 Å. This spreading pattern can be understood as follows. The high local resistivity at the channel end due to the nonabrupt profile leads to a very high value of R_{sp} if the current spreads into the source/drain bulk immediately. A less resistive path is provided by the surface accumulation layer in the gate-to-source/drain overlap region. The current prefers at first to confine itself to this surface layer and later spread into the bulk once the local resistivity drops to a reasonably small value. The resistance component R_{ac} is attributed to this surface layer effect. In Ref. 48, a minimum for the sum of $R_{sp} + R_{ac}$ is used as a criterion to determine the point of current spreading. In practical non-LDD devices, the distance for R_{ac} is on the order of 200 Å. Notice that this distance is usually much smaller than the total gate-to-S/D overlap region. These two resistance components are gate-voltage dependent, and neglect of this dependence reduces the accuracy of most methods in the literature for measuring the actual effective channel length of a MOSFET.[49,50] Attempts to correct this situation have begun.[51]

Figure 1.20a presents a quantitative comparison of different contributions from different resistance components as a function of channel length for one particular scaling scenario.[44] The dominant contribution is $R_{sp} + R_{ac}$ and it does reduce with scaling. The impact of series resistance on MOSFET scaling also is reflected in the g_m prediction as a function of channel length in Fig. 1.20b. It shows that g_m increases as devices are scaled down and that the resistance has little impact on g_m in the saturation region.

A very important and interesting question to be answered is how much resistance a MOSFET can tolerate before the circuit performance is affected. Figure 1.21 shows the simplified output characteristics for $V_G = V_{DD}$ with and without series resistance. Notice that the impact in the triode region is much more pronounced than in the saturation region. Analytically, the following relationship can be used[52]:

$$g_m = \frac{g_{mi}}{1 + (R_s + R_d)g_{di} + R_s g_{mi}}, \tag{43}$$

where g_{mi} is the intrinsic transconductance without source and drain resistances (R_s and R_d) and g_{di} is the intrinsic drain conductance. In the triode region, it can be shown that the g_m or current is degraded by a factor of $[1 + (R_s + R_d)/R_{ch}]$, where R_{ch} is the channel resistance in the triode region and is equal to I/g_{di} (or V_{SAT}/I_{SAT}). In the saturation region, g_m is degraded by a factor of $(1 + R_s g_m)$. For submicron devices, velocity saturation occurs and $g_m \approx I_{SAT}/V_G$. Furthermore, V_{SAT} is approximately one-third of V_G for practical devices. With these relationships, it can be deduced that the degradation of g_m and current in the triode region is approximately six times as much as that in the saturation region.

The estimated tolerable series resistance on each side of source and drain is shown in Fig. 1.22 as a function of channel length. The absolute values should be taken cautiously, as they change with different scaling laws. Generally, these values should be chosen to be around 10% of the channel resistance in the triode region. By the previous argument, both g_m and current in the triode region would be degraded by $\approx 20\%$, whereas those in the saturation region would be degraded by $\approx 3.5\%$. With these resistance values, both noise margins and charging speed are affected. In NMOS circuits, the low-voltage point is increased by 20% for the same load current. In CMOS circuits, noise margin is not sensitive to series resistance. Speed is degraded by $\approx 10\%$ for both NMOS and CMOS circuits. This

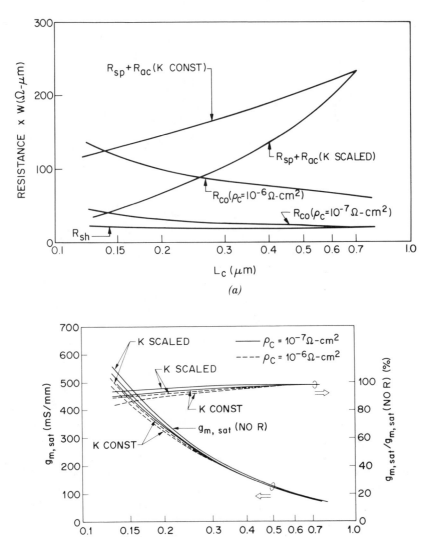

Fig. 1.20 (a) The predicted contributions of each resistance component as a function of channel length for n-MOSFETs. (b) Predicted g_m in the saturation region with and without series resistance. (After Ref. 44. Reprinted with permission, copyright 1987, IEEE.)

number is derived in the following manner. From Fig. 1.21, the pull-down time of an inverter can be divided into t_1 (from V_{DD} to V_{SAT}) and t_2 (from V_{SAT} to ≈ 0), with $t_1 \approx C_L(V_{DD} - V_{SAT})/I_{SAT}$ and $t_2 \approx C_L V_{SAT}/I_{SAT}$ (one time constant, $R = V_{SAT}/I_{SAT}$), where C_L is the nodal load capacitance. Since $V_{SAT} \approx 1/3 V_G$, $t_1 \approx 2t_2$. The overall speed degradation is approximated by $20\%(1/3) + 3.5\%(2/3) = 9\%$.

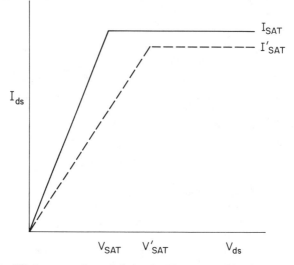

Fig. 1.21 Simplified output characteristics for $V_G = V_{DD}$ with (dotted line) and without (solid line) series resistance. The impact in the triode region is shown to be approximately six times that in the staturation region.

1.6 TRANSPORT

In this section, we want to sketch some developments in two areas: (a) modeling the velocity–field relationship and (b) modeling the hot-carrier distribution function. For transport in bulk silicon, these calculations are not simple, but for the MOSFET an additional complication is the influence of the Si–SiO$_2$ interface. At the outset, a major problem must be faced, namely, that the interfacial scattering effects on transport in the channel are not well known. No microscopic model exists that does not involve empirical parameters to be found by fitting measurements. In addition to unclear scattering mechanisms, the confinement of the carriers near the interface introduces quantum effects. The shape of the potential well confining the carriers to the interface is not known and fluctuates in shape from point to point across the interfacial plane due to variations in the field normal to the interface. Such variations are perhaps due to surface roughness, localized variations in trapped charge near the interface, or more macroscopically, to the two-dimensional nature of the potential and the transport as the drain is approached. In fact, as the carriers approach the drain, transport may shift from surface to bulk behavior, requiring some knowledge of how the interfacial scattering effects attenuate with depth away from the interface.

With all these uncertainties, some empiricism must creep into our modeling. The aim is to keep empiricism under control.

1.6.1 Velocity–Field Models

Many velocity–field expressions exist, and applications to circuit modeling, yield analysis, or process monitoring have their own demands on such models. However, in

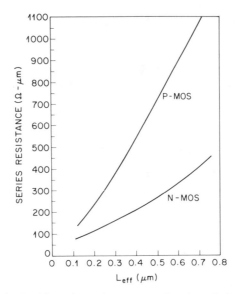

Fig. 1.22 Estimated tolerable series resistance as a function of channel length, on each side of source and drain. The scaling law according to Ref. 44 is assumed.

this section it is physically motivated models that are important. It is mechanism that is to be emphasized, not computational ease nor database efficiency.

In modeling transport near the Si–SiO$_2$ interface, two different electric field dependencies arise. The first is due to carrier drift in the source/drain field, usually called the parallel field dependence. The second is due to the action of the gate electrode in pinning the carriers to the interface, usually called the normal field dependence.

1.6.1.1 Normal Field Dependence. Carriers are confined to the Si–SiO$_2$ interface in a potential well created by the gate electrode. As the normal field at the interface increases, the well deepens and narrows. Consequently, the interaction of the carriers with the interface is modulated by the normal field.

It was shown by Schrieffer[53] that an increase in the normal field reduced the mobility. Since then, many authors have tried to predict this normal field dependence quantitatively. For strong normal fields at room temperature, Sabnis and Clemens[54] have shown experimentally that the channel mobility is fairly independent of interfacial properties. They showed that a variety of samples prepared with different oxide thicknesses and doping levels exhibit the same normal field dependence, provided the normal field is computed as the "effective" field seen by the carriers. This effective field is considered intuitively to reflect the tendency of the carriers to screen themselves from the interface, and is computed as

$$E_{\text{eff}} = \left(\frac{q}{\kappa_s \varepsilon_0}\right)\left(N_{\text{sub}}w + \frac{1}{2}N_{\text{inv}}\right), \tag{44}$$

with w = depletion layer width. If we accept this view and search for a normal field dependence via the "effective" field, then Schwartz and Russek[55] showed that a combined classical and quantum model (an entirely quantum model is too cumbersome) leads to a mobility given by

$$\frac{1}{\mu} = \frac{1}{\mu_s} + \frac{\overline{P}m^*v_{th}}{2q(Z_{QM} + Z_{CL})}, \tag{45}$$

where $\mu_s \simeq 1150(T/300)^{2.5}$ cm^2/V · s, independent of doping, and

$$Z_{QM} = \left(\frac{9\hbar^2}{4m^*qE}\right)^{1/3} \tag{46}$$

is the quantum-mechanical channel width based on the uncertainty principle, with $\hbar = h/(2\pi)$, h = Planck's constant, and

$$Z_{CL} = \left(\frac{3kT}{2qE}\right), \tag{47}$$

the classical channel width. Other parameters in (45) are the Fuchs scattering factor $\overline{P}$, the thermal velocity $v_{th} = (3kT/m^*)^{1/2}$, and the effective mass $m^* = 0.92\ m_0$, where m_0 is the free electron mass. This result is quasi-empirical, involving the crude assumption that the channel depth in the direction normal to the interface can be approximated by the sum of a classical and a quantum width.

At lower inversion layer carrier densities just above threshold, a dopant level dependence of the mobility is evident in addition to that contained in E_{eff}, a consequence of scattering by dopant ions. At still lower gate biases, where the normal field is determined by the gate and the ionized dopant ions in the depletion layer, localized charges near the interface cause fluctuations in carrier density. That is, for moderate to high interface charge densities, due to interface traps or to oxide charge, a "swiss-cheese" model results for the inversion layer. As a result, conduction follows tortuous paths to avoid "bubbles" in the inversion layer once the inversion layer carrier density is reduced sufficiently. Hence, at low gate biases, in the subthreshold region, the apparent mobility suddenly drops with decreasing carrier density. Such behavior was observed by Chen and Muller,[56] and agrees with a theory described by Brews.[57,58] This carrier density fluctuation model has been pursued further by other authors.[59–61]

The form chosen for the normal field dependence varies from author to author. Hiroki et al.[62] suggest

$$\frac{1}{\mu_\perp} = \frac{1}{\mu(N)} + \alpha\frac{E_\perp}{1 + \beta E_\perp^{2/3}}, \tag{48}$$

with $\mu(N)$ the zero-field bulk mobility of Scharfetter and Gummel as a function of dopant density N. Equation (48) can be interpreted as an example of Matthiessen's rule for combining bulk and surface scattering, as proposed by Schwartz and Russek.[55] For agreement with data of Cooper and Nelson,[63] Hiroki et al.[62] found α and β to be 8.2×10^{-9} s/cm and 3.18×10^{-4} $(cm/V)^{2/3}$. This expression is another version of the Schwartz–Russek mobility (45) if $E_\perp$ is taken as E_{eff} from (44) and μ_s replaces $\mu(N)$. However, Hiroki et al.[62] interpret $E_\perp$ as a microscopic field, not as E_{eff}, and hence incorporate the doping dependence through $\mu(N)$ instead of through E_{eff}.

1.6.1.2 Parallel Field Dependence. Some interesting ideas about modeling carrier velocity stem from the work of Thornber,[64] who furthered the goal of parameterizing the carrier velocity in terms of (a) the low-field mobility, (b) the saturation velocity, and (c) the electric field. Rather than strictly empirical fitting, Thornber examined the transformation properties of the Boltzmann equation when scattering rates were scaled. The resulting scaling of the electron distribution function places restrictions on the possible forms for the drift velocity. If now it is supposed that bulk transport differs from surface transport by such a scaling, then bulk velocity–field relations can be related to surface velocity–field relations.

It was shown by such scaling that the velocity–field relation of Yamaguchi[65] probably was too restricted, and that this relation could lead to an exaggerated normal field dependence of the mobility. This conclusion is supported by work of Cham et al. (Ref. 2, pp. 88–90) and Hiroki et al.[62] However, universal agreement has yet to be reached.

The form for the mobility in bulk silicon proposed empirically by Scharfetter and Gummel, and reinterpreted in terms of basic scattering mechanisms by Thornber, is used by many authors for the parallel field dependence, namely,

$$\frac{1}{\mu} = \frac{1}{\mu_\perp}\left[1 + \frac{(\mu_\perp E_\parallel/v_c)^2}{\mu_\perp E_\parallel/v_c + G} + \left(\frac{\mu_\perp E_\parallel}{v_s}\right)^2\right]^{1/2}, \qquad (49)$$

where v_s is saturation velocity ($\sim 1.04 \times 10^7$ cm/s) and v_c and G are related to acoustic phonon scattering (4.9×10^6 cm/s and 8.8, respectively). Here, $\mu_\perp$ is the mobility at low $E_\parallel$, [e.g., from Eq. (48)], and it depends on the normal field, which modulates the amount of interfacial scattering.

Equations (48) and (49) are very different from the form proposed by Yamaguchi,[65] and also differ from Thornber's proposal[64] that (49) be used with $\mu_\perp = \mu(N)g(E_\perp)$ [$g(E_\perp)$ an empirical function independent of doping], possibly supplemented by making v_s and v_c weakly dependent on $E_\perp$.

1.6.1.3 Extention to Detailed Computer Modeling. The mobility expressions above refer in some average way to all the carriers in the channel. However, in a MOSFET the carriers contributing to the measured source/drain current need not remain near the interface as they approach the drain. In general, one can raise the question of how the mobility is to be modeled for carriers at a general position somewhere between the surface region and a position more remote from the surface.

Any computer model that does not solve the transport problem using an interfacial scattering model for individual electrons, but instead uses a channel mobility or a channel velocity–field relation, must resort to expediency. For example, Selberherr[66] has suggested an empirical approach to fade out the interface effect with depth. Thornber's approach seems to suggest a spatially varying scaling factor that would interpolate the scattering from surface to bulk. Hiroki et al.[62] suggest using the mobility expressions quoted above with normal and parallel fields interpreted as microscopic field components normal and parallel to the directions of current flow, rather than effective channel fields normal and parallel to the interface. For the parallel field, this interpretation seems valid, but for the normal field no fundamental argument has been advanced to support this view. Clearly, this difficulty requires more attention.

A more basic approach is to solve the Boltzmann equation using a Monte Carlo method. For example, Chu-Hao et al.[67] modeled interfacial scattering in terms of surface roughness scattering (introducing the parameters $\Delta \cdot L$ to describe the deviation of the surface from flatness) and interfacial charge scattering (introducing a parameter GN_I to describe the "effective" density of interface charges per unit area). Fitting the results to experimental data, they found $\Delta \cdot L \sim 0.5$–$18 \ \text{Å}^2$ and $GN_I \sim 3.5$–75×10^{10} charges/cm^2, depending on the sample fitted. Because their model is that of a two-dimensional electron gas, the depth dependence of interfacial scattering referred to above is evidenced in their model only indirectly as a dependence of the scattering upon the energy sub-band of the carrier — higher sub-bands have wave functions extending deeper into the silicon away from the interface. On this basis, their model for surface roughness scattering has no depth dependence (it is energy independent) and the interface charge scattering falls off with distance from the surface (it has a dependence on 1/energy). Because of the need to treat many sub-bands, it may be difficult to test these basic approaches against the quasi-empirical approaches of Thornber and Schwartz–Russek, especially at room temperature.

1.6.2 Distribution Function for Electrons

Let us now discuss why device simulation programs are expected to fail as miniaturization proceeds. These programs solve Poisson's equation and the current continuity equations,

$$\varepsilon_s \nabla \cdot \mathbf{E} = -\varepsilon_s \nabla^2 \psi = -q(n - p + N_D^+ - N_A^+), \tag{50a}$$

$$\frac{\partial n}{\partial t} - \frac{1}{q} \nabla \cdot \mathbf{J}_n = G - R, \tag{50b}$$

$$\frac{\partial p}{\partial t} + \frac{1}{q} \nabla \cdot \mathbf{J}_p = G - R, \tag{50c}$$

in two or three dimensions, using as input the device description from a process simulator (see Fig. 1.4). Here, G and R are generation and recombination rates. Equation (50a) is Poisson's equation, and (50b) and (50c) are the continuity equations for electrons and holes, respectively. To these equations must be added relations between the carrier densities, currents, and potential, and expressions for G and R.

Of particular interest is the expression for electron current. (We emphasize electrons, but many of the same ideas apply to holes as well.) Traditionally, the current is expressed in the "drift–diffusion" approximation as two terms: a drift term proportional to the electric field and a diffusion term proportional to the gradient of the carrier density. However, this approximation is not valid in regions of rapid spatial variation nor for rapid time variation.

To extend device simulation to these regimes requires an approach more basic than "drift-diffusion," such as the Boltzmann transport equation for the electron distribution. This starting point will be examined further, although it too may be too crude for the smallest channel lengths, that is, below 0.1 μm. The problems in this extreme regime, where scattering events are no longer localized, individual events, have been discussed by Barker and Ferry.[68] The basis for the Boltzmann equation, a study in itself, is not pursued here.

Assuming the adequacy of the Boltzmann transport equation, one still faces a formidable problem in trying to find the electron distribution function. Three basic approaches have evolved: (a) Monte Carlo methods, (b) moment methods, and (c) combination or fitted Monte Carlo methods.

Our discussion begins with a statement of the Boltzmann transport equation, and then discusses each of these three approaches in turn.

1.6.2.1   The Boltzmann Transport Equation.   We introduce the distribution function $f(\mathbf{k}, \mathbf{r}, t)$ that describes the probable number of electrons with crystal momentum $\mathbf{k}$ at spatial position $\mathbf{r}$ at time instant t. Using this function, we can find the average carrier density at position $\mathbf{r}$ as

$$n(\mathbf{r}, t) = \int d^3k f(\mathbf{k}, \mathbf{r}, t), \tag{51}$$

the average velocity as

$$n(\mathbf{r}, t)\langle \mathbf{v}(\mathbf{r}, t)\rangle = \int d^3k \frac{1}{\hbar}\nabla_k \varepsilon(\mathbf{k}) f(\mathbf{k}, \mathbf{r}, t), \tag{52}$$

where $\varepsilon(\mathbf{k})$ is the energy vs. $\mathbf{k}$-vector relation for the carriers, and the average energy as

$$n(\mathbf{r}, t)\langle w(\mathbf{r}, t)\rangle = \int d^3k \, \varepsilon(\mathbf{k}) f(\mathbf{k}, \mathbf{r}, t). \tag{53}$$

If several bands or band minima are involved, multiple distribution functions and multiple $\varepsilon(\mathbf{k})$ relations must be introduced. Here, only one band minimum is considered.

The distribution function $f(\mathbf{k}, \mathbf{r}, t)$ satisfies the Boltzmann transport equation under certain restrictions not explored here. However, it should be mentioned that, as usually employed for semiconductor transport, the Boltzmann equation is drastically simplified. Thus, basic reservations about the usual derivation of this equation are almost irrelevant. For example, correlations between pairs of particles (important in assessing the effect of interactions between particles) are often neglected altogether. If so done, it becomes moot to discuss the validity of the approximation expressing the two-particle distribution function. $f(\mathbf{k}, \mathbf{r}, \mathbf{k}', \mathbf{r}', t)$ in terms of the single-particle distribution function, for example, as

$$f(\mathbf{k}, \mathbf{r}, \mathbf{k}', \mathbf{r}', t) \sim f(\mathbf{k}, \mathbf{r}, t)f(\mathbf{k}', \mathbf{r}', t),$$

to obtain a closed equation for the distribution function. See, for example, Ferziger and Kaper.[69]

In any event, the Boltzmann transport equation is as follows,

$$\frac{\partial f}{\partial t} + \frac{1}{\hbar}\mathbf{E} \cdot \nabla_k f + \frac{1}{\hbar}\nabla_k \varepsilon \cdot \nabla_r f = S, \tag{54}$$

where S represents the effects of scattering, and could be calculated if the details of all the scattering mechanisms were known, and multiple-particle correlations were treated using an ansatz such as that just described.

How may we solve this equation? Perhaps the most detailed solution is found using a Monte Carlo approach. A less detailed, but perhaps more illuminating approach, is to restrict one's attention to some gross averages and try to understand their interrelations. Each approach is examined in turn.

1.6.2.2   The Monte Carlo Approach.

Reggiani[70] describes the Monte Carlo approach in detail. He also describes a proof that the Monte Carlo procedure does solve the time-varying Boltzmann equation for a spatially uniform distribution. Baccarani et al.[71] discuss a proof that the Monte Carlo method solves the Boltzmann equation in steady state (no time dependence) for spatially varying fields. In neither case are electron–electron correlations included, which means energy exchange between carriers must be mediated by the other scattering mechanisms. Thus, while the theoretical equivalence of the Monte Carlo approach and the simplified Boltzmann equation is not fully established, it applies for a case general enough to provide a test of some even more drastically simplified approaches, to be discussed next.

In view of the extensive discussion of Monte Carlo methods already available, we confine ourselves here to two points receiving some attention at present. One point is the adaptation of Monte Carlo methods to include particle–particle correlation. At present, this work is confined to a kind of self-consistent field approach, which couples the electron in flight to an average field as modified by its fellow electrons via Poisson's equation. One approach is to excise regions of a device where a Boltzmann transport equation seems necessary and use a Monte Carlo approach in these regions. To find the boundary conditions on these regions, and the overall behavior, moment equations are solved for the entire device, with explicit moments of the distribution function being computed in the excised regions. Such an approach has been outlined by Bandyopadhyay et al.[72]

Another application of Monte Carlo techniques is to problems involving the high-energy tail of the distribution function. The tail of the distribution is of particular importance in estimating gate current, hot-carrier degradation of the interface, substrate current, and so forth. Because the tail of the distribution is rather unrelated to gross averages such as the overall average carrier energy, a technique that does not prejudge the form of the tail region, such as the Monte Carlo method, is desirable. On the other hand, the tail region involves sensitivity to some rather unknown aspects of electron transport.

One basic point that could have major importance for modeling hot-carrier injection into SiO_2 is the role of collisional broadening. Collisional broadening is important for hot carriers because the scattering rate can be so large that a carrier remains only briefly in a given energy state. Under these circumstances, the energy of a quasi-particle is not strictly conserved in a phonon collision, and the final kinetic energy of a carrier may be greater than the initial kinetic energy even if an emission process has occurred. For electron energies above ~ 2 eV in GaAs, Kim et al.[73] find large increases in the number of hot electrons when collisional broadening is taken into account. About a factor of ten increase in injected current from Si into SiO_2 due to collisional broadening was found by Tang and Hess[74] for substrate biases above 11 V. At lower voltages, these effects could be much more important on a percentage basis, because other contributions to the injected current are much smaller. A basic method for inclusion of these effects in a Monte Carlo calculation was presented by Reggiani et al.[75]

Just how these effects are to be treated still is under development, and there are other aspects of hot-electron transport that also require investigation. Nonetheless, we cannot wait for all these questions to be resolved before seeking some guidance to practical problems. Hence, application of Monte Carlo methods in advance of a solid foundation is proceeding, with enough success to support hope that not all these complications are important. For example, see Sangiorgi et al.[76]

1.6.2.3 Moment Methods (Hydrodynamic Models). Moment methods attempt to simplify the transport problem by restricting attention to just a few parameters, such as those of Eqs. (52–53), rather than looking for all the detail contained in the distribution function itself. This approach is patterned after the treatment of liquids and gases as described, for example, by Ferziger and Kaper,[69] Montgomery,[77] or Harris.[78] A typical derivation is given by Blotekjaer,[79] and has been developed by Cook and Frey,[80] Rudan and Odeh,[81] and Forghieri et al.,[82] among others. The moment equations are obtained by multiplying the Boltzmann equation by unity, $\mathbf{v} = 1/\hbar \nabla_k \varepsilon$ and $\varepsilon(\mathbf{k})$ and then integrating over the Brillouin zone. Denoting tensors by bold face symbols in square brackets, the resulting equations can be written as

$$\frac{\partial}{\partial t} n + \mathbf{\nabla}_r \cdot (n\langle \mathbf{v}\rangle) = 0, \qquad (55)$$

$$\frac{\partial}{\partial t} (n\langle \mathbf{v}\rangle) - \mathbf{E} \cdot \langle[\mathbf{m}^{-1}]\rangle n - \mathbf{\nabla}_r \cdot (\langle[\mathbf{vv}]\rangle n) = \langle \mathbf{v}S\rangle, \qquad (56)$$

$$\frac{\partial}{\partial t} (n\langle w\rangle) - \mathbf{E} \cdot n\langle \mathbf{v}\rangle + \mathbf{\nabla}_r \cdot (n\langle \mathbf{v}\varepsilon\rangle) = \langle \varepsilon S\rangle, \qquad (57)$$

where the reciprocal effective mass tensor is given by

$$n\langle[\mathbf{m}^{-1}]\rangle = \int d^3k \frac{1}{\hbar}\mathbf{\nabla}_k \frac{1}{\hbar}\mathbf{\nabla}_k \varepsilon f. \qquad (58)$$

It remains to interpret the expressions given by

$$n\langle \mathbf{v}\varepsilon\rangle \equiv \int d^3k \, \varepsilon \frac{1}{\hbar}\mathbf{\nabla}_k \varepsilon f = n\mathbf{w}, \qquad (59)$$

$$\langle[\mathbf{vv}]\rangle n \equiv \int d^3k \left(\frac{1}{\hbar}\mathbf{\nabla}_k \varepsilon\right)\left(\frac{1}{\hbar}\mathbf{\nabla}_k \varepsilon\right) f = k_B[\mathbf{T}]. \qquad (60)$$

Until these quantities are related to n, $\langle w\rangle$, and $\langle \mathbf{v}\rangle$, the moment equations are not a closed set. Of course (cf. Harris[78]), one could continue with higher-order moments, multiplying the Boltzmann equation by $[\mathbf{vv}]$ and $\mathbf{v}\varepsilon$, for example. Then one finds additional equations which couple $n\mathbf{w}$ and $k_B[\mathbf{T}]$ at the expense of introducing new entities.

When do we stop adding moments, and how do we stop? The basic strategy is to introduce macroscopic laws, such as Fourier's law of heat conduction relating heat flow to temperature gradient via the thermal conductivity, and Newton's law relating the pressure tensor to the rate of shear via the bulk and shear viscosities. For example, Fourier's law could be introduced in the form

$$\mathbf{w} = k_B\mathbf{\nabla}[\text{Trace}([\mathbf{T}])]. \qquad (61)$$

These basic laws must be incorporated correctly because an incorrect implementation could force-fit the underlying distribution function to a drifted Maxwellian or some other unrealistic form. We will not realize that this forcing has happened because we never find the actual distribution function itself.

To a large extent, the left sides of the above moment equations contain little information about the real, physical system at hand: greater information is contained in the various moments of the scattering term on the right. For example, a common approach is to introduce momentum and energy relaxation times using the definitions

$$\tau_m \equiv \frac{\langle \mathbf{v}\rangle - \langle \mathbf{v}\rangle_0}{\langle \mathbf{v}S\rangle}; \qquad \tau_\varepsilon \equiv \frac{w - w_0}{\langle \varepsilon S\rangle}. \qquad (62)$$

Then most of the physics is buried in the times τ_m and τ_ε. How are these times to be estimated? The answer to this point brings us to the hybrid methods which attempt to use Monte Carlo estimates for these times.

1.6.2.4 Hybrid Methods. A good deal of work has been done using the moment equations with approximate relaxation times. For example, Nougier et al.[83] used the idea of a steady-state determination of these times as a function of the average energy of the carriers for spatially independent problems. Assuming that the relaxation times can continue to be approximated as these known functions of average energy even in transient conditions, the moment equations (55–57) become a closed set of equations in a spatially uniform system. These steady-state functions are found using a Monte Carlo analysis of the steady-state problem.[83,84]

Baccarani and Wordeman[85] and Rudan and Odeh[81] treated steady-state overshoot in spatially nonuniform systems, but they introduced crude approximations to the relaxation times in place of the Monte Carlo results. The thrust of their work was the role of the truncation assumptions like (61) in spatially varying systems. This matter is not yet settled because of uncertainties in the introduction of macroscopic laws, the role of steady-state macroscopic laws in a transient analysis, and the uncertainties in the boundary conditions to be used with all the moment equations.

Although not expressed in terms of a moment method, Thornber[86] also has suggested that static, uniform field dependencies suffice to characterize overshoot outside the ballistic regime, and his results apply to spatially varying systems as well as time-varying ones. His approach has been tried by Kizilyalli and Hess[87] with good results.

1.7 HOT-CARRIER DEGRADATION

As the physical dimensions of MOSFETs are miniaturized for both speed and density, the internal field inside the device increases. This increase is caused by the increased doping concentration needed to avoid punchthrough, as well as by the use of a supply voltage that is not scaled for constant field. This partial scaling is due to a need to make the chip voltage compatible with that of other system components, to a desire to obtain high noise margins, and, in some cases, to gain circuit speed. The net result is higher fields that create hot carriers, which in turn cause long-term instability. As a result of this phenomenon, MOSFETs under operational voltages degrade with time and the $I–V$ characteristics change. These changes include a change of threshold voltage V_T, a degradation of transconductance g_m and of current, and a degradation of subthreshold slope that causes the off-current at $V_G = 0$ to increase. These undesirable effects eventually cause malfunction of the circuits. The problem of degradation is severe because it causes downtime in the system, and it is also extremely costly to replace malfunctional circuits once they are in the field.

One origin of instability is high gate voltage (or high oxide field on the order of 5 MV/cm) which results in carriers crossing the oxide layer. Carriers in the oxide can create interface traps and oxide charge and eventually lead to the phenomenon of time-dependent-dielectric-breakdown (TDDB).[88–90] TDDB occurs as electrons falling into deep oxide or interface traps release energy sufficient to cause oxide or interface damage. Once defects form, there is a tendency toward "decoration," that is, further dendritic damage near the same site.[88] When these weak spots link together across the oxide thickness, they short out the capacitively stored energy and breakdown occurs. The time to failure is related to the time required for a time integral of the current through the oxide to equal a critical total injected charge density to breakdown. This charge-to-breakdown is fairly constant at low injected current levels, but can decrease abruptly at larger current levels.[88] That is, it is injection level dependent.

One way to study this behavior is to use MOS capacitors, injecting carriers across the entire gate area. In contrast to capacitors, in MOSFETs the carriers can gain energy from the lateral field caused by the applied drain bias. Thus, the hot-carrier distribution in energy in an MOS capacitor normally differs from that in a MOSFET, which might change the degradation behavior. A possibly more serious complication in the MOSFET is that hot carriers are distributed spatially in a very nonuniform manner, due to both transport and fields. This spatial nonuniformity does not exist in an MOS capacitor, and one can speculate whether nonuniform injection leads to a spatially nonuniform charge-to-breakdown density in MOSFETs. In this section, only hot carriers resulting from the high channel field caused by the drain bias are considered.

In the case of an n-channel MOSFET, electrons gain energy from the field as they move down the channel. Before they lose energy through collisions, they possess high kinetic energy and are considered to be hot carriers (hot electrons). It generally is accepted that hot carriers cause device degradation, even though the exact mechanism is controversial. Hot carriers are known also to produce two parasitic currents, namely, gate current and substrate current. Thus, understanding of these two currents may lead to correlations with degradation. These two currents can be simulated through numerical calculations such as Monte Carlo techniques, but only analytical approaches are given here.

1.7.1 Substrate Current

For an n-MOSFET, electrons in the channel experience a very large field near the drain. The high field can cause impact ionization, and additional electrons and holes are generated by avalanche multiplication. The generated electrons are attracted to the drain, adding to the channel current, while holes are collected by the substrate contact, resulting in a substrate current (Fig. 1.23). This channel-initiated multiplication is different from pn junction avalanche breakdown in two aspects. First, the initiating current is fed from a third terminal and, second, the avalanche current is much smaller than the channel current. Hence, it is considered a low-level avalanche multiplication.

Analytically, the substrate current can be expressed as[91]

$$I_{SUB} = (M - 1)I_{CH}, \tag{63}$$

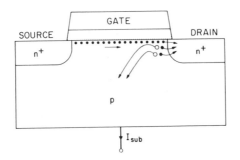

Fig. 1.23 Schematic diagram showing the generation of substrate current in an *n*-channel MOSFET.

where I_{CH} is the channel current and M the conventional multiplication factor given by

$$M = \frac{1}{1 - \int \alpha \, dx} , \tag{64}$$

and α is the ionization coefficient. For low-level multiplication, $M \simeq 1$ and the substrate current is given by

$$I_{SUB} = I_{CH} \int_0^L \alpha \, dx , \tag{65}$$

where α is a strong function of the field E and can be expressed as[91]

$$\alpha(E) = Ae^{-B/E(x)}, \tag{66}$$

where A and B are constants. For a spatially constant field,

$$I_{SUB} = I_{CH} \alpha \, \delta L , \tag{67}$$

where δL is the length of the high-field region. For a spatially varying field, the substrate current can be approximated by[92]

$$I_{SUB} = \frac{I_{CH} A E_{max}^2 e^{-B/E_{max}}}{B [dE(x)/dx]|_{E_{max}}} . \tag{68}$$

As a first approximation, the critical field near the drain E_{max} can be estimated using some simple device model.

A typical substrate current vs. gate bias curve for a submicron channel *n*-MOSFET is shown in Fig. 1.24. For a given drain voltage, E_{max} decreases as V_G increases,

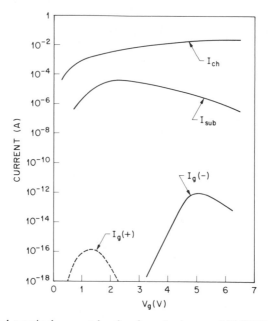

Fig. 1.24   Typical terminal current levels of a submicron n-MOSFET ($L_{\text{eff}} \approx 0.5\ \mu$m, $W \approx 10\ \mu$m) for a drain bias of ≈ 5 V. The channel current I_{CH} increase monotonically. The substrate current I_{SUB} has a characteristic peak at low V_G, whereas the gate current $I_G(-)$ peaks at $V_G \approx V_D$. The dotted line represents gate current $I_G(+)$ due to avalanched hot holes as observed by only a few experimentalists.[94–97]

but I_{CH} increases monotonically. Therefore, I_{SUB} has a maximum at low gate voltages where a significant number of carriers are available, but E_{max} has not been reduced significantly. The substrate current is usually two to five orders of magnitude smaller than the channel current. The main characteristics are its bell shape and its maximum occurring at a low gate voltage ($\approx 0.4\ V_D$). The whole curve shifts upward or downward depending on the drain bias.

The generated holes can be hot carriers (hot holes), depending on the amount of energy they gain from the hot-electrons and from the field. Because of this avalanche effect, an n-MOSFET can have both hot holes and channel hot electrons.

1.7.2   Gate Current

Gate current results from hot carriers, either channel electrons or avalanched holes, that possess sufficient energy to surmount the Si–SiO$_2$ barrier. The energy band diagram in the direction perpendicular to the Si–SiO$_2$ interface is shown in Fig. 1.25a. The barriers for electrons and holes emitted to the gate are 3.1 eV and 4.9 eV, respectively. Also shown here is the electron barrier lowering by image force effect. The net barrier height is changed to[93]

$$\phi_b = \phi_{bo} - CE_{\text{ox}}^{1/2} - DE_{\text{ox}}^{2/3}, \qquad E_{\text{ox}} > 0, \qquad (69)$$

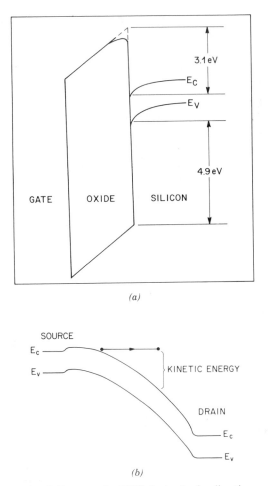

(a)

(b)

Fig. 1.25 (*a*) Energy band diagram of a MOS device in the direction perpendicular to the surface. The barriers on the Si–SiO$_2$ interface are approximately 3.1 eV and 4.9 eV for electrons and holes, respectively. (*b*) Energy band diagram of an *n*-MOSFET at Si surface above threshold with drain bias. Excess energy is indicated by the amount above E_c.

where E_{ox} is the oxide field (in the direction from gate to substrate), C is calculated as 2.6×10^{-4} (cm V)$^{1/2}$ and D is obtained by fitting experiment as 10^{-5}(cm^2 V)$^{1/3}$. The last term accounts for tunneling. For an opposing oxide field, the barrier is increased to

$$\phi_b = \phi_{bo} + E_{ox}t_{ox}, \qquad E_{ox} < 0, \tag{70}$$

where t_{ox} is the oxide thickness. In an *n*-channel MOSFET, electrons gain energy from the field in the direction of the channel. The extra energy gained is indicated by the additional energy above the conduction band edge as shown in Fig. 1.25*b*.

The typical characteristics of electron gate current $[I_G(-)]$ are shown in Fig. 1.24. The peak gate current occurs at $V_G \approx V_D$, which is distinctly different from the peak of the substrate current. At the left side of the peak ($V_G < V_D$), the oxide field near the drain hinders electron injection and the barrier is increased according to Eq. (70). The gate current thus rises exponentially with V_G. Also in this region, the gate current is found to be quite independent of V_D. Beyond the gate current peak ($V_G > V_D$), the MOSFET is driven toward the linear region of operation and the maximum field in the channel decreases. This field decrease results in less energetic electrons, reducing the gate current in spite of a rise in channel current.

Gate current can arise also from avalanched hot holes, as shown in Fig. 1.24. This current is extremely small and has been observed only by a few experimentalists,[94–97] using very specialized techniques on very thin gate oxides. As expected, the shape of this gate current follows that of the substrate current. This gate current due to hot holes is less probable than that due to electrons for the following reasons:

1. Substrate current in MOSFETs is due to low level multiplication. Hot holes are fewer than hot electrons.
2. Hot holes require a higher energy to be emitted to the gate.
3. The mean free path for holes is smaller than that of electrons. They tend to scatter and lose energy more rapidly.
4. Hole mobility in the oxide is smaller. This lower mobility will actually increase the trapping efficiency.

The two common models for gate current are the lucky-electron model and the electron-temperature model. In the lucky-electron model, the probability of an electron suffering a collision within a certain distance is analyzed. If all possible collisions were considered, this probability ultimately would determine the energy distribution of carriers. However, collisions resulting in increased carrier energy for some electrons are ignored. The "lucky" electron simply escapes energy loss events and gains energy only from the applied field. The gate current is given by[98,99]

$$I_G = I_{CH} \int_0^L P_\phi P_{ox} \frac{dx}{\lambda_r}, \tag{71}$$

where P_ϕ is the probability of an electron gaining the required energy to surmount the barrier and P_{ox} the probability of that carrier reaching the gate without suffering further collision. λ_r is the redirectional scattering mean free path, a fitting parameter related to long wavelength and intervalley acoustic phonon scattering, $\lambda_r \approx 92$ nm.[99] P_ϕ in essence describes the motion parallel to the Si–SiO$_2$ interface, whereas P_{ox} is associated with the motion perpendicular to the interface. Furthermore, P_ϕ is calculated to be

$$P_\phi = \frac{E\lambda}{4\phi_b} e^{-\phi_b/E\lambda}, \tag{72}$$

where λ is the scattering mean free path, another fitting parameter dominated by optical phonon scattering, $\lambda \approx 9.2$ nm at $T = 300$ K.[99] P_{ox} is the product of P_1, the probability that a hot electron travels to the Si–SiO$_2$ interface without any collision, and P_2, the probability that no scattering occurs in the oxide image-potential well. According to Ref. 99,

$$P_1 = \frac{\int_0^\infty n(y)e^{-y/\lambda}\, dy}{\int_0^\infty n(y)\, dy}, \tag{73}$$

and

$$P_2 = e^{-\sqrt{q/(16\pi E_{ox}\varepsilon_{ox}\lambda_{ox}^2)}}, \tag{74}$$

where λ_{ox} is the mean free path in the oxide, $\lambda_{ox} \approx 3.2$ nm.[99]

Using (71) the measured ratio I_G/I_{CH} has been fitted successfully at large drain voltages ($V_{DS} > 6$ V).[99] The main shortcoming of the lucky-electron model is that it fails to describe the observed gate current under small drain bias. That is, even the "luckiest" electron which escapes all collisions attains a maximum energy given by its initial energy plus the applied drain voltage. Experimentally, however, electron gate current can be measured with drain biases as low as 2.3 V[100] and 1.4 V.[101] The high-energy tail of an equilibrium Boltzmann distribution does not contain enough carriers to explain the observed gate current. For example, with $T = 77$ K, $V_D = 2.7$ V, $q\phi_b = 3.1$ eV, the concentration of electrons with energies above the barrier is only ≈ 0.04/cm^3. For an entire gate area 1 μm long and 10 μm wide with a velocity (corresponding to a kinetic energy of 3.1 eV) of 10^8 cm/s, the current is calculated to be only $\approx 10^{-19}$ A, far below the measured value.[†]

[†]The number of hot-electrons above in the barrier is given by

$$n = \int_{E_c + q(\phi_b - V_D)}^{\infty} N(E)F(E)\, dE,$$

where $N(E)$ and $F(E)$ are the density of states and Fermi–Dirac distribution. Since $q(\phi_b - V_D)$ is 0.4 eV above E_c and $E_c \approx E_f$, $F(E)$ can be approximated by $\exp[-(E - E_f)/kT]$. Substituting $\xi = (E - E_c)/kT$ and $a = q(\phi_b - V_D)/kT$,

$$n = \frac{2N_c}{\sqrt{\pi}}e^{-(E_c - E_f)/kT}\int_a^\infty \xi^{1/2}e^{-\xi}\, d\xi,$$

where N_c is the effective density of states in the conduction band. The above integral can be shown by integration by parts to be equal to

$$\sqrt{a}\,e^{-a}\left[1 + \sum_{n=1}^{\infty}\frac{(-1)^{n+1}|2n - 3|!!}{2^n a^n}\right].$$

Since $a(\approx 60) \gg 1$,

$$n \approx N_c e^{-(E_c - E_f)/kT}\sqrt{a}\,e^{-a}.$$

$(E_c - E_f)$ can be estimated to be ≈ -0.08 eV $(-12\,kT)$ using N_c (77 K) $\approx 3.6 \times 10^{18}$/cm^3, and a surface concentration of 10^{20}/cm^3.

Too large a gate current in the substrate injection experiments of Ning and Yu[102] was accounted for by the barrier-lowering terms in Eq. (69), not included in the estimate of the footnote. However, in the case of channel injection considered here, the maximum gate current occurs for the bias condition $V_G \sim V_D$. This bias condition leads to a low oxide field over the part of the channel where the lucky electrons have the most energy. Toward the source, the oxide field increases and the barrier-lowering effect becomes large, but near the source the lucky electrons have little energy, so injection does not occur. Hence, the discrepancy between the low theoretical estimate and the large experimental gate current at low drain bias cannot be explained by the barrier-lowering or tunneling terms of (69). What is more, Troutman[103] has shown that the lucky electrons are a minor contributor to injection for the channel injection case. He advocated the electron-temperature model, in which carrier heating by the field leads to a form of thermionic emission rather than injection of lucky electrons. This model is described next.

The electron-temperature model applies the theory of thermionic emission of "hot" electrons into the oxide from the channel, with the major assumption that the electron temperature T_e is not in equilibrium with the lattice. The higher electron temperature increases the number of electrons in the high-energy tail of the distribution. The gate current density is given by[102,104]

$$J_G = qn_s \left(\frac{kT_e}{2\pi m^*} \right) e^{-q\phi_b/kT_e},\tag{75}$$

where n_s is the surface carrier density. The parameter T_e is related to the electric field along the channel by different theories. In a model with phonon and ionization scattering, the electron temperature is given by the Bartelink–Moll–Meyer formula[94,105,106]

$$T_e = \frac{[(qE\lambda)^2/r\varepsilon_r](1/k)}{\frac{1}{2} + [\frac{1}{4} + (qE\lambda/r\varepsilon_r)^2]^{1/2}}\tag{76}$$

where ε_r is the Raman optical phonon energy ($= 0.063$ eV), $r (= 3.2$ for Si) is the ratio of the mean free paths for ionization and phonon scattering, E is the electric field, and λ is the mean free path for phonon emission ($\lambda \approx 6$ nm[106]). At very high fields ($E \gg r\varepsilon_r/q\lambda \approx 3 \times 10^5$ V/cm), T_e approaches $q\lambda E/k$. Another approach is based on the energy balance equation[79-85,104,107] in a relaxation time approximation and leads to the formula

$$T_e = T_0 + \frac{2qv_{sat}\tau E}{3k},\tag{77}$$

where v_{sat} is the saturation velocity ($\sim 10^7$ cm/s for $E \gtrsim 10^5$ V/cm), τ the energy relaxation time ($\approx 6-8 \times 10^{-14} s$[104]), and T_0 the lattice temperature. With the constants substituted, the above two approaches lead to similar T_e values given approximately by $6-9 \times 10^{-3} E$ (K · cm/V). The electron temperature has been correlated with the photon emission spectrum.[107]

Both the lucky-electron model and the electron-temperature model have an exponential dependence on ϕ_b. If λ in Eq. (72) is replaced using the high-field limit of (76), namely, $\lambda = kT_e/qE$, they provide similar results.[100] A major difficulty with such models is that an electron temperature that adequately characterizes the majority of the hot-electron distribution has little value in predicting the high-energy tail of the electron distribution because the distribution usually is not Maxwellian.[76,82,84,85] It is this tail that governs the gate current, weak avalanche current, and oxide degradation.

Both the lucky-electron and the thermionic emission models fail to describe the injection current at low values of drain bias. It is possible that the collisional broadening mechanism described by Refs. 73–75 can do so, but for low drain bias the emission current might be too low (too improbable) to compute in reasonable times using Monte Carlo methods. It also would be useful if careful experiments were made to eliminate weak spots in the oxide or numerous microscopic regions of low barrier height as explanations. Another suggestion is that Auger recombination is responsible for some electrons having higher energy than the applied drain bias.[108]

1.7.3 Cause of Degradation

It has been shown that there are basic inadequacies in simple theories of gate current, and it will be shown that there are even more uncertainties about the real cause of degradation. Since degradation generally is believed to be due to hot carriers, four mechanisms can be considered.

1. Interactions of channel hot electrons (in an n-MOSFET) with the interfacial region (not necessarily related to I_G or I_{SUB}).
2. Interactions during passage of channel hot electrons through the gate oxide, probably correlated with an electron gate current $I_G(-)$.
3. Interactions of avalanched hot holes with the interfacial region, perhaps correlated with the substrate current I_{SUB}. This correlation will differ for different device geometries.
4. Interactions during passage of avalanched hot holes through the gate oxide, probably correlated with a hole gate current $I_G(+)$.

Avalanched hot electrons are neglected because the channel hot electrons are more numerous due to the low level of multiplication. Besides, the field between the pinch-off point and the drain tends to repel avalanched electrons from the surface into the bulk (Fig. 1.23), so they should have less effect on the oxide interface than channel hot electrons. It also has been suggested that hot carriers can be emitted into the oxide and subsequently reemitted to the Si substrate because of the curved field lines in the oxide. Even though this process may not be correlated with the measured gate current, it can cause degradation.

Mechanism #3 and partially #1 may be correlated with the substrate current, and mechanisms #2 and #4 with the gate current. Early MOS capacitor experiments show that degradation does occur upon the passage of gate current. One could hope that if the applied voltage were scaled below the barrier value, gate current and thus degradation would disappear. However, no threshold voltage for cutting off the gate current has been observed. Gate current simply decreases steadily with reduced applied bias until it disappears in the measurement noise. Degradation measured at small drain biases also suggests that there is no threshold voltage for degradation.[108,109] The unfortunate fact that gate current never can be eliminated complicates a clear distinction between effects correlated with gate current and those correlated with substrate current.

General threshold degradation data as a function of stressing gate bias for fixed drain voltages are shown in Fig. 1.26. The amount of degradation, both of threshold voltage and transconductance, tends to follow the bias dependence of the substrate current (compare with Fig. 1.24). It is this correlation that leads some to believe that degradation can be monitored using the substrate current (#1 or #3). In contrast, in this bias range the gate current $I_G(-)$ increases exponentially with V_G, suggesting that degradation may not require the passage of gate current through the oxide. Because of these observations, comparisons are sometimes made between different structures stressed under conditions of equal substrate current rather than equal gate current. This view could be supported if the structures were not different as to injection conditions, for example, provided the hot carriers origi-

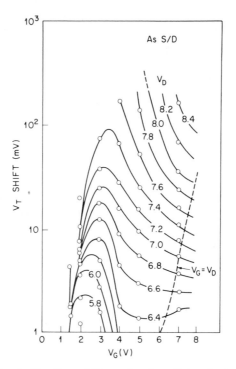

Fig. 1.26 Typical threshold voltage shift as a function of stressing gate voltage for a fixed drain bias. The shape duplicates the substrate current rather than the gate current. (After Ref. 94. Reprinted with permission, copyright 1986, IEE.)

nated at the same depth from the interface. One might think a correlation with I_{SUB} meant degradation was caused by the hot-hole gate current that accompanies substrate current (#4), but it seems more plausible that it is mechanism #1 or #2, with I_{SUB} correlated with E_{max} [see Eq. (68)].

As mentioned before, the type of hot carriers causing degradation is unclear. Part of the reason is that the measured gate current or substrate current does not adequately monitor degradation. For instance, different bias conditions can cause different degrees of degradation, even at the same gate or substrate current level. One explanation is that the measured I_G or I_{SUB} incorporates hot carriers of all energies of which only some may have energies appropriate to cause degradation. Perhaps a more important factor is that degradation often is localized in a small region and hence is caused by the injected current in that region. This local component of injected current is not related simply to the overall average current. It also is clear that any correlation can vary with changes in device structure that change the injection probabilities, for example, (73) or (74). An extreme example of poor correlation between a measured current and degradation would be a case comprised of compensating gate currents due to both hot electrons and hot holes, where both currents could cause degradation, but only the net gate current was measured.

Hot-carrier effects in p-channel MOSFETs are studied to a lesser extent. The measured terminal currents are shown in Fig. 1.27. The substrate current is smaller

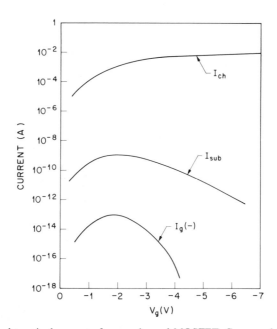

Fig. 1.27   Typical terminal currents for a p-channel MOSFET. Compared to n-MOSFETs, the substrate current is smaller by 4–5 orders of magnitude. Only gate current $I_G(-)$ due to avalanched electrons has been measured. The substrate current I_{SUB} also is due to electrons.

than that in n-MOSFETs by 4–5 orders of magnitude due to the smaller hole ionization coefficient in Eq. (65). Hot electrons resulting from the substrate current contribute to the gate current and so the shapes of the gate current and substrate current are similar. No gate current resulting from channel hot holes has been reported in the literature. It is generally observed that after stressing, negative charge is created near the drain. This leads to an increase of MOSFET current due to reduction of effective channel length.[110–113] Nevertheless, instability in p-channel MOSFETs is considered a minor problem because the shift tends to saturate at an acceptable percentage, even though it has been cautioned that a reduction in punchthrough voltage may result.[114]

Hot-carrier instability in depletion mode devices also is not considered a major problem because the current travels through the buried channel where gate current or interaction with the Si–SiO_2 interface is less probable. Second, for a given amount of charges or interface traps, the impact on buried channel device behavior is less dramatic than for the enhancement, surface channel devices. Third, in most real circuits, the gate voltage is always tied to the source, a condition which is not the worst case for degradation.

In a given case, whether degradation of I–V characteristics is caused by fixed charges or interface traps (differentiated by whether charge is varied by Fermi level) also is not clear. A decision is complicated by the fact that the spatial extent of the affected area is unknown. Generally, charge pumping measurements indicate that interface traps are responsible, but fixed charges localized to only part of the channel also can explain the observed degradation.[111,115–117]

To simulate devices in real operation, MOSFETs have been stressed under ac biases.[118–121] It generally is indicated that degradation is as much or more in the ac mode as under dc stress.[118,119,121] It also is found that the greatest degradation is caused by the falling edge of the gate voltage when the drain bias is still high.[121–123] This phenomenon may shed some light on the real cause of degradation. On the other hand, it is also reported that in a real circuit inverter operation, the degradation is much less than that under dc conditions.[124]

The structural effect due to different channel width has also been the subject of some discussion. It has been reported that both hot-carrier currents and degradation are enhanced in small-width devices,[125,126] but the effects are absent in other reports.[127]

In the midst of controversies around hot-carrier degradation, the matter in least dispute is probably the impact of hydrogen or water content in the oxide. Water has been observed to enhance degradation since the early work with MOS capacitors,[128] and nitride capping or hydrogen anneal has been found to have similar effects.[129,130]

1.8 DEVICE DESIGNS

Some of the constraints upon submicrometer MOSFETs have been described, largely in terms of the traditional HMOS structure of Fig. 1.8. Although it is not intended here to describe processing details, we would like to discuss some of the new ideas for structures designed to operate in the submicron regime.

Most of these designs target some particular difficulty of submicron devices for improvement. Examples are: hot-electron reliability, driving ability, and maximum allowable voltage. It is noteworthy that these objectives are directly related to circuit behavior — they impact the viability of the device in a real circuit application.

Once an objective is selected at least one device parameter is affected. For example, to increase the maximum allowable voltage, one may suggest a modification of junction depth. Naturally, modification of any device parameter automatically affects all aspects of device operation that physically depend on this parameter. For example, junction depth affects series resistance, breakdown voltage, barrier lowering, and bulk punchthrough. It is the aim of physical device models to describe these direct dependencies.

However, the situation is far more complicated than this direct physical coupling alone. The device processing that adjusts targeted parameters also couples with other device parameters, often in an unexpected way. As an example, a shallow junction may be achieved by a process that also leads to higher junction leakage or to a parasitic device related to an edge or corner. These side effects may well not be part of any process simulation package, and their identification or anticipation requires insight and experience. If a process-related correlation is unsuspected, it can lead to an erroneous reputation for an approach or a physically incorrect model.

In addition to these influences of processing, the viability of a new design depends on the variability of device behavior introduced by variations in the process. For example, the effective electrical length of the gate varies from device to device for the same nominal processing because of factors such as variability in etching, oxidation, junction implants, drive-in, and so on. Thus, the correlations between parameters introduced by fabrication imply correlated variations in all the coupled parameters. If the resulting variations in electrical behavior are not acceptable in a certain circuit, the circuit or the device must be changed.

With these sobering thoughts, let us examine some of proposals for structures designed for survival in the submicron regime.

1.8.1 Designs for Hot-Carrier Reliablity

Hot-carrier reliability can be controlled by (a) reducing the number of hot carriers produced, (b) reducing the number that reach the interface or enter the oxide, or (c) confining the damage to a region that has little electrical impact.

To reduce the number of hot carriers generated, one can reduce the applied voltage as a partial solution. A voltage reduction by itself implies less drive current and, hence, slower circuits and poorer noise margins. Thus, it must be accompanied by some miniaturization to reduce load capacitance, lower thresholds, and decrease channel length. We have described some aspects of this miniaturization process already. There are a few predictions for the highest voltage allowed for each effective channel length, and they are summarized in Fig. 1.28.[131-135] These extrapolations are based on short-term degradation results as outlined by Ref. 133. Even though these curves do not agree exactly with one another, partially because of the assumption of different scaling laws, they do indicate, for

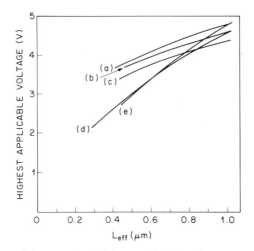

Fig. 1.28 A survey of the predicted highest applicable voltage as a function of channel length. The curves are from (a) Mikoshiba et al.,[131] (b) Takeda et al.,[132] (c) Takeda et al.,[133] (d) Hu,[134] and (e) Chen et al.[135] (Reprinted with permission, copyrights 1983, 1985, and 1986, IEEE.)

example, the urgency of reduction of supply voltage below 3.5 V for effective channel lengths less than 0.4 μm.

Another way to reduce the number of hot carriers is to change the device structure to reduce the fields inside or to limit the spatial extent of the high-field regions, so that carriers do not gain enough energy from the field to cause difficulty. Since high fields occur only near the drain, the drain doping profile can be tailored such that the potential gradient is more gradual. There exist two structures, namely, the LDD (lightly doped drain) and DDD (doubly diffused drain) devices, which are shown schematically in Fig. 1.29. The LDD device can be obtained by two independent implantations, with one of them offset by an oxide sidewall on the gate structure. The DDD structure uses simpler processing based on the difference in diffusion coefficients for phosphorus and arsenic. Alternatively, one can use different implant energies.

The field distribution in a LDD device is indicated in Fig. 1.30a, whereas the reduction of field in a DDD device is shown in Fig. 1.30b. It has been cautioned that degradation of a LDD device can be worse than for a normal device.[136,137] Since the deposited oxide on the sidewall has higher trapping efficiency and lightly doped drains are more easily depleted by oxide charge, the parasitic series resistance of the LDD drain can increase due to degradation.[138]

A number of engineers recognized that (a) the amount of field reduction, (b) the role of series resistance, and (c) the location of damage due to hot carriers all depend on the geometry of the drain near the gate edge. The three main parameter of the LDD structure are the length of the n^- region L_n, the n^- dose, and the gate overlap length L_g. See Fig. 1.31. If the length L_n is less than the depletion width w_n in the n^- region, the lateral field is dropped over L_n. If the depletion

a. CONVENTIONAL

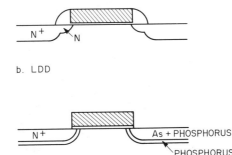

b. LDD

c. DDD

Fig. 1.29 Schematic diagram showing the drain doping profile of (*a*) a conventional MOSFET, (*b*) a LDD device, and (*c*) a DDD device.

width w_n is less than L_n, then the depletion width governs the field reduction. Hence, there is a maximum length of L_n that is useful, as pointed out by Izawa and Takeda,[139] namely,

$$L_n \sim w_n . \tag{78}$$

In the region where the gate overlaps the n^- region, the two-dimensional influence of the gate causes a lengthening of the depletion width w_n, allowing greater lateral field reduction. The gate overlap is therefore advantageous. In a more complex process, Izawa et al.[140] have made the gate overlap length an independent parameter, which they anticipate will allow the LDD structure to be used down to 0.3 μm.

A simple model was developed by Mayaram et al.[3,141] to describe the field reduction factor and series resistance as a function of the amount of overlap of the gate over the n^- drain region. The field-reduction factor FRF is defined in terms of the maximum lateral field E_m as

$$FRF = \frac{E_m(\text{LDD})}{E_m(\text{conventional})}, \tag{79}$$

and is typically greater than 0.60. However, as the hot-carrier generation is exponential in E_m [see Eq. (68)], this improvement is significant. Using their model, it is found that the best FRF for a given series resistance occurs when the n^- region is entirely under the gate electrode. However, this situation also increases the gate-to-drain parasitic capacitance.

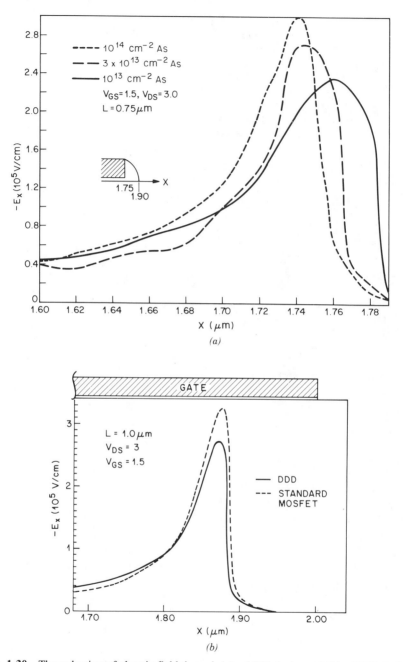

Fig. 1.30 The reduction of electric field through (*a*) a LDD device and (*b*) a DDD device.

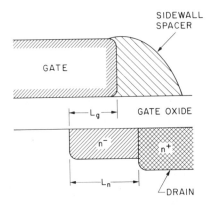

Fig. 1.31 Cross section of the drain end of an LDD device showing the sidewall spacer, the gate overlap length L_g, and the length of the lightly doped drain region L_n.

Of course, the LDD consciously increases the series resistance of the drain/source, degrading g_m [see Eq. (43)] in order to lower the maximum lateral electric field. At one extreme, when the n^- dose is so light as to be negligible, the LDD becomes a normal MOSFET (perhaps with a problem that the gate does not reach the drain) and the maximum lateral field occurs at the n^+ junction. At the other extreme, when the n^- doping is so strong that the n^- region becomes just an extension of the n^+ region, the maximum lateral field occurs at the tip of the n^- region. If the n^- drain is shallower than the n^+ drain, this case results in the highest field. An example is shown in Fig. 1.32 from Tsang et al.[142] (Actually, this case with the gate edge between the n^- tip and the n^+ drain can lead to a maximum field in one of three positions:[141] namely, at the n^- tip, at the gate edge, or at the n^+ drain. It appears that the maximum field position is near the gate edge in Fig. 1.32.) We see then, that hot-carrier injection can be moved between the two extremes: from the n^+ drain edge to the n^- drain edge. For intermediate cases, as the n^- dose is increased from a low value to a high value, there is a maximum n^- dose above which the peak field shifts position from the n^+ region to the n^- region. There is less degradation if injection is located over the n^+ drain, because the n^+ drain is doped heavily enough that it will not deplete in response to trapped charge, so trapping causes no increase in series resistance. It may, however, accentuate gate-to-drain leakage currents, as observed by Chang and Lien[143] and Chan et al.[144] at large oxide fields, and in any case may lead to long-term gate-to-drain shorts.

It would appear, then, that the optimal LDD configuration employs (a) the maximum n^- dose that allows a peak field near or within the n^+ drain and (b) a gate aligned with the n^+ drain. A maximum n^- dose minimizes series resistance within an LDD design. The gate alignment provides the greatest field-reduction factor, and automatically ensures that the sidewall spacer lies entirely over the n^+ drain. Of course, gate alignment is not precise. For example, the sidewall spacer thickness may vary with fabricational variations. If the spacer is too thick, the n^+

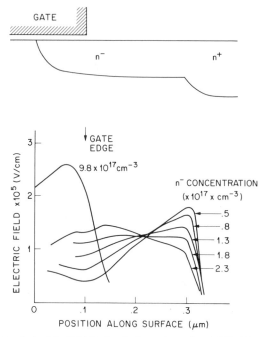

Fig. 1.32 The lateral electric field in the drain region of an LDD. The volume concentration of dopant in the lightly doped drain is the parameter. Bias conditions assumed are $V_{DS} = V_{GS} = 8.5$ V. Device parameters are $t_{ox} = 45$ nm, $r_j(n^+) = 0.45$ μm, $r_j(n^-) = 0.30$ μm. Note that for n^- doping levels below 2.3×10^{17} cm^{-3} the field peaks near the n^+ boundary, whereas for higher doping it peaks near the gate edge. (After Tsang et al.[142] Reprinted with permission, copyright 1982, IEEE.)

region will not reach the gate edge under drive-in, and hot-carrier injection into the sidewall oxide will occur. To avoid this, the process must be centered so the typical device has a nonzero overlap, setting a lower bound on how low the gate-to-drain parasitic capacitance can become (see Chan et al.[145]).

Besides the LDD and DDD structures, other types of drain modification have been proposed using retrograde or buried drains. Takeda et al.[146] proposed a Gaussian drain profile peaked below the interface. The advantage of this structure is that the peak electric field occurs below the interface, and as the hot-carrier density decays exponentially with distance [see Eq. (73)], slower degradation is expected. A somewhat more complex idea of this sort was advanced by Grinolds et al.,[147] with the buried LDD structure.

1.8.2   Designs to Improve Response Time

Let us accept that the supply voltage has been chosen as some industry standard that ensures general compatibility with other circuitry on the marketplace. With this given voltage, we set ourselves the object of designing a device with good response. Response time is governed by charging of a capacitance C by a MOSFET. If the current delivered by the MOSFET is I_{SAT} and the capacitor must charge to the supply voltage V_{DD}, then the response time T is

$$T = \frac{CV_{DD}}{I_{SAT}}. \tag{80}$$

The capacitive load to be driven is roughly the sum of some parasitic capacitances and the gate capacitance of the next stage [perhaps times some fan-out (FO) supplied by the circuit designer]. That is,

$$C \approx C(\text{parasitic}) + FO \cdot C_{ox}WL. \tag{81}$$

The saturation current is related to the saturated value of the small signal transconductance $g_m(\text{sat})$,

$$g_m(\text{sat}) \sim \frac{2I_{SAT}}{V_G - V_T} \quad (\text{mobility regime}),$$

$$\sim \frac{I_{SAT}}{V_G - V_T} \quad (\text{velocity saturated}). \tag{82}$$

Hence, the response time T is roughly (see also the discussion at the end of Section 1.5)

$$T \sim \frac{CV_{DD}}{g_m(V_G - V_T)}$$

$$\sim \frac{C}{g_m(\text{sat})}. \tag{83}$$

On the other hand, g_m is given by

$$g_m(\text{sat}) \sim C_{ox}W\left(\frac{\mu(V_G - V_T)}{L}\right) \quad (\text{mobility regime}),$$

$$\sim C_{ox}Wv_{sat} \quad (\text{velocity saturated}). \tag{84}$$

Assuming that $C(\text{parasitic}) \ll FO \cdot C_{ox}WL$, the response time improves if L is reduced. If $C(\text{parasitic}) \gg FO \cdot C_{ox}WL$ the response time improves if t_{ox} is reduced, because the parasitic capacitance is weighted less in the response time. In addition, if t_{ox} is reduced, short-channel effects are reduced, allowing a reduction in L. Thus, our response time can be improved by increasing the transconductance (e.g., by decreasing the series resistance [see discussion following Eq. (43)], or reducing the capacitance (e.g., by reducing parasitic drain-to-substrate junction capacitance or gate to drain overlap capacitance or by reducing the channel length to reduce $C_{ox}WL$).

1.8.2.1 Reduction of Series Resistance. The subject of series resistance was discussed in detail in Section 1.5. A major decrease in series resistance has been made possible using self-aligned silicides. Using silicides, one would like to form shallow junctions that still have low resistances. For example, Horiuchi and Yamaguchi[148] proposed the SOLID (silicide on lightly doped drain) structure in which the only n^+ region is a very thin layer introduced by a segregation phenomena during silicidation. Fabrication was not reported. Schmitz and Chen[149] made CMOS devices as small as 0.3 μm using Pt silicide and a 10 s E-beam transient anneal to activate junctions with $r_j \sim 0.1$ μm. The resulting $g_m \sim 100$ mS/mm, a low enough value to suggest that series resistance still was high. Ford and Stemple[150] also made Pt silicide MOSFETs down to 0.5 μm. All these authors had difficulty with junction leakage. Some precautions that can help reduce leakage were discussed by Liu et al.[151] A better approach may be to implant dopant into preformed silicide.[31,152]

In Section 1.5, it was pointed out that steeper (more abrupt) junctions improve series resistance. Study is needed to assess the trade-offs here with higher fields and hot-carrier generation.

1.8.2.2 Reduction of Parasitic Capacitance. Improvement of gate-to-drain overlap capacitance has been discussed in connection with LDD devices. The limitations here seem to be that device reliability suffers if the gate is not aligned with the drain and that too little overlap leads to too much process variability of hot-carrier injection.

Drain-to-substrate junction capacitance can be reduced, most commonly by reducing the area of the drain-to-substrate junction. This area reduction is accomplished by contacting over field oxide, leaving only a small-area, local interconnect of polysilicon or TiN over a reduced-area drain.[152] The parasitic capacitance is then the parallel combination of the local interconnect (which includes a contact pad over field oxide) and a reduced drain-to-substrate junction capacitance. There is no need for deep junctions to avoid spiking. This approach can be incorporated with a raised source and drain, reducing the effective junction depth as well. Series resistance still must be watched.

An example of this sort of structure are the devices based on raised polysilicon electrodes, such as the buried oxide MOSFET (BOMOS) of Sakurai[153] and Shinchi and Sakurai[154] (see Fig. 1.33a). With a refinement of this structure, NMOS ring oscillators were made with 0.5 μm gate lengths, $g_m \sim 115$ mS/mm, and a delay of 59 ps/stage at 5 V by Inokawa et al.[155] Series resistance was a limitation, which the authors hope can be improved using silicide. A very similar (but much larger) structure was reported by Mieno et al.[156] using different processing. Both of these devices are built on top of the substrate, after patterning of the field insulator by anisotropic etching, using epitaxially grown silicon for the active device area.

An LDD structure with polysilicon electrodes but using a recessed field oxide and no epitaxial growth was described by Misawa et al.[157] as shown in Fig. 1.33b. Chiu et al.[158] reported a similar structure for CMOS application that showed 82 ps delay at 0.6 μm L_{eff} at 5 V. The butting of the junctions against the recessed oxide

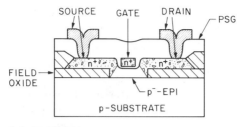

SOURCE   GATE   DRAIN

PSG

FIELD OXIDE

p⁻-EPI

p-SUBSTRATE

(a) EPITAXIAL

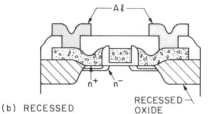

(b) RECESSED

(c) UPMOS

Fig. 1.33 Three different self-aligned structures intended to reduce parasitic drain-to-substrate capacitance: (a) a structure using epitaxial silicon for the device body continuous with raised polysilicon source/drain contacts (after Mieno et al.[156]); (b) a structure using a recessed isolation oxide and self-aligned polysilicon contacts to source and drain (after Misawa et al.[157]); (c) self-aligned polysilicon contacts over a patterned field oxide, compatible with silicide and symmetric CMOS technology (after Lynch[152]). (Reprinted with permission, copyright 1987, IEEE.)

is considered to further reduce parasitic junction capacitance. A number of such recessed oxide isolation techniques have been discussed by Tsai et al.[159] and Coppee and Van de Wiele.[160] As shown in Fig. 1.33b, the resulting structure has a sidewall capacitative coupling of the gate to the source and drain polysilicon.

Another approach compatible with a symmetric CMOS technology is the UP-MOS process of Fig. 1.33c, as proposed by Lynch et al.[161] The interested reader can find a critique of the processing trade-offs involved in these structures in the review by Lynch.[152]

A different way to reduce the junction capacitances is to move to silicon-on-insulator (SOI) technology. Colinge et al.[162] demonstrated CMOS ring oscillators with gate lengths of 0.5 μm (effective lengths of 0.2 μm) delays of 52 ps/stage and $g_m \sim 80$ mS/mm. A number of experiments and simulations show that the disadvantages of partially depleted SOI devices such as kink effect disappear for fully depleted devices made by thinning the silicon to the order of 50 to 100 nm.[163] The trade-offs in these structures are quite different from HMOS.[164,165] The SOI approach will become more attractive as various SOI processing problems are overcome, such as poor mobility, trapping, and parasitics, and as miniaturization of HMOS, LDD, etc., places more demands on process control.

1.8.2.3 Reduction of Channel Length.

The remaining approach to decrease response time is to decrease channel length. How can we make the device shorter than an HMOS device without making hot-electron stability or short-channel effects intolerable?

Let us review the constraints imposed on HMOS. The minimum oxide thickness as set by Fowler–Nordheim tunneling or by gate-induced drain leakage[143,144] is below the limits set at present by processing. Hence, t_{ox} currently is set by fabricational difficulties for voltages below ~3.5 V. Although devices have been made with oxide thicknesses as low as 25 Å, a reasonable estimate for near-future process-limited oxide thicknesses is about 50 Å (see Tanaka and Fukuma[166]). This figure is below that of many present designs. In addition, it appears that the junction depth currently is set by series resistance and junction leakage constraints at a value $r_j > 0.1$ μm. With these values for these parameters, for a symmetric CMOS technology, I_{OFF} requirements limit the amount of barrier lowering that can be allowed. This limit on barrier lowering in turn sets the minimum allowable channel length, and the targeted length must be above this minimum by an amount given by the product (barrier-lowering sensitivity) $\times$ (process variation in L). Thus, for a given variation in L, the targeted L can be reduced either by reducing the barrier lowering or reducing sensitivity of the barrier lowering to variations in L.

The most obvious way to reduce barrier lowering and its sensitivity is to minimize short-channel effects by reducing oxide thickness. It is notable that HMOS devices with thinner oxides are being made. For example, Fichtner et al.[167] have described a design for a supply voltage of 1.5 V, $V_T \sim 0.5$ V, $L_{eff} = 0.22$ μm, using a fairly standard NMOS process modified as follows. (1) Oxide thickness was reduced to 80 Å. (2) Junction depths were reduced to 0.09 μm by reducing the thermal budget after implant to 20 min at 900°C. (3) Channel doping was increased to 10^{18} cm^{-3}. (4) Polysilicon gate thickness was reduced to 0.15 μm. (5) All levels were defined by electron-beam writing. Excellent I–V characteristics were obtained, with saturation $g_m \sim 300$ mS/mm.

As another example, for a 2.5 V supply and t_{ox} of 50 Å, Kobayashi et al.[168] achieved $g_m \sim 260$ mS/mm at $L_{eff} \sim 0.28$ μm, $r_j \sim 0.15$ μm. These devices were series-resistance limited. Ring oscillator delay was 50 ps/stage. In later work,[169] devices with $t_{ox} \sim 25$ Å were made. In this case, a 1 V supply voltage is appropriate, $L_{eff} \sim 0.15$ μm, $r_j \sim 0.23$ μm, and $g_m \sim 500$ mS/mm. Both g_m values agree with the prediction using the scaling algorithm of Ng and Lynch[48] (see Fig. 1.20). Gate current was high, about 0.13 mA/cm^2 at 1.2 V.

A final example of thin-oxide devices is that of Sai-Halasz et al.[170] for 77 K, a continuation of the work by Baccarani et al.,[19] Wordeman et al.,[24] and Sun et al.[35] The advantages and disadvantages of operation at 77 K are discussed by Sun et al.[35] The main advantages are a reduced off-current and *pn* junction leakage, a reduced temperature variation of threshold, and a lower interconnect resistance. Among the disadvantages at a device level are carrier freezeout (which affects LDD design and depletion devices) and increased oxide trapping.

To explore 77 K operation, Sai-Halasz et al.[170] made devices with 45 Å oxides and effective channel lengths in the range 0.07 μm $< L_{eff} < 0.25$ μm. Shallow source/drain extensions with $r_j = 0.05$ μm were connected to main junctions with $r_j = 0.1$ μm. Supply voltage has $V_{DD} = 0.6$ V with $V_T = 0.15$ V. For $L_{eff} = 0.1$ μm, $g_m(\text{sat}) \sim 760$ mS/mm at 77 K, and the large signal $G_m = I(V_D = V_G)/V_G = 440$ mS/mm at $V_G = 0.6$ V. Room temperature values were one-half to two-thirds of the 77 K values. Room temperature field-effect mobility was 370 cm^2/V · s.

A feature of this design is a channel implant that is peaked below the surface. This implant controls punchthrough while keeping impurity scattering low in the channel region to improve the mobility. A strong process dependence of the device parameters meant that modeling was too uncertain to allow any conclusions about velocity overshoot.

In general, these reports for thinner oxide structures estimate hot-electron effects using measurements of I_{SUB}, and circuit and processing sensitivity issues are given only superficial attention. These attempts to miniaturize MOSFETs within the HMOS framework may have to be supplemented by more radical designs.

One of these designs is the so-called double implanted LDD or "halo" structure proposed by Ogura et al.,[171] Codella and Ogura,[172] and Rathnam et al.,[173] as shown in Fig. 1.34*a* and *b*. It has been applied to buried-channel devices by Odanaka et al.,[174] as shown in Fig. 1.34*c*. The midchannel portion of this structure resembles HMOS, and the dependence of I_{OFF} on the channel profile is the same. However, a shorter L can be used than for HMOS on the same substrate because the substrate doping is increased only locally, on the inside walls of the source/drain junctions. Because of this so-called "halo" doping, bulk punchthrough occurs at shorter L than for HMOS. Thus, the punchthrough limitation is relaxed and, for example, the simplified constraint of Eq. (10), expressed also in Fig. 1.18, can be lowered. Also (of greater interest for NMOS enhancement–depletion logic than for symmetric CMOS), the reverse bias sensitivity of threshold is lower, governed by the lightly doped part of the substrate.

A penalty in sidewall junction capacitance is paid, but this cost is less than that incurred should the bottom area of the junctions be over heavily doped substrate. Another penalty is an increase of fields in the bulk, and the maximum halo doping is determined by the onset of low-level avalanche at the chosen supply voltage and

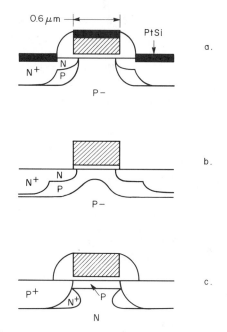

Fig. 1.34 Three designs for submicron MOSFETs as described in the text: (*a*) Ogura et al.[171,172] (*b*) Rathnam et al.,[173] (*c*) Odanaka et al.[174] (Reprinted with permission, copyrights 1983, 1985, and 1986, IEEE.)

channel length. The impact of these increased fields on the generation of hot carriers is controlled by the LDD-like *n*-extensions of the junctions.

Codella and Ogura[172] report a 0.35 μm structure like that of Fig. 1.34*a* operating near 3V with 15 nm t_{ox}. Junction depth of the *n*-extensions to the junctions was $r_j \sim 0.14$ μm, and the *p*-halo peak doping was $\sim 3 \times 10^{17}$ cm^{-3} at a depth of ~ 0.2 μm. This doping did increase the channel doping slightly at the ends of a 0.5 μm channel device.

Some authors (e.g., Bauer et al.[175]) have argued that the ultimate improvement over HMOS is possible only at the cost of an asymmetrical structure. Such structures are common in high-voltage applications, for example, the lateral insulated gate transistor (LIGT)[176] or conductivity-modulated FET (COMFET)[177] which combines MOS and bipolar transistors, and the double diffused (DMOS) transistor.[178] In an asymmetrical structure, the source region can be protected from the drain by tailoring a heavily doped substrate region surrounding the source, whereas the drain region can be designed using a buried LDD or retrograde Gaussian profile to keep hot electrons far from the surface and reduce peak fields. It seems clear that an asymmetrical structure does allow greater physical independence in adjusting these trade-offs; a processing independence is less clear. A symmetrical structure is easier to make and to miniaturize, and also offers more flexibility in circuit design. Hence, effort will continue to realize shorter devices that still are symmetrical but more complex.

Some processing issues in making small devices like LDD or the double implanted LDD just discussed have been examined by Hamada et al.[179] These authors suggest that careful implant control is essential to avoid compensation effects due to overlapping *n*- and *p*-type dopants. Otherwise, series resistance problems become severe. In addition, leakage control becomes a basic problem when *pn* junctions become too highly doped, leading to dynamic circuit constraints. Process modeling also requires greater accuracy.

1.9   CLOSING REMARKS

Optimal miniaturization remains somewhat a trial and error process, and it is tightly coupled to the entire IC design and fabrication procedure. There is need for good organization of the many steps involved, and it has been suggested that Warnier–Orr diagrams might help. A key objective in such organization is a disciplined flow of information and iteration. If feedback is undisciplined, besides causing slowdown and wasted resources, the entire process may be abortive.

This need for a more organized approach stems from pushing the technology to its limits. That is, a coupling of all facets occurs as each limitation is extended until a trade-off with another is encountered. Many examples of the trade-offs involved in design have been discussed. It was proposed that simple models of these trade-offs could be useful in guiding the overall flow of a design. As an exploratory attempt of this sort, a design procedure based on off-current as a basic parameter was presented. This initial design approach demonstrated how channel doping profile was dictated by off-current and bulk punchthrough, leaving the final compromises in terms understandable at a circuit or system level.

In an overall examination of miniaturization, several limitations stand out. One is the problem of fabricational tolerances. As devices shrink, variations in their parameters due to limited process control become larger on a percentage basis. Hence, device variability increases, and circuit design has become in part a game of statistics. One adjusts distributions of parameters—massaging the centroids and variances for maximum yield. A device design is successful only if its sensitivity to fabricational variations meshes with the statistical requirements.

To reduce these fabricational sensitivities, one recourse is to improve process control. Close monitoring and maintenance of process is one aspect of this reduction. Better processes are another. From the device standpoint, design of devices with less process sensitivity is a third.

Here is a second major limitation. To model device sensitivity, process modeling must be accurate. This accuracy requires progress from a knowledge standpoint—we do not understand the processes well—and from a computational standpoint because three-dimensional computation of great accuracy is needed. To understand processing better, models will have to be tested using test structures and test processes outside the normal course of IC fabrication, because experiments to test the models are insufficiently wide-ranging in the constrained IC environment.

Of course, besides accurate process modeling, to model device sensitivity requires accurate device modeling. In the area of hot-electron transport, this requirement has yet to be met. Even without considering overshoot and ballistic transport the modeling of electron transport is largely empirical because of uncertainties about interfacial scattering. Additional uncertainties surround hot-carrier transport for the calculation of gate current and hot-carrier degradation. Some of these issues have been discussed.

Series resistance and hot-electron reliability illustrate the above generalities. Both of these subjects lead to redesign of the source/drain junction geometry. For the series resistance problem, an abrupt profile is desirable, and deep junctions are better. For the hot-electron problem, LDD and DDD structures have been proposed that increase the series resistance. Again, deep junctions are better. However, deep junctions conflict with punchthrough and barrier-lowering constraints, limiting channel length reduction. Thus, we see efforts to redesign the junctions to be deep and yet not aggravate punchthrough. These designs again illustrate the birth of trade-offs and of increasing demands for good process control, good process models, and good computational abilities.

In summary, miniaturization pushes simultaneously against the limits of device physics and process chemistry, computational abilities, and manufacturing organization itself. With no exaggeration, manufacture now is *built* upon accurate simulation, which implies an unprecedented parallel pursuit in real time of new knowledge of devices and processes, new numerical methods, and new control of manufacture. The successful integration of these factors may well become the real meaning of "integration" in ICs.

ACKNOWLEDGMENTS

The authors are pleased to thank their colleagues for their comments, particularly S. J. Hillenius, W. T. Lynch, S. M. Sze, and M. R. Pinto.

REFERENCES

1. R. W. Hamming, *Numerical Methods for Scientists and Engineers*, Chapter N+1, McGraw-Hill, New York, 1962.

2. K. M. Cham, S.-Y. Oh, D. Chin, and J. L. Moll, *Computer-Aided Design and VLSI Device Development*, Kluwer Academic, Boston, MA, 1986.

3. J. Lee, K. Mayaram, and C. Hu, "A Theoretical Study of Gate/Drain Offset in LDD MOSFETs," *IEEE Electron Device Lett.*, **EDL-7**(3), 152 (1986).

4. S. M. Sze, *Physics of Semiconductor Devices*, 2nd ed., Chapter 8, Wiley, New York, 1981.

5. P. Richman, *MOS Field Effect Transistors and Integrated Circuits*, Wiley, New York, 1973.

6. D. K. Ferry, "Physics and Modeling of Submicron IGFETs. I," in N. G. Einspruch, Ed., *VLSI Electronics: Microstructure Science*, Vol. 1, Chapter 6, Academic, New York, 1981.

7. J. R. Brews, "Physics of the MOS Transistor," in D. Kahng, Ed., *Silicon Integrated Circuits, Applied Solid State Science,* Suppl. 2, Part A, Chapter 1, Academic, New York, 1981.

8. H. J. Rood, *Logic and Structured Design for Computer Programmers,* Chapter 6, Prindle, Weber, & Schmidt, Boston, MA, 1985.

9. D. V. Steward, *Software Engineering with Systems Analysis and Design,* Brooks/ Cole, Monterey, CA, 1987.

10. A. J. Strojwas and S. W. Director, "The Process Engineer's Workbench," *IEEE J. Solid-State Circuits,* **SC-23**(2), 377 (1988).

11. M. A. Styblinski and L. J. Opalski, "Algorithms and Software Tools for IC Yield Optimization Based on Fundamental Fabrication Parameters," *IEEE Trans. Comput.-Aided Des.,* **CAD-5**(1), 79 (1986).

12. P. Yang, D. E. Hocevar, P. F. Cox, C. Machala, and P. K. Chatterjee, "An Integrated and Efficient Approach for MOS VLSI Statistical Circuit Design," *IEEE Trans. Comput.-Aided Des.,* **CAD-5**(1), 5 (1986).

13. S. Tazawa, T. Takeda, K. Yokoyama, M. Tomizawa, and A. Yoshii, "Application of Two-Dimensional Process/Device Simulation for Evaluating MOSFET Fabrication Processes," *Solid-State Electron.,* **30**(4), 375 (1987).

14. S. W. Director, "Manufacturing-Based Simulation: An Overview," *IEEE Circuits Devices Mag.,* **3**(5), 3 (1987).

15. C. J. B. Spanos and S. W. Director, "Parameter Extraction for Statistical IC Process Characterization," *IEEE Trans. Comput.-Aided Des.,* **CAD-5**(1), 66 (1986).

16. I. Chen and A. J. Strojwas, "A Methodology for Optimal Test Structure Design for Statistical Process Characterization and Diagnosis," *IEEE Trans. Comput.-Aided Des.,* **CAD-6**(4), 592 (1987).

17. T.-K. Yu, S. M. Kang, I. N. Haji, and T. N. Trick, "Statistical Performance Modeling and Parametric Yield Estimation of MOS VLSI," *IEEE Trans. Comput.-Aided Des.,* **CAD-6**(6), 1013 (1987).

18. Y. Aoki, H. Masuda, S. Shimada, and S. Sato, "A New Design-Centering Methodology for VLSI Device Development," *IEEE Trans. Comput.-Aided Des.,* **CAD-6**(2), 452 (1987).

19. G. Baccarani, M. R. Wordeman, and R. H. Dennard, "Generalized Scaling Theory and its Application to a 1/4 Micrometer MOSFET Design," *IEEE Trans. Electron Devices,* **ED-31**(4), 452 (1984).

20. R. H. Dennard, F. H. Gaensslen, H.-N. Yu, V. L. Rideout, E. Bassous, and A. R. LeBlanc, "Design of Ion-Implanted MOSFETs with Very Small Physical Dimensions," *IEEE J. Solid-State Circuits,* **SC-9**(5), 256 (1974).

21. J. R. Brews, W. Fichtner, E. H. Nicollian, and S. M. Sze, "Generalized Guide to MOSFET Miniaturization," *IEEE Eelectron Device Lett.,* **EDL-1**(1), 2 (1980).

22. P. K. Chatterjee, W. R. Hunter, T. C. Holloway, and V. T. Lin, "The Impact of Scaling Laws on the Choice of n-Channel or p-Channel for MOS VLSI," *IEEE Electron Device Lett.,* **EDL-1**(10), 220 (1980).

23. J. R. Pfiester, J. D. Shott, and J. D. Meindl, "Performance Limits of CMOS ULSI," *IEEE J. Solid-State Circuits,* **SC-20**(1), 253 (1985).

24. M. R. Wordeman, A. M. Schweighart, R. H. Dennard, G. Sai-Halasz, and W. W. Molzen, "A Fully Scaled Submicron NMOS Technology Using Direct Write E-Beam Lithography," *IEEE Trans. Electron Devices,* **ED-32**(11), 2214 (1985).

25. J. M. Shannon, J. Stephen, and J. H. Freeman, "MOS Frequency Soars with Ion-Implanted Layers," *Electronics,* **42**(Feb. 3), 96 (1969).

26. R. Pashley, K. Kokonnen, E. Boleky, R. Jecmen, S. Liu, and W. Owen, "H-MOS Scales Traditional Devices to Higher Performance Level," *Electronics,* **50**(Aug. 18), 94 (1977).

27. F. M. Klaassen, "Design and Performance of Micron-Sized Devices," *Solid-State Electron.,* **21**(3), 565 (1978).

28. H. Shichijo, "A Re-examination of Practical Performance Limits of Scaled n-Channel and p-Channel MOS Devices for VLSI," *Solid-State Electron.,* **26**(10), 969 (1983).

29. C. Mead and L. Conway, *Introduction to VLSI Systems,* p. 9, Addison-Wesley, Reading, MA, 1980.

30. N. H. E. Weste and K. Eshraghian, *Principles of CMOS VLSI Design,* p. 168, Addison-Wesley, Reading, MA, 1985.

31. S. J. Hillenius, R. Liu, G. E. Georgiou, R. L. Field, D. S. Williams, A. Kornblit, D. M. Boulin, R. L. Johnston, and W. T. Lynch, "A Symmetric Submicron CMOS Technology," *Tech. Dig. — Int. Electron Devices Meet.,* p. 252 (1986).

32. S. J. Hillenius and W. T. Lynch, "Gate Material Work Function Considerations for 0.5 Micron CMOS," *IEEE Proc. Int. Conf. Comput. Des., 1985,* Port Chester, NY, p. 147, (1985).

33. T. Noguchi, Y. Asahi, M. Nakahara, K. Maeguchi, and K. Kanzaki, "High Speed CMOS Structure with Optimized Gate Work Function," *Symp. VLSI Technol. Dig. Tech. Pap.,* p. 19 (1986).

34. G. J. Hu and R. H. Bruce, "Design Trade-Offs Between Surface and Buried-Channel FETs," *IEEE Trans. Electron Devices,* **ED-32**(3), 584 (1985).

35. J. Y.-C. Sun, Y. Taur, R. H. Dennard, and S. P. Klepner, "Submicrometer-Channel CMOS for Low-Temperature Operation," *IEEE Trans. Electron Devices,* **ED-34**(1), 19 (1987).

36. W.-S. Feng, T. Y. Chan, and C. Hu, "MOSFET Breakdown Voltage," *IEEE Electron Device Lett.,* **EDL-7**(7), 449 (1986).

37. J. Chen, T. Y Chan, I. C. Chen, P. K. Ko, and C. Hu, "Subbreakdown Drain Leakage Current in MOSFET," *IEEE Electron Device Lett.,* **EDL-8**(11), 515 (1987).

38. J. R. Brews, "Threshold Shifts Due to Nonuniform Doping Profiles in Surface Channel MOSFETs," *IEEE Trans. Electron Devices,* **ED-26**(11), 1696 (1979).

39. T. Skotnicki, G. Merckel, and T. Pedron, "New Punchthrough Current Model Based on the Voltage-Doping Transformation," *IEEE Trans. Electron Devices,* **ED-35**(7), 1076 (1988).

40. S. C. Jain and P. Balk, "A Unified Analytical Model for Drain-Induced Barrier Lowering and Drain-Induced High Electric Field in a Short-Channel MOSFET," *Solid-State Electron.,* **30**(5), 503 (1987).

41. T. Skotnicki, G. Merckel, and T. Pedron, "The Voltage-Doping Transformation: A New Approach to the Modeling of MOSFET Short-Channel Effects," *IEEE Electron Device Lett.,* **EDL-9**(3), 109 (1988).

42. J. D. Kendall and A. R. Boothroyd, "A Two-Dimensional Analytical Threshold Voltage Model for MOSFETs with Arbitrarily Doped Substrates," *IEEE Electron Device Lett.,* **EDL-7**(7), 401 (1986).

43. V. Marash and R. W. Dutton, "Methodology for Submicron Device Model Development," *IEEE Trans. Comput.-Aided Des.,* **CAD-7**(2), 299 (1988).

44. K. K. Ng and W. T. Lynch, "The Impact of Intrinsic Series Resistance on MOSFET Scaling," *IEEE Trans. Electron Devices*, **ED-34**, 503 (1987).

45. H. Murrmann and D. Widmann, "Current Crowding on Metal Contacts to Planar Devices," *IEEE Trans. Electron Devices*, **ED-16**, 1022 (1969).

46. H. H. Berger, "Models for Contacts to Planar Devices," *Solid-State Electron.*, **15**, 145 (1972).

47. D. B. Scott, W. R. Hunter, and H. Shichijo, "A Transmission Line Model for Silicided Diffusions: Impact on the Performance of VLSI Circuits," *IEEE Trans. Electron Devices*, **ED-29**, 651 (1982).

48. K. K. Ng and W. T. Lynch, "Analysis of the Gate-Voltage-Dependent Series Resistance of MOSFETs," *IEEE Trans. Electron Devices*, **ED-33**, 965 (1986).

49. K. Terada and H. Muta, "A New Method to Determine Effective MOSFET Channel Length," *Jpn. J. Appl. Phys.*, **18**, 953 (1979).

50. J. Whitfield, "A Modification on 'An Improved Method to Determine MOSFET Channel Length'," *IEEE Electron Device Lett.*, **EDL-6**(3), 109 (1985).

51. G. J. Hu, C. Chang, and Y.-T. Chia, "Gate Voltage Dependent Effective Channel Length and Series Resistance of LDD MOSFETs," *IEEE Trans. Electron Devices*, **ED-34**(12), 2469 (1987).

52. S. Y. Chou and D. A. Antoniadis, "Relationship Between Measured and Intrinsic Transconductances of FET's," *IEEE Trans. Electron Devices*, **ED-34**, 448 (1987).

53. J. R. Schrieffer, "Effective Carrier Mobility in Surface Space Charge Layers," *Phys. Rev.*, **97**, 641 (1955); in R. H. Kingston, Ed., *Semiconductor Surface Physics*, p. 55, University of Pennsylvania, Philadelphia, PA, 1957.

54. A. G. Sabnis and J. T. Clemens, "Characterization of the Electron Mobility in the Inverted (100)-Si Surface," *Tech. Dig. — Int. Electron Devices Meet.*, p. 18, (1979).

55. S. A. Schwartz and S. E. Russek, "Semiempirical Equations for Electron Velocity in Silicon. Part II. MOS Inversion Layer," *IEEE Trans. Electron Devices*, **ED-30**, 1634 (1983).

56. J. T. C. Chen and R. S. Muller, "Carrier Mobilities at Weakly Inverted Silicon Surfaces," *J. Appl. Phys.*, **45**, 828 (1974).

57. J. R. Brews, "Theory of the Carrier-Density Fluctuations in an IGFET Near Threshold," *J. Appl. Phys.*, **46**, 2181 (1975).

58. J. R. Brews, "Carrier-Density Fluctuations and the IGFET Mobility Near Threshold," *J. Appl. Phys.*, **46**, 2193 (1975).

59. V. A. Gergel and R. A. Suris, "Fluctuations of the Surface Potential in MIS Structures," *Sov. Phys. — JETP (Engl. Transl.)*, **48**, 95 (1978).

60. A. A. Guzev and V. A. Gurtov, "Spatial Scale of Statistical Fluctuations of the Potential in MIS Structures," *Sov Phys. — Semicond. (Engl. Transl.)*, **18**, 856 (1984).

61. G. J. Declerck and R. J. Van Overstraeten, "Characterization of the MOSFET Operating in Weak Inversion," in L. Marton, Ed., *Adv. Electron. Electron Phys.*, vol. 47, p. 197, Academic, NY, 1978.

62. A. Hiroki, S. Odanaka, K. Ohe, and H. Esaki, "A Mobility Model for Submicrometer MOSFET Device Simulations," *IEEE Electron Device Lett.*, **EDL-8**, 231 (1987).

63. J. A. Cooper and D. F. Nelson, "High-Field Drift Velocity of Electrons at the Si SiO_2 Interface as Determined by a Time-of-Flight Technique," *J. Appl. Phys.*, **54**, 1445 (1983).

64. K. K. Thornber, "Relation of Drift Velocity to Low-Field Mobility and High-Field Saturation Velocity," *J. Appl. Phys.*, **51**, 2127 (1980).

65. K. Yamaguchi, "Field-Dependent Mobility for Two-Dimensional Numerical Analysis of MOSFETs," *IEEE Trans. Electron Devices*, **ED-26**, 1068 (1979).

66. S. Selberherr, *Analysis and Simulation of Semiconductor Devices*, p. 101, Springer-Verlag, New York, 1984.

67. Chu-Hao, J. Zimmermann, M. Charef, R. Fauquembergue, and E. Constant, "Monte Carlo Study of Two-Dimensional Electron Gas Transport in Si-MOS Devices," *Solid-State Electron.*, **28**, 733 (1985).

68. J. R. Barker and D. K. Ferry, "On the Physics and Modeling of Small Semiconductor Devices," *Solid-State Electron.*, **23**, 519 (1980).

69. J. H. Ferziger and H. G. Kaper, *Mathematical Theory of Transport Processes in Gases*, North-Holland, Amsterdam, 1972.

70. L. Reggiani, "General Theory," in L. Reggiani, Ed., *Hot Electron Transport in Semiconductors*, p. 11, Springer-Verlag, New York, 1985.

71. G. Baccarani, C. Jacaboni, and A. M. Mazone, "Current Transport in Narrow-Base Transistors," *Solid-State Electron.*, **20**, 5 (1977).

72. S. Bandyopadhyay, M. E. Klausmeier-Brown, C. M. Maziar, S. Datta, and M. S. Lundstrom, "A Rigorous Technique to Couple Monte Carlo and Drift-Diffusion Models for Computationally Efficient Device Simulation," *IEEE Trans. Electron Devices*, **ED-34**, 392 (1987).

73. K. Kim, B. A. Mason, and K. Hess, "Inclusion of Collision Broadening in Semiconductor Electron-Transport Simulations," *Phys. Rev., B: Condens. Matter* [3], **36**(12), 6547 (1987).

74. J. Y. Tang and K. Hess, "Theory of Hot Electron Emission from Silicon Into Silicon Dioxide," *J. Appl. Phys.*, **54** (9), 5145 (1983).

75. L. Reggiani, P. Lugli, and A. P. Jauho, "Quantum Kinetic Equation for Electronic Transport in Nondegenerate Semiconductors," *Phys. Rev., B: Condens. Matter* [3] **36**(12), 6602 (1987).

76. E. Sangiorgi, M. R. Pinto, F. Venturi, and W. Fichtner, "A Hot-Carrier Analysis of Submicron MOSFETs," *IEEE Electron Device Lett.*, **EDL-9**(1), 13 (1988).

77. D. C. Montgomery, *Theory of the Unmagnetized Plasma*, p. 15, Gordon & Breach, New York, 1971.

78. S. Harris, *An Introduction to the Theory of the Boltzmann Equation*, pp. 26 and 113, Holt, Rinehart, & Winston, New York, 1971.

79. K. Blotekjaer, "High Frequency Conductivity, Carrier Waves and Acoustic Amplification in Drifted Semiconductor Plasma," *Ericsson Technics*, **22**, Annex A, 125 (1966).

80. R. K. Cook and J. Frey, "An Efficient Technique for Two-Dimensional Simulation of Velocity Overshoot Effects in Si and GaAs Devices," *COMPEL — Int. J. Comput. Math. Electr. Electron. Eng.*, **1**, 65 (1982).

81. M. Rudan and F. Odeh, "Multi-Dimensional Discretization Scheme for the Hydrodynamic Model of Semiconductor Devices," *COMPEL — Int. J. Comput. Math. Electr. Electron. Eng.*, **5**, 149 (1986).

82. A. Forghieri, R. Giuerreri, P. Ciampolini, A. Gnudi, M. Rudan, and G. Baccarani, "A New Discretization Strategy for the Semiconductor Equations Comprising Mo-

mentum and Energy Balance," *IEEE Trans. Comput.-Aided Des.*, **CAD-7**(2), 231 (1988).

83. J. P. Nougier, J. C. Vaissiere, D. Gasquet, J. Zimmermann, and E. Constant, "Determination of Transient Regime of Hot Carriers in Semiconductors Using the Relaxation Time Approximations," *J. Appl. Phys.*, **52**, 825 (1981).

84. D. L. Woolard, R. J. Trew, and M. A. Littlejohn, "Hydrodynamic Hot-Electron Transport Model with Monte Carlo-Generated Transport Parameters," *Solid-State Electron.*, **31** (3/4), 571 (1988).

85. G. Baccarani and M. R. Wordeman, "An Investigation of Steady-State Velocity Overshoot in Silicon," *Solid-State Electron.*, **28**, 407 (1985).

86. K. K. Thornber, "Current Equations for Velocity Overshoot," *IEEE Electron Device Lett.*, **EDL-3**, 69 (1982).

87. I. C. Kizilyalli and K. Hess, "Simplified Device Equations and Transport Coefficients for GaAs Device Modeling," *IEEE Trans. Electon Devices*, **ED-34**(11), 2352 (1987).

88. D. R. Wolters and J. J. van der Schoot, "Dielectric Breakdown in MOS Devices, Parts I–III," *Philips J. Res.*, **40**, 115, 137, 164 (1985).

89. T. Kusaka, Y. Ohji, and K. Mukai, "Time-Dependent Dielectric Breakdown of Ultra-Thin Silicon Oxide," *IEEE Electron Device Lett.*, **EDL-8**(2), 61 (1987).

90. M. Davis and R. Lahri, "Gate Oxide Charge-to-Breakdown Correlation to MOSFET Hot-Electron Degradation," *IEEE Electron Device Lett.*, **EDL-9**(4), 183 (1988).

91. Y. A. El-Mansy, and D. M. Caughey, "Modeling Weak Avalanche Multiplication Currents in IGFETs and SOS Transistors for CAD," *Tech. Dig. — Int. Electron Devices Meet.*, p. 31 (1975).

92. S. Tam, P. K. Ko, C. Hu, and R. S. Muller, "Correlation Between Substrate and Gate Currents in MOSFETs," *IEEE Trans. Electron Devices*, **ED-29**, 1740 (1982).

93. T. H. Ning, C. M. Osburn, and H. N. Yu, "Emission Probability of Hot Electrons from Silicon into Silicon Dioxide," *J. Appl. Phys.*, **48**, 286 (1977).

94. E. Takeda, Y. Nakagome, H. Kume, and S. Asai, "New Hot-Carrier Injection and Device Degradation in Submicron MOSFETs," *Proc. Inst. Electr. Eng.*, **130**, 144 (1983).

95. K. R. Hofmann, C. Werner, W. Weber, and G. Dorda, "Hot-Electron and Hole-Emission Effects in Short n-Channel MOSFETs," *IEEE Trans. Electron Devices*, **ED-32**, 691 (1985).

96. Y. Nissan-Cohen, "A Novel Floating-Gate Method for Measurement of Ultra-Low Hole and Electron Gate Currents in MOS Transistors," *IEEE Electron Device Lett.*, **EDL-7**, 561 (1986).

97. N. S. Saks, P. L. Heremans, L. Van den Hove, H. E. Maes, R. F. De Keersmaecker, and G. J. Declerck, "Observation of Hot-Hole Injection In NMOS Transistors Using a Modified Floating-Gate Technique," *IEEE Trans. Electron Devices*, **ED-33**, 1529 (1986).

98. C. Hu, "Lucky-Electron Model of Channel Hot Electron Emission," *Tech. Dig. — Int. Electron Devices Meet.*, p. 22 (1979).

99. S. Tam, P. K. Ko, and C. Hu, "Lucky-Electron Model of Channel Hot-Electron Injection in MOSFETs," *IEEE Trans. Electron Devices*, **ED-31**, 1116 (1984).

100. S. Tam, F. C. Hsu, C. Hu, R. S. Muller, and P. K. Ko, "Hot-Electron Currents in Very Short Channel MOSFETs," *IEEE Electron Device Lett.*, **EDL-4**, 249 (1983).

101. B. Ricco, E. Sangiorgi, and D. Cantarelli, "Low Voltage Hot-Electron Effects in Short Channel MOSFETs," *Tech. Dig.—Int. Electron Devices Meet.*, p. 92 (1984).

102. T. H. Ning and H. N. Yu, "Optically Induced Injection of Hot Electrons into SiO_2," *J. Appl. Phys.*, **45**, 5373 (1974).

103. R. R. Troutman, "Silicon Surface Emission of Hot Electrons," *Solid-State Electron.*, **21**(1), 283 (1978).

104. E. Takeda, H. Kume, T. Toyabe, and S. Asai, "Submicrometer MOSFET Structure for Minimizing Hot-Carrier Generation," *IEEE Trans. Electron Devices*, **ED-29**, 611 (1982).

105. D. J. Bartelink, J. L. Moll, and N. I. Meyer, "Hot-Electron Emission from Shallow p-n Junctions in Silicon," *Phys. Rev.*, **130**, 972 (1963).

106. C. Bulucea, "Avalanche Injection into the Oxide in Silicon Gate-Controlled Devices (i) Theory," *Solid-State Electron.*, **18**, 363 (1975).

107. A. Toriumi, M. Yoshimi, M. Iwase, Y. Akiyama, and K. Taniguchi, "A Study of Photon Emission from n-Channel MOSFETs," *IEEE Trans. Electron Devices*, **ED-34**, 1501 (1987).

108. E. Sangiorgi, B. Ricco, and P. Olivo, "Hot Electrons and Holes in MOSFETs Biased Below the Si-SiO_2 Interfacial Barrier," *IEEE Electron Device Lett.*, **EDL-6**, 513 (1985).

109. E. Takeda, N. Suzuki, and T. Hagiwara, "Device Performance Degradation Due to Hot-Carrier Injection at Energies Below the Si-SiO_2 Energy Barrier," *Tech. Dig.—Int. Electron Devices Meet.*, p. 396 (1983).

110. D. J. Coe, "Changes in Effective Channel Length Due to Hot-Electron Trapping in Short-Channel M.O.S.T.s," *IEE J. Solid-State Electron Devices*, **2**, 57 (1978).

111. K. K. Ng and G. W. Taylor, "Effects of Hot-Carrier Trapping in n- and p-Channel MOSFETs," *IEEE Trans. Electron Devices*, **ED-30**, 871 (1983).

112. H. Hara, Y. Okamoto, and H. Ohnuma, "A New Instability in MOS Transisitor Caused by Hot Electron and Hole Injection from Drain Avalanche Plasma into Gate Oxide," *Jpn. J. Appl. Phys.*, **9**, 1103 (1970).

113. E. Takeda, Y. Nakagome, H. Kume, N. Suzuki, and S. Asai, "Comparison of Characteristics of n-Channel and p-Channel MOSFETs for VLSIs," *IEEE Trans. Electron Devices*, **ED-30**, 675 (1983).

114. M. Koyanagi, A. G. Lewis, R. A. Martin, T. Y. Huang, and J. Y. Chen, "Hot-Electron-Induced Punchthrough (HEIP) Effect in Submicrometer PMOSFETs," *IEEE Trans. Electron Devices*, **ED-34**, 839 (1987).

115. A. Kalnitsky and S. Sharma, "The Effect of Channel Hot Electron Stress on a.c. Device Characteristics of MOSFETs," *Solid-State Electron.*, **29**, 1053 (1986).

116. P. E. Cottrell and E. M. Buturla, "Steady State Analysis of Field Effect Transistors via the Finite Element Method," *Tech. Dig—Int. Electron Devices Meet.*, p. 51 (1975).

117. C. Lombardi, P. Olivo, B. Ricco, E. Sangiorgi, and M. Vanzi, "Hot Electrons in MOS Transistors: Lateral Distribution of the Trapped Oxide Charge," *IEEE Electron Device Lett.*, **EDL-3**, 215 (1982).

118. W. Weber, C. Werner, and G. Dorda, "Degradation of n-MOS-Transistors after Pulsed Stress," *IEEE Electron Device Lett.*, **EDL-5**, 518 (1984).

119. K. L. Chen, S. Saller, and R. Shah, "The Case of AC Stress in the Hot-Carrier Effect," *IEEE Trans. Electron Devices*, **ED-33**, 424 (1986).

120. B. S. Doyle, M. Bourcerie, J. C. Marchetaux, and A. Boudou, "Dynamic Channel Hot-Carrier Degradation of MOS Transistors by Enhanced Electron-Hole Injection into the Oxide," *IEEE Electron Device Lett.*, **EDL-8,** 237 (1987).

121. J. Y. Choi, P. K. Ko, and C. Hu, "Hot-Carrier-Induced MOSFET Degradation Under AC Stress," *IEEE Electron Device Lett.*, **EDL-8,** 333 (1987).

122. T.-C. Ong, K. Seki, P. K. Ko, and C. Hu, "Hot-Carrier-Induced Degradation in p-MOSFETs Under ac Stress," *IEEE Electron Device Lett.*, **EDL-9**(5), 211 (1988).

123. R. Bellens, P. Heremans, G. Groeseneken, and H. E. Maes, "Hot-Carrier Effects in n-Channel MOS Transistors Under Alternating Stress Conditions," *IEEE Electron Device Lett.*, **EDL-9**(5), 232 (1988).

124. T. Horiguchi, H. Mikoshiba, K. Nakamura, and K. Hamano, "Hot-Carrier Induced Degradation of n-MOSFETs in Inverter Operation," *Proc. Symp. VLSI Technol.*, p. 104 (1985).

125. V. Srinivasan and J. J. Barnes, "Small Width Effects on MOSFET Hot-Electron Reliability," *Tech. Dig — Int. Electron Devices Meet.*, p. 740 (1980).

126. L. A. Akers, M. A. Holly, and C. Lund, "Hot Carriers in Small Geometry CMOS," *Tech. Dig.—Int. Electron Devices Meet.*, p. 80 (1984).

127. T. C. Ong, S. Tam, P. K. Ko, and C. Hu, "Width Dependence of Substrate and Gate Currents in MOSFETs," *IEEE Trans. Electron Devices*, **ED-32,** 1737 (1985).

128. E. H. Nicollian, C. N. Berglund, P. F. Schmidt, and J. M. Andrews, "Electrochemical Charging of Thermal SiO_2 Films by Injected Electron Currents," *J. Appl. Phys.*, **42,** 5654 (1971).

129. R. A. Gdula, "The Effects of Processing on Hot Electron Trapping in SiO_2," *J. Electrochem. Soc.*, **123,** 42 (1976).

130. F.-C. Hsu, J. Hui, and K. Y. Chiu, "Effect of Final Annealing on Hot-Electron-Induced MOSFET Degradation," *IEEE Electron Device Lett.*, **EDL-6,** 369 (1985).

131. H. Mikoshiba, T. Horiguchi, and K. Hamano, "Comparison of Drain Structures in n-Channel MOSFETs," *IEEE Trans. Electron Devices*, **ED-33,** 140 (1986).

132. E. Takeda, H. Kume, Y. Nakagome, T. Makino, A. Shimizu, and S. Asai, "An As-P(n^+ − n) Double Diffused Drain MOSFET for VLSIs," *IEEE Trans. Electron Devices*, **ED-30,** 652 (1983).

133. E. Takeda, Y. Ohji, and H. Kume, "High Field Effects in MOSFETs," *Tech. Dig.— Int. Electron Devices Meet.*, p. 60 (1985).

134. C. Hu, "Hot-Electron Effects in MOSFETs," *Tech. Dig.—Int. Electron Devices Meet.*, p. 176 (1983).

135. K. L. Chen, S. A. Saller, I. A. Groves, and D. B. Scott, "Reliability Effects on MOS Transistors Due to Hot-Carrier Injection," *IEEE Trans. Electron Devices*, **ED-32,** 386 (1985).

136. F.-C. Hsu and K.-Y. Chiu, "Evaluation of LDD MOSFETs Based on Hot-Electron-Induced Degradation," *IEEE Electron Device Lett.*, **EDL-5,** 162 (1984).

137. F.-C. Hsu and H. R. Grinolds, "Structure-Enhanced MOSFET Degradation Due to Hot-Electron Injection," *IEEE Electron Device Lett.*, **EDL-5,** 71 (1984).

138. E. Takeda and N. Suzuki, "An Empirical Model for Device Degradation Due to Hot-Carrier Injection," *IEEE Electron Device Lett.*, **EDL-4,** 111 (1983).

139. R. Izawa and E. Takeda, "The Impact of n^- Drain Length and Gate-Drain/Source Overlap on Submicrometer LDD Devices for VLSI," *IEEE Electron Device Lett.*, **EDL-8**(10), 480 (1987).

140. R. Izawa, T. Kure, S. Iijima, and E. Takeda, "The Impact of Gate-Drain Overlapped LDD (GOLD) for Deep Submicron VLSIs," *Tech. Dig.—Int. Electron Devices Meet.*, p. 38 (1987).

141. K. Mayaram, J. C. Lee, and C. Hu, "A Model for the Electric Field in Lightly Doped Drain Structures," *IEEE Trans. Electron Devices*, **ED-34**(7), 1509 (1987).

142. P. J. Tsang, S. Ogura, W. W. Walker, J. F. Shepard, and D. L. Critchlow, "Fabrication of High-Performance LDD FETs with Oxide Sidewall Spacer Technology," *IEEE Trans. Electron Devices*, **ED-29**(4), 590 (1982).

143. C. Chang and J. Lien, "Corner-Field Induced Drain Leakage in Thin Oxide MOS-FETs," *Tech. Dig.—Int. Electron Devices Meet.*, p. 714 (1987).

144. T. Y. Chan, J. Chen, P. K. Ko, and C. Hu, "The Impact of Gate-Induced Drain Leakage Current on MOSFET Scaling," *Tech. Dig.—Int. Electron Devices Meet.*, p. 718 (1987).

145. T. Y. Chan, A. T. Wu, P. K. Ko, and C. Hu, "Effects of the Gate-to-Drain/Source Overlap on MOSFET Characteristics," *IEEE Electron Device Lett.*, **EDL-8**(7), 326 (1987).

146. E. Takeda, T. Makino, and T. Hagiwara, "The Impact of Drain Impurity Profile and Junction Depth on Submicron MOSFETs," *Ext. Abstr. 15th Conf. Solid State Devices Mater., 1983*, p. 261 (1983).

147. H. R. Grinolds, M. Kinugawa, and M. Kakumu, "Reliability and Performance of Submicron LDD NMOSFETs with Buried As n⁻ Impurity Profiles," *Tech. Dig.—Int. Electron Devices Meet.*, p. 246 (1985).

148. M. Horiuchi and K. Yamaguchi, "SOLID: High-Voltage, High-Gain 300 nm Channel-Length MOSFETs—(i) Simulation," *Solid-State Electron*, **28**(5), 465 (1985).

149. A. E. Schmitz and J. Y. Chen, "Design, Modeling and Fabrication of Sub-Half Micrometer CMOS Transistors," *IEEE Trans. Electron Devices*, **ED-33**(1), 148 (1986).

150. J. M. Ford and D. K. Stemple, "The Effects of As Drain Profile on Submicron Silicide MOSFETs," *IEEE Trans. Electron Devices*, **ED-35**(3), 302 (1988).

151. R. Liu, D. S. Williams, and W. T. Lynch, "Mechanisms for Process Induced Leakage in Shallow Silicided Junctions," *Tech. Dig.—Int. Electron Devices Meet.*, p. 58 (1986).

152. W. T. Lynch, "Self-Aligned Contact Schemes for Source-Drains in Submicron Devices," *Tech. Dig.—Int. Electron Devices Meet.*, p. 354 (1987).

153. J. Sakurai, "A New Buried-Oxide Isolation for High-Speed High Density MOS Integrated Circuits," *IEEE Trans. Solid-State Circuits*, **SC-13**(4), 468 (1978).

154. O. Shinchi and J. Sakurai, "The Buried-Oxide MOSFET—A New Type of High-Speed Switching Device," *IEEE Trans. Electron Devices*, **ED-23**, 1190 (1976).

155. H. Inokawa, T. Kobayashi, and K. Kiuchi, "A Submicrometer Lifted Diffused-Layer MOSFET," *IEEE Electron Device Lett.*, **EDL-8**(3), 98 (1987).

156. F. Mieno, A. Shimizu, S. Nakamura, T. Deguchi, N. Haga, I. Matsumoto, Y. Furumura, T. Yamaguchi, K. Inayoshi, M. Maeda, and K. Yanagida, "Novel Selective Poly- and Epitaxial-Silicon Growth (SPEG) Technique for VLSI Processing," *Tech. Dig.—Int. Electron Devices Meet.*, p. 16 (1987).

157. Y. Misawa, H. Homma, K. Sato, and N. Momma, "A Self-Aligning Polysilicon Electrode Technology (SPEL) for Future LSIs," *Tech. Dig.—Int. Electron Devices Meet.*, p. 32 (1987).

158. T.-Y. Chiu, G. M. Chin, M. Y. Lau, R. H. Hanson, M. D. Morris, K. F. Lee, A. M. Voshchenkov, R. G. Swartz, V. D. Archer, and S. N. Finegan, "A High Speed Super

Self-Aligned Bipolar CMOS Technology," *Tech. Dig.— Int. Electron Devices Meet.*, p. 24 (1987).

159. H.-H. Tsai, S.-M. Chen, and C-Y. Wu, "A New Fully Recessed Oxide (FUROX) Field Isolation Technology for Scaled VLSI Circuit Fabrication," *IEEE Electron Device Lett.*, **EDL-7**(2), 124 (1986).

160. J.-L. Coppee and F. Van de Wiele, "SILO Isolation Technique: A Study of Active and Parasitic Device Characteristics with Semi-Recessed and Fully-Recessed Field Oxides," *Solid-State Electron*, **31**(5), 887 (1988).

161. W. T. Lynch, P. D. Foo, R. Liu, J. Lebowitz, K. J. Orlowsky, G. E. Georgiou, and S. J. Hillenius, "UPMOS — A New Approach to Submicron VLSI," *Eur. Solid State Device Res. Conf. Proc.* (1987).

162. J.-P. Colinge, K. Hashimoto, T. Kamins, S.-Y. Chiang, E.-D. Liu, S. Peng, and P. Riseman, "High-Speed, Low-Power, Implanted Buried Oxide CMOS Circuits," *IEEE Electron Device Lett.*, **EDL-7**(5), 279 (1986).

163. J.-P. Colinge, "Reduction of Kink Effect in Thin-Film SOI MOSFETs," *IEEE Electron Device Lett.*, **EDL-9**(2), 97 (1988).

164. J.-P. Colinge, "Hot-Electron Effects in SOI n-Channel MOSFETs," *IEEE Trans. Electron Devices*, **ED-34**(10), 2173 (1987).

165. M. Yoshimi, T. Wada, K. Kato, and H. Tango, "High Performance SOI MOSFET Using Ultrathin SOI Film," *Tech. Dig.—Int. Electron Devices Meet.*, p. 640 (1987).

166. K. Tanaka and M. Fukuma, "Design Methodology for Deep Submicron CMOS," *Tech. Dig.—Int. Electron Devices Meet.*, p. 628 (1987).

167. W. Fichtner, E. N. Fuls, R. L. Johnston, R. K. Watts, and W. W. Weik, "Optimized MOSFETs with Sub-Quarter Micron Channel Lengths," *Tech. Dig.—Int. Electron Devices Meet.*, p. 384 (1983).

168. T. Kobayashi, S. Horiguchi, and K. Kiuchi, "Deep-Submicron Characteristics with 5 nm Gate Oxide," *Tech. Dig.—Int. Electron Devices Meet.*, p. 414 (1984).

169. S. Horiguchi, T. Kobayashi, M. Miyake, M. Oda, and K. Kiuchi, "Extremely High Transconductance (above 500 nS^/^mm) MOSFET with 2.5 nm Gate Oxide," *Tech. Dig.—Int. Electron Devices Meet.*, p. 761 (1985).

170. G. A. Sai-Halasz, M. R. Wordeman, D. P. Kern, E. Ganin, S. Rishton, D. S. Zicherman, H. Schmid, M. R. Polcari, H. Y. Ng, P. J. Restle, T.-H. P. Chang, and R. H. Dennard, "Design and Experimental Technology for 0.1 μm Gate Length Low-Temperature Operation FETs," *IEEE Electron Device Lett.*, **EDL-8**(10), 463 (1987).

171. S. Ogura, C. F. Codella, N. Rovedo, J. F. Schepard, and J. Riseman, "A Half Micron MOSFET Using Double Implanted LDD," *Tech. Dig.—Int. Electron Devices Meet.*, p. 718 (1982).

172. C. F. Codella and S. Ogura, "Halo Doping Effects in Submicron DI-LDD Device Design," *Tech. Dig.—Int. Electron Devices Meet.*, p. 230 (1985).

173. S. Rathnam, H. Bahramian, D. Laurent, and Y.-P. Han, "An Optimized 0.5 Micron LDD Transistor," *Tech. Dig.—Int. Electron Devices Meet.*, p. 237 (1983).

174. S. Odanaka, M. Fukumoto, G. Fuse, M. Sasago, T. Yabu, and T. Ohzone, "A New Half-Micrometer p-Channel MOSFET with Efficient Punchthrough Stops," *IEEE Trans. Electron Devices*, **ED-33**(3), 317 (1986).

175. F. Bauer, S. C. Jain, J. Korec, V. Lauer, M. Offenberg, and P. Balk, "Incompatibility of Requirements of Optimizing Short Channel Behavior and Long Term Stability in MOSFETs," *Solid-State Electron*, **31**(1), 27 (1988).

176. D. N. Pattanayak, A. L. Robinson, T. P. Chow, M. S. Adler, B. J. Baliga, and E. J. Wildi, "N-Channel Lateral Insulated Gate Transistors. Part 1. Steady-State Characteristics," *IEEE Trans. Electron Devices,* **ED-33**(12), 1956 (1986).

177. J. P. Russell, A. M. Goodman, L. A. Goodman, and J. M. Nelson, "The COMFET — A New High Conductance MOS-Gated Device," *IEEE Electron Device Lett.,* **EDL-4,** 63 (1983).

178. M. R. Claessen and P. Van der Zee, "An Accurate DC Model for High-Voltage Lateral DMOS Transistors Suited for Computer-Aided Circuit Design," *IEEE Trans. Electron Devices,* **ED-33**(12), 1964 (1986).

179. A. Hamada, Y. Igura, R. Izawa, and E. Takeda, "N-Source/Drain Compensation Effects in Submicrometer LDD MOS Devices," *IEEE Electron Devices Lett.,* **EDL-8**(9), 398 (1987).

2 Scaling the Silicon Bipolar Transistor

DENNY D. TANG

IBM Thomas J. Watson Research Center
Yorktown Heights, New York

2.1 INTRODUCTION

Miniaturization has been the way to improve the performance and cost of semiconductor integrated circuits. In the course of miniaturization, scaling serves as the guiding principle for device design. In recent years, the down-scaling of the bipolar transistor has led to a very significant decrease in the switching delay of circuits. The progress can be measured using the switching delay of the ECL (emitter-coupled-logic) ring oscillators, representing the capability of the transistors. Figure 2.1 shows the gate delay data collected from several industrial laboratories. The gate delay becomes shorter year after year and is reaching into the sub-50 ps range.

There are two inherent device attributes that make bipolar devices suitable for high-speed integrated circuits. First, the exponential $I-V$ relation gives the bipolar transistor the highest transconductance among the semiconductor devices. Second, compared with other field-effect devices, the "turn-on" voltage of bipolar transistors is least sensitive to process variation. Thus, with good tracking among transistors, the circuit can be operated at a very low power supply voltage and very small logic swing. These two unique properties make the bipolar transistor an excellent driver of the wire capacitance in integrated circuit chips; in other words, its gate delay is least sensitive to the wire capacitance in the LSI and VLSI environments.

Bipolar circuits, however, dissipate constant power. High-level integration chips must be mounted in sophisticated thermal packages to remove the heat. Retrospectively, the down-scaling of the bipolar transistor has been more successful in increasing the circuit speed than the integration level. While today's CMOSs (complementary metal-oxide-semiconductor field-effect transistors) become the synonyms of VLSI (very large scale integration) circuits, the bipolar ECL circuits become the key to the high-speed electronics of the mainframe computers and supercomputers. These very high-speed systems are made of bipolar LSI chips, which are driven to the speed limit and dissipate several watts of power. The kind

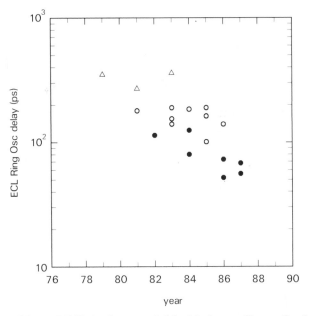

Fig. 2.1 Gate delays of ECL (emitter-coupled-logic) ring oscillators. Symbols: triangles for 2 μm or greater designs, circles for 1–2 μm designs, and dots for submicron designs.

of evolution in MOSFET technology from the high-power NMOS to the low-power CMOS has not happened in the bipolar technology. A practical, CMOS-like, low-power complementary bipolar circuit is yet to be found. The power dissipation issue in the bipolar chip remains.

This chapter serves to review the scaling properties and the salient features of silicon bipolar transistors. Vertical *npn* transistors will be used as examples in most discussions unless otherwise mentioned. In Section 2.2, we will first review the design and scaling of silicon bipolar transistors. The evolving physical charac-ters of the scaled transistors and the properties that impose limitations to scaling will be discussed in Section 2.3. The evolution of the modern transistor structures will be given in Section 2.4. In Section 2.5, we will describe the state-of-the-art circuit capabilities. We will look at trends and future possibilities in Section 2.6.

2.2   BIPOLAR TRANSISTOR DESIGN AND SCALING

In the literature, one can find two approaches to the bipolar scaling. One approach is to analyze the properties of a scaled-down transistor with certain assumed dop-ing profile and calculate its circuit performance.[1,2] The other approach is based on synthesis rather than analysis,[3,4] where device design is chosen to circumvent limits as it is scaled down. The design and scaling described in this session are based on the synthesis approach. Device impurity profiles (thus the associated parameters) are tailored at each level of the lithographic scaling until the physical limit is reached. To continue the scaling process, the device structure may have to be modified. We will first describe the design procedure and then the scaling process and rules.

2.2.1 Bipolar Design

By the very nature of the bipolar action, more "free" parameters are involved in the design of bipolar devices as compared with MOSFETs. In addition, some parameters are related to the physical structure and the contact electrodes of the bipolar transistor. Figure 2.2[5] shows two most commonly used elements of integrated bipolar circuits: a vertical *npn* transistor and a resistor. The emitter and the base region of the transistor are formed by successively counter-doping the lightly doped *n*-epitaxial collector layer with opposite type of doping impurities. The active (intrinsic) region of the device is only the region under the n^+ emitter. Around the emitter is the inactive (extrinsic) region that connects the base and collector to the surface. The collector current is much larger than the base current, a thick n^+ subcollector provides a low resistive path to the surface. Further outside is the p^+ isolation region that electrically separates the n^+ subcollectors on the same wafer. The size of the transistor is generally much larger than its active region.

   In the transistor design, the vertical impurity profile must satisfy the power supply condition before being considered for performance optimization. The collector–base junction breakdown voltage and the collector-to-emitter "punchthrough" voltage must be greater than the power supply voltage. The latter is the collector voltage at which the depletion layer of the base–collector junction on the base side reaches the emitter. Other than for the special high-voltage application, digital bipolar circuits are operated with a supply of 5 V or less. Thus, the collector–base junction

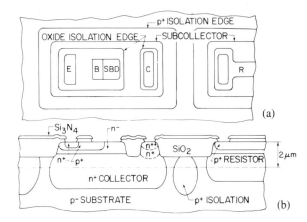

Fig. 2.2 (*a*) The topview and (*b*) cross section of a junction-isolated planar Schottky diode-clamped transistor and a resistor. (From Ref. 5. Reprinted with permission, copyright 1979, IEEE.)

breakdown is not a concern; instead, the punchthrough voltage of the thin base transistor should be considered. The widths of the depletion layer of the collector–base junction on the base side and collector side are

$$x_b = \left[\frac{2(V_c + \phi_{bi})\varepsilon N_c}{q(N_c + N_b)N_b}\right]^{0.5}, \tag{1}$$

$$x_c = \left[\frac{2(V_c + \phi_{bi})\varepsilon N_b}{q(N_c + N_b)N_c}\right]^{0.5}, \tag{2}$$

where q and ε are the electronic charge and the dielectric permittivity, respectively, ϕ_{bi} is the junction built-in potential, and N_c and N_b are the epi-collector and base doping density, respectively. If the doping density of the epi is much lower than the base doping, or $N_c \ll N_b$, these relations become

$$x_b = \left[\frac{2(V_c + \phi_{bi})\varepsilon N_c}{q N_b^2}\right]^{0.5}, \tag{3}$$

$$x_c = \left[\frac{2(V_c + \phi_{bi})\varepsilon}{q N_c}\right]^{0.5}, \tag{4}$$

and $x_c \gg x_b$. The width x_b expands as V_c increases; base punchthrough occurs when it is equal to the $w_b - x_e$, where x_e is the depletion layer width of the emitter–base junction and w_b is the base width. Equation (3) is valid as long as the depletion layer in the epitaxial collector does not reach the subcollector, or

$$W_{epi} \geq x_c \tag{5}$$

for V_c at the punchthrough voltage.

   The collector current flow modifies the depletion layer width of the collector–base junction. The electrons, which drift across the collector–base junction, compensate for the positively charged donor atoms in the depletion layer. The density of electrons n, with velocity v, carries a current density of $J_c = qnv$. The collector doping density N_c is effectively reduced and the depletion layer width increases to

$$x_c = \left[\frac{2(V_c + \phi_{bi})\varepsilon}{q(N_c - n)}\right]^{0.5}. \tag{6}$$

When $n > N_c$, the epi region becomes negatively charged. Holes enter the collector region to neutralize the negative charge and rapidly "stretch" the neutral base beyond the metallurgical junction. It is known as Kirk effect[6] or base stretching. The extra charge stored in the "stretched" base slows the transistor action. To prevent base stretching, one may design the epi doping density N_c to be greater than J_c/qv_1, where J_c is the desired operating current density and v_1 is the saturation velocity of electrons (10^7 cm/s); the electric field in the collector junction of modern transistors is large enough to bring carriers close to saturation velocity under the typical power supply voltage. The epilayer under the collector–base junction should be slightly wider than the depletion layer width. Another way to circumvent base stretching is the elimination of the lightly doped epilayer. Rather than having a $n^+pn^-n^+$ profile, one has a n^+pn^+ profile. This approach requires a thicker base to stand the same punchthrough voltage. Besides, the collector–base junction capacitance is larger and thus less desirable.

Once the supply voltage condition is satisfied, the profile optimization process can proceed. The sheet resistance of the base region is an important parameter. It is related to the collector current and device performance. The intrinsic-base sheet resistance is

$$R_{sb} = \frac{1}{q\mu_p Q_b},\tag{7}$$

where μ_p is the hole mobility, and Q_b is the integral base charge $Q_b = p_b w_b$. The density of the collector saturation current is

$$J_{co} = \frac{qn_i^2 D_n}{Q_b},\tag{8}$$

where n_i and D_n are the intrinsic carrier density and the electron diffusion coefficient, respectively. The ratio of the collector saturation current and the base sheet resistance is therefore

$$\frac{J_{co}}{R_{sb}} = q^2 D_n \mu_p n_i^2.\tag{9}$$

It is a function of the doping density of the base region and is independent of the base geometry. Figure 2.3 shows that the relation of R_{sb} and the collector current density appears quite linear over a wide doping range[7,8] until the doping density in the base exceeds the 10^{19} cm^{-3} range. This point will be reiterated when discussing scaling properties. For emitter–base junction biased at 0.66 V, the ratio of J_c/R_{sb} is 1.25×10^{-3} (A/$\Omega \cdot$ cm^2) at room temperature.

For most circuit applications, a current gain β of 100 is adequate. Thus, the base current density can be related to the R_{sb} as

$$J_b = \frac{J_c}{\beta} = \frac{CR_{sb}}{\beta}.\tag{10}$$

The relation between the collector current density J_c and the emitter stripe width W_e can be examined by looking into the lateral voltage drop in the base region due

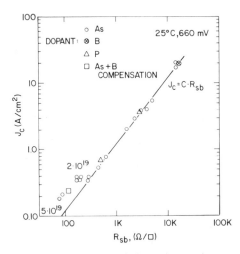

Fig. 2.3 Collector current density vs. intrinsic-base sheet resistance measured at 25°C and at $V_{be} = 0.660$ V. (From Ref. 8. Reprinted with permission, copyright 1980, IEEE.)

to the base current.[4] For a transistor with two base contacts, the lateral voltage drop in the base is

$$I_b R_b = I_b \left(\frac{R_{sb} W_e}{g L_e} \right), \quad \text{and} \quad I_b = J_b W_e L_e , \tag{11}$$

where L_e is the emitter stripe length and g is the base geometric factor with a value between 12 and 20. To ensure uniform injection across the emitter from the edge to the center, the lateral base voltage drop should be less than KT/q. From Eqs. (10) and (11), one gets

$$R_{sb} J_c W_e^2 = \beta g \left(\frac{KT}{q} \right) ,$$

and from (9),

$$W_e^2 J_c^2 = \text{constant.} \tag{12}$$

Thus, as a result of the decrease in the base resistance, transistors with a narrower emitter stripe width can carry more current density without emitter crowding effects. High-current capability is important, and this point will become clear in the latter part of this section.

At high collector currents, the electron density in the base can be equal to or greater than the base doping density. More holes are injected from the base contacts to neutralize the electrons; and that effectively raises Q_b, reduces R_b, and lowers J_{co}. This is called the base-conductivity modulation. Clearly, thin-base transistors suffer least from the base-conductivity modulation since the doping density is scaled up as the base is thinned down.

The switching delay of the bipolar circuit is made up of three major components. Figure 2.4 shows a schematic of switching delay of bipolar circuits vs. the collector current. In the low-current range, the long switching delay is equal to the charging (and discharging) time of the parasitic capacitors. The delay is inversely proportional to the switching current. In the high-current range, the base stored-charge time constant τ_b and the base resistance R_b limit the performance. The base stored-charge time constant is slowly varying until the current density exceeds the Kirk effect limit. Thus, if the operating current per gate is a limiting factor, for example, in a VLSI chip, the transistor design should be focused on the reduction of parasitic capacitance and the wire capacitance. More often, speed is the only concern; the focus will be on the formation of a very thin base, a low base-resistance vertical profile and a high current density capability.[9]

The switching delay contributed by each part of the transistor is different, depending on the type of circuit in question. To talk about the optimization of the transistor profile, one should start from the electrical parameters, more specifically, the equivalent circuits (Fig. 2.5a). We will show a design example of the transistor vertical profiles for ECL circuits (Fig. 2.5b). The key circuit parameters that influence the transistor design are the switch current I_s and the logic swing voltage V_l. The load resistor R_l is therefore equal to V_l/I_s. A delay equation can be derived from the sensitivity of the gate delay to the RC time constants of the device parameters.[10] It is not as accurate as the detailed circuit simulation, but it provides more insight into the interrelationship between the circuit delay and the transistor layout and vertical profile.[3] For a given device layout, the depletion

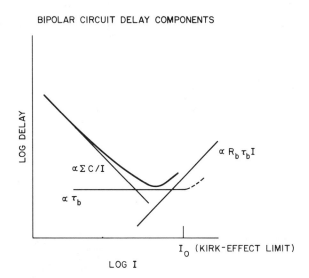

BIPOLAR CIRCUIT DELAY COMPONENTS

Fig. 2.4 Variation of various delay components of a bipolar circuit vs. switching current I or the collector current of the switching transistor.

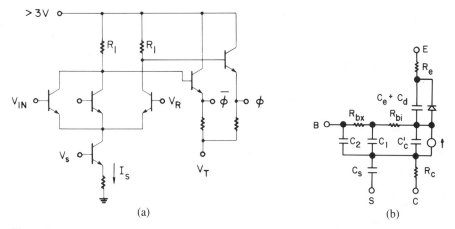

Fig. 2.5 A schematic of (*a*) the equivalent circuit of a bipolar transistor and (*b*) an ECL circuit.

layer capacitance and the base resistance can easily be estimated. The base stored-charge time constant is represented by the diffusion capacitance C_d. Based on the charge control concept,[11] it is equal to the total charge $(\tau_b I_c)$ divided by the changes in V_{be} (= half of the logic swing of the ECL gate), or $C_d = \tau_b I_c/(0.5\ V_l)$. Let's consider the design of a transistor for an ECL gate optimized for the following circuit conditions:

Circuit Parameters

Switch current	$I_s = 0.5$ mA
Logic swing	$V_l = 0.5$ V $\rightarrow R_l = 1$ kΩ
Wire capacitance	$C_w = 0.2$ pF (including resistor capacitance)
Circuit fan-out	$FO = 3$

For a transistor layout with

$$A_e = 2.5 \times 5.0\ \mu m^2, \qquad A_c = 7.5 \times 8.5\ \mu m^2, \qquad A_s = 8.5 \times 14.75\ \mu m^2,$$

one can establish the following:

Transistor Parameters

Emitter capacitance	$C_e = 22$ fF
$I_c = I_s = 0.50$ mA $\rightarrow$	$J_c = 0.04$ mA/μm^2
	$N_c = 5 \times 10^{15}$ cm^{-3}
	$W_{epi} = 0.5\ \mu m$ (under CB junction)
Collector capacitance	$C_c = 13$ fF
Isolation capacitance	$C_s = 29$ fF
Choose:	
Diffusion capacitance	$C_d = 26$ fF ($\tau_b = 13$ ps)
Base resistance R_{sb}	$R_b = 400\ \Omega$ (10 kΩ/$\square$)

One can estimate the contribution of each device parameter to the switching delay, as shown in Table 2.1.

TABLE 2.1   Delay Components of ECL Circuits

Weighing Factor	Component	ps
$2.50 + 0.413 \times FO$	$R_l C_c$	47
$0.24 \times FO$	$R_l C_e$	16
$1.1 + 0.19 \times FO$	$R_l C_d$	44
1.27	$R_b C_c$	7
2.6	$R_b C_d$	27
1.10	$R_l C_s$	32
0.233	$R_l C_w$	47
Total		*220*

As shown in Table 2.1, the three major delay components from the transistor $R_l C_c$, $R_l C_d$ and $R_b C_d$ are comparable at this current value. Further reduction in base width will reduce the terms $R_l C_d$ and $R_b C_d$ or improve the circuit delay. At a certain stage, the return will diminish. Other criteria, such as yield, may become a more important design consideration. On the other hand, the thin base is important if one intends to operate the transistor at a higher current density or to reduce the parasitic capacitance by shrinking the lateral geometry.

Using the current density as the center of focus, one can easily synthesize the transistor parameters and derive the optimal profile for a given layout. The procedure can be summarized as follows:

1. Since current density must be supported by the device structure, it should be first evaluated.
2. The collector epi doping density and epi thickness can be obtained from the Kirk effect limit and the power supply voltage.
3. From transistor layouts, the key depletion layer capacitances (C_c, C_s) and C_w become available, and then one can estimate the delay associated with these components.
4. Base profile, characterized by the R_b and τ_b (or N_b and w_b), should yield delay components that are comparable to the delay components of the capacitances. Self-consistency checks are necessary against the current gain and base punchthrough voltage, which may impose additional constraints.

It is noted that the requirements on the collector doping density for low capacitance and for supporting high current density are conflicting. The current density should be considered as a reference for a balanced design. So should the requirements on the base profile for low base resistance and short base stored-charge time constant. The current gain and junction leakage set the upper bound for the base doping, and the punchthrough voltage the lower bound.

This synthesis approach can provide a device design that is optimized at each current level. It can give a result that is not obvious to those using analysis ap-

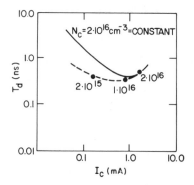

Fig. 2.6 ECL circuit delays vs. switching current, solid line; same transistor, only circuit load resistor is varied, dashed line. Both the collector doping profile of the transistor and circuit load resistor are changed according to switching current. (From Ref. 3. Reprinted with permission, copyright 1979, IEEE.)

proach. Figure 2.6 illustrates the difference in ECL performance from the two approaches. The analysis approach will yield a performance shown in the solid line; the synthesis approach will yield a performance shown in the dashed line, in which the collector doping (thus capacitance) is adjusted according to the switching current density.

The synthesis design represents the first pass design that establishes the target profile. Knowing the key device parameters, one can proceed to do detailed evaluation of the device profile that is obtainable from processing. The nonlinear behaviors of the device parameters should be examined using numerical simulation tools[12,13] for creating models and fine-tuning.

In vertical transistors, the stored charge is confined in the intrinsic base; the stored-charge time constant is mainly a function of the base width ($w_b^2/\eta D_n$, $\eta \geq 2$) and can easily be calculated. In the lateral transistors (see Fig. 2.7), the stored charge spreads from the intrinsic base to the parasitic diode under the emitter region. Detailed modeling is therefore needed to compute the charge and the stored-charge time constant.[14] The synthesis procedure remains the same.

2.2.2 Bipolar Scaling

As mentioned in the previous section, more bipolar device parameters are involved in the design; the bipolar scaling is also considerably more complicated

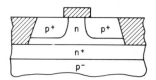

Fig. 2.7 Lateral *pnp* transistor on n^+ subcollector.

than the MOSFET scaling.[15] Also, in bipolar down-scaling, power supply voltage and the logic swing cannot be reduced much since they are already close to their lower limit at room temperature.[16] So that, unlike the MOSFET case, the electric field strength and current densities increase as one scales down. Thus, rather than scaling the transistor equations, the transistor is represented by an equivalent circuit and the circuit equation is scaled.[17]

In general, semiconductor circuits can be modeled basically as RC networks. In scaling the circuits, one therefore needs only to change all of the important capacitors by a common factor; then the circuit delay is scaled by the same factor. Although the same kind of scaling factor can be applied to all resistive elements, the scaling down of resistors would mean scaling up the operating current. Since the logic swing is $I_s R_l$, scaling resistors would lead to a prohibitively rapid increase in current density unless one reduces the logic swing, for example, by operating the circuit at a lower temperature.[16] Thus, capacitance scaling is more practical. In bipolar transistors, this means scaling both C_d (the e-b diffusion capacitance) and C_c, among others, where the former is dictated by the vertical profile (or base stored-charge time constant) and the latter, mainly the lateral dimension (or lithographic level). Thus, scaling bipolar means not only the shrinking of the lateral dimension, but also the vertical profile of the transistor.

A set of rules that govern the relation between the vertical profile of the transistor and the lithographic level for ECL circuits[17] is given in Table 2.2. The key device parameters are the same as the key design parameters: the emitter width, the base doping and width, the collector doping, and the current density. The base and collector dopings N_b and N_c should be increased such that $N_b \sim W_b^{-2}$ to avoid punchthrough, $N_c \sim J_c$ to avoid base stretching. Included in the exponents is the dependence of mobilities on doping in the 10^{17}–10^{18} cm^{-3} range.

Figure 2.8 shows the projected gate delay of the scaled ECL circuit for lithographic levels from 2.5 to 0.25 μm. It is shown that in order to achieve circuit delay in proportion to the lithographic dimension, the current and, hence, the power dissipation, of the circuit remains essentially constant in scaling. Continued improvement in circuit performance is expected from scaling at least down to 0.25 μm dimensions. It also shows that the integration level is always limited by the cooling capability of the package, since the reduction in the operating current will add more wiring delay, which is a large portion of the total circuit delay in such a speed regime. It also indicates that a trade-off can be made between the integration level and the circuit speed, depending on the product objective.

TABLE 2.2 Scaling Rules[a]

Base doping N_b	w_b^{-2}
Collector doping N_c	J
Base width w_b	$a^{0.8}$
Collector current density J	a^{-2}
Circuit delay	a

[a]Lithographic level (emitter stripe width).

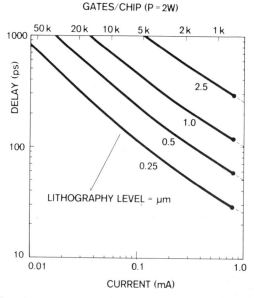

Fig. 2.8 The projected gate delay of the scaled ECL circuit for a 2.5–0.25 μm lithographic level.[17]

So far, we have focused on the scaling of the vertical transistors. Lateral *pnp* transistors scale differently. Its stored charges are not only a function of the base width but also functions of the abruptness of the transition region from epi to the n^+ subcollector and the depth of the p^+ emitter, etc. Most lateral transistors perform poorly; thus, there is little incentive to scale them.

Schottky barrier diodes (SBD) are frequently used in bipolar circuits. They should be scaled to operate at current densities compatible with the scaled transistors. At high injections, the minority carriers also inject into the lightly doped n–epi region and slow down the switching speed.[18] The concept of transistor scaling can be extended to SBDs.

2.2.3 Nonscalable Components

The resistive voltage drop along the current path should rise as the device is scaled down.[17] Specific contact resistance at the metal–semiconductor interface does not change. Reduction of contact size will raise the contact resistance. The resistance at the emitter contact is the most critical parameter since it affects the current injection, and will be discussed in Section 2.3. The thickness of the subcollector layer cannot be reduced, since the collector current remains unchanged. In addition, the wire capacitance associated with the metal interconnects is dictated by a very different set of conditions.

To scale the wire capacitance of a circuit, the length and the width of the wire should be reduced with the lithographic level. This can be realized if the number of circuit on a chip does not increase. Figure 2.9 shows the wire capacitance.[19] For narrow wire pitch, the fringe capacitance becomes significant.[20] The average wire capacitance of wires on the multiple-plane wiring grid is greater than 0.2 pF/mm in length due to cross-coupling. For constant current scaling, the current density and IR drop along the wire increases as the metal line becomes narrower. A typical 1 μm thick Al line has a sheet resistance of 30 mΩ/$\square$. For chips with increasing circuit counts, the average number of wire per gate[21] and the average wire length[22] both increase. As a result, the wire capacitance always increases. The situation can only be improved by adding another layer of metal wire to the chip.[23]

2.3 SCALING PROPERTIES AND LIMITS

When scaling down the bipolar transistor, the impurity doping density in every region of the transistor is raised. For devices with base width near 100 nm, the peak doping is in the 10^{18} cm^{-3} range. The increase in doping density alters the silicon electrical properties, which eventually impose limits to the scaling[16,19,24] or force a deviation from the ideal scaling. Limitations also arise from a specific engineering practice. We will come back to discuss the engineering issues in the later sections. Here, we will focus on the properties of the thin base and shallow emitter of the scaled transistor.

2.3.1 The Thin Base

Several interesting properties of the thin base are related to the high doping density and its consequent bandgap narrowing, high junction field, and leakage currents.

The bandgap narrowing in the heavily doped silicon was well accepted. In a simple picture, the bandgap is reduced by the electron exchange energy which

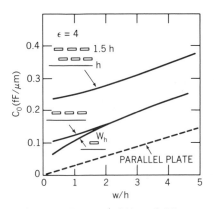

Fig. 2.9 Wire capacitance, metal thickness/width = 0.25, separation between wires = wire width. (From Ref. 19. Reprinted with permission, copyright 1982, IEEE.)

scales with the interelectron distance. The amount of narrowing can therefore be written as $\Delta E_g \sim N^{1/3}$, where N is the electron density or equivalently, the doping density.[25] Data of the measured ΔE_g using different methods spread widely.[26-28] Figure 2.10 shows the results from three different measurement techniques. In general, data from photoluminescence should give a more accurate bandgap value, whereas injection current measurement on transistors gives the pn product, or n_i^2.[29] The n_i^2 should increase as the bandgap narrows with the increase in doping density. The heavy doping effect in the emitter reigon is very significant. In fact, the heavy doping effect in the emitter has been one factor that lowers the emitter efficiency of bipolar transistors. Due to the bandgap difference in the emitter region and base region, the current gain $\beta \sim \exp\{[-E_g(\text{emitter}) + E_g(\text{base})]/KT\}$ has a positive temperature dependence.[30] A typical transistor with base doping in the 10^{17} cm^{-3} range will suffer a current gain reduction about a factor of 40 or more from room temperature to the liquid nitrogen temperature. For the scaled transistor with base doping raised above 2×10^{18} cm^{-3}, it is expected that the bandgap-narrowing effect in the base should reduce the temperature sensitivity of the current gain. From room temperature to the liquid nitrogen temperature, a gain reduction of about 3–5 times was observed in pnp transistors[8] with base doping in the 10^{18} cm^{-3} range.

As one raises the base doping density, the carrier mobility decreases due to the increase in impurity scattering. Recently, it was suggested that the minority carrier is scattered more often by the majority carriers than by the impurity ions.[31] Carrier–carrier scattering is weaker than the heavier impurity-ion scattering; thus, the mobility of a carrier can be larger when it is a minority carrier (e.g., hole in n-Si) than when it is a majority carrier (e.g., hole in p-Si). The diffusion coefficient of the minority carrier is related to the low-field mobility through the Einstein rela-

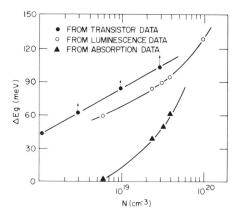

Fig. 2.10 Collection of representative values of gap reductions extracted from transistor data,[28] luminescence data,[27] and optical absorption data[26] plotted against impurity concentration. (From Ref. 29. Reprinted with permission, copyright 1985, Pergamon Journals, Ltd.)

tion, $D = (KT/q)\mu$. Several measurements of the diffusion coefficient of minority carriers using current injection techniques[32,33] and time-of-flight technique[34] showed that in the doping range of 10^{18} to mid-10^{19} cm^{-3}, the μ_n (minority) is slightly greater than μ_n (majority) and μ_p (minority) is almost twice μ_p (majority) (Fig. 2.11). The combination of the reduced bandgap and difference in the minority-carrier and majority-carrier mobility in the heavily doped silicon can explain the constant value observed in the J_c/R_{sb} ratio [Eq. (9)], which is $q^2\mu_p D_n n_i^2$. The roll-off of $\mu_p D_n$ at higher doping densities is offset by the increases in n_i^2.

Another aspect of the heavy doping effect is the high junction field, namely, the high built-in field of the emitter–base junction. The maximum electric field is $2(\phi_{bi} - V_{be})/x_e \sim N_b^{0.5}$, where x_e is the depletion layer width of the emitter–base junction and V_{be} is positive in forward bias. Increasing the base doping density will raise the field, and the collector saturation current density decreases [Eq. (8)]. To scale up the collector current density, V_{be} must be raised, and that in turn lowers back the junction field at the operating point. However, the field at zero and reverse bias increase with scaling. For an emitter–base junction with intersect doping density in the low-10^{18} cm^{-3} range, the reverse current of the junction current starts to behave like a "backward diode."[35] Although the forward current remains ideal, the reverse current is substantially greater than the Shockley–Read–Hall (SRH) recombination current[36] and increases rapidly with the voltage; it is attributed to the tunneling. Figure 2.12 shows the junction current of various background doping levels. If the base doping is further increased above mid-10^{18} cm^{-3} range, the "trap-assisted" tunneling current may also exist in the forward bias.[37,38] The electron in the conduction band can tunnel into a state in the forbidden band and later recombine with a hole. The necessary states for this process may be the Si–Si oxide interface state. This current is an exponential function of the emitter–base bias voltage but is insensitive to temperature. Figure 2.13 shows the current density vs. depletion layer width; the currents start to rise as the depletion layer width at zero bias becomes less than 40 nm.

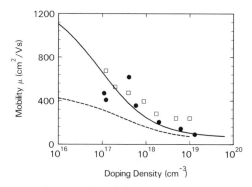

Fig. 2.11 Mobilities of electron in p-Si (Squares) and holes in n-Si (dots) vs. doping concentration at 300 K. The mobility is obtained from diffusion coefficient at zero-field condition. For comparison, electron mobility in n-Si, solid line; hole mobility in p-Si, dashed line. (From Ref. 32. Reprinted with permission, copyright 1979, *Applied Physics Letters*.)

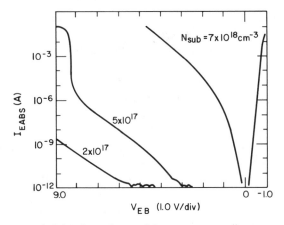

Fig. 2.12   Substrate resistivity dependence of the reverse tunneling currents of n^+p diode. The scale of vertical axis is log current. (From Ref. 36. Reprinted with permission, copyright 1983, IEEE.)

It is expected that when the base width is scaled to 25 nm, the average doping density in the base reaches 7×10^{18} cm^{-3}.[17] Even at this doping level, the transistor is still functional at high currents, though it may lose its current gain at low currents. The transistor will not be suitable for the leakage-sensitive circuits, for example, the dynamic memory. However, such a low-level leakage is of little concern for the operation of the high-speed ECL circuits.

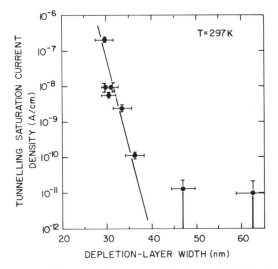

Fig. 2.13   Forward tunneling saturation current density vs. depletion layer width of the p^+n^+ junction at zero bias $w_d(0)$. Tunneling current density is $J_{oT} \simeq k_1 \exp[-k_2 w_d(0)]$. (From Ref. 38. Reprinted with permission, copyright 1986, IEEE.)

Ultimately, in a very thin base, the electrons encounter finite number of scattering when they diffuse across the base. The current transport in the base becomes more like a thermionic emission.[39] The base stored-charge time constant would be proportional to the base width rather than the square of the base width. The calculation using Monte Carlo technique for a base width down to 20 nm[40] showed that there is no significant departure from the diffusion transport. It is also found that when the electrons pass through the collector–base junction, under the acceleration of the junction field, an overshoot in the carrier velocity happens at the collector–base junction and lowers the carrier density there.[40,41]

When we try to keep the base punchthrough voltage unchanged in scaling, the total base dopant is decreasing. This is due to the fact that the punchthrough voltage is set by maintaining a constant $N_b w_b^2$, whereas the base dopant in a cubic is $N_b w_b^3$. Due to the random fluctuation in the dopant distribution, there are local fluctuations in the depletion layer width of the emitter–base junction, resulting in variations of the base width.[16] Undoped layer at the emitter–base junction can be used to eliminate this problem and the associate limitation.[19] It also reduces the maximum electric field at the emitter–base junction by a factor of 2, and thus the junction leakage current.

2.3.2   The Shallow Emitter

Ideally, the emitter–base junction depth is independent of the base width, which is the difference between the collector–base junction depth and the emitter–base junction depth. In practice, the depth of either junction cannot be controlled perfectly and some tolerance exists. The tolerance is proportional to the amount of dopant movement or the junction depth. To maintain the control and reproducibility of a thin base width, both the emitter–base and collector–base junction must be scaled. The emitter–base junction depth of the modern high-speed transistor is made shallower than the diffusion length of the holes that are injected into the emitter. Thus, the base saturation current density is affected by the surface of the emitter as given by

$$J_{bo} = \frac{qn_i^2}{N_e(w_e/D_p + 1/s)}, \tag{13}$$

where w_e is the emitter depth, s the surface recombination velocity, and N_e the emitter doping density. The metal contact to the emitter surface has a high recombination velocity ($s \simeq 10^6$ cm/s) and forms an almost perfect sink for the injected minority carriers. The shallow emitter raises the gradient of the minority-carrier density and hence the base current, resulting in a decrease in current gain. For a typical emitter with doping density of 2.10^{20} cm^{-3}, the hole diffusion length is around 250–300 nm.

A polysilicon buffer layer can be added between the metal contact and the surface of the shallow emitter to suppress the base current and therefore bring back the current gain.[42] The polysilicon layer effectively moves the metal contact farther away from the emitter–base junction.[43] The high dopant diffusivity in the polysilicon also shortens the heat cycle for forming an emitter of equivalent junction "depth." Cross-sectional TEM studies[44–46] showed that a thin and discontinuous layer of native oxide exists at the poly–mono silicon interface. During the subsequent annealing, the polysilicon recrystallizes epitaxially except over the regions covered by the interfacial oxide. In the high-temperature anneals, the interfacial oxide breaks up and the recrystalization process continues. Usually, dopants (e.g., As) pile up at the interface when the polysilicon receives little anneal and the pile-up disappears when the polysilicon is recrystallized.[47]

Intensive studies have been conducted to understand the effects of the polysilicon contact on the base current. Possible mechanisms are the reduced carrier mobility in the polysilicon,[43] tunneling through the interfacial oxide,[48] finite recombination at the poly–mono silicon interface,[46] and the thermionic injection over the interfacial barrier.[49] Accumulated evidences are that to be effective, the polysilicon layer has to be thicker than 50 nm and the doping density must be greater than 10^{20} cm^{-3}.[43,46]

The surface cleaning and the post-polysilicon deposition heat cycle are important.[50,51] Both the current gain and the emitter series resistance are a strong function of the surface preparations.[45,50,51] A thicker (or "intentionally" grown) oxide layer at the interface brings about a higher current gain, but gives an unacceptably high emitter resistance.

In scaling, the parasitic-resistor value is assumed unchanged; however, the contact resistance actually increases since the contact size shrinks. The ideal contact resistance between the heavily doped silicon and Al is $10 \ \Omega \cdot \mu m^2$. For an emitter of $0.25 \times 0.25 \ \mu m^2$, the emitter resistance is at least 160 Ω! The polysilicon emitter has one additional interface; higher emitter resistance is expected. One needs to reduce or completely eliminate the interfacial oxide and thus its resistance for further scaling. The metal–polysilicon contact resistance is reduced further by making the metal–polysilicon contact area larger than the emitter opening on the monocrystalline silicon.

Low emitter resistance is particularly important for the bipolar transistor, since it acts as a feedback resistor that reduces the effective transconductance of the transistor, and therefore the ac small signal gain of an amplifier as well as the noise margins of a logic circuit. By setting a criterion that the voltage drop across the emitter contact resistance is to be less than 10% of the logic swing (or say, 50 mV) of a logic circuit, one can estimate the maximum current density through an emitter to be $J \simeq 5$ mA/μm^2.[17]

2.4 TRANSISTOR STRUCTURES

Generically, there are two classes of bipolar transistors; namely, the planar structure and the polysilicon-base transistor.[52] The planar structure (Fig. 2.2) that brought bipolar transistor integrated circuits into dominance in the 1960s is still being produced in the 1980s. Its vertical profile is scaled; the diffusion process used to form the base region is gradually being replaced by the more controllable ion-implantation process. It is a common practice to add a polysilicon emitter to the planar structure when the vertical profile is scaled.[42] This device structure is relatively simple by today's process standard and lacks surface topography (thus, the name "planar"). However, the size of this device structure is very sensitive to the alignment accuracy between masking levels. For a two-base contact transistor, the collector–base junction is at least five times the minimal lithographic line width. The reduction of the collector–base capacitance can be done primarily through the improvement in the lithographic processes. The extrinsic-base resistance can be reduced by adding an extra dose of implant to the extrinsic base. Using ion implantation to form the extrinsic base, the sheet resistance can be reduced to less than 200 $\Omega/\square$, with a junction depth under 0.35 μm. The gate delay of an ECL ring oscillator using planar devices with a 0.8 μm wide emitter is in the range of 50 ps.[53]

The polysilicon-base transistor was developed in the late 1970s.[54–56] The metal contact to the base is not over the monocrystalline silicon but over the polysilicon layer that bridges the two. As a result, the collector–base junction area is not limited by the metal pitch and is therefore reduced substantially. These structures readily lend themselves to self-aligned schemes.[54,55] Those schemes are designed to overcome the large alignment inaccuracy of the lithographic tools. In most of the self-aligned structures, the emitter-to-base contact spacing is determined by the sidewall thickness, which is the CVD (chemical vapor deposition) film thickness. Thus, a submicron spacing is typical for self-aligned devices. Figure 2.14 shows the cross section of one of the "double-poly" self-aligned bipolar structures. The overall parasitic capacitance of this transistor is much smaller than that of the planar device. Therefore, circuits built with polybase transistors show lower power-delay products. 50 ps switching delays had been reported for unloaded ECL ring oscillators (fan-in = 1).[57,58] Further reduction of the collector area can be made by butting the emitter edge against the oxide isolation on two sides.[9]

The minimal spacing between the emitter and base contact is determined by the electrical property of emitter perimeter. In principle, the depletion layer of the emitter–extrinsic-base junction must be covered entirely by the oxide sidewall, so

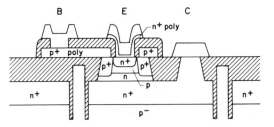

Fig. 2.14 Schematic of a "double-poly" bipolar device showing its three key features, namely, (a) self-aligned base contact, (b) deep-trench isolation, and (c) polysilicon emitter contact.[52]

that the junction is not exposed to the polysilicon. Since the emitter junction diffuses laterally, the minimum spacing should be at least the sum of the emitter lateral diffusion, the extrinsic-base lateral diffusion and the depletion layer width. The arsenic lateral diffusion is somewhat larger than the depth of the emitter.[59] If the base is in contact with the metal, the emitter perimeter should be a diffusion length (of carriers) away from the base metal contact, so that the laterally injected carriers do not recombine immediately at the base contact and cause a current gain loss. The doping density at the emitter–extrinsic base intercept should leave a depletion layer of 40 nm or greater[38] so that tunneling around the emitter perimeter is avoided. The minimal spacing between the emitter and base contact is thus about 120 nm, assuming the emitter and base dopant each diffuse laterally 40 nm.

All the self-alignment schemes rely on the surface topography to create a thin sidewall "spacer" that separates the emitter and base contacts. Thus, self-aligned devices have a much higher surface topography than the planar devices and require more process control, especially when the lateral dimension is reduced to the same as the step height. One should notice that the dominant base resistance comes from the intrinsic-base region under the emitter. While the extrinsic-base resistance is proportional to the emitter-to-base contact spacing, the intrinsic-base resistance is proportional to the emitter stripe width and cannot be reduced by using the self-aligned structure; rather, it can only be reduced by narrowing the emitter stripe width. In general, the advantage of a self-alignment structure becomes less significant as the alignment accuracy of lithographic systems improves.

The ideal structure is the so-called symmetrical structure[60] or the sidewall-base-contact structure (SICOS),[61] shown schematically in Fig. 2.15. With a buried oxide underneath the extrinsic-base region, the *npn* transistor has the least base–collector junction capacitance. It also has comparable upward and downward *I–V* characteristics. In reality, the *npn* has higher speed in the normal downward mode than in the upward mode. High-speed lateral *pnp* transistors with f_t at 3 GHz have been realized with this structure,[62] since the buried oxide under the *pnp* emitter eliminates the stored charge that slows down the typical lateral *pnp* transistors (shown in Fig. 2.7).

Deep-trench isolation is replacing the traditional junction isolation. The deep-trench is lined with an insulator, such as oxide, filled with silicon or other material and penetrates the subcollector down to the substrate. Its biggest leverage lies in the area reduction of the transistor. This is why the early applications were mostly

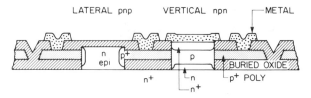

Fig. 2.15 Schematic of the symmetrical bipolar device structure. (From Ref. 60, copyright 1980, IEEE.)

to memory design.[63] However, it also reduces the isolation capacitance and thus improves logic circuit speed.

A comparison of the layout and the electrical parameters of the planar transistor with junction isolation and the self-aligned transistor with deep-trench isolation is shown in Fig. 2.16. For the same active area (emitter area), the planar device with junction isolation has much larger parasitic capacitances and, in general, the circuits must dissipate a rather large power in order to have a speed approaching that limited by the base width. With the large reduction of parasitics, the vertical profile affects the device performance more. Optimizing the device profile for higher current density operation becomes more important. Experimental observations of Kirk effects in logic circuits are increasingly common.[64,65]

In all modern transistors, the impurity profile of the active region and the inactive region of the transistor are formed and optimized separately. Generally, the intrinsic base that is formed late in the fabrication process is extendable to very shallow emitter and base profiles. Processes in which there are long thermal cycles after the intrinsic base formation have problems yielding very small and reproducible base widths.

2.5   BIPOLAR CIRCUITS

Logic and memory applications have been the key driving forces behind the bipolar technology development. The performance and integration level are good measurements of the technology. Here we will give a survey of the recent advances in this area, which should represent the overall capability of the bipolar technology.

One important attribute of the bipolar scaling is the constant power scaling. While circuit speed improves through scaling, the chip power increases with the integration level. Thus, the integration level is more often gauged by the package technology and not by the bipolar technology alone. Packages for bipolar VLSI circuits should be capable of removing more heat and should have a large number of signal input/output pins. The latter allows more communication between chips, an important feature of the high-speed VLSI logic. Integration level sometimes has little to do with the speed of the system performance. A very high-speed electronic system can be made with chips of low-level integration, as mentioned at the beginning of this chapter. Nevertheless, we will focus on the capability of today's logic- and memory-chip technology in this section.

2.5.1   Logic

The impact of scaling on the speed of the bipolar circuits is most direct. A literature survey of the operating current density[4] is shown in Fig. 2.17. The operating current density continues to increase over the years and is reaching 60 kA/cm^2 today. The evolution of the integration level of the bipolar logic products[66,67] is shown in Fig. 2.18. The integration level of the bipolar gate array follows the progress of the multiple-level metal process for interconnects, since the number

CONVENTIONAL TRANSISTOR WITH NON-BUTTED EMITTER

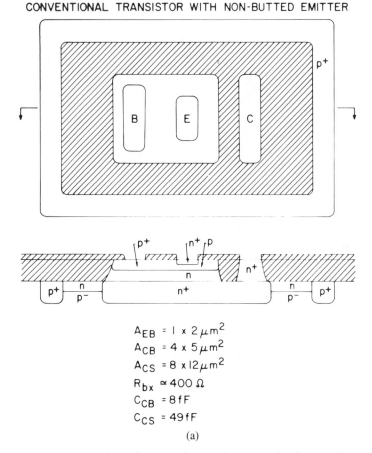

$$A_{EB} = 1 \times 2\,\mu m^2$$
$$A_{CB} = 4 \times 5\,\mu m^2$$
$$A_{CS} = 8 \times 12\,\mu m^2$$
$$R_{bx} \approx 400\,\Omega$$
$$C_{CB} = 8\,fF$$
$$C_{CS} = 49\,fF$$

(a)

Fig. 2.16 A comparison of the planar transistor with nonbutted emitter and junction isolation and self-aligned transistor with butted emitter and deep-groove isolation. (From Ref. 67. Reprinted with permission, copyright 1985, IEEE.)

of wireable circuits on a gate array is determined by the availability of wiring tracks.[21] The polysilicon and diffused lines extensively used in MOSFET circuits for circuit interconnection are too resistive for the bipolar application. A large number of gate arrays utilize three levels of metal for wiring. In designing the basic gate layout of a gate array, the emphasis is not to minimize the area occupied by a gate, but to maximize the accessibility of logic terminals of the gate to the wiring tracks. Continuing demand for more interconnect wires through either more levels of metal or much finer metal line pitch is expected for the future VLSI gate-array products. A 10,000-gate gate array has been reported using four levels of metal.[68]

Currently, most gate arrays are designed with ECL circuits for their superior circuit speed, noise immunity, and the capacitance driving capability. An ECL cir-

SELF-ALIGNED TRANSISTOR
WITH BUTTED EMITTER

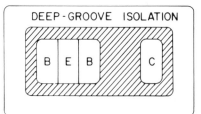

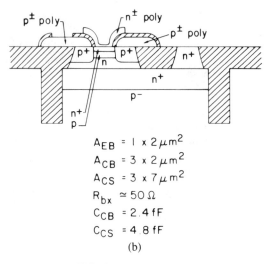

$$A_{EB} = 1 \times 2 \,\mu m^2$$
$$A_{CB} = 3 \times 2 \,\mu m^2$$
$$A_{CS} = 3 \times 7 \,\mu m^2$$
$$R_{bx} \simeq 50 \,\Omega$$
$$C_{CB} = 2.4 \,fF$$
$$C_{CS} = 4.8 \,fF$$

(b)

FIG. 2.16 (Continued)

cuit consumes much more power than other circuits due to its higher power supply
voltage (>3 V) and constant current dissipation. On the other hand, it has more
logic functional capability per gate. For example, it provides OR and NOR functions, and also the combination of both by cascoding the inputs and "dotting" the
outputs. Thus, its "functional" delay is rather short.

To reduce the power of the circuit without substantially reducing the current of
the circuit (thus the wiring delay), the choices of the future VLSI circuits appear
to be the circuits with lower supply voltage and possibly smaller logic swing.
There are many bipolar circuits, such as NTL, ISL, TTL, STL, and MTL (or
I^2L) (Fig. 2.19), that operate with a power supply less than 2 V and can achieve a
power-delay product of less than 0.2 pJ at 1 μm design rules.[69,70] The fastest of
these circuits appear to be NTL with emitter-follower. NTL as fast as 30 ps at
1.48 mW has been demonstrated.[71] The MTL circuit has a better wiring efficiency
because it doesn't need intragate wiring. However, gate delay is very sensitive to

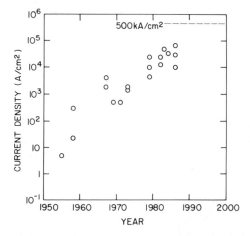

Fig. 2.17   A survey of the operating current density of bipolar transistors. The theoretical limitation is set by the contact resistance.[9] [From Ref. 4. Reprinted with permission, IBM Research report, 1987 (unpublished).]

the transistor structure,[72,73] due to the inverse operation of the *npn* transistors. Nevertheless, MTL circuits built with planar bipolar technology have been used in a high-performance microprocessor.[74]

2.5.2   Memory

Memory performance benefits directly from the lithographic linewidth/overlay reduction alone. This is because most circuits on a memory chip are heavily loaded

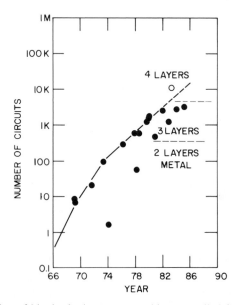

Fig. 2.18   Evolution of bipolar logic gate array chips, compiled from Refs. 66 and 67.

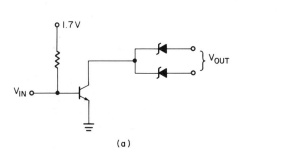
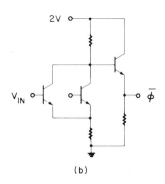
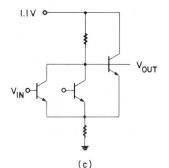
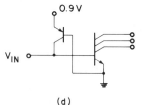

Fig. 2.19 Schematics of lower power bipolar gates: (*a*) STL (Schottky transistor logic) (*b*) NTL (nonthreshold logic) with emitter-follower, (*c*) CML (current-mode logic), (*d*) MTL (merged-transistor logic), and (*e*) ISL (integrated Schottky logic).

with the capacitance of the long line that connects a row of memory cells. The load capacitance is far greater than the internal capacitance of the gate. A smaller cell size immediately means smaller load capacitance and therefore higher circuit speed. There is a practical concern about the scaling of the transistor vertical profile for memory cells. Thin-base transistors are more vulnerable to the emitter-to-collector shorts caused by crystalline defects or "pipes," especially when the total area of the emitters in a memory chip is also much larger than that in a logic chip. Thus, wide-base transistors are frequently used for memory cells for yield consideration. Further improvement in memory performance can be realized by scaling the vertical profile of the transistors in the small number of peripheral circuits, since a thin-base transistor can deliver more current in a shorter time to charge the capacitance of the long lines.[75]

Excellent accounts of the history of the memory development have been given in Refs. 76 and 77. The evolution of memory development is shown in Fig. 2.20. The integration level of the bipolar memory remains at about 1/16 the level of the MOSFETs. Nevertheless, there are several highlights in the memory development. First, the access time of the high-speed bipolar memory is already in the sub-nanosecond[78] regime promised by the GaAs MESFET technology. The reason that bipolar can compete against GaAs in this area is again due to its superior capacitance driving capability and its excellent voltage tracking that allows the circuits to be operated with a very small logic swing. The switching delay of unloaded bipolar circuits may not be faster, but in an LSI environment, where the wiring capacitance is heavy, they can do better than the present GaAs circuits.

A key step in the development of the bipolar VLSI memory is the introduction of the merged *pnp*-load memory cell.[79] In the past, high-speed memories had been designed with resistor-load memory cell (usually called ECL cell, but not related to ECL logic). Due to the finite value of the resistor, a typical standby current of tens of μA is dissipated by the cell, which is too large for VLSI memories. Presently, most of the RAMs of 16 kbit capacity or greater are implemented with the *pnp* load cells.[75,77,80] The standby current is reduced down to the nanoampere range, although theoretically the cell remains stable in the picoampere level, which is comparable to the CMOS memory cells. The *pnp* load cell is also considerably smaller than the resistor-load cell since the lateral *pnp* can be merged into the vertical *npn* as shown in Fig. 2.21 and is therefore called merged-transistor logic (MTL) cell. Together with the polysilicon intracell wiring and the deep-groove isolation, MTL memory is potentially very competitive in the VLSI memory

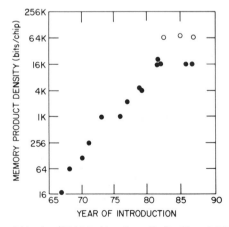

Fig. 2.20 Evolution of bipolar SRAM chips from Refs. 67 and 76. Darkened circles are product chip and open circles are chips made in laboratories.

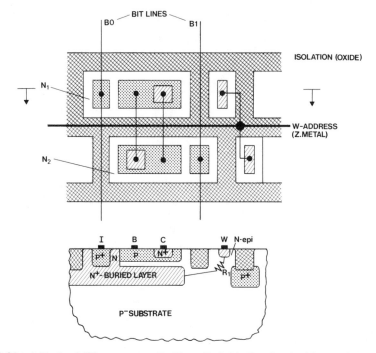

Fig. 2.21 4-Device MTL memory cell. (From Ref. 81. Reprinted with permission, copyright 1981, IEEE.)

arena.[81] One most important property of the MTL memory cell is its extraordinary resistance to the α-particle-induced soft errors.[82] The n^+ sublayer, acting as a common word line, drains completely the carriers generated in the substrate by the ionizing α-particles and keeps them away from the sensitive storage nodes: the top n^+ pocket of the *npn* transistor. This property allows MTL memory cell to be scaled down more than all other bipolar and MOSFET memory cells in this respect.

2.6   FUTURE POSSIBILITIES

It is undoubted that Si bipolar transistors will continue to be scaled down to sub-half micron. A review of the future trend has been given in Refs. 83 and 84. Driven by the silicon MOSFET dynamic RAM developments, lithographic tools for patterning half and sub-half microns will become a reality.[85] Self-aligned device structures of some sort will continue to play key roles until the alignment accuracy of the lithographic tools is improved down below ± 0.1 μm. It is in position to challenge the GaAs MESFET technology in the high-speed LSI environment.

To fabricate transistors with a base width of 25 nm or less in a controllable manner, a new way to form the thin base is needed. A typical ion-implanted base profile is shown in Fig. 2.22.[86] The shallow profile formed by low-energy ion-

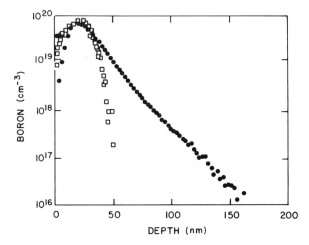

Fig. 2.22 Ion-implanted boron profile at 5 keV. A channeling tail exists. SIMS measurements, dots, Monte Carlo calculations, squares.[87] (From Ref. 86. Reprinted with permission, copyright 1984, *Applied Physics Letters*.)

implantation resembles to an exponential function rather than a Gaussian profile due to channeling effects. The slope of the as-implanted boron profile is not steep enough to begin with and is further reduced by the boron diffusion in the subsequent heat cycle. Recent advances in lower-temperature processes, such as the molecular-beam epitaxy (MBE),[88] and the photo-epitaxy[89] have demonstrated the feasibility of forming heavily doped and thin-base region. Figure 2.23 shows the thin-base profile achieved by the photo-epitaxy process. To lower the emitter contact resistance, the elimination of the poly–mono silicon interface resistance will be necessary. Similar to the migration from NMOS to CMOS in the evolution of the MOSFET technology, increasing sophistication and maturity in the bipolar fabrication process can accommodate both high-speed *npn* and *pnp* transistor and make high-performance complementary circuits possible, such as CBL (charge buffer logic[90]).

Both the package and the multilayer interconnect metallurgy, rather than the transistor will continue to set the pace of the increase in the integration level of bipolar logic circuits. Although progress is being made in those areas, more pressure is on finding circuit techniques for reducing the power dissipation at the functional level without compromising the performance. Mixing both the high-power high-speed and low-power moderate-speed circuit is one method.[91] More advanced CAD (computer-aided design) tools are needed to discriminate the capacitance associated with circuit nets, so that higher power circuit is placed to drive the high-capacitance circuit node and vice versa. Such a practice is common in the MOSFET circuit design and is relatively easy to implement to the bipolar design. Another possible circuit scheme will be the selective power-on circuit blocks. This scheme is used in memory design. Circuits are in the power-down state during standby and the power-up state when activated through an external control, for example, the chip select circuitry.

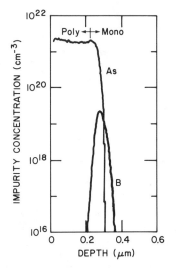

Fig. 2.23 SIMS depth profiles of an intrinsic transistor region. The base is formed by photo-epitaxy process at 650°C.[89] The base width is 65 nm and the peak base doping density 10^{19} cm^{-3}. (From Ref. 89. Reprinted with permission, copyright 1987, IEEE.)

Along the memory front, the MTL memory with very competitive density, low power dissipation, and high resistance to α-particle-induced soft error has been the bipolar entry into the high-performance VLSI. The scaling of the MTL memory does not have to include the reduction of the base width, at least in the memory cells, since the transistor is operated in the inverse mode. Scaling lithographic dimension alone will directly increase its integration level.

Looking farther ahead, bipolar actions do not have to be restricted to the homojunction bipolar transistors; heterojunction transistors are other possibilities.[92] An emitter with energy bandgap larger than the base creates an energy barrier to the minority-carrier injection to the emitter. The advantage of the heterojunction bipolar transistor is that its base can be doped heavily to reduce the base resistance. These heterojunction bipolar transistors are made generally using molecular-beam epitaxy technique. Their speed advantage at room temperature is limited. However, they create the possibility of low-temperature operation. With a larger bandgap emitter, the current gain should increase as temperature decreases, opposite to the homojunction Si bipolar transistor. Although a larger bias voltage is needed to turn on the *pn* junction at lower temperatures, the logic voltage swing can be scaled down with kT/q, while keeping the same "on-" and "off-" current ratio. It is expected that at the liquid nitrogen temperature, the overall power-delay product of a loaded circuit ($\simeq CV_s V_l$, where V_s is the power supply voltage and V_l the logic swing) should decrease by close to a factor of 4 compared with room temperature. Another heterojunction device is called BICFET[93] (bipolar inversion channel field-effect transistor). A very thin base layer can be formed by inverting the surface, rather by doping the substrate. However, all these new devices need a strong material development effort.

Today's bipolar technology is under intense competition with other rival technologies. The most notable are the GaAs technology in the high-speed LSI area and the CMOS and BiCMOS technology in the VLSI area. Future possibilities can only be realized with continuing efforts, and the final outlook is strongly dependent on the relative rate of the progress.

REFERENCES

1. S. Gaur, "Performance Limitations of Silicon Bipolar Transistors," *IEEE Trans. Electron Devices,* **ED-26,** 415 (1979).

2. P. A. H. Hart, T. Van't Hof, and F. M. Klassen, "Device Down Scaling and Expected Circuit Performance," *IEEE Trans. Electron Devices,* **ED-26,** 421 (1979).

3. D. D. Tang and P. M. Solomon, "Bipolar Transistor Design for Optimized Power-Delay Logic Circuits," *IEEE J. Solid-State Circuits,* **SC-14,** 679 (1979).

4. R. W. Keyes, *Bipolar Transistors: Trends and Limitations,* IBM Res. Rep. RC12843, IBM, Yorktown Heights, NY, 1987.

5. R. J. Blumberg and S. Brenner, "A 1500 Gate Random Logic LSI Masterslice," *IEEE J. Solid-State Circuits,* **SC-14**(5), 818 (1979).

6. C. T. Kirk, "A Theory of Transistor Cutoff Frequency Fall-off at High Current Density," *IEEE Trans. Electron Devices,* **ED-9,** 114 (1962).

7. H. H. Heimier and H. H. Berger, "Evaluation of Electron Injection Current Density in P-Layers for Injection Modeling of I^2L," *IEEE J. Solid-State Circuits,* **SC-13**(2), 205 (1977).

8. D. D. Tang, "Heavy Doping Effects in PNP Bipolar Transistors," *IEEE Trans. Electron Devices,* **ED-27,** 563 (1980).

9. D. D. Tang, P. M. Solomon, T. H. Ning, R. D. Isaac, and R. E. Burger, "1.25 μm Deep-Groove-Isolated Self-Aligned Bipolar Circuits," *IEEE J. Solid-State Circuits,* **SC-17,** 925 (1982).

10. G. D. Hachtel, M. R. Lignter, and H. J. Kelly, "Application of Optimization Program AOP to Design of Memory Cell," *IEEE Trans. Circuits Syst.,* **CAS-22,** 496 (1975).

11. H. K. Gummel and H. C. Poon, "An Integral Charge Control Model of Bipolar Transistors," *Bell Syst. Tech. J.,* **49,** 827 (1970).

12. E. M. Buturla, P. E. Cottrell, B. M. Grossman, and K. A. Salsburg, "Finite-Element Analysis of Semiconductor Devices: The FIELDAY Program," *IBM J. Res. Dev.,* **25,** 218 (1981).

13. R. W. Knepper, S. P. Gaur, F. Y. Chang, and G. R. Srinivasen, "Advanced Bipolar Transistor Modeling: Process and Device Simulation Tools for Today's Technology," *IBM J. Res. Dev.,* **29**(3), 218 (1985).

14. H. H. Berger, "The Injection Model — A Structure-Oriented Model for Merged Transistor Logic (MTL)," *J. Solid-State Circuits,* **SC-5**(5), 218 (1974).

15. R. H. Dennard, F. H. Gaensslen, L. Kuhn, and H. N. Yu, "Design of Micron MOS Switch Devices," *Tech. Dig. — Int. Electron Devices Meet.,* p. 168 (1972).

16. R. W. Keyes, "Physical Limits of Digital Electronics," *Proc. IEEE,* **63,** 740 (1975).

17. P. M. Solomon and D. D. Tang, "Bipolar Circuit Scaling," *Int. Solid-State Circuit Conf., Dig. Tech. Pa.,* p. 86 (1979).

18. S. M. Sze, *Physics of Semiconductor Physics,* 2nd ed., p. 265, Wiley, New York, 1981.

19. P. M. Solomon, "A Comparison of Semiconductor Devices for High-Speed Logic," *Proc. IEEE,* **70**(5), 489 (1982).

20. P. E. Cottrell and E. M. Butula, "VLSI Wiring Capacitance," *IBM J. Res. Dev.,* **29**(1), 227 (1985).

21. W. R. Heller, C. G. Hsi, and W. F. Mikhail, "Wireability — Designing Wiring Space for Chips and Chip Packages," *IEEE Des. Test,* **1**, 43 (1984).

22. W. E. Doanth, "Wire Length Distribution for Placements of Computer Logic," *IBM J. Res. Dev.,* **25**(3), 152 (1981).

23. K. Klein, E. F. Miersch, R. Remschardt, H. Schettler, U. Schultz, and R. Zuhlke, "A Study on Bipolar VLSI Gate-Arrays Assuming Four Layers of Metal," *IEEE J. Solid-State Circuits,* **SC-17**, 472–480 (1982).

24. B. Hoeneissen and C. A. Mead, "Limitations in Microelectronics II. Bipolar Technology," *Solid-State Electron.,* **15**, 891 (1972).

25. R. W. Keyes, "The Energy Gap of Impure Silicon," *Commun. Solid State Phys.,* **7**, 149 (1977).

26. A. A. Vol'fson and V. K. Subashlev, "Fundamental Absorption Edge of Silicon Heavily Doped with Donor or Acceptor Impurities," *Sov. Phys. — Semicond. (Engl. Transl.),* **1**(3), 327 (1967); M. Balkanski, A. Aziza, and E. Amallag, "Infrared Absorption in Heavily Doped N-Type Si," *Phys. Status Solidi,* **31**, 323 (1969).

27. P. E. Schmidt, M. L. W. Thewalt, and W. P. Dumke, "Photoluminescence in Heavily Doped Si: B and As," *Solid State Commun.,* **38**, 1091 (1981); W. P. Dumke, "Comparison of Band-Gap Shrinkage Observed in Luminescence from n^+-Si with that from Transport and Optical Absorption Measurements," *Appl. Phys. Lett.,* **42**, 196 (1983).

28. W. Slotboom and H. C. de Graaff, "Measurements of Band-Gap Narrowing in Si Bipolar Transistors," *Solid-State Electron.,* **19**, 857 (1962).

29. S. Pantelides, A. Selloni, and R. Car, "Energy-gap Reduction in Heavily Doped Silicon: Causes and Consequences," *Solid-State Electron.,* **28**(1/2), 17 (1985).

30. N. L. Kauffman and A. A. Bergh, "Temperature Dependence of Ideal Gain in Double Diffused Silicon Transistor," *IEEE Trans. Electron Devices,* **ED-15**, 732 (1968); D. Buhanan, "Investigation of Current Gain Temperature Dependence in Silicon Transistors," *ibid.,* **ED-16**, 117 (1969).

31. H. S. Bennett, "Hole and Electron Mobility in Heavily Doped Silicon: Comparison of Theory and Experiment," *Solid-State Electron.,* **26**(12), 1157 (1983).

32. J. Ziewior and D. Silber, "Minority Carrier Diffusion Coefficients in Highly Doped Silicon," *Appl. Phys. Lett.,* **35**, 170 (1979).

33. J. del Alamo, S. Swirhun, and R. M. Swanson, "Simultaneous Measurement of Hole Lifetime, Hole Mobility and Bandgap Narrowing in Heavily Doped n-Silicon," *Tech. Dig. — Int. Electron Devices Meet.,* p. 290 (1985); S. E. Swirhun, Y.-H. Kwark, and R. M. Swanson, "Measurement of Electron Lifetime, Electron Mobility and Bandgap Narrowing in Heavily Doped p-Type Silicon," *ibid.,* p. 24 (1986).

34. D. D. Tang, F. F. Fang, M. Scheuermann, T. C. Chen, and G. A. Sai-Halasz, "Minority Carrier Transport in Silicon," *Tech. Dig. — Int. Electron Devices Meet.,* p. 20 (1986).

35. S. M. Sze, *Physics of Semiconductor Devices,* 2nd ed., p. 537. Wiley, New York, 1981.

36. J. M. C. Stork and R. D. Isaac, "Tunnelling in Base-Emitter Junctions," *IEEE Trans. Electron Devices*, **ED-30**, 1527 (1983).

37. A. G. Chynoweth, W. L. Feldman, and R. A. Logan, "Excess Tunneling Current in Silicon Esaki Junction," *Phys. Rev.*, **121**(3), 684 (1961).

38. J. del Alamo and R. M. Swanson, "Forward-Bias Tunneling: A Limitation to Bipolar Device Scaling," *IEEE Electron Devices Lett.*, **EDL-7**(11), 629 (1986).

39. P. Rohr, F. A. Lindholm, and K. R. Allen, "Questionability of Drift-Diffusion Transport in the Analysis of Small Semiconductor Devices," *Solid-State Electron.*, **17**, 729 (1974).

40. G. Baccarani, C. Jacoboni, and A. M. Mazzone, "Current Transport in Narrow-Base Transistors," *Solid-State Electron.*, **30**, 5 (1977).

41. R. K. Cook, "Numerical Simulation of Hot-Carrier Transport in Silicon Bipolar Transistors," *IEEE Trans. Electron Devices*, **ED-30**, 1103 (1983); Y. J. Park, D. H. Navon, and T.-W. Tang, "Monte Carlo Simulation of Bipolar Transistors," *ibid.*, **ED-31**(12), 1724 (1984).

42. J. Graul, A. Glasl, and H. Murrmann, "High-Performance Transistors With Arsenic-Implanted Polysilicon Emitter," *IEEE J. Solid-State Circuits*, **SC-11**, 491 (1976).

43. T. H. Ning and R. D. Isaac, "Effects of Emitter Contact on Current Gains of Bipolar Transistors," *Tech. Dig.—Int. Electron Devices Meet.*, p. 473 (1979).

44. N. Natsuaki, M. Tamura, and T. Miyazaki, "Rapid-thermal Annealing of a Polysilicon-Stack Emitter Structure," *Tech. Dig.—Int. Electron Devices Meet.*, p. 662 (1983).

45. M. C. Wilson, N. Jorgensen, G. R. Booker, and P. C. Hunt, "Electrical and Microstructural Investigation Into the Effect of Arsenic Emitter Concentration on the Enhanced Gain Polysilicon Emitter Bipolar Transistor," *Symp. VLSI Technol., Dig. Tech. Pap.*, p. 46 (1985).

46. G. L. Patton, J. C. Bravman, and J. D. Plummer, "Physics, Technology and Modeling of Polysilicon Emitter Contacts for VLSI Bipolar Transistors," *IEEE Trans. Electron Devices*, **ED-33**(11), 1754 (1986).

47. A. E. Michel, R. H. Kastl, and S. R. Mader, "Ion-implanted Polysilicon Diffusion Source," *Nucl. Instrum. Methods*, **209/210**, 719 (1983).

48. Z. Yu, B. Ricco, and R. W. Dutton, "A Comprehensive Analytical and Numerical Model of Polysilicon Emitter Contacts in Bipolar Transistors," *IEEE Trans. Electron Devices*, **ED-31**, 773 (1984).

49. C. C. Ng and E. S. Yang, "A Thermionic Diffusion Model of Polysilicon Emitter," *Tech. Dig.—Int. Electron Devices Meet.*, p. 32 (1986).

50. J. M. C. Stork, C. Y. Wong, and M. Arienzo, "Polysilicon Emitter Resistance and Carrier Transport Studies," *Symp. VLSI Technol. Dig. Tech. Pap.*, p. 44 (1985).

51. P. Ashburn and B. Soerowirdjo, "Comparison of Experimental and Theoretical Results on Polysilicon Emitter Bipolar Transistors," *IEEE Trans. Electron Devices*, **ED-31**, 852 (1984); A. Neugroschel, M. Arienzo, Y. Komem, and R. D. Isaac, "Experimental Study of the Minority-Carrier Transport at the Polysilicon-Monosilicon Interface," *ibid.*, **ED-32**, 807 (1985).

52. D. D. Tang, G.-P. Li, C. T. Chuang, D. A. Danner, M. B. Ketchen, J. L. Mauer, M. J. Smyth, M. P. Manny, J. D. Cressler, B. J. Ginsberg, E. J. Petrio, and T. H. Ning, "73 ps Si Bipolar ECL Circuits," *Int. Solid-State Circuit Conf., Dig. Tech. Pap.*, p. 104 (1986).

53. T. Chen, D. Tang, C. Chuang, J. Cressler, J. Warnock, G. Li, P. Biolsi, D. Danner, M. Polcari, T. Ning, "Sub-50 ps Single Poly Bipolar Technology," *Tech. Dig.—Int. Electron Devices Meet.*, p. 748 (1988).

54. T. Sakai and M. Suzuki, "Super Self-Aligned Bipolar Technology," *Symp. VLSI Technol., Dig. Tech. Pap.*, p. 16 (1983).

55. T. H. Ning, R. D. Isaac, P. M. Solomon, D. D. Tang, and H. N. Yu, "Self-Aligned npn Bipolar Transistors," *Tech. Dig—Int. Electron Devices Meet.*, p. 823 (1980).

56. H. Nakashiba, I. Ishida, K. Aomura, and T. Nakamura, "An Advanced PSA Technology for High-Speed Bipolar LSI," *IEEE J. Solid-State Circuits*, SC-15, 455 (1980).

57. H. Takemura, T. Kamiya, S. Ohi, M. Sugiyama, T. Tashiro, and M. Nakamae, "Submicron Epitaxial Layer and RTA Technology for Extremely High Speed Bipolar Transistors," *Tech. Dig.—Int. Electron Devices Meet.*, p. 424 (1986).

58. Hee K. Park, K. Boyer, C. Clawson, G. Eiden, A. Tang, T. Yamaguchi, and J. Sachitano, "High-speed Polysilicon Emitter-Base Bipolar Transistor," *IEEE Electron Device Lett.*, EDL-7(12), 658 (1986).

59. Y. Tamaki, F. Murai, K. Sagara, and A. Anzai, "A 100 nm Emitter Transistor Fabricated with Direct EB Writing for High-Speed Bipolar LSIs," *Symp. VLSI Technol. Dig. Tech. Pap.*, p. 31 (1987).

60. D. D. Tang, V. J. Silvestri, H. N. Yu, and A. Resiman, "A Symmetrical Bipolar Transistor," *Tech. Dig.—Int. Electron Devices Meet.*, p. 58 (1980).

61. T. Nakamura, T. Miyazaki, S. Takahashi, T. Kure, T. Okabe, and M. Nagata, "Self-Aligned Transistor with Sidewall Base Electrode," *IEEE Trans. Electron Devices*, ED-29, 596 (1982).

62. K. Nakazato, T. Nakamura, and M. Kato, "A 3GHz Lateral PNP Transistor," *Tech. Dig.—Int. Electron Devices Meet.*, p. 416 (1986).

63. H. Goto, T. Takada, R. Abe, Y. Kawabe, K. Oami, and M. Tanaka, "An Isolation Technology for High Performance Bipolar Memories—IOP II," *Tech. Dig.—Int. Electron Devices Meet.*, p. 58 (1982).

64. D. D. Tang, K. P. MacWilliams, and P. M. Solomon, "Effects of Collector Epitaxial Layer on Switching Speed of High-Performance Bipolar Transistors," *IEEE Electron Device Lett.*, EDL-4, 17 (1983).

65. T. Tashiro, H. Takamura, T. Kamiya, F. Tokuyoshi, S. Ohi, H. Shiraki, M. Nakamae, and T. Nakamura, "An 80-ps ECL Circuit with High Current Density Transistor," *Tech. Dig.—Int. Electron Devices Meet.*, p. 686 (1984).

66. E. J. Rymasewski, J. L. Walsh, and G. W. Wheeler, "Semiconductor Logic Technology in IBM," *IBM J. Res. Dev.*, 25, 603 (1981).

67. D. D. Tang and T. H. Ning, "Advances in Bipolar IC Technology," *Proc. IEEE Custom Integr. Circuits Conf.*, p. 20 (1985).

68. S. Brenner, T. A. Bartush, D. J. Swietek, D. C. Banker, F. J. Crispi, D. J. Delotto, D. L. Merrill, J. P. Norsworthy, M. N. Shen, and C. D. Waggoner, "A 10,000 Gate Bipolar VLSI Masterslice Utilizing Four Levels of Metal," *Int. Solid State Circuits Conf., Dig. Tech. Pap.*, p. 152 (1983).

69. S. K. Wiedmann, "Advancements in Bipolar VLSI Circuits and Technologies," *IEEE J. Solid-State Circuits*, SC-19, 282 (1984).

70. J. Lohstroh, "Devices and Circuits for Bipolar (V)LSI," *Proc. IEEE*, 69, 265 (1981).

71. S. Konaka, Y. Yamamoto, and T. Sakai, "A 30 ps Si Bipolar IC Using Super Self-Aligned Process Technology," *Ext. Abstr. Int. Conf. Solid-State Devices Mater., 16th, 1984*, p. 209 (1984).

72. D. D. Tang, T. H. Ning, R. D. Isaac, G. C. Feth, S. K. Wiedmann, and H. N. Yu, "Subnanosecond I²L/MTL Circuits," *IEEE Trans. Electron Devices*, **ED-27**, 1379 (1980).

73. T. Nakamura, K. Nakazato, T. Miyazaki, M. Ogivima, T. Okabe, and M. Nagata, "290-psec I²L Circuits with Five-Fold Self-Alignment," *Tech. Dig.—Int. Electron Devices Meet.*, p. 684 (1982).

74. M. H. Hingarh, M. Vora, D. Wilnai, and T. Longo, "A 16b Microprocessor for Real-time Applications," *Int. Solid-State Circuit Conf., Dig. Tech. Pap.*, p. 28 (1983).

75. T. Awaya, K. Toyoda, O. Normura, Y. Nakaya, K. Tanaka, and H. Sugawara, "A 5 ns Access Time 64kb ECL RAM," *Int. Solid-State Circuit Conf., Dig. Tech. Pap.*, p. 130 (1987).

76. W. E. Harding, "Semiconductor Manufacturing in IBM, 1957 to the Present: A Perspective," *IBM J. Res. Dev.*, **25**, 647 (1981).

77. K. Ogiue, S. Ohwaki, and Y. Katoo, "16 Kbit High Speed Bipolar Static RAM," *Hitachi Rev.*, **33**, (2), 77 (1984).

78. T. Sakai, H. Miyanaga, S. Konaka, and Y. Yamamoato, "A 0.85 ns 1Kb Bipolar RAM," *Int. Conf. Solid-State Devices Mater. Dig. Tech. Pap.*, p. 225 (1984) C. T. Chuang, D. D. Tang, G. P. Li, E. Hackbarth, and R. R. Boedeker, "A 1.0-ns 5-kbit ECL RAM," *IEEE J. Solid-State Circuits*, **SC-21**(5), 670 (1986).

79. S. K. Wiedmann and H. H. Berger, "Small-Size, Low Power Bipolar Memory Cell," *IEEE J. Solid-State Circuits*, **SC-6**, 283 (1971).

80. S. K. Wiedmann and K. Heuber, "A 25 ns 8K × 8 static MTL/I²L RAM," *Int. Solid-State Circuit Conf., Dig. Tech. Pap.*, p. 220 (1987).

81. S. K. Wiedmann, D. D. Tang, and R. Beresford, "High-Speed Split-Emitter MTL/I²L Memory Cell," *IEEE J. Solid-State Circuits*, **SC-16**, 429 (1981).

82. G. A. Sai-Halasz and D. D. Tang, "Soft-Error Rates in Bipolar Static RAM's," *Tech. Dig.—Int. Electron Devices Meet.*, p. 344 (1983).

83. T. H. Ning, "Possibilities and Limitations of Bipolar Devices," *Ext. Abstr. Int. Conf. Solid-State Devices Mater., 16th, 1984*, p. 205 (1984).

84. T. H. Ning and D. D. Tang, "Bipolar Trends," *Proc. IEEE*, **24**(12), 1669 (1986).

85. H. L. Stover, "Stepping into the 80's With Die-By-Die Alignment," *Solid State Technol.*, **24**, 112 (1981).

86. A. E. Michel, R. H. Kastl, S. R. Mader, B. J. Masters, and J. A. Gardiner, "Channeling in Low Energy Boron Ion Implantation," *Appl. Phys. Lett.*, **44**(4), 404 (1984).

87. J. F. Ziegler, J. P. Biersack, and U. Littmark, *Stopping and Ranges of Ions in Matter*, Vol. 1, Pergamon, New York, 1983.

88. R. G. Swarts, J. H. Mcfee, P. Grabbe, and S. N. Finegan, "An Uncompensated Silicon Bipolar Junction Transistor Fabricated Using Molecular Beam Epitaxy," *IEEE Electron Device Lett.*, **EDL-2**(11), 293 (1981).

89. T. Sugii, T. Yamazaki, T. Fukano, and T. Ito, "Thin Base Bipolar Technology By Low-Temperature Photo-Epitaxy," *Symp. VLSI Technol., Dig. Tech. Pap.*, p. 35 (1987).

90. S. K. Wiedmann, "Charge Buffer Logic (CBL)—A New Complementary Bipolar Circuit Concept," *Symp. VLSI Technol., Dig. Tech. Pap.*, p. 38 (1985).

91. M. Usami, S. Hososaka, A. Anzai, K. Otsuka, A. Masaki, S. Murta, M. Ura, and M. Nakagawa, "Status and Prospects for Bipolar ECL Gate Arrays," *Int. Circuit Conf. Devices, Dig. Tech. Pap.*, p. 272 (1983); M. Suzuki, H. Okamoto, and S. Horiguchi, "Advanced 5K-Gate Bipolar Gate Array with a 267 ps Basic Gate Delay" *IEEE J. Solid-State Circuits*, **SC-19**(6), 1038 (1984).

92. K. Shohno and H. Otake, "Boron Monophosphide on Si and Device Applications," *Tech. Dig.—Int. Electron Devices Meet.*, p. 490 (1977).

93. G. W. Taylor and J. G. Simmons, "The Bipolar Inversion Channel Field-Effect Transistor (BICFET)—A New Field-Effect Solid State Devices," *IEEE Trans. Electron Devices*, **ED-32**(11), 2345 (1985).

3 Submicron GaAs, AlGaAs/GaAs, and AlGaAs/InGaAs Transistors

MICHAEL SHUR

Department of Electrical Engineering
University of Minnesota
Minneapolis, Minnesota

3.1 INTRODUCTION

I inferred that the number of vibrations of the shorter string must also be twenty times as great.

—Marin Mersenne of Paris (1588–1648)[†]

A never-ending quest for higher-speed, lower-power, and higher-integration scale demands scaling down device size. The minimum device size in silicon integrated circuits has reduced from about 20 microns in the early 1960s to submicron dimensions in the late 1980s.

In shorter devices, carrier transit times are reduced, leading to shorter propagation delays and higher operating frequencies. In addition, a smaller active device volume means lower power required for a switching event.

Whereas silicon is the undisputed king in conventional electronics, submicron devices made of compound semiconductors successfully compete in the area of microwave and ultrafast digital circuits. By combining elements from columns 3 and 5 of the periodic table (e.g., Ga and As), one creates compound semiconductors with the same number of valence electrons per atom as in Si (four). Compounds such as GaAs, InP, InAs, InSb, AlAs, and others have semiconducting properties and band structure somewhat similar to "classic" elemental semiconductors, such as silicon or germanium. At the same time, they provide a gamut of materials with different bandgaps (direct and indirect), different lattice constants, and other physical properties for the semiconductor device designer. Moreover, some of these elements can be combined to form solid-state solutions such as $Al_xGa_{1-x}As$, where the composition x may vary continuously from 0 to 1 with a corresponding change in physical properties from those of GaAs into those of AlAs.

[†]Quoted by J. A. Zahm, *Sound and Music*, McClurg, Chicago, IL, 1892.

Gallium arsenide is the most studied and understood compound semiconductor material. It has been proven indispensable for many device applications from ultrahigh speed transistors to lasers and solar cells. Its lattice constant (5.6533 Å) is very close to that of AlAs (5.6611 Å) and the heterointerface between the two materials has a very small density of interface states. Newer technologies, such as molecular-beam epitaxy (MBE) and metal organic chemical vapor deposition (MOCVD), allow us to grow these materials with very sharp and clean heterointerfaces and have very precise control over doping and composition profiles, literally (in the case of MBE) changing these parameters within an atomic distance.

Other compound semiconductors, important for applications in ultrahigh speed submicron devices, include InP, $In_xGa_{1-x}As$, AlN, GaP, $Ga_xIn_{1-x}As$, etc.

There are several advantages of GaAs and related compound semiconductors for applications in submicron devices. First of all, the effective mass of electrons in GaAs is much smaller than in Si ($0.067\,m$ in GaAs compared with $0.98\,m$ longitudinal effective mass and $0.19\,m$ transverse effective mass in Si, where m is the free electron mass). This lead to much higher electron mobility in GaAs: approximately 8500 $cm^2/V \cdot s$ in pure GaAs at room temperature compared with 1500 $cm^2/V \cdot s$ in silicon. Moreover, in high electric fields, the electron velocity in GaAs is also larger (see Fig. 3.1),[1,2] which is even more important in submicron devices where electric fields are high. Light electrons in GaAs are much more likely to experience so-called ballistic transport, that is, to move through the short active region of a high-speed device without having any collisions (with lat-

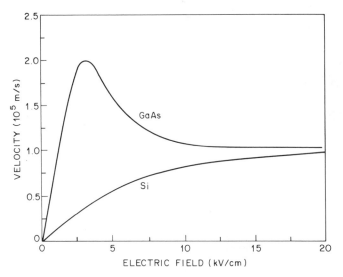

Fig. 3.1 Electron drift velocity vs. electric field for GaAs and Si: the curve for Si is calculated for undoped materials at 300 K using the equation given by Yu and Dutton[1]; The curve for GaAs is calculated for undoped materials at 300 K using the equation given by Xu and Shur.[2]

tice imperfections or lattice vibration quanta). This may boost their velocity far beyond the values expected for long devices that are shown in Fig. 3.1. Such ballistic transport may become important in very short devices with sizes on the order of 0.1 μm or less. It was first predicted by Shur and Eastman[3] and was recently observed by Heiblum et al.[4] and Levy et al.[5] In somewhat longer GaAs devices (with dimensions between 0.1 and 1.5 μm or so), so-called overshoot effects are important. These effects (first predicted by computer simulations of Ruch[6]) are related to the finite time that it takes for an electron to change its energy and may also result in boosting the electron velocity to considerably higher levels than the equilibrium values shown in Fig. 3.1. The increase of electron velocity in short samples is illustrated by Fig. 3.2 (from Shur[7]), where the evolution of electron velocity as a function of distance in a constant electric field is shown. Very close to the injecting contact, electrons are moving ballistically and electron velocity is proportional to time. Further from the contact, the velocity reaches a peak value and then decreases. However, due to overshoot effects, the peak value of velocity is far higher than the equilibrium value reached as the distance increases further.

In silicon, ballistic and overshoot effects may also be important. However, they are much less pronounced because of a larger electron effective mass. To summarize, electrons in GaAs, and many related compounds, are faster than in silicon. This increase of effective velocity in short-channel field-effect transistors is illustrated by Fig. 3.3 (from Cappy et al.[8]), which shows computed electron velocities vs. distance for field-effect transistors fabricated from different materials. Also shown, for comparison, are the values of peak velocities in long-channel devices fabricated from the same materials.

Another important advantage of materials such as GaAs and InP for applications in high-speed submicron devices is the availability of semi-insulating sub-

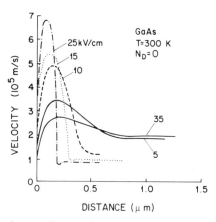

Fig. 3.2 Electron velocity vs. distance for constant electric field. The curves were calcualted by solving the equations describing the electron transport using effective energy-dependent momentum and energy relaxation times. (From Shur.[7] Reprinted with permission, copyright 1976, *Electronics Letters.*)

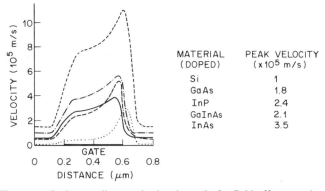

MATERIAL (DOPED)	PEAK VELOCITY (x 10⁵ m/s)
Si	1
GaAs	1.8
InP	2.4
GaInAs	2.1
InAs	3.5

Fig. 3.3 Electron velocity vs. distance in the channel of a field-effect transistor for different semiconductor materials: ...., Si; — — — , InP; ———, GaAs; – · – · –, GaInAs; ----------, InAs. Peak velocities in these materials are shown for comparison. (From Cappy et al.[8] Reprinted with permission, copyright 1980, IEEE.)

strates that eliminate parasitic capacitances related to junction isolation in silicon circuits and allow fabrication of microstrip lines with small losses (the latter being especially important for applications in microwave monolithic integrated circuits). We may speculate that with the development of high T_c superconductors, which may be used in such microstrip lines, this advantage will become especially important.

It is also worth mentioning that GaAs is a direct gap material widely used in optoelectronic applications. The direct gap makes possible a monolithic integration of ultrahigh-speed submicron transistors and lasers or light emitting diodes on the same chip for applications in optical communications. This and the resulting faster recombination rates may also lead to a better radiation hardness.

Last but not least, the availability of excellent heterostructure systems, such as AlGaAs/GaAs, GaInAs/InP, InGaAs/AlGaAs, etc., and new technologies, such as MBE and MOCVD, open up unlimited opportunities for experimentation with new devices, such as modulation doped field-effect transistors (MODFETs), heterojunction bipolar transistors (HBTs), hot-electron transistors (HETs), induced base transistors (IBTs), permeable base transistors (PBTs), vertical ballistic transistors (VBTs), planar doped barrier transistors (PDBTs), and many novel quantum devices (the latter will be considered in Chapter 5).

There are also drawbacks of compound semiconductor technology. Silicon is blessed with an excellent native oxide. Silicon nitride is also used as an excellent insulator in silicon field-effect transistors. A poor quality of oxide on GaAs and a corresponding high density of interface states make it difficult to fabricate a GaAs metal oxide semiconductor field-effect transistor. A wide bandgap AlGaAs and, more recently, AlN may substitute for such an insulator, albeit in a limited way.

Silicon is an elemental semiconductor. Of course, silicon purity determines the device quality, but one does not have to worry about composition. In a compound semiconductor, such as GaAs, the material composition is of utmost importance. For example, defects may be caused by a deficiency of arsenic atoms.

Silicon is abundantly available in nature. Gallium is a relatively rare element. Arsenic is toxic. As a consequence, GaAs wafers cost many times more than comparable silicon wafers and may be in short supply.

GaAs thermal conductivity at 300 K is three times less than that for silicon (0.46 W/cm $\cdot$ °C for GaAs compared with 1.5 W/cm $\cdot$ °C).

That is why compound semiconductor transistors are widely used for microwave and ultrahigh speed applications where their high-speed properties are the most important. Other possible uses include high-temperature electronics (because of a larger energy gap in GaAs), power devices (because of higher breakdown field and an ability to speed up their turn-on by light), optoelectronics applications, and radiation-hard electronics (because of a direct bandgap). However, microwave and ultrahigh speed applications constitute the mainstream of compound semiconductor technology. Hence, scaling down the device sizes in order to enhance the electron velocity by ballistic and/or overshoot effects and reduce the transit time is especially important for compound semiconductor transistors.

Submicron devices can be lateral, such as standard field effect transistors, or vertical, such as typical bipolar junction transistors. A lateral device is a planar structure with a critical device dimension controlled by a lithography process. In a vertical device, the critical device dimension is controlled by a deposition process rather than by lithography. In this chapter, we first discuss general scaling considerations for lateral devices, that is, field-effect transistors. We analyze how the current–voltage characteristics of GaAs metal semiconductor field-effect transistors (MESFETs) and AlGaAs/GaAs MODFETs are expected to change with the gate size. We also consider short-channel effects in field-effect transistors, such as gate length modulation, substrate injection, and technological problems related to the dopant diffusion into the channel and stresses from the passivating dielectric that are extremely important in submicron field-effect transistors. This will be followed by the discussion of novel submicron vertical structures, such as HBTs, PBTs, IBTs, ballistic transistors, etc.

3.2   SCALING CONSIDERATIONS FOR FIELD-EFFECT TRANSISTORS

A typical field-effect transistor can be considered as a capacitor that has the gate as one plate and the conducting channel between the source and drain as the other plate. In considering how the speed of a field-effect transistor scales with the gate length, we should study the time required to charge the gate capacitance C_0. This capacitance can be estimated as

$$C_0 = \frac{\varepsilon W L}{d}, \tag{1}$$

where ε is the dielectric permittivity of the material separating the channel from the gate, W the device width, L the gate length, and d the separation between the gate and the channel. The total gate charge Q_0 is proportional to C_0:

$$Q_0 = C_0 \Delta V, \qquad (2)$$

where ΔV is the voltage swing. The minimum time required for removing this charge (i.e., for a switching event) is proportional to

$$t_0 = \frac{Q_0}{I_0} \qquad (3)$$

where

$$I_0 = \frac{Q_0 v_0}{L} \qquad (4)$$

is the device current and v_0 the effective carrier velocity. Substituting Eq. (4) into Eq. (3), we obtain

$$t_0 = \frac{L}{v_0} = t_{tr} \qquad (5)$$

where t_{tr} is the carrier transit time. Equation (5) seems to imply that scaling the length from one micron down to a quarter micron should increase the speed by four times or even more, if ballistic and overshoot effects enhancing the electron velocity in short structures are taken into account.

However, Eq. (5) must be modified to account for fringing and interconnect capacitances, C_f and C_{int}, that play an import role in determining the overall circuit speed. Replacing charge Q_0 by

$$Q_t = (C_0 + C_f + C_{int}) \Delta V, \qquad (6)$$

we obtain for the characteristic switching time

$$t_c = \frac{(C_0 + C_f + C_{int}) \Delta V}{I_0}. \qquad (7)$$

This equation may be rewritten as follows:

$$t_c \approx t_{tr} \left[1 + \frac{C_f + C_{int}}{C_0} \right] \qquad (8)$$

The second term in the brackets on the right-hand side of Eq. (8) may be several times larger than one. For example, for a 0.3 μm gate field-effect transistor, $t_{tr} \approx 0.3 \times 10^{-6}$ m/2×10^5 m/s ≈ 1.5 ps. In fact, the smallest propagation delay achieved for 0.3 μm gate AlGaAs/GaAs MODFET ring oscillators is 5.8 ps (see Shah et al.[9]).

The ring oscillator circuit has fan-in and fan-out of one and very short interconnects. More practical circuits, such as divide-by-two and divide-by-four circuits, multiplexers, multipliers, etc., have substantially larger fan-outs, FO, (typically two to three) and correspondingly larger propagation delays.

The fringing capacitances scale with the gate length L and device width W:

$$C_f \approx a_{f1}\varepsilon W + a_{f2}\varepsilon L, \tag{9}$$

where a_{f1} and a_{f2} are numerical constants of the order of unity. The interconnect capacitance is proportional to the average interconnect length L_{int}:

$$C_{int} \approx a_i \varepsilon L_{int}, \tag{10}$$

where a_i is a numerical constant on the order of unity. We may assume that W and L_{int} scale with the gate length:

$$W \approx S_w L, \tag{11}$$

$$L_{int} \approx S_{int} L, \tag{12}$$

where S_w and S_{int} are scaling factors that depend on circuit design and layout. Substituting Eqs. (9)–(12) into (8), we obtain for the propagation delay

$$t_c = \frac{L}{v_0(1 + K_t d/L)}, \tag{13}$$

where K_t is a constant that depends on L_{int} and W. This scaling relation clearly indicates the importance of having a small separation d between the gate and the channel. This separation determines the driving capabilities of a device. Let us consider, for example, a typical MODFET with the gate dimensions of 0.7 $\times$ 10 μm (see Cirillo et al.[10]). The effective electron velocity at room temperature is on the order of 1.5×10^5 m/s and the gate-to-channel distance is on the order of 400 Å. The shortest propagation delays observed for ring oscillators ($FO = 1$) are on the order of 10 ps. Substituting these numbers into Eq. (13) we find $K_t \approx 20$. Using this value for K_t, we can plot the propagation delay t_c vs. L for different values of d (see Fig. 3.4).

The scaling relation (13) is derived assuming that the effective electron velocity in the channel v_0 is independent of the gate length. In fact, this is not true in short-channel devices.

First of all, as mentioned in the introduction, overshoot effects become important in short-channel devices where the transit time becomes comparable with the electron energy relaxation time. As a consequence, the electron velocity may exceed its steady-state value achieved in long-channel devices. In even shorter sam-

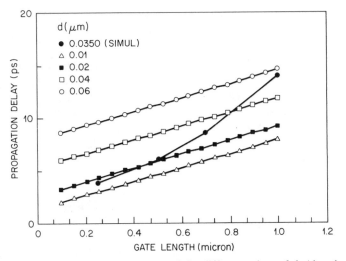

Fig. 3.4 Propagation delay per fan-out t_c vs. L for different values of d. Also shown is the propagation delay for a DCFL MODFET ring oscillator vs. gate length. At each value of L, the load current was optimized to minimize the propagation delay.

ples (shorter than 0.1 μm or so), the transit time becomes comparable with the electron momentum relaxation time so that the electron transport becomes ballistic, increasing the electron velocity even more.

Second, the electron velocity in the channel is not uniform. In a linear regime of operation (i.e., at small drain-to-source voltages V_D), the electron velocity is inversely proportional to the device length

$$v_0 = \frac{\mu V_D}{L} .$$ (14)

Here, μ is the low-field mobility and V_D/L is the electric field in the channel. As the drain-to-source voltage increases, the electric field in the channel becomes nonuniform. It is larger at the drain side of the gate, where the carrier concentration in the channel is smaller, because the drain voltage adds a reverse bias to the gate-to-channel capacitance. As a consequence, the electron velocity also becomes nonuniform. It reaches larger values closer to the drain end of the channel, where the electron concentration is smaller. If the electron velocity in high electric fields is saturated, the saturation first occurs at the drain side of the gate, where the electric field is the highest. In short-channel devices, the velocity saturation region occupies a larger fraction of the channel. Hence, in short-channel devices, the average electron velocity in the channel is closer to the saturation velocity. In long-channel devices, the average electron velocity in the channel in the saturation regime may be considerably smaller than the saturation velocity. This effect is more pronounced when the carrier mobility is low (see also Shur[11]). Hence, this is very important for p-channel devices because the hole mobility in compound semiconductors is much smaller than electron mobility.

The scaling equation (13) implies that the device width is scaled proportionally to the gate length. If the device width is kept constant, however, the speed advantage of short-channel devices increases. To illustrate the effects of all these factors, the propagation delay for a DCFL MODFET ring oscillator vs. gate length is shown in Fig. 3.4. At each value of L, the load current was optimized to minimize the propagation delay. The gate width was kept constant. The effective value of the electron saturation velocity was assumed to be dependent on the gate length using an approximate equation for velocity given by Shur and Long.[12] The model used in this simulation was described by Hyun et al.[13] As can be seen from Fig. 3.4, the computed dependence of τ vs. L is somewhat different from that predicted by Eq. 43.

The scaling considerations (corrected for the increase of the electron velocity in short-channel transistors due to the overshoot and ballistic effects) seem to indicate that propagation delays as low as 1 ps or less may be achieved in 0.1 μm gate GaAs or AlGaAs/GaAs field-effect transistors. Let us assume that such a device is 1 μm wide, that it has a gate-channel separation $d = 100$ Å and a voltage swing of 500 mV. Then the gate capacitance C_0 is on the order of femtofarad, and the power-delay product $P\tau \approx C_0 \Delta V^2 \approx 0.3$ fJ, corresponding to the power dissipation of 0.3 mW per gate. The total number of electrons in the channel $C_0 \Delta V/q \approx 3500$, is still large enough to prevent large statistical fluctuations.

In microwave applications, cutoff frequency

$$f_T = \frac{1}{2\pi\tau} \tag{15}$$

(frequency where the short-circuit current gain is unity) and maximum oscillation frequency f_{max} are important figures of merit. The dependences of f_T and f_{max} on the gate length calculated by Das[14] for GaAs MESFETs and MODFETs are shown in Fig. 3.5. Also shown in Fig. 3.5 are experimental data from Van Zeghbroeck et al.,[15] Henderson et al.,[16] Drummond et al.,[17] and Chen et al.[18] According to the calculations by Das,[14] similar performance may be expected for vertical device structures, such as permeable base transistors. Heterojunction bipolar transistors are other very promising devices that yielded maximum oscillation frequency as high as 105 GHz (see Chang et al.[19]).

According to predictions by Eastman,[20] within the next three years one or more of these devices will be capable of operating at about 500 GHz with logic switching times under 3 ps. This can be only done by scaling the device sizes down. However, the difficulties involved in this approach are considerable. Better fabrication technology, more complete understanding of device physics, better modeling capabilities, and more sophisticated characterization techniques are required to make GaAs technology widespread and successful.

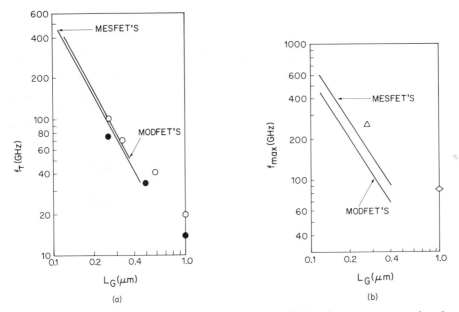

Fig. 3.5 (*a*) Cutoff frequency f_T, and (*b*) maximum oscillation frequency vs. gate length for GaAs MESFETs and MODFETs: ○, f_T for MESFETs (after Van Zeghbroeck et al.[15]); ●, f_T (after Drummond et al.[17]); △, f_{max} for MODFETs (after Henderson et al.[16]); ◇, f_{max} for MODFETs (after Chen et al.[18]). (After Das.[14] Reprinted with permission, copyright 1987, IEEE.)

3.3   CURRENT–VOLTAGE CHARACTERISTICS OF SUBMICRON GaAs AND AlGaAs/GaAs FIELD-EFFECT TRANSISTORS

A GaAs metal semiconductor field-effect transistor (MESFET) is a workhorse of compound semiconductor electronics. The absence of a stable native oxide on GaAs makes GaAs MESFETs or junction field-effect transistors (JFETs) the devices of choice for compound semiconductor electronics. More recently, different heterojunction field-effect transistors have emerged and become the fastest semiconductor transistors.

Schematic structures of different GaAs transistors are compared in Fig. 3.6. In a GaAs MESFET or JFET, the gate voltage modulates channel conductance by varying the depth of the depletion region. In heterojunction transistors, such as modulation dooped field-effect transistors (MODFETs), gate voltage modulates the channel conductance by changing the density of two-dimensional electron gas (or two-dimensional hole gas in *p*-channel devices). An exception is the doped channel modulation doped field-effect transistor, where the channel is created by both a doped GaAs layer and by an induced two-dimensional gas. Different heterostructure transistors shown in Fig. 3.6 differ primarily in how the threshold voltage is controlled. In a conventional modulation doped field-effect transistor [also called a high electron mobility transistor (HEMT); see Mimura et al.[21]], the

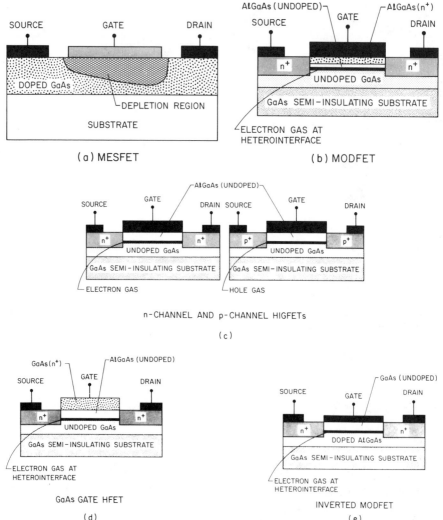

Fig. 3.6 Schematic structures of GaAs field-effect transistors: (*a*) GaAs Metal semiconductor field-effect transistor (MESFET); (*b*) modulation doped field-effect transistor (MODFET); (*c*) *n*-channel and *p*-channel heterostructure insulated gate field-effect transistors (HIGFETs); (*d*) GaAs gate Heterostructure field-effect transistor (GaAs gate HFET); (*e*) inverted modulation doped field-effect transistor (IMODFET); (*f*) superlattice modulation doped field-effect transistor (SL-MODFET); and (*g*) doped channel modulation doped field-effect transistor (DCMODFET).

threshold voltage is controlled by changing doping in the doped AlGaAs layer (Fig. 3.6*b*). Doping in the AlGaAs layer frequently leads to the creation of deep traps that cause numerous deleterious effects, such as a threshold voltage shift with temperature, persistence photoconductivity, etc. (see, e.g., Drummond et al.[17]).

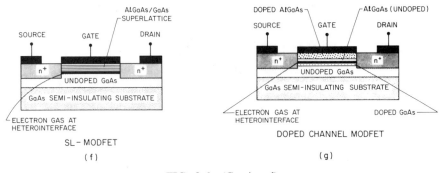

SL-MODFET

(f)

DOPED CHANNEL MODFET

(g)

FIG. 3.6   (Continued)

In n-channel and p-channel heterostructure insulated gate field-effect transistors (Fig. 3.6c, Cirillo et al.[22]), an AlGaAs layer is undoped. This leads to a well-defined threshold voltage which is only weakly dependent on temperature and has only small light sensitivity. However, the value of the threshold voltage in HIGFETs is simply determined by the difference between the Schottky barrier height and the conduction band discontinuity and is fairly large ($\sim$0.8 V or so for n-channel devices).

In a GaAs gate heterostructure field-effect transistor proposed by Solomon et al.[23] (Fig. 3.6d), a highly doped n^+ GaAs layer is used instead of the metal gate. This reduces the threshold voltage to values close to zero. However, it also reduces the turn-on voltage V_{on} for the gate current.

An inverted modulation doped field-effect transistor (Fig. 3.6e) has the channel closer to the metal gate than a conventional MODFET. This leads to better gate control, that is, to higher device transconductance. Extremely high transconductance values have been observed in such devices (1820 mS/mm at 77 K and 1180 mS/mm at 300 K; see Cirillo et al.[24]).

In superlattice modulation doped field-effect transistors (Fig. 3.6f), dopants in the superlattice layer separating the gate from the heterointerface are located in narrow, highly doped quantum wells. This decreases the number of traps and leads to better device performance.

In doped channel modulation doped field-effect transistors (Fig. 3.6g), dopants controlling device threshold voltage are located in the GaAs near the heterointerface. Hence, this device can be considered as a combination of a GaAs MESFET and an AlGaAs/GaAs MODFET.

The principle of operation of all these devices is closely related to that of a GaAs MESFET and an AlGaAs/GaAs MODFET. Band diagrams of a GaAs MESFET and an AlGaAs/GaAs MODFET are compared in Fig. 3.7. In a GaAs MESFET, the conducting channel is in doped GaAs, whereas in a MODFET it is spatially separated from the donors located in AlGaAs. Also, in MODFETs, electrons in the channel are confined very close to the heterointerface (within $\sim$80Å or so). As a consequence, the electron motion in the direction perpendicular to the heterointerface is quantized. Two lowest (and most important) sub-bands are de-

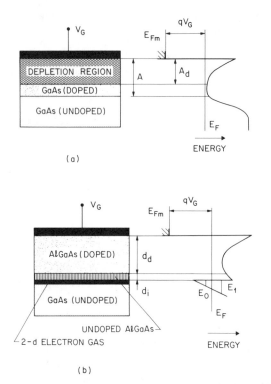

Fig. 3.7   Band diagrams of a (a) GaAs MESFET and (b) an AlGaAs/GaAs MODFET.

noted in Fig. 3.7b as E_0 and E_1. For typical electron concentrations in the two-dimensional gas, the electron quasi-Fermi level at the heterointerface is fairly high in the conduction band and the Fermi velocity is also high. The high Fermi velocity, high carrier concentration in the two-dimensional gas (resulting in an effective screening of Coulomb potentials), and spatial separation of electrons from donors lead to a higher electron mobility in MODFET channels, especially at low temperature and low electric fields (see Dingle et al.[25]). However, in higher electric fields, the electron velocity in MODFET channels is probably close to that in GaAs MESFETs (see, e.g., Tomizawa et al.[26] and Masselink et al.[27]).

Another important advantage of MODFETs is the higher Schottky barrier height for AlGaAs (≈ 1–1.1 eV compared with ≈ 0.75 V for GaAs). This leads to a larger gate voltage swing that is primarily limited by the gate leakage current.

The physics of compound semiconductor devices is quite complex, especially for devices with submicron feature sizes. In addition to the overshoot and ballistic effects mentioned in Section 3.1, the nonuniformity of the channel doping (most GaAs MESFETs are ion-implanted devices), the dependence of electron mobility and velocity in the channel on the distance from the surface, surface effects, effects of passivating layers, diffusion of dopants from the heavy implanted ohmic regions into the channel, and effects of traps leading to the dependence of device parameters on time, bias history, illumination dependence, and to the parasitic interaction of devices on a wafer (so-called backgating and sidegating) all play an important role; they must be understood and accounted for in developing reliable submicron GaAs technology. Some of these effects will be considered in the following sections. In this section, the behavior of "ideal" devices is considered in order to establish how their characteristics scale with gate length and in order to compare GaAs MESFETs and AlGaAs/GaAs MODFETs.

Our comparison of current–voltage characteristics of these devices will be based on a so-called pinned velocity saturation model, first considered by Grebene and Ghandhi[28] and Pucel et al.[29] According to this model, the drain current saturation as a function of drain-to-source voltage occurs when electron velocity at the drain side of the channel reaches the saturation velocity v_s. In fact, as can be seen from Fig. 3.3, the velocity profile in short-channel devices is more complex. The peak velocity in the channel may depend on channel length, doping profile, and drain-to-source voltage. A complete velocity saturation in the channel may not even occur. It is well established, however, that the electron velocity in high electric fields is much less dependent on the electric field than in low electric fields. Moreover, this simplistic velocity saturation model does provide a fairly good agreement with the results of the numerical calculations and experimental data and, hence, may be used in order to establish how device characteristics scale with the gate length.

As shown by Shur,[11] the dependence of the drain saturation current on the gate voltage predicted by this model for GaAs MESFETs may be approximated by a "square law" expression:

$$(I_{ds})_{\text{sat}} = \beta(V_g - V_T)^2, \tag{16}$$

where

$$\beta = \frac{2\varepsilon\mu v_s W}{A(\mu V_{\text{po}} + 3v_s L)}, \tag{17}$$

$$V_T = V_{\text{bi}} - V_{\text{po}} \tag{18}$$

is the threshold voltage, V_{bi} the built-in voltage, ε the dielectric permittivity of GaAs, L the gate length, W the gate width, v_s the electron saturation velocity, A the channel thickness, N_d the doping concentration in the channel, and

$$V_{\text{po}} = \frac{qN_dA^2}{2\varepsilon} \tag{19}$$

is the pinch-off voltage. Equation (16) is applicable for devices with relatively small pinch-off voltages ($V_{\text{po}} \leq 1.5$–2 V). For devices with larger pinch-off voltages, and empirical expression proposed by Statz et al.,[30]

$$(I_{ds})_{\text{sat}} = \frac{\beta(V_g - V_T)^2}{1 + b(V_g - V_T)}, \tag{20}$$

provides a much better fit to the experimental data.

The dependence of the drain saturation current on the gate voltage predicted by the unpinned velocity saturation model for AlGaAs/GaAs MODFETs is given by (see Drummond et al.[31])

$$I_{ds} = \beta_0 V_{sl}^2 \left\{ 1 + \left[\frac{V_g - V_T}{V_{sl}} \right]^2 \right\}^{1/2} - \beta_0 V_{sl}^2, \tag{21}$$

where

$$\beta_0 = \frac{\varepsilon \mu W}{(d + \Delta d)L} \tag{22}$$

in which d is the total thickness of the AlGaAs layer separating the two-dimensional gas from the gate, $\Delta d \approx 80$ Å is the correction term related to the quantization of the two-dimensional electron gas,

$$V_T \approx \phi - \Delta E_c - \frac{qN_d d^2}{2\varepsilon} \tag{23}$$

is the threshold voltage, ϕ the Schottky barrier height, and N_d the doping concentration in the doped AlGaAs layer.

This is an approximate model that does not account for gate current and parallel conduction in AlGaAs layer (see Ruden et al.[32] for further discussion).

From Eqs. (16) and (21), we can obtain the following expressions for the device transconductances, g_{mes} and g_{mod}, in the saturation regime:

$$g_{mes} = 2\beta(V_g - V_T), \tag{24}$$

$$g_{mod} = \frac{\beta_0(V_g - V_t)}{\{1 + [(V_g - V_t)/V_{sl}]^2\}^{1/2}}. \tag{25}$$

These transconductances increase with gate voltage and reach the maximum values when $V_g \approx V_{on}$, where V_{on} is the gate turn-on voltage (when the gate current becomes appreciable). At higher gate voltages, the gate leakage current leads to a rapid drop in device transconductance. Hence, the maximum device transconductances are given by

$$(g_{mes})_{max} \approx 2\beta(V_{on} - V_T), \tag{26}$$

$$(g_{mod})_{max} \approx \frac{\beta_0(V_{on} - V_t)}{\{1 + [(V_{on} - V_t)/V_{sl}]^2\}^{1/2}}. \tag{27}$$

The maximum drain-to-source current obtained can be estimated as

$$(I_{mes})_{max} \approx \beta(V_{on} - V_T)^2, \tag{28}$$

$$(I_{mod})_{max} \approx \beta_0 V_{sl}^2 \left\{ 1 + \left[\frac{V_{on} - V_t}{V_{sl}} \right]^2 \right\}^{1/2} - \beta_0 V_{sl}^2 . \qquad (29)$$

If we approximately estimate the maximum voltage swing V_{on} in GaAs MESFETs as

$$V_{on} \approx V_{po} + V_t , \qquad (30)$$

then in the limiting case of a very short-channel $(L \Rightarrow 0)$, Eq. (28) yields

$$(I_{mes})_{max} \approx q N_d A v_s W , \qquad (31)$$

that is, the expression for the full channel current, as should be expected.

In the same limiting case $(L \Rightarrow 0)$, Eq. (29) yields

$$(I_{mod})_{max} \approx q n_s v_s W , \qquad (32)$$

where

$$n_s = \frac{\varepsilon (V_g - V_t)}{q(d + \Delta d)} \qquad (33)$$

is the density of the two-dimensional gas induced into the channel. The simple model leading to Eq. (29) does not take into account the fact that the density of two-dimensional electron gas in the channel is limited. This is because at large gate voltage swings and, consequently, large values of n_s, the minimum of the conduction band in the AlGaAs layer (see Fig. 3.7b) may approach the quasi-Fermi level of the AlGaAs, leading to a parallel conduction path in the low-mobility AlGaAs and limiting the maximum drain current (see Lee et al.[33] and Ruden et al.[32]). Hence,

$$(I_{mod})_{max} \leq q n_{smax} v_s W , \qquad (34)$$

where n_{smax} is the maximum concentration of the two-dimensional gas.

Equations (26)–(34) can be used to determine the dependence of the device transconductance on gate length, channel doping, electron mobility, and saturation velocity. In Figs. 3.8 and 3.9, we compare the calculated dependences of the maximum transconductance and maximum current swing on the gate length for GaAs MESFETs and AlGaAs/GaAs MODFETs. In this calculation, we account for the dependence of the effective saturation velocity on the gate length assuming that for $L \geq 0.2 \ \mu m$,

$$v_{seff} \approx (0.22 + 1.39 L) \times 10^5 / L , \qquad (35)$$

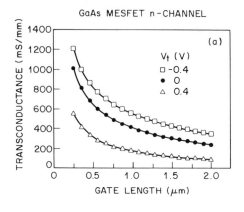

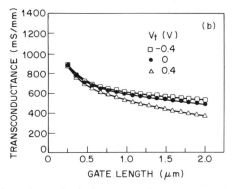

Fig. 3.8 Calculated maximum intrinsic transconductance vs. gate length for n-channel GaAs MESFETs (a) and AlGaAs/GaAs MODFETs (b). Parameters used in the calculation: dielectric permittivity of GaAs $\varepsilon = 1.14 \times 10^{-10}$ F/m, dielectric permittivity of AlGaAs $\varepsilon = 1.1 \times 10^{-10}$ F/m, MESFET channel depth $A = 6 \times 10^{-8}$ m, gate turn voltage for GaAs MESFETs $V_{on} = 0.6$ V, gate turn voltage for AlGaAs MODFETs $V_{on} = 0.9$ V, low-field mobility for GaAs MESFETs $\mu = 4000$ cm^2/V·s, low-field mobility for AlGaAs MODFETs $\mu = 5000$ cm^2/V·s, gate-to-channel spacing for AlGaAs MODFETs $d = 2 \times 10^{-8}$ m, effective thickness of the two-dimensional electron gas $\Delta d = 80$ Å. The effective saturation velocity was assumed to be dependent on the gate length and given by Eq. (35).

where v_{seff} is in m/s and L in μm. This semiempirical expression was obtained by Shur and Long[12] based on transient calculations of the electron velocity in short $n^+ n n^+$ structures.

As can be seen from Fig. 3.8, submicron GaAs MESFETs with very short channels may have even larger transconductances than AlGaAs/GaAs MODFETs, even though a relatively small gate-to-channel spacing (200 Å) was used in the calculation of the MODFET transconductance. The reason is that the maximum transconductance in GaAs MESFETs is reached when the conducting channel boundary is very close to the gate, corresponding to a very small effective gate-to-

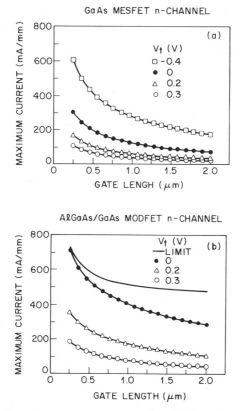

Fig. 3.9 Calculated maximum current vs. gate length for n-channel GaAs MESFETs (a) and AlGaAs/GaAs MODFETs (b) Parameters used in the calculation are given in the caption for Fig. 3.8. Also shown in Fig. 3.9b is the maximum limiting value of the MODFET current obtained using Eqs. (18) and (19) for $n_{smax} = 2 \times 10^{12}$ cm^{-2}.

channel spacing. Longer-channel MODFETs have higher transconductance than GaAs MESFETs with comparable threshold voltages. Also, MODFET transconductances are much less sensitive to the device threshold voltages.

The curves shown in Fig. 3.8 are especially important for analyzing a general trend of the transconductance dependence on the gate length, because the numerical value of the maximum transconductance is determined primarily by the gate-to-channel separation. The fact that the device transconductance is controlled by the gate-to-channel spacing is clearly indicated by the experimental data for inverted AlGaAs/GaAs MODFETs (see Cirillo et al.[24]). Extrinsic transconductances as high as 1820 mS/mm gate at 77 K and 1180 mS/mm at 300 K were obtained for 2μm gate inverted MODFETs by decreasing the effective gate-to-channel spacing to 60–80 Å. For all other devices, transconductance may also be increased by decreasing the gate-to-channel separation. If the threshold voltage is kept constant, the doping level should be increased proportionally to d^2 and A^2, in MODFETs and MESFETs respectively. Small effective separation between the gate and conducting channel also allows minimization of short-channel effects, which become quite noticeable when $L/A < 3$ (see Dambkes et al.[34]).

MODFET devices have larger maximum current than GaAs MESFETs with comparable threshold voltages. This may be explained by a larger maximum gate voltage swing in MODFETs (directly related to a higher Schottky barrier height in AlGaAs and also to the conduction band discontinuity at the heterointerface). The conduction band discontinuity at the heterointerface acts as a "retaining wall" limiting the gate current at high gate voltages. This increases the maximum gate voltage swing (see Baek et al.[35]). Also, the MODFET transconductance increases more rapidly with the gate voltage at gate voltages close to the threshold. This is more pronounced in the MODFETs with small or negative gate voltages and leads to a higher maximum current. Gate voltage dependences of the device transconductances for a GaAs MESFET and an AlGaAs/GaAs MODFET are compared in Fig. 3.10. (In this figure, we do not show the transconductance dependences at large gate voltage where the transconductance sharply decreases with the gate voltage because of an increase in gate current.)

As was mentioned above, the models used in this calculation do not take into account many important effects occurring in short-channel field-effect transistors. Nevertheless, they describe correctly the scaling of device characteristics with the gate length, channel depth, doping, etc. As an example, in Fig. 3.11 the normalized dependence of the drain current on the gate length calculated numerically by Hess and Kizilyalli[36] is compared with the analytical calculation. As can be seen, the agreement is quite good.

However, the simple models do not predict the correct electric field, energy, and velocity distributions in the channel. They are not even applicable for the de-

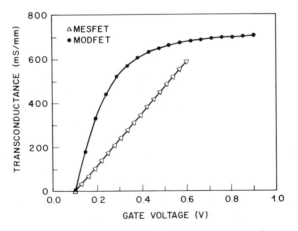

Fig. 3.10   The gate-voltage dependences of the device transconductances for a GaAs MESFET and an AlGaAs/GaAs MODFET. Parameters used in the calculation: threshold voltage $V_t = 0.1$ V for both devices and the other parameters are the same as given in the caption for Fig. 3.8.

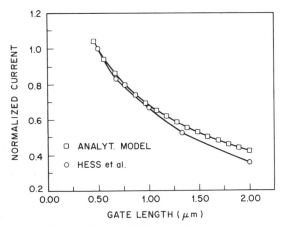

Fig. 3.11 Comparison of dependences of MODFET drain current on gate length calculated numerically by Hess and Kizilyalli[36] with the analytical calculation. The drain current is normalized with respect to its value for 0.5 μm devices.

scription of the device in the saturation regime. Two-dimensional numerical simulations, such as reported by Hess and Kizilyalli[36] for MODFETs and by Snowden and Loret[37] for GaAs MESFETs, allow some insight into the actual physics of the electron transport in short-channel field-effect transistors. We speculate that AlGaAs/GaAs MODFETs may have a lower effective electron temperature in the channel because the electron distribution may be "cooled off" by the hot-electron "real space transfer" over the conduction band discontinuity into AlGaAs (see Hess et al.[38]). Experimental data by Chen et al.[39] seem to indicate that the electron temperature in the channel may be as low as 400 K. This may explain the better noise performance of MODFETs.

Equations (26)–(34) show that the low-field mobility becomes less important in low pinch-off and short-channel devices. As was shown by Shur,[11] the higher values of low-field mobility in GaAs (up to 4500 cm^2/V · s in highly doped active layers) should lead to substantially improved performance compared with silicon devices, even for short-channel enhancement mode devices where velocity saturation effects are very important. The higher mobility that may be achieved in modulation doped field-effect transistors at room temperature leads to even better performance. However, a further increase in low-field mobility, say, to 20,000 cm^2/ V · s (such as in modulation doped structures at the temperature of liquid nitrogen), should not lead to much further improvement. The higher values of low-field mobility may still help reduce the series source resistance, especially in devices with non-self-aligned gates.

As also can be seen from Eqs. (26)–(34) (and as can be expected from the device physics), the relative benefits of scaling down the device size are especially important for low-mobility devices. In this regard, we should mention AlGaAs/GaAs and GaAs complementary, CMOS-like technologies involving both n-channel and p-channel devices (see, e.g., Cirillo et al.,[22] Daniels et al.,[40] Baier et al.,[41] and Kiehl et al.[107]). These technologies have a promise of ultrahigh-speed, low-power circuits. However, the overall speed is slowed down by low-mobility *p*-channel devices leading to relatively low transconductances. As shown in Fig. 3.12, the transconductance of *p*-channel devices can be substantially increased by scaling down the device size. The relative improvement with scaling down the device size is especially pronounced in high-threshold voltage devices, such as *p*-channel HIGFETs, where the decrease in channel length from 1 to 0.25 μm is expected to increase device transconductance by more than a factor of 3.

Even further increase may be achieved by increasing the hole mobility. The dependence of the maximum MODFET transconductance on low-field mobility is shown in Fig. 3.13 for different channel lengths. As can be seen, even an increase in hole mobility to approximately 1500 cm^2/V · s for 0.25 μm gate MODFETs is expected to achieve transconductances close to that of *n*-channel devices. One way

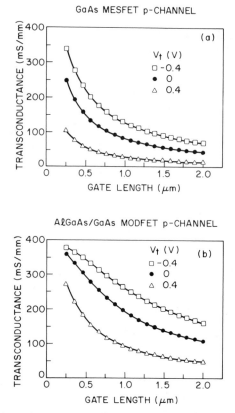

Fig. 3.12 Calculated maximum transconductance vs. gate length for *p*-channel AlGaAs/GaAs MODFETs. Parameters used in the calculation: hole mobility $\mu = 800$ cm^2/V · s, and the other parameters are the same as given in the caption for Fig. 3.8.

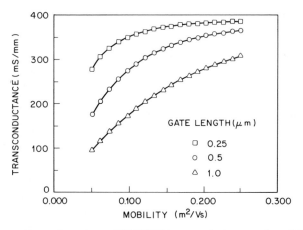

Fig. 3.13 Dependence of maximum MODFET transconductance on low-field mobility for AlGaAs/GaAs MODFETs with different gate lengths. Parameters used in the calculation are the same as given in the caption for Fig. 3.8.

to increase hole mobility is to use strain in order to split the light and heavy hole sub-bands in GaAs (see Fig. 3.14 and Drummond et al.[108]). This technique was utilized by Drummond et al.[42] in their high-speed p-channel transistor. Recently, Sweeny et al.[43] pointed out that a similar increase in hole mobility may be expected in quantum well devices, especially in one-dimensional quantum wires where the confinements of holes lift the symmetry responsible for the degeneracy of light and heavy holes in cubic semiconductors, leading to a substantial decrease in hole effective mass. This seems to suggest that a submicron quantum well HIGFET may become a leading contender for high-speed p-channel devices and that quantum well n-channel and p-channel submicron HIGFETs can be used for ultrahigh-speed, very low power complementary technology.

   Let us now discuss some experimental data obtained for submicron devices.

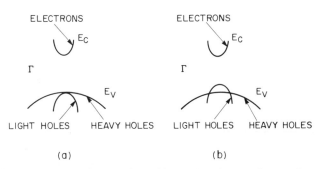

Fig. 3.14 Schematic band diagram for cubic semiconductors for small wave vectors: (a) without strain, (b) with strain.

Mishra et al.[44] measured transconductances as high as 600 mS/mm gate width for 0.1 μm gate GaAs MESFETs with very high doping (6×10^{18} cm^{-3}) in the active layer. The high doping decreased the depletion layer thickness, minimizing short-channel effects that depend on the ratio of gate length over channel depth. As a consequence, output drain conductance in the saturation region g_d was relatively low with the g_m/g_d ratio $\approx$ 20. The expected cutoff frequency for these devices was about 80 GHz.

Van Zeghbroeck et al.[45] reported transconductances as high as 442 mS/mm for 0.5 μm gate GaAs MESFETs with the threshold voltage of 0.29 V. This is in agreement with values expected from the results shown in Fig. 3.8a. They also used high channel doping (9×10^{18} cm^{-3}) to reduce the short-channel effects.

Jaeckel et al.[46] presented the results of a systematic study of GaAs MESFETs with gate lengths from 0.1 μm to 1 μm. In this particular work, they did not report very high values of device transconductance. However, their results, reproduced in Fig. 3.15 and Fig. 3.16, allow us to see a trend in dependences of the MESFET parameters on device length.

The highest values of transconductance for GaAs submicron MESFETs (up to 1750 mS/mm) were measured by Van Zeghbroeck et al.[47] These very high values are related, however, to a parasitic bipolar action in these devices and not to a standard MESFET action.

Sauner and Lee[48] measured transconductance of 400 ms/mm in 0.5 μm AlGaAs/ GaAs n-channel MODFETs. They also reported maximum drain currents in 0.4 μm AlGaAs/GaAs n-channel MODFETs as high as 800 mA/mm gate width (for multiple-channel structures, see Sauner and Lee[49]).

Daniels[50] obtained transconductances as high as 471 mS/mm for InGaAs doped channel MODFETs with maximum drain current in excess of 660 mA/mm gate width.

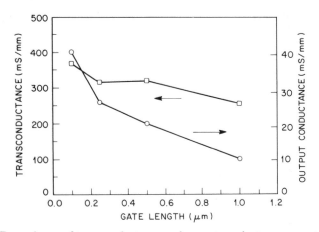

Fig. 3.15 Dependence of transconductance and output conductance on gate length for GaAs MESFETs. (Data from Jaeckel et al.[46] Reprinted with permission, copyright 1986, IEEE and IBM Corp.)

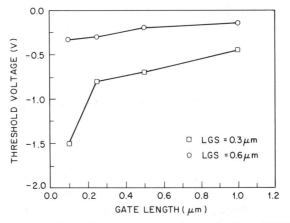

Fig. 3.16 Dependence of threshold voltage on gate length for GaAs MESFETs for different gate-to-source spacings. (Data from Jaeckel et al.[46] Reprinted with permission, copyright 1986, IEEE and IBM Corp.)

Mishra et al.[51] fabricated AlGaAs/InGaAs MODFETs with 0.3 μm gates and obtained transconductances over 650 mS/mm at room temperature.

Baier et al.[41] used tailored-barrier-height (TBH) implants in order to fabricate complementary GaAs MESFETs. They obtained a 33 mS/mm transconductance for 1 μm gate p-channel devices.

Lee et al.[52] reported transconductances of 17.8 mS/mm and 89 mS/mm for p-channel 1 μm gate InGaAs/AlGaAs MODFETs. They also speculated that transconductances greater than 200 mS/mm can be obtained from these p-channel devices.

These experimental data allow us to reach certain conclusions regarding submicron devices. First of all, it is apparent that effective gate length may be quite different from nominal (metallurgical) gate length. It may be larger because gate fringing capacitance modulates regions adjacent to the gate, increasing the effective gate length by approximately 1.5–2 d_{eff} where d_{eff} is the effective channel depth in non-self-aligned gate devices. d_{eff} is determined by the channel depth A in GaAs MESFETs and by the gate-to-channel separation d in AlGaAs/GaAs MODFETs. d_{eff} may even be larger due to carrier injection into the substrate. In self-aligned devices, the effective gate length may be shorter than the metallurgical gate length because of the diffusion of donors from implanted n^+ regions into the channel, effectively pushing the contact regions under the edges of the gate. (This effect will be considered in detail in Section 3.4.) As a consequence, devices with the same nominal gate length and threshold voltage may have maximum transconductances that differ by a factor of 2 or 3.

Pseudomorphic AlGaAs/InGaAs MODFETs generally exhibit better characteristics than AlGaAs/GaAs MODFETs. That may be related to higher mobility in InGaAs and larger band discontinuity at the heterointerface.

The behavior of short-channel devices may be quite different from that of longer channel devices. This is illustrated by Fig. 3.15 and 3.16 showing that higher transconductances in submicron devices may be offset by higher output conductances and may be accompanied by changes in the threshold voltage. This may be caused by short-channel effects that will be considered in detail in the next section.

3.4   SHORT-CHANNEL EFFECTS IN SUBMICRON GaAs AND AlGaAs FIELD-EFFECT TRANSISTORS

3.4.1   Output Conductance

Several important so-called short-channel effects play an important role in devices with short gate lengths (on the order of a micron or less). In short-channel devices, there is a noticeable increase in drain-to-source current in the saturation region (see, e.g., Fig. 3.17, where we show the current–voltage characteristics of a 0.55 μm gate GaAs MESFET (from Sadler and Eastman[53] and Chen[54]). A fairly large output conductance in the saturation regime may be primarily related to two effects: gate length modulation and space charge injection into a semi-insulating substrate. Let us first consider the gate length modulation effect. This effect, which was first considered in the frame of the velocity saturation model by Grebene and Ghandhi,[28] can be visualized as an extension of the region near the drain, where electron velocity is saturated, toward the source (see Fig. 3.18).

The exact analysis of this effect requires two-dimensional modeling. However, to the first order it can be explained as a reduction of the effective channel length in the saturation regime:

$$L_{\text{eff}} = L - \Delta L . \qquad (36)$$

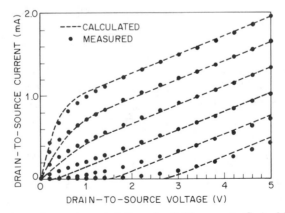

Fig. 3.17 Current–voltage characteristics of a 0.55 μm gate GaAs MESFET. (From Sadler and Eastman[53] and Chen.[54] Reprinted with permission, copyright 1983, IEEE, and T. H. Chen.)

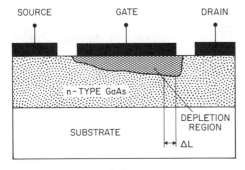

(a)

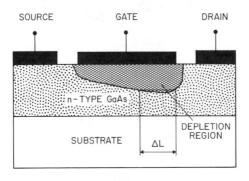

(b)

Fig. 3.18 Gate length modulation in GaAs MESFET: (a) drain-to-source voltage slightly higher than the saturation voltage, and (b) much higher drain-to-source voltage. ΔL is the length of the velocity saturation region under the gate.

Here, L_{eff} is the effective channel length such that the voltage drop across the corresponding section of the channel is equal to the drain-to-source saturation voltage V_{dsat} and ΔL is the length of the velocity saturation region. This section of the channel supports the remainder of the drain-to-source voltage $V_D - V_{Dsat}$. With the increase in V_D, the length of the velocity saturation section ΔL increases leading to a shorter effective channel length L_{eff} (i.e., channel length modulation) and, hence, to a higher drain saturation current and finite output conductance dI_{dsat}/dV_{ds} in the saturation region.

The theory of this effect was developed by Pucel et al.[29] for GaAs MESFETs and by Chang and Fetterman[55] for AlGaAs/GaAs MODFETs.

There are also two mechanisms of space charge injection that may play a role in determining the output conductance. The first mechanism, which may be especially important in self-aligned devices, is illustrated by Fig. 3.19a. It is space charge injection from the source contact into a semi-insulating substrate. According to calculations by Chen[54] (see also Shur,[56] p. 386), this mechanism is respon-

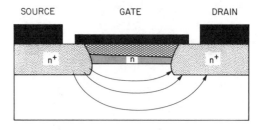

(a)

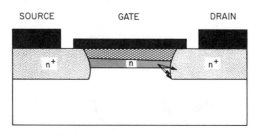

(b)

Fig. 3.19   Two mechanisms of the space charge injection that play a role in determining the output conductance: (*a*) streamlines of current for space charge injection from the source contact, (*b*) streamlines of current for space charge injection from the velocity saturation region in the channel into the drain ohmic implant.

sible for the large output conductance in the 0.55 μm GaAs MESFET (characteristics shown in Fig. 3.17).

In the linear region (when the velocity is proportional to the electric field, i.e., $\mu_n V_D / L_c < v_s$, where L_c is the separation between the contacts, i.e., $L_c \sim L$), this current is proportional to V_D^2 (Mott–Gurney law, after Mott and Gurney[57]):

$$I_{sp} = \frac{9\varepsilon_s \mu_n W d_c V_D^2}{8 L_c^3}. \tag{37}$$

Here, d_c is the depth of the n^+ contacts. In the saturation region where electron velocity is approximately equal to v_s,

$$I_{sp} = \frac{2\varepsilon_s v_s W d_c V_d}{L^2} \tag{38}$$

The following interpolation formula provides a good approximation for space charge injection current for arbitrary drain voltages:

$$j_{sc} = \frac{2}{3} u^2 - \frac{2}{27} u^3, \qquad u \le 3, \tag{39}$$

and

$$j_{sc} = 2(u - 1), \qquad u > 3, \tag{40}$$

where $u = \mu_n V_D/(v_s L_c)$ and $j_{sc} = I_{sc}\mu_n L/(\varepsilon_s v_s W d_c)$ (see Chen[54] and Shur[56], p. 385).

The second mechanism of space charge current may be important in the saturation regime (Fig. 3.19b). In this case, the space charge limited current is caused by injection of electrons with high velocity from the channel into the velocity saturation region near the drain. The cross section of the region, where the concentration of injected carriers is large, increases toward the drain contact. A similar mechanism was considered by Hanafi[58] for Si MOSFETs.

Finally, traps may play an important role in determining output conductance. This may be proven by establishing that output conductance in most devices is actually frequency-dependent at low frequencies (see Fig. 3.20a from Canfield et al.[59]).

Studies of the drain-voltage dependence of threshold voltage (see Chen et al.[60]) also confirm the importance of traps in determining both this dependence and the

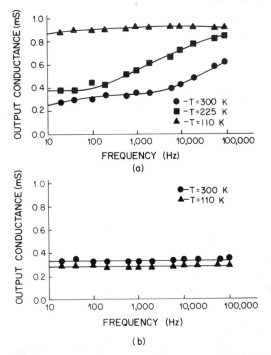

Fig. 3.20 MESFET output conductance vs. frequency: (a) standard MESFET, (b) buried-channel MESFET with p-type implantations on the surface above the channel and in the substrate behind the channel. (From Canfield et al.[59] Reprinted with permission, copyright 1987, IEEE.)

output conductance. Finally, studies of the transient response of drain current by McCamant[61] also show traps strongly affect output conductance at low frequencies.

The variety of different effects contributing to output conductance makes it difficult to develop a universal quantitative model. However, based on the theoretical work quoted above and on experimental data, we can mention three factors determining output conductance. First, it is ratio of the effective channel depth (or of gate-to-channel separation in MODFETs) over the gate length. The smaller ratio leads to a reduction in output conductance. The key to controlling short-channel effects seems to be the scaling of the channel depth and the commensurate increase in the doping levels (proportionally to A^2 in GaAs MESFETs). For example, we notice much smaller output conductance and much higher transconductance in very highly doped 0.1 μm GaAs MESFETs fabricated by Mishra et al.[44] compared with the devices having nominally the same gate length but much smaller doping fabricated by Jaeckel et al.[46]

The second factor is the localization of the channel that prevents electron injection into the substrate, thereby reducing the space charge limited current (see Eastman and Shur[62]) and effects related to traps. Such a localization can be achieved by using an AlGaAs buffer layer (see Eastman and Shur[62]), buried p layers (see Canfield and Forbes[63]), or quantum well structures (see Ketterson et al.[64]).

As an example, we show in Fig. 3.20 (from Canfield and Forbes[63]) how the dependence of MESFET output conductance on frequency (related to traps) is dramatically reduced when MESFET channel is localized using buried implanted p-type layers.

Finally, in self-aligned structures, T-gate or sidewall technology (see Section 3.4.2) may be used to suppress the space charge limited current injection from the contacts. Contact implant depth is also a factor in determining this current. More shallow contact implants reduce the space charge limited current by reducing the effective injection cross section [see Eq. (37)].

3.4.2 Gate Length Dependence of the Threshold Voltage

According to simple one-dimensional device models, the threshold voltage V_t should be independent of channel length. In fact, the threshold voltage is a strong function of device length in short-channel devices (see Fig. 3.16).

Another related short-channel effect is dependence of the parameters of GaAs FETs on gate orientation on the wafer with respect to the crystallographic directions. Such an orientation effect was reported by Lee et al.[65] and by Yokoyama et al.[66] Asbeck et al.[67] suggested that this orientation dependence could be related to a piezoelectric stress effect. The dielectric passivation layers (typically silicon nitride) in the gate-source and gate-drain regions cause compressive or tensile stress, leading to piezoelectric charges induced in the channel and in the substrate. The stress and stress-induced electric field depend on gate orientation on the wafer, thickness of the passivating dielectric film-, and gate length. This leads to the dependence of the device parameters, such as threshold voltage and transconductance, on gate orientation on the wafer, thickness of the passivating dielectric film, and gate length (see Figs. 3.21[68,69] and 3.22[68,70]).

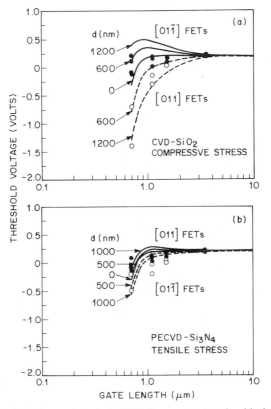

Fig. 3.21 Threshold voltage of GaAs MESFETs vs. gate length with the dielectric overlayer thickness as parameter: (a) silicon dioxide overlayer, (b) silicon nitride overlayer. Calculated curves from Chen et al.[68] Measured data from Onishi et al.[69] (Reprinted with permission, copyrights 1985 and 1987, IEEE.)

Another effect determining gate-length dependence of the threshold voltage in submicron FETs is the effect of donor diffusion from ion-implanted contact regions (see Fig. 3.23). In a self-aligned device, the gate acts as an implantation mask for the n^+ source and drain contacts. During implant annealing, donors diffuse from the n^+ contacts into the channel, thereby changing the carrier concentration in the channel.

Matsumoto et al.[71] fabricated GaAs MESFETs with gate lengths of 0.8, 0.6 μm, and 0.4 μm. Se ions were implanted to form ohmic contacts for the drain and source with no implantation for the n-channel. The device with a 0.8 μm gate acted as a normal FET. The output conductance and current level of the device with a 0.6 μm gate at the same drain and gate voltages were larger than those of the 0.8 μm device. The MESFET with a 0.4 μm gate exhibited a much larger pinch-off voltage and small controllability of the drain current by the gate bias, suggesting that the two n^+ regions were effectively in contact with each other. These results clearly show that n^+ dopants diffused heavily into the channel.

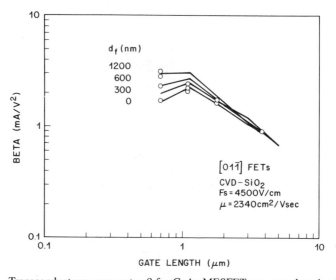

Fig. 3.22 Transconductance parameter β for GaAs MESFETs vs. gate length with the dielectric overlayer thickness as parameter. Calculated curves from Chen et al.[68] Measured data from Onodera et al.[70] Device transconductance $g_m \approx 2\beta(V_{gs} - V_T)$, where V_T is the threshold voltage (this equation is valid only at relatively small gate-to-source voltages when the gate current may be neglected). (Reprinted with permission, copyrights 1986 and 1987, IEEE.)

More recently, Ueto et al.[72] demonstrated that, by using 0.2 μm thickness of SiO_2 sidewall (see Fig. 3.24), the threshold voltages of submicron devices could be made to be about the same as those of long gate length (≥ 3 μm) FETs. Kato et al.[73] achieved a similar improvement by using a T-gate structure (see Fig. 3.25). These results and theoretical analysis given by Chen et al.[68] show that, for a realistic description of gate length dependence of threshold voltage of self-aligned MESFETs, the lateral spread of the n^+ layer should be taken into account.

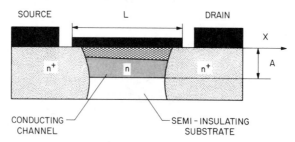

Fig. 3.23 Schematic representation of dopant diffusion from n^+ ohmic regions into MESFET.

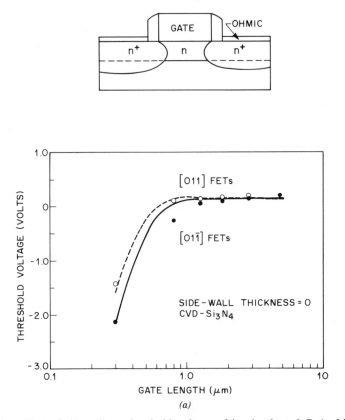

Fig. 3.24 Effect of sidewall on threshold voltage of ion-implanted GaAs MESFETs. (a) Schematic drawing of the doping implants made without the sidewall (the sidewall was made after n^+ implantation) and the dependence of the threshold voltage on the gate length for two different gate orientations for this structure. (b) Schematic drawing of the doping implants with sidewall structure (sidewall is shown by shaded region) and the dependence of the threshold voltage on the gate length for two different gate orientations for this structure. Sidewall thickness is 0.2 μm. (c) Schematic drawing of the doping implant without the n^+ implants (the sidewall was added for the process convenience) and the dependence of the threshold voltage on the gate length for two different gate orientations for this structure. Calculated curves from Chen et al.[68] Measured data from Ueto et al.[72] (Reprinted with permission, copyrights 1985 and 1987, IEEE.)

3.5   VERTICAL DEVICE STRUCTURES

Field-effect transistors are lateral devices with a critical device dimension determined by a lateral dimension. This dimension is controlled by a lithography process. An alternative approach is to use bipolar and unipolar vertical structures such as heterostructure bipolar junction transistors, tunneling emitter bipolar transistors, permeable base transistors, vertical ballistic transistors, and induced base transis-

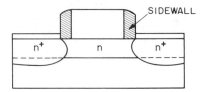

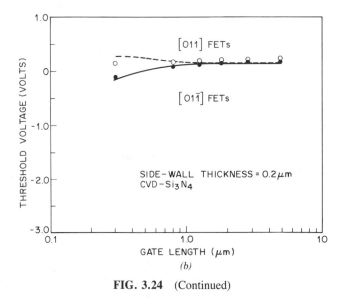

FIG. 3.24   (Continued)

tors. There are also different vertical quantum devices such as resonant tunneling diodes and resonant bipolar transistors. These quantum devices are considered in Chapter 5.

3.5.1   Heterojunction Bipolar Junction Transistors

The idea of a heterojunction bipolar transistor (HBT) was proposed by Shockley[74] and was later developed by Kroemer.[75,76] The principle of operation of this device is described below.

The largest current components in the forward active mode of a bipolar junction transistor are the electron components of the emitter and the collector current, I_{ne} and I_{nc}. They are determined by electron diffusion current in the base at the emitter–base and collector–base junctions, respectively. The common emitter current gain β of a conventional transistor may be estimated as

$$\beta = \frac{I_c}{I_b} < \beta_{\max} = \frac{D_n N_{de} x_e}{D_p N_{ab} W} \exp\left(\frac{-\Delta E_g}{k_B T}\right), \tag{41}$$

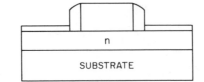

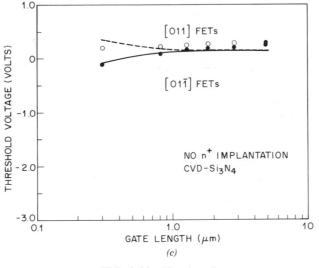

FIG. 3.24   (Continued)

where $I_c \approx I_{cn}$ is the collector current and I_b is the base current. The collector current is proportional to $D_n/(N_{ab}W)$, where D_n is the effective electron diffusion coefficient, N_{ab} the density of acceptors in the base region, and W the base width. The base current is primarily determined by the hole component of the emitter current (caused by hole injection from the base into the emitter region). This current component is proportional to $D_p/(N_{de}x_e)$, where D_p is the effective hole diffusion coefficient, N_{de} is the density of donors in the emitter region, and x_e is the emitter width. The exponetial factor $\exp(-\Delta E_g/k_B T)$ that reduces gain is caused by bandgap narrowing in the highly doped emitter region. It takes into account that, in a semiconductor with a narrower energy gap, the concentration of minority carriers increases exponentially.

In a heterojunction bipolar junction transistor (HBT), the emitter region has a wider bandgap than the base. If for illustrative purposes the discontinuity of the bandgaps is assumed to be entirely related to the valence band discontinuity, then $\beta_{\max}$ for a wide-gap emitter device is given by

$$\beta_{\max} = \frac{D_n N_{de} x_e}{D_p N_{ab} W} \exp\left(\frac{\Delta E_g}{k_B T}\right).$$ (42)

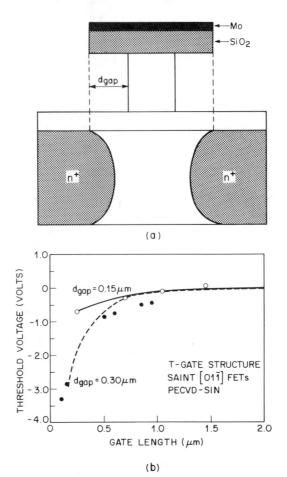

Fig. 3.25 Effect of T-gate structure on threshold voltage of ion-implanted GaAs MES-FETs: (a) schematic drawing of the T-gate structure, (b) threshold voltage vs. gate length for two different T-gate structures. Calculated curves from Chen et al.[68] Measured data from Kato et al.[73] (Reprinted with permission, copyrights 1983 and 1987, IEEE.)

As a result, a very high value of β_{max} may be achieved even when N_{de} is smaller than N_{ab}.

In fact, the band diagram of a heterojunction looks like that shown in Fig. 3.26 and, as a consequence, Eq. (42) should be rewritten as

$$\beta_{max} = \frac{D_n N_{de} x_e}{D_p N_{ab} W} \exp\left(\frac{\Delta E_v}{k_b T}\right), \tag{43}$$

where $\Delta E_v = \Delta E_g - \Delta E_c$. This leads to a less dramatic increase in β_{max}, but the basic principle remains intact. Moreover, the decrease in maximum gain related to

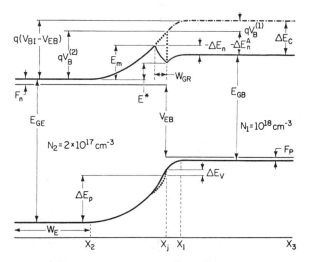

Fig. 3.26 Energy band diagrams of emitter–base heterojunction: dotted line, abrupt emitter–base heterojunction; solid line, emitter–base heterojunction with graded composition. (After Grinberg et al.[77] Reprinted with permission, copyright 1984, IEEE.)

the decrease of the potential barrier for holes may be partially compensated for by the increase in electron velocity in the base caused by the conduction band spike. As pointed out by Kroemer,[76] electrons in this "spike-notch" structure enter the base with a very large energy (close to ΔE_c) and, as a consequence, may have very high velocities (on the order of several times 10^5 m/s). Because of the directional nature of the dominant polar optical scattering, they may traverse the base region ballistically or near ballistically, maintaining a very high velocity. Kroemer described the conduction band spike as a "a launching pad" for ballistic electrons. Also, the magnitude of the spike can be reduced by grading the composition of the wide bandgap emitter near the heterointerface (see Kroemer[76]), as shown in Fig. 3.26. Such a grading plays an important role in optimizing the emitter injection efficiency of HBTs (see, e.g., Grinberg et al.[77]).

This discussion shows that the common-emitter current gain of an HBT may be made very high. In practical devices, the transistor gain may be limited by the recombination current I_r:

$$\beta = \frac{I_{ne}}{I_r}, \tag{44}$$

and could be as high as a few thousand or more if the heterojunction interface is relatively defect-free, so that I_r is not excessively high.

If the recombination current is caused primarily by recombination in the base, it can be estimated as

$$I_{br} \sim \frac{qcn_p(0)W}{\tau}, \tag{45}$$

where c is a numerical constant on the order of unity, $n_p(0)$ the concentration of minority carriers (electrons) at the emitter end of the base, and τ the lifetime. The electron component of emitter current is given by

$$I_{ne} \sim q n_p(0) v_{nb}, \tag{46}$$

where v_{nb} is the effective electron velocity in the base. Then we obtain

$$\beta \approx \frac{I_{ne}}{I_{br}} \approx \frac{\tau}{t_{TR}}, \tag{47}$$

where t_{TR} is the electron transit time across the base. For a sufficiently short base (say, $W \sim 0.1 \ \mu m$), $\beta > 1000$ can be obtained even if the lifetime is only on the order of a nanosecond. In a conventional homojunction silicon transistor, the dependence of the energy gap on the doping level leads to a shrinkage of the energy gap in the emitter region. This energy gap shrinkage represents one of the dominant performance limitations for conventional Si BJTs (see, e.g., Sze[78]).

In addition to high injection efficiencies and, as a consequence, high common-emitter current gains, HBTs have a number of other advantages over conventional bipolar transistors. As a consequence of higher base doping, base spreading resistance is smaller. Because of relatively low doping of the emitter region, emitter–base capacitance could be made quite small. All these factors result in a higher speed of operation.

Recent advances in molecular-beam epitaxy (MBE) technology have made it possible to obtain abrupt or graded heterojunctions with a high degree of reproducibility. Common-emitter gains in excess of 1600 have been achieved for AlGaAs-GaAs bipolar junction transistors.

The equations presented above merely illustrate the principle of operation of a HBT. More detailed and accurate analysis of an HBT operation should be based on a more detailed model (see, e.g., Tiwari[79]). Such an analysis shows that both bulk recombination in the depletion regions and surface recombination, such as at the emitter–base junction periphery limit the common-emitter current gain. It also shows that the propagation delay of an HBT logic gate is primarily limited by the time constant equal to the product of collector capacitance and load resistance. The propagation delay can be analyzed using the equation proposed by Dumke et al.[80] for an HBT common mode logic (CML) gate:

$$\tau_s = 2.5 R_B C_{BC} + \frac{R_b \tau_B}{R_L} + (3 C_{BC} + C_L) R_L \tag{48}$$

where τ_B is the base transit time, R_B is the base resistance, R_L and C_L are the load resistance and load capacitance, and C_{BC} is the collector–base capacitance.

A schematic structure of a high-speed AlGaAs/GaAs HBT is shown in Fig. 3.27 (after Chang et al.[19]). This structure has a proton-implanted area that reduces the collector–base capacitance by as much as 60%, leading to an increase in cutoff

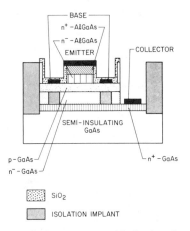

Fig. 3.27   Schematic diagram of a heterostructure bipolar junction transistor (HBT) (emitter-on-top configuration). (After Chang et al.[19] Reprinted with permission, copyright 1987, IEEE.)

frequency f_T and maximum oscillation frequency f_{max} compared with unimplanted devices.

Recently, the NTT group[81,82] reported a 5.5 ps/gate propagation delay for AlGaAs/GaAs HBTs with nonalloyed compositionally graded InGaAs/GaAs emitter ohmic contacts. They reduced the specific emitter contact resistance to as low as 1.4×10^{-7} $\Omega \cdot cm^2$. Emitter and collector dimensions were 2×5 and 4×7 μm^2, respectively. For the collector current of 12 mA, Nagata et al.[81,82] reported $f_{max} = 60$ GHz and $f_T = 80$ GHz.

Another heterostructure system suitable for HBTs is InGaAs/InP (see Nottenburg et al.[83,84]). The schematic structure of a InGaAs/InP double heterostructure bipolar transistor is shown in Fig. 3.28. The potential advantages of such a device are related to small built-in emitter junction potential and high mobility in InGaAs, high electron drift velocity in InP, and low surface recombination velocity (see Nottenburg et al.[83]). Nottenburg et al.[84] reported high current gain (>500) over eight decades of collector current for InGaAs/InP double heterostructure bipolar transistors.

Using high-mobility InGaAs as the base material for HBTs can also improve performance compared with AlGaAs/GaAs HBTs. A schematic band structure and layer structure for such a device are shown in Figs. 3.29 and 3.30 respectively (after Asbeck et al.[85]). Cutoff frequency as high as 40 GHz (see Asbeck et al.[85]) and propagation delay as short as 19.7 ps (for 19 stage CML ring oscillators) (see Asbeck et al.[85] and Sullivan et al.[86]) were obtained.

Even though the base width in HBTs determines transit time across the base and, hence, device speed for this vertical device structure, in practical transistors the lateral dimensions of the emitter stripes are very important. These dimensions determine parasitic capacitances, resistances, and power dissipation. This is clearly demonstrated by Chang et al.[19]). They obtained a 14.5 ps gate propagation

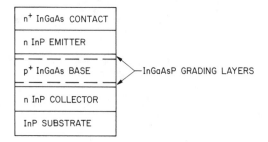

	THICKNESS (μm)	DOPING LEVEL (cm^{-3})
CONTACT	0.2	$> 10^{19}$
EMITTER	0.3	2×10^{17}
GRADING LAYER	0.02	2×10^{17}
BASE	0.12	2×10^{18}
COLLECTOR	0.3	2×10^{17}

Fig. 3.28   The schematic structure of a InGaAs/InP double heterostructure bipolar transistor. (From Nottenburg et al.[83] Reprinted with permission, copyright 1986, IEEE.)

delay in HBT common mode logic ring oscillators using devices with an emitter width of only 1.2 μm. These devices exhibited cutoff frequency $f_T = 67$ GHz and maximum oscillation frequency $f_{\max} = 105$ GHz. Another important feature of these devices was a very high base-region doping (up to 10^{20} cm^{-3}), leading to a small base spreading resistance.

Ishibashi and Yamauchi[87] reported an HBT with a new device structure. They replaced an n-type GaAs collector layer by a double layer that included a rela-

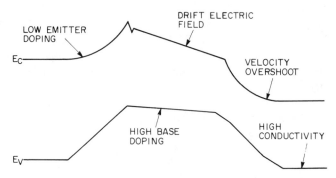

Fig. 3.29   Schematic band structure of AlGaAs/InGaAs/GaAs HBT. (After Asbeck et al.[85] Reprinted with permission, copyright 1986, IEEE.)

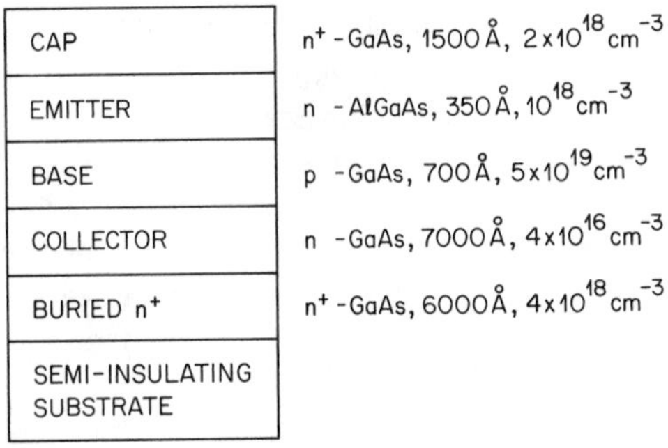

CAP	n^+-GaAs, 1500 Å, 2×10^{18} cm^{-3}
EMITTER	n -AlGaAs, 350 Å, 10^{18} cm^{-3}
BASE	p -GaAs, 700 Å, 5×10^{19} cm^{-3}
COLLECTOR	n -GaAs, 7000 Å, 4×10^{16} cm^{-3}
BURIED n^+	n^+-GaAs, 6000 Å, 4×10^{18} cm^{-3}
SEMI-INSULATING SUBSTRATE	

Fig. 3.30 Schematic layer structure of AlGaAs/InGaAs/GaAs HBT. (After Asbeck et al.[85] Reprinted with permission, copyright 1986, IEEE.)

tively thick i-GaAs layer (2000 Å in their devices) and a thin p^+ GaAs layer (200 Å thick doped at 2×10^{18} cm^{-3}). The p^+ layer is totally depleted and introduces a potential drop and electric field in the i-layer, resulting in a near-ballistic collection of electrons in a certain voltage range. They obtained a very high cutoff frequency of 105 GHz.

Most HBTs are emitter-on-top structures (see Fig. 3.27). However, as was pointed out by Kroemer,[76,88] the collector capacitance may be substantially smaller for the collector-on-top configuration. This was recently confirmed by the circuit simulation of HBTs reported by Akagi et al.,[89] who predicted a higher speed for collector-on-top configurations when comparing non-self-aligned structures.

A new type of a heterojunction bipolar transistor, a tunneling emitter bipolar transistor (TEBT) was recently proposed by Xu and Shur[90] and fabricated by Najar et al.[91] (see Fig. 3.31). In this device, a wide bandgap AlGaAs emitter is replaced by a conventional n^+ GaAs emitter, but a thin compositionally graded AlGaAs layer is inserted between the emitter and base regions. This layer has vastly differ-

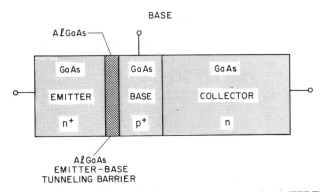

Fig. 3.31 Schematic diagram of tunneling emitter bipolar transistor (TEBT). (From Xu and Shur.[90] Reprinted with permission, copyright 1986, IEEE.)

ent tunneling rates for electrons and holes, so electrons can easily go through but holes are prevented from being injected into the emitter region. This is reminiscent of a "mass filtering" idea proposed by Capasso et al.[92] for superlattice devices. The TEBT should have smaller emitter contact resistance, higher gain, fewer traps, and higher speed of operation than a conventional HBT.

The progress achieved in developing HBTs is illustrated in Fig. 3.32, where cutoff frequencies and gate propagation delays are shown as functions of time. These results show that HBTs present a serious challenge to GaAs MESFETs, AlGaAs/GaAs MODFETs, and other high-speed devices.

3.5.2   Unipolar Vertical Device Structures

Heterojunction bipolar transistors are the most studied vertical devices. However, different unipolar submicron compound semiconductor vertical device structures have been proposed to take advantage of ballistic and overshoot effects and graded and abrupt heterostructures — an approach sometimes referred to as energy band

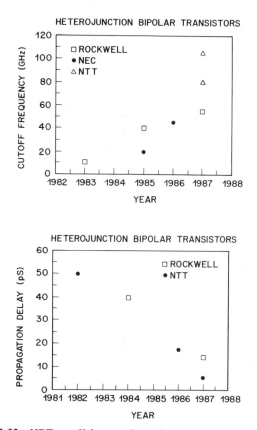

Fig. 3.32   HBT cutoff frequencies and gate propagation delays.

engineering. These devices include vertical field-effect transistors (VFETs), permeable base transistors (PBTs), vertical ballistic transistors (VBTs), induced base transistors (IBTs), hot-electron transistors (HETs), planar doped barrier transistors (PDBTs), and others.

Vertical FETs were first proposed by Shockley[93] and further developed by Zuleeg[94] and Nishizawa (see, e.g., Nishizawa et al.[95]). Different types of GaAs and AlGaAs/GaAs vertical FETs are shown in Fig. 3.33 (from Rav-Noy et al.[96]). Rav-Noy et al. achieved transconductances as high as 280 mS/mm.

Frensley et al.[97] studied microwave properties of GaAs VFETs. The maximum oscillation frequency obtained was not very high ($f_{max} = 12$ GHz). However, the microwave performance was primarily limited by parasitics, and much higher cutoff and maximum oscillation frequency may be achieved.

In a permeable base transistor (Fig. 3.34), a thin tungsten grating is embedded into a single crystal of gallium arsenide, forming a transistor base (see Bozler et al.[98] and Bozler and Alley[99]). The device consists of four layers: the n^+ substrate, the n-type emitter layer, the thin film tungsten grid, and the n-type collector layer. Tungsten forms a Schottky barrier with GaAs. The doping concentration N_D in the n-layer is such that the zero-bias depletion width around the Schottky barrier is larger than the space between the tungsten strips. As a result, the current is zero at zero emitter–base voltage. When a positive bias is applied to the base, the depletion layer shrinks and a conductive path forms between the collector and emitter.

The effect of doping density and device dimensions on the cutoff frequency f_T is shown in Fig. 3.35 (from Bozler and Alley[99]). In this calculation, the device dimensions were scaled down with N_D, so the threshold voltage V_T remains nearly constant. As can be seen from the figure, f_T increases rapidly with doping and with a decrease in device size.

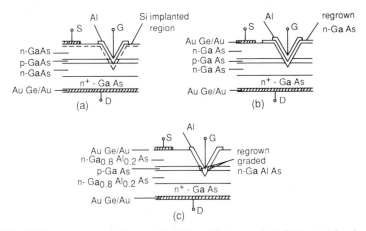

Fig. 3.33 Different types of GaAs and AlGaAs/GaAs vertical FETs: (*a*) ion-implanted vertical GaAs MESFET, (*b*) vertical GaAs MESFET with regrown *n*-type GaAs, (*c*) heterojunction vertical FET. (From Rav-Noy et al.[96] Reprinted with permission, copyright 1984, IEEE.)

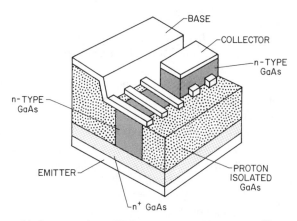

Fig. 3.34 Permeable base transistor (PBT). (After Bozler and Alley.[99] Reprinted with permission, copyright 1980, IEEE.)

Recently, Shur[100] proposed a bipolar version of this device where one of the injecting contacts is *p*-type and the other is *n*-type. As a consequence, the electron-hole plasma is injected into the spacings between the fingers of the grid. The density of plasma is modulated by changing the grid potential. The possibility of field-effect modulation of the electron-hole plasma was recently demonstrated by Hack et al.[101] in amorphous silicon double injection field-effect transistors.

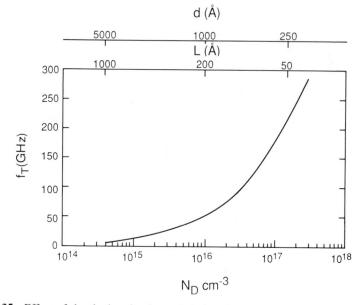

Fig. 3.35 Effect of the doping density and device dimensions on cutoff frequency of a PBT. Dimensions *d* and *L* are shown in Fig. 3.34. (From Bozler and Alley.[99] Reprinted with permission, copyright 1980, IEEE.)

A schematic diagram of a vertical ballistic transistor (VBT) is shown in Fig. 3.36 (from Eastman et al.[102]). The principle of operation of this device is similar to that of a PBT. The main advantage of this device is that the gate contact is on the top of the structure and embedded into a GaAs layer. That is why this structure is easier to fabricate.

Luryi[103] proposed a new interesting device concept. In his device, called in the induced base transistor (IBT), the base conduction occurs in the induced two-dimensional gas (see Fig. 3.37). Chung et al.[104] fabricated IBTs and obtained common-base current gain α as high as 0.96.

A planar doped barrier transistor (PDBT) was proposed by Malik et al.[105] A schematic diagram of this device is shown in Fig. 3.38. Barrier heights in planar doped barrier devices are controlled by the charge of the plane of dopants (see Fig. 3.39). This allows us to control the band diagram by changing this charge, thereby optimizing the device characteristics.

Recently, Heiblum et al.[4] and Levy et al.[5] used a hot-electron transistor (HET) and a planar doped barrier transistor, respectively, for the first direct observations of the ballistic transport.

A schematic band diagram of the HET used by Heiblum et al.[4] is shown in Fig. 3.40. By changing the collector voltage, the number of the collected electrons can be changed so that the energy distribution of the collected electrons is proportional to the derivative of collector current with respect to collector–base voltage. This measured derivative is shown in Fig. 3.41. The sharp peak that increases with emitter current corresponds to ballistic electrons. At the present time, this device is primarily used as a hot-electron spectrometer that yields information about the energy distribution of hot electrons. However, if parasitics are minimized and base spreading resistance is kept low, it may be also used for ultrahigh frequency applications.

Xu and Shur[106] proposed a different version of this device, a double base hot-electron transistor (DBHET), where the first (doped and/or graded) base region acts as an "electron gun," accelerating electrons, and as a "lens", providing a better focused ballistic electron beam that is injected into the second base where an input signal is applied. The same concept may be applied to form a beam of ballistic electrons in other structures.

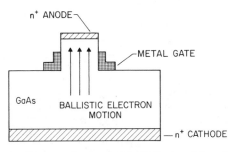

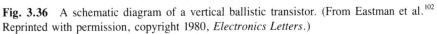

Fig. 3.36 A schematic diagram of a vertical ballistic transistor. (From Eastman et al.[102] Reprinted with permission, copyright 1980, *Electronics Letters*.)

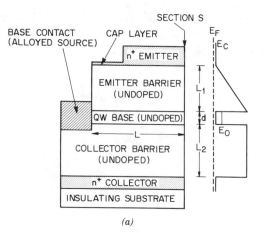

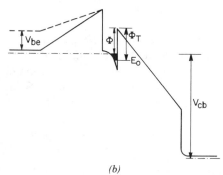

(b)

Fig. 3.37 Schematic band diagram of induced base transistor (*a*) under equilibrium conditions, (*b*) under positive bias V_{cb} applied to the collector. (From Luryi[103] and Chung et al.[104] Reprinted with permission, copyrights 1985 and 1986, IEEE.)

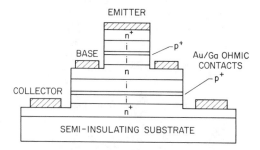

Fig. 3.38 Schematic diagram of planar doped barrier transistor (PDBT). (After Malik et al.[105] Reprinted with permission, copyright 1981, IEEE.)

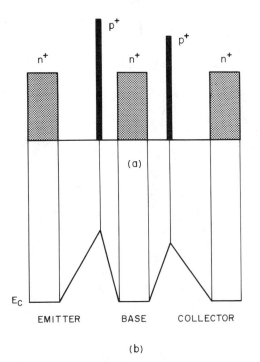

(a)

(b)

Fig. 3.39 (a) Schematic doping profile and (b) band diagram of planar doped barrier transistor (PDBT). (After Malik et al.[105] Reprinted with permission, copyright 1981, IEEE.)

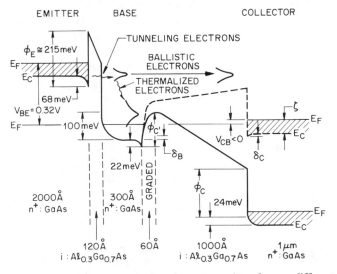

Fig. 3.40 Schematic band diagram of a hot-electron transistor for two different collector–base voltages. (From Heiblum et al.[4] Reprinted with permission, copyright 1985, *Physical Review Letters*.)

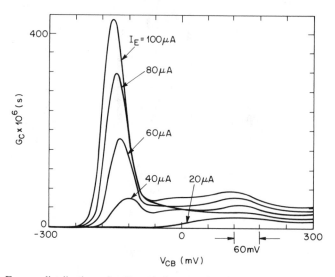

Fig. 3.41 Energy distribution of collected electrons in a hot-electron transistor for different emitter currents. $G_c = \partial I_c / \partial V_{cb}$, where I_c is the collector current and V_{cb} the base–collector voltage. (From Heiblum et al.[4] Reprinted with permission, copyright 1985, *Physical Review Letters*.)

3.6 CONCLUSION

Submicron compound semiconductor devices, such as field-effect transistors and heterostructure bipolar junction transistors, have great promise for ultrahigh-speed and millimeter wave operation. New improvements in speed, fabrication process, reliability, integration scale, material composition, etc., are reported almost daily, increasing both our knowledge base and our hopes for the future.

New ideas, or new implementations of old ideas, have led to many new submicron devices, such as vertical transistors, permeable base transistors, ballistic transistors, induced base transistors, planar doped barrier transistors, hot-electron transistors, tunneling emitter barrier transistors, etc., that may compete for higher speed and better performance.

The physical basis for this achieved and expected improvement is ballistic and near ballistic transport in submicron structures. The existence of this mode of transport has been recently proven, and now its advantages are to be utilized in order to achieve switching speeds and operation frequencies that have never been possible in the past.

ACKNOWLEDGMENTS

I would like to thank Jun-ho Baek, Chung-Hsu Chen, Tzu-hung Chen, A. Grinberg, Choong Hyun, Kwyro Lee, Kang Lee, Paul Ruden, P. C. Chao, Andy Peczalski, Nick

Cirillo, Jr., Robert Daniels, David Arch, Obert Tufte, John Abrokwah, Phil Jenkins, Tim Drummond, Hadis Morkoç, Steve Baier, and Jingming Xu, who worked with me on the development of GaAs and AlGaAs/GaAs device models. I am also grateful to my graduate students, Michael Norman, Jun-ho Baek, Yung Byun, Phil Jenkins, and Byung Moon, for the critical reading of the manuscript.

This work was partially supported by the Microelectronics and Information Sciences Center at the University of Minnesota, and by Honeywell, Inc.

REFERENCES

1. Z. Yu and R. W. Dutton, *Sedan III — A General Electronic Material Device Analysis Program*, Program Manual, Stanford University, Stanford, CA, July 1985.

2. J. Xu and M. Shur, "Velocity-Field Dependence in GaAs," *IEEE Trans. Electron Devices*, **ED-33**, 1831 (1987).

3. M. Shur and L. F. Eastman, "Ballistic Transport in Semiconductors at Low Temperatures for Low Power High Speed Logic," *IEEE Trans. Electron Devices*, **ED-26**, 1677 (1979).

4. M. Heiblum, M. I. Nathan, D. C. Thomas, and C. M. Knoedler, "Direct Observation of Ballistic Transport in GaAs," *Phys. Rev. Lett.*, **55**, 2200 (1985).

5. A. F. J. Levy, J. R. Hayes, P. M. Platzman, and W. Wiegmann, "Injected Hot Electron Transport in GaAs," *Phys. Rev. Lett.*, **55**, 2071 (1985).

6. J. G. Ruch, "Electronics Dynamics in Short Channel Field-Effect Transistors," *IEEE Trans. Electron Devices*, **ED-19**, 652 (1972).

7. M. Shur, "Influence of the Non-Uniform Field Distribution in the Channel on the Frequency Performance of GaAs FETs," *Electron. Lett.*, **12**, 615 (1976).

8. A. Cappy, B. Carnes, R. Fauquembergues, G. Salmer, and E. Constant, "Comparative Potential Performance of Si, GaAs, GaInAs, InAs Submicrometer-gate FETs," *IEEE Trans. Electron Devices*, **ED-27**, 2158 (1980).

9. N. J. Shah, S. S. Pei, C. W. Tu, and R. C. Tiberio, "Gate-Length Dependence of the Speed of SSI Circuits Using Submicron Selectively Doped Heterostructure Transistor Technology," *IEEE Trans. Electron Devices*, **ED-33**, 543 (1986).

10. N. C. Cirillo, Jr., J. K. Abrokwah, and M. Shur, "A Self-Aligned Process for IC's Based on Modulation-Doped (Al,Ga)As/GaAs FET's," *IEEE Trans. Electron Devices*, **ED-31**, 1963 (1984).

11. M. Shur, "Low Field Mobility, Effective Saturation Velocity, and Performance of Submicron GaAs FETs," *Electron. Lett.*, **18**(21), 909 (1982).

12. M. Shur and D. Long, "Performance Prediction for Submicron GaAs SDFL Logic," *IEEE Electron Device Lett.*, **EDL-3**, 124 (1982).

13. C. H. Hyun, M. Shur, and N. C. Cirillo, Jr., "Simulation and Design Analysis of AlGaAs/GaAs MODFET Integrated Circuits," *IEEE Trans. Comput.-Aided Des.*, **CAD-5**, 284 (1986).

14. M. B. Das, "Millimeter Wave Performance of Ultrasubmicrometer-Gate Field Effect Transistors: A Comparison of MODFET, MESFET, and PBT Structures," *IEEE Trans. Electron Devices*, **ED-34**, 1429 (1987).

15. B. J. Van Zeghbroeck, W. Patrick, H. Meier, P. Vettiger, and P. Wolf, "High Performance GaAs MESFET's," *Tech. Dig. — Int. Electron Devices Meet.*, p. 832 (1986).

16. T. S. Henderson, M. I. Aksun, C. K. Peng, H. Morkoç, P. C. Chao, P. M. Smith, K. H. G. Duh, and L. F. Lester, "Power and Noise Performance of the Pseudomorphic Modulation Doped Field Effect Transistor at 60 GHz, *"Tech. Dig.—Int. Electron Devices Meet.,* p. 464 (1986).

17. T. J. Drummond, T. W. Masselink, and H. Morkoç, "Modulation-Doped GaAs/ (Al,Ga) As Heterojunction Field-Effect Transistors: MODFETs," *Proc. IEEE,* **74,** 733 (1986).

18. Y. K. Chen, D. C. Radulescu, G. W. Wang, A. N. Lepore, P. J. Tasker, L. F. Eastman, and E. Strid, "Bias-Dependent Microwave Characteristics of an Atomic Planar-Doped AlGaAs/InGaAs/GaAs Double Heterojunction MODFET," *Proc. IEEE Microwave Theory Tech. Symp. 1987,* Las Vegas, NV, p. 871.

19. M. F. Chang, P. M. Asbeck, K. C. Wang, G. J. Sullivan, N. H. Sheng, J. A. Higgins, and D. L. Miller, "AlGaAS/GaAs Heterojunction Bipolar Transistors Fabricated Using a Self-Aligned Dual-Lift-Off Process," *IEEE Electron Device Lett.,* **EDL-8,** 303 (1987).

20. L. F. Eastman, "Tunneling, Ballistic, and Hot Electron Devices," *Program, 45th Annu. Device Res. Conf., 1987,* Santa Barbara, CA, p. I-3 (1987).

21. T. Mimura, S. Hiyamizu, T. Fukii, and K. Nambu, "A New Field Effect Transistor with Selectively Doped GaAs/n-Al$_x$Ga$_{1-x}$As heterostructures," *Jpn. J. Appl. Phys.,* **19,** L225 (1980).

22. N. C. Cirillo, Jr., M. Shur, P. J. Vold, J. K. Abrokwah, R. R. Daniels, and O. N. Tufte, "Realization of n-Channel and p-Channel High Mobility (Al,Ga)As-GaAs Heterostructure Insulated Gate FETs on a Planar Wafer Surface," *IEEE Electron Device Lett.,* **EDL-6,** 645 (1985).

23. P. Solomon, C. M. Knoedler, and S. L. Wright, "A GaAs Gate Heterojunction FET," *IEEE Electron Device Lett.,* **EDL-5,** 379 (1984).

24. N. C. Cirillo, Jr., M. Shur, and J. K. Abrokwah, "Inverted GaAS/AlGaS Modulation Doped Transistors with Extremely High Transconductances," *IEEE Electron Device Lett.,* **EDL-7,** 71 (1986).

25. R. Dingle, H. L. Stormer, A. C. Gossard, and W. Wiegman, "Electron Mobilities in Modulation-Doped Semiconductor Heterojunction Superlattices," *Appl. Phys. Lett.,* **33,** 665 (1978).

26. M. Tomizawa, K. Yokoyama, and Y. Yoshi, "Hot-Electron Velocity Characteristics at AlGaAs/GaAs Heterostructures," *IEEE Electron Device Lett.,* **EDL-5,** 464 (1984).

27. T. W. Masselink, N. Braslau, D. LaTulipe, W. Wang, and S. Wright, "Electron Velocity at High Electric Fields in AlGaAs/GaAs Modulation-Doped Heterostructures," *Proc. Int. Conf. Hot Carriers Semicond. 5th,* Boston, MA (1987).

28. A. B. Grebene and S. K. Ghandhi, "General Theory for Pinched Operation of the Junction Gate FET," *Solid-State Electron.,* **12,** 573 (1969).

29. R. A. Pucel, H. Haus, and H. Statz, "Signal and Noise Properties of Gallium Arsenide Microwave Field-Effect Transistors," *Adv. Electron. Electron Phys.,* **38,** 195 (1975).

30. H. Statz, P. Newman, I. W. Smith, R. A. Pucel, and H. A. Haus, "GaAs FET Device and Circuit Simulation in SPICE," *IEEE Trans. Electron Devices,* **ED-34,** 160 (1987).

31. T. J. Drummond, H. Morkoç, K. Lee, and M. Shur, "Model for Modulation Doped Field Effect Transistor," *IEEE Electron Device Lett.,* **EDL-3,** 338 (1982).

32. P. P. Ruden, C. J. Han, and M. Shur, "Gate Current in MODFETs," *J. Appl. Phys.*, **64**(3), 1541 (Aug. 1, 1988).

33. K. Lee, M. Shur, T. J. Drummond, and H. Morkoç, "Parasitic MESFET in (Al,Ga)As/GaAs Modulation Doped FET," *IEEE Trans. Electron Devices*, **ED-31**, 29 (1984).

34. H. Dambkes, W. Brokerhoff, and K. Heime, "GaAs MESFETs with Highly Doped (10^{18} cm^{-3}) Channels — An Experimental and Numerical Investigation," *Tech. Dig. — Int. Electron Devices Meet.*, p. 621 (1983).

35. J. H. Baek, M. Shur, R. R. Daniels, D. K. Arch, J. K. Abrokwah, and O. N. Tufte, "New Mechanism of Gate Current in Heterostructure Insulated Gate Field-Effect Transistors," *IEEE Electron Device Lett.*, **EDL-7**, 519 (1986).

36. K. Hess and C. Kizilyalli, "Scaling and Transport Properties of High Electron Mobility Transistors," *Tech. Dig. — Int. Electron Devices Meet.*, p. 556 (1986).

37. C. M. Snowden and D. Loret, "Two-Dimensional Hot-electron Models for Short-Gate-Length GaAs MESFET's," *IEEE Trans. Electron Devices*, **ED-34**, 212 (1987).

38. K. Hess, H. Morkoç, H. Shichijo, and B. G. Streetman, "Negative Differential Resistance Through Real Space Transfer," *Appl. Phys. Lett.*, **35**, 459 (1979).

39. C. H. Chen, S. Baier, D. K. Arch, and M. Shur, "A New and Simple Model for GaAs Heterojunction FET Gate Characteristics," *IEEE Trans. Electron Devices*, **ED-35**, 570 (1987).

40. R. R. Daniels, R. MacTaggart, J. K. Abrokwah, O. N. Tufte, M. Shur, J. Baek, and P. Jenkins, "Complementary Heterostructure Insulated Gate FET Circuits for High-Speed, Low-Power VLSI," *Tech. Dig. — Int. Electron Devices Meet.*, p. 448 (1986).

41. S. M. Baier, G. Y. Lee, H. K. Chung, B. J. Fure, and R. MacTaggart, "Complementary GaAs MESFET Logic Gates," *IEEE Electron Device Lett.*, **EDL-8**, 260 (1987).

42. T. J. Drummond, T. E. Zipperian, I. J. Fritz, J. E. Schiber, and T. A. Plut, "p-Channel, Strained Quantum Well, Field-Effect Transistor," *Appl. Phys. Lett.*, **49**, 461 (1986).

43. M. Sweeny, J. Xu, and M. Shur, "Light and Heavy Holes in One-Dimensional Systems," *Proc. Int. Conf. Superlattices Microstruct. 1987*, Chicago, IL (to be published).

44. U. K. Mishra, R. S. Beubien, M. J. Delaney, A. S. Brown, and L. H. Hacket, "MBE Grown GaAs MESFETs with Ultra-High g_m and f_T," *Tech. Dig. — Int. Electron Devices Meet.*, p. 829 (1986).

45. B. J. Van Zeghbroeck, W. Patrick, H. Meier, and P. Vettiger, "Submicrometer GaAs MESFET with Shallow Channel and Very High Transconductance," *IEEE Electron Device Lett.*, **EDL-8**, 118 (1987).

46. H. Jaeckel, V. Graf, B. J. Van Zeghbroeck, P. Vettiger, and P. Wolf, "Scaled GaAs MESFET's with Gate Length Down to 100 nm," *IEEE Electron Device Lett.*, **EDL-7**, 522 (1986).

47. B. J. Van Zeghbroeck, W. Patrick, H. Meier, and P. Vettiger, "Parasitic Bipolar Effects in Submicrometer GaAs MESFETs," *IEEE Electron Device Lett.*, **EDL-8**, 188 (1987).

48. P. Sauner and J. W. Lee, "High-Efficiency Millimeter Wave GaAs/AlGaAs Power HEMT's," *IEEE Electron Device Lett.*, **EDL-7**, 503 (1986).

49. P. Sauner and J. W. Lee, "Multiple HEMT with One Watt per Millimeter Power Density at 21 GHz," *Program, 45th Annu. Device Res. Conf. 1987,* Santa Barbara, CA, p. II-A8 (1987).

50. R. R. Daniels, "Doped Channel FETs," private communication (1987).

51. U. K. Mishra, A. S. Brown, L. M. Jelloian, L. H. Hacket, and M. J. Delaney, "High Performance Submicron AlInAs-GaInAs HEMT's," *Program, 45th Annu. Device Res. Conf., 1987,* Santa Barbara, CA, p. II-A6 (1987).

52. C. P. Lee, H. T. Wang, G. J. Sullivan, N. H. Sheng, and D. L. Miller, "High-transconductance p-Channel InGaAs/AlGaAs Modulation-Doped Field Effect Transistors," *IEEE Electron Device Lett.,* **EDL-8,** 85 (1987).

53. R. A. Sadler and L. F. Eastman, "High-Speed Logic at 300 K with Self-Aligned Submicron Gate GaAs MESFET's," *IEEE Electron Device Lett.,* **EDL-4,** 215 (1983).

54. T. H. Chen, "High Speed GaAs Device and Integrated Circuit Modeling and Simulation," PhD. thesis, University of Minnesota, Minneapolis, MN (1984).

55. C. S. Chang and H. R. Fetterman, "An Analytical Model for HEMT's Using New Velocity-Field Dependence," *IEEE Trans. Electron Devices,* **ED-34,** 1456 (1987).

56. M. Shur, *GaAs Devices and Circuits,* Plenum, New York, 1987.

57. N. F. Mott and R. W. Gurney, "Electronic Processes in Ionic Crystals, Oxford Univ. Press (Clarendon), London and New York, 1940.

58. H. I. Hanafi, "Current Modeling for MOSFET," in A. E. Ruehli, Ed., *Circuit Analysis, Simulation, and Design. Advances in CAD for VLSI,* Vol. 3 p. 71, North-Holland, Amsterdam, 1986.

59. P. Canfield, J. Medinger, and L. Forbes, "Buried Channel GaAs MESFET's with Frequency-Independent Output Conductance," *IEEE Eelectron Device Lett.,* **EDL-8,** 88 (1987).

60. C. H. Chen, A. Peczalski, and M. Shur, "Trapping-Enhanced Temperature Variation of the Threshold Voltage of GaAs MESFET's," *IEEE Trans. Electron Devices,* **ED-33,** 792 (1986).

61. A. McCamant, "TriQuint Semiconductor, Beaveron, OR," private communication, referenced by Canfield et al.[59].

62. L. F. Eastman and M. Shur, "Substrate Current in GaAs MESFET's," *IEEE Trans. Electron Devices,* **ED-26,** 1359 (1979).

63. P. Canfield and L. Forbes, "Suppression of Drain Current Transients, Drain Current Oscillations, and Low Frequency Generation-Recombination Noise in GaAs FET's Using Buried Channels," *IEEE Trans. Electron Devices,* **ED-33,** 925 (1986).

64. A. A. Ketterson, W. T. Masselink, J. S. Gedymin, J. Klem, C. K. Peng, W. F. Kopp, and K. R. Gleason, "Characterization of InGaAs/AlGaAs Pseudomorphic Modulation-Doped Field-Effect Transistors," *IEEE Trans. Electron Devices,* **ED-33,** 564 (1986).

65. C. P. Lee, R. Zucca, and B. M. Welch, "Orientation Effect on Planar GaAs Schottky Barrier Field Effect Transistors," *Appl. Phys. Lett.,* 37(3), 311 (1980).

66. N. Yokoyama, H. Onodera, T. Ohnishi, and A. Shibatomi, "Orientation Effect of Self-Aligned Source/Drain GaAs Schottky Barrier Field-Effect Transistor," *Appl. Phys. Lett.,* 42(3), 270 (1983).

67. P. M. Asbeck, C. P. Lee, and F. M. Chang, "Piezoelectric Effects in GaAs FETs and Their Role in Orientation Dependent Device Characteristic," *IEEE Trans. Electron Devices,* **ED-31,** 1377 (1984).

68. C. H. Chen, A. Peczalski, M. Shur, and H. K. Chung, "Orientation and Ion-Implanted Transverse Effects in Self-Aligned GaAs MESFETs," *IEEE Trans. Electron Devices,* **ED-34,** 1470 (1987).

69. T. Onishi, T. Onodera, N. Yokoyama, and H. Nishi, "Comparison of the Orientation Effect of SiO_2 and Si_3N_4-Encapsulated GaAs MESFET's," *IEEE Electron Device Lett.,* **EDL-6,** 172 (1985).

70. T. Onodera, T. Onishi, N. Yokoyama, and H. Nishi, "Improvement in GaAs MESFET Performance due to Piezoelectric Effect," *IEEE Trans. Electron Devices,* **ED-32,** 2314 (1986).

71. K. Matsumoto, N. Hashizume, N. Atoda, K. Tomizuwa, T. Kurosu, and M. Ioda, "Submicron-Gate Self-Aligned GaAs FET by Ion Implantation," *Int. Symp. GaAs Related. Comp., 10th,* New Mexico, p. 317 (1982).

72. K. Ueto, T. Furutsuka, H. Toyoshima, M. Kanamori, and A. Higashisaka, "A High Transconductance GaAs MESFET with Reduced Short Channel Effect Characteristics," *Tech. Dig., Electron Devices Meet.,* p. 82 (1985).

73. N. Kato, Y. Matsuoka, K. Ohwada, and S. Moriya, "Influence of n^+ Layer-Gate Gap on Short-Channel Effects of GaAs Self-Aligned MESFETs (SAINT)," *IEEE Electron Device Lett.,* **EDL-4,** 417 (1983).

74. W. Shockley, U.S. Pat. 2,569,347 (1951).

75. H. Kroemer, "Theory of a Wide-Gap Emitter for Transistors," *Proc. IRE,* **45,** 1535 (1957).

76. H. Kroemer, "Heterostructure Bipolar Transistors and Integrated Circuits," *Proc. IEEE,* **70,** 13 (1982).

77. A. A. Grinberg, M. Shur, R. J. Fisher, and H. Morkoç, "Investigation of the Effect of Graded Layers and Tunneling on the Performance of AlGaAs/GaAs Heterojunction Bipolar Transistors," *IEEE Trans. Electron Devices,* **ED-31**(12), 1758 (1984).

78. S. M. Sze, *Physics of Semiconductor Devices,* 2nd ed., Wiley, New York, 1981.

79. S. Tiwari, "GaAlAs/GaAs Heterostructure Bipolar Transistors: Experiment and Theory" *Tech. Dig.—Int. Electron Devices Meet.,* p. 262, (1986).

80. W. P. Dumke, J. M. Woodall, and V. L. Rideout, "GaAs-AlGaAs Heterojunction Transistor for High frequency Operation," *Solid-State Electron.,* **15,** 1339 (1972).

81. O. Nakajima, K. Nagata, Y. Yamauchi, H. Ito, and T. Ishibashi, "High-Speed AlGaAs/GaAs HBTs with Proton-Implanted Buried Layers," *Tech. Dig.—Int. Electron Devices Meet.,* 266 (1986).

82. K. Nagata, O. Nakajima, Y. Yamauchi, H. Ito, T. Nittono, and T. Ishibashi, "High Speed Performance of AlGaAs/GaAs Heterojunction Bipolar Transistor with Non-Alloyed Emitter Contacts," *Program, 45th Annu. Device Res. Conf., 1987,* Santa Barbara, CA, p. IV-A2 (1987).

83. R. N. Nottenburg, J. C. Bischoff, J. H. Abeles, M. B. Panish, and H. Temkin, "Base Doping Effects in InGaAS/InP Double Heterostructure Bipolar Transistors," *Tech. Dig.—Int. Electron Devices Meet.,* p. 278 (1986).

84. R. N. Nottenburg, H. Temkin, M. B. Panish, R. Bhat, and J. C. Bischoff, "High Speed InGaAS/InP Double-Heterostructure Bipolar Transistors," *IEEE Electron Device Lett.,* **EDL-8,** 282 (1987).

85. P. M. Asbeck, M. F. Chang, K. C. Wang, G. J. Sullivan, and D. L. Miller, "GaAlAs/GaInAs/GaAs Heterojunction Bipolar Technology for sub-35 ps Current-Mode Logic

Circuits," *Proc. Bipolar Circuits Technol. Meet., IEEE, 1986,* Minneapolis, MN, p. 25 (1986).

86. G. J. Sullivan, P. M. Asbeck, M. F. Chang, D. L. Miller, and K. C. Wang, "High Frequency Performance of AlGaAs/InGaAs/GaAs Strained Layer Heterojunction Bipolar Transistors," *IEEE Trans. Electron Devices,* **ED-33**(11), 1845 (1986).

87. T. Ishibashi, and Y. Yamauchi, "A Novel AlGaAs/GaAs HBT Structure for Near Ballistic Collection," *Program, 45th Annu. Device Res. Conf., 1987,* Santa Barbara, CA, p. IV-A6 (1987).

88. H. Kroemer, "Heterostructure Bipolar Transistors: What Should We Build," *J. Vac. Sci. Technol.,* **13**[2], 1(2), 112 (1983).

89. J. Akagi, J. Yoshida, and M. Kurata, "A Model-Based Comparison of Switching Characteristics Between Collector-Top and Emitter-Top HBT's," *IEEE Trans. Electron Devices,* **ED-34**, 1413 (1987).

90. J. Xu and M. Shur, "Tunneling Emitter Bipolar Junction Tansistor," *IEEE Electron Device Lett.,* **EDL-7**, 416 (1986).

91. F. E. Najar, D. C. Radulescu, Y. K. Chen, G. W. Wicks, P. J. Tasker, and L. F. Eastman, "DC Characterization of the AlGaAS/GaAs Tunneling Emitter Bipolar Transistor," *Appl. Phys. Lett.,* **50**(26), 1915 (1987).

92. F. Capasso, K. Mohammed, A. Y. Cho, R. Hull, and A. L. Hutchinson, "Effective Mass Filtering: Giant Quantum Amplification of the Photocurrent in a Semiconductor Superlattice," *Appl. Phys. Lett.,* **47**(4), 420 (1985).

93. W. Shockley, "A Unipolar 'Field-Effect' Transistor," *Proc. IRE,* **40**, 1289 (1952).

94. R. Zuleeg, *Solid-State Electron.,* **10**, 449 (1967).

95. J. I. Nishizawa, T. Terasaki, and K. Shibata, Field-Effect Transistor Versus Analog Transistor (Static Induction Transistor), " *IEEE Trans. Electron Devices,* **ED-22**, 185 (1975).

96. Z. Rav-Noy, U. Schreter, S. Mukai, E. Kapon, J. S. Smith, L. C. Chiu, S. Margalit, and A. Yariv, "Vertical FET's in GaAs," *IEEE Electron Device Lett.,* **EDL-5**, 228 (1984).

97. W. R. Frensley, B. Bayraktaroglu, S. E. Campbell, H. D. Shih, R. E. Lehmann, and R. E. Williams, "Microwave Operation of Gallium Arsenide Vertical MESFET," *Proc. Bienn. IEEE Cornell Bienn., 1983,* Ithaca, NY, p. 87 (1983).

98. C. O. Bozler, G. D. Alley, D. C. Flanders, and W. T. Lindley, "Fabrication and Microwave Performance of the Permeable Base Transistor," *Tech. Dig.—Int. Electron Devices Meet.,* p. 384 (1979).

99. C. O. Bozler and G. D. Alley, "Fabrication and Numerical Simulation of the Permeable Base Transistor," *IEEE Trans. Electron Devices,* **ED-27**, 1128 (1980).

100. M. Shur, "Bipolar Permeable Base Transistor," *Proc. IEEE Bipolar Conf., 1987,* Minneapolis, MN, p. 37 (1987).

101. M. Hack, M. Shur, and W. Czubatyj, "Double Injection Field Effect Transistor—A New Type of Solid State Device," *Appl. Phys. Lett.,* **48**(20), 1386 (1986).

102. L. F. Eastman, R. Stall, D. Woodard, N. Dandekar, C. Wood, M. Shur, and K. Board, "Ballistic Motion in GaAs at Room Temperature," *Electron. Lett.,* **16**, 524 (1980).

103. S. Luryi, "An Induced Base Hot-Electron Transistor," *IEEE Electron Device Lett.,* **EDL-6**, 178 (1985).

104. C. H. Chung, W. C. Liu, M. S. Jame, Y. H. Wang, S. Luryi, and S. M. Sze, "Induced Base Transistor Fabricated by Molecular Beam Epitaxy," *IEEE Electron Device Lett.*, **EDL-7,** 497 (1986).

105. R. J. Malik, M. A. Hollis, L. F. Eastman, D. W. Woodard, C. E. C. Wood, and T. R. AuCoin, "GaAs Planar-Doped Barrier Transistors Grown by Molecular Beam Epitaxy," *Proc. 8th Bienn. IEEE, Cornell Bienn., 1981,* p. 87, Ithaca, NY, (1981).

106. J. Xu and M. Shur, "Ballistic Transport in Hot Electron Transistors," *J. Appl. Phys.*, **62**(9), 3816 (Nov. 1, 1987).

107. R. A. Kiehl, D. A. Frank, S. L. Wright, and J. H. Magerlein, "Device Physics of Quantum-Well Heterostructure M1#SFET's," *Tech. Dig.—Int. Electron Devices Meet.* p. 70 (1987).

108. T. J. Drummond, T. E. Zipperian, I. J. Fritz, J. E. Schiber, and T. A. Plut, "p-channel, Strained Quantum Well, Field-Effect Transistor," *Appl. Phys. Lett.*, **49,** 461 (1986).

4 Ultrahigh-Speed HEMT LSI Circuits

MASAYUKI ABE, TAKASHI MIMURA, KAZUO KONDO, and
JUNJI KOMENO
Fujitsu Laboratories Ltd.
Fujitsu Limited
Atsugi, Japan

4.1 INTRODUCTION

Nine years have now passed since the 1980 announcement of the high electron mobility transistor (HEMT),[1] and HEMT technology has certainly opened the door to new possibilities for ultrahigh-speed large-scale integration (LSI)/very large scale integration (VLSI) applications.[2] The evolution of high-speed GaAs-based HEMT integrated circuits (ICs) is the result of continuous technological progress utilizing the superior electronic properties due to the supermobility of GaAs/AlGaAs heterojunction structure. Electron mobility in the conventional GaAs metal semiconductor field-effect transistor (MESFET) channel with typical donor concentrations of around 10^{17} cm^{-3} ranges from 4000 to 5000 cm^2/V · s at room temperature. The mobility in the channel at 77 K is not too much higher than at room temperature due to ionized impurity scattering. In undoped GaAs, however, electron mobility of $2-3 \times 10^5$ cm^2/V · s has been obtained at 77 K. The mobility of GaAs with feasibly high electron concentrations for facilitating the fabrication of devices was found to increase through modulation-doping techniques demonstrated in GaAs/AlGaAs superlattices.[3] As the first application of this electron-mobility-enhanced phenomenon to the new transistor approach, a high electron mobility transistor (HEMT), based on modulation-doped GaAs/AlGaAs single heterojunction structures, was invented[1] and was demonstrated to improve the 77 K channel mobility greatly.

HEMT technology has been demonstrated to promise for ultrahigh-speed LSI/ VLSI applications.[1,4-6] Due to the supermobility GaAs/AlGaAs heterojunction structure, the HEMT is especially attractive for low-temperature operations at liquid nitrogen temperature. In 1981, a HEMT ring oscillator with a gate length of 1.7 μm demonstrated a 17.1 ps switching delay with 0.96 mW power dissipation per gate at 77 K, indicating that switching delays below 10 ps will be achievable with 1 μm gate devices.[4] A switching delay of 8.5 ps with 2.59 mW power dissipation per gate has already been obtained with a 0.5 μm gate device at 77 K.[7] Furthermore, 5.8 ps with 1.76 mW power dissipation per gate at 77 K and 10.2 ps with 1.03 mW power dissipation per gate at 300 K have been achieved for a 0.35 μm gate device.[8] Even at room temperature, 9.2 ps with 4.2 mW per gate has been obtained with a 0.28 μm gate device.[9]

More complex circuits were achieved by divide-by-two operations of HEMT frequency dividers with direct coupled FET logic (DCFL) circuits, demonstrating maximum clock frequencies of 8.9 GHz[10,11] and 13 GHz[12] at 77 K. The maximum clock frequency achieved with HEMT technology is roughly two times higher than that of its GaAs MESFET counterpart with comparable geometry.

For LSI-level complexity, HEMT technology has made it possible to develop a 4 kbit static RAM,[13,14] 16 kbit static RAM[15] as memory circuits, and a 4.1-kgate gate array with a 16 × 16 bit parallel multiplier[16] as a logic circuit. The 4 kbit static RAM, designed for an ECL-compatible level, has an address access time of 500 ps with a power dissipation of 5.7 W per chip.[14] The 16 × 16 bit parallel multiplier designed on 4.1-kgate gate array has a multiply time of 4.1 ns with a power dissipation of 6.2 W. HEMT has already jumped into the LSI/VLSI application field.

This chapter first presents the performance advantages of HEMT approaches, with the focus on scaled-down device structure in the submicron dimensional range. Next, a HEMT technology for VLSIs is described, including material and self-alignment device fabrication technology. The current status and recent advances in HEMT logic and memory LSI circuit implementation are reviewed and the future HEMT VLSI prospects for ultrahigh-speed computer applications are projected.

4.2 PERFORMANCE ADVANTAGES OF HEMT APPROACHES

HEMT technology has new possibilities for LSI/VLSI with high-speed and low-power dissipation. This section describes the principles of the HEMT, its technological advantages compared with other technologies for high-speed devices, and HEMT performances scaled-down in the submicron dimensional range.

4.2.1 HEMT Principle

A cross-sectional view of the basic structure of a HEMT, with a selectively doped GaAs/AlGaAs heterojunction structure, is shown in Fig. 4.1. An undoped GaAs layer and Si-doped n-type AlGaAs layer are successively grown on a semi-insulating GaAs substrate by molecular-beam epitaxy (MBE). Because of the higher electron affinity of GaAs, free electrons in the AlGaAs layer are transferred to the undoped GaAs layer, where they form a two-dimensional high-mobility electron gas within 10 nm of the interface. The n-type AlGaAs layer of the HEMT is completely depleted in two depletion mechanisms: (1) the surface depletion results from the trapping of free electrons by surface states, and (2) the interface depletion re-

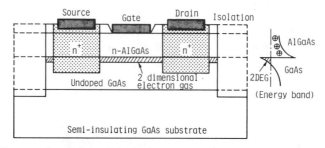

Fig. 4.1 Cross-sectional view of the basic structure of a HEMT, with a selectively doped GaAs/AlGaAs heterostructure.

sults from the transfer of electrons into the undoped GaAs. The Fermi level of the gate metal is matched to the pinning point, which is 1.2 eV below the conduction band. With the reduced AlGaAs layer thickness, the electrons supplied by donors in the AlGaAs layer are insufficient to pin the surface Fermi level. Therefore, the space-charge region extends into the undoped GaAs layer and, as a result, band bending results in the upward direction, and the two-dimensional electron gas does not appear. When a positive voltage higher than the threshold voltage is applied to the gate, electrons accumulate at the interface and form a two-dimensional electron gas (2DEG).

Thus, we can control the electron concentration to achieve depletion (D) mode and enhancement (E) mode HEMT operation.[1,17] Electron mobility and sheet electron concentration N_s in the heterostructure are shown as a function of temperature in Fig. 4.2.[18] As temperature decreases, the electron mobility, which was about 8×10^3 cm²/V · s at 300 K, increases dramatically and reaches 2×10^5 cm²/V · s at 77 K due to reduced phonon scattering. A further increase with a considerable gradient occurred even below 50 K, and a maximum value of 1.5×10^6 cm²/V · s in the dark and 2.5×10^6 cm²/V · s under light illumination was attained at 4.2 K. Sheet electron concentration decreases with decreasing temperature until it becomes constant below 150 K. The almost constant value of about 3.5×10^{11} cm⁻² below 150 K corresponds to that of 2DEG at the interface, since this value agrees well with the value of N_s determined by Shubnikov–de Haas measurement at 4.2 K. Apparent excess carriers above 150 K are attributed to free electrons which are thermally excited from relatively deep donors in n-type $Al_{0.3}Ga_{0.7}As$.

In HEMT structures, the AlGaAs layer heavily doped with donors such as Si contains DX centers[19] which behave as electron traps at low temperatures. Some anomalous behaviors at low temperatures are believed to be related to these traps. These include distortion of drain I–V characteristics, an unexpected threshold voltage shift at low temperatures, and highly sensitive persistent photoconduction. We have found that the distortion of drain I–V characteristics is related to the type of device structures. In a conventional partial gate HEMT structure that has relatively long (>100 μm) exposed surfaces of AlGaAs at both sides of the gate, electrons heated by the drain field have sufficient energy to transfer from a GaAs channel to

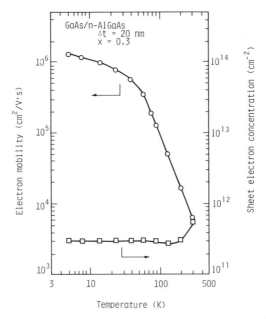

Fig. 4.2 Electron mobility and sheet electron concentration in GaAs/n-AlGaAs with a 20 nm thick undoped AlGaAs spacer layer, as a function of a temperature.

the AlGaAs region, where they are trapped at DX centers. As a result, space charges build up at the drain side of the gate. This eventually increases the drain output resistance in the linear region of operation, leading to drain current collapse. To eliminate drain current collapse at low temperature, we adopt a self-aligned gate structure, as shown in Fig. 4.1. The n-AlGaAs layer is completely covered by the n-GaAs top layer. There are no exposed surfaces at the drain end of the gate. In the structure, high-energy electrons can easily pass through the thin n-AlGaAs layer (30 nm) without being trapped and can reach the n-GaAs top layer, eliminating the anomalous drain I–V characteristics at low temperature. Concerning the threshold voltage shift with temperature, a 0.2 V to 0.3 V shift for the temperature range from 300 K to 77 K has been reported in a HEMT. We have succeeded in eliminating the threshold voltage shift and sensitive persistent photoconduction effect by optimizing device parameters with a reduced AlAs mole fraction of 0.2.

4.2.2 Comparison With Other High-Speed Device Approaches

The technological advantages of HEMTs are compared with various competing approaches to high-speed device design in Table 4.1. It is difficult to compare their optimized performances fairly. Here, we have the criteria based on 0.5– 1 μm device technology. The switching delay of GaAs MESFETs is two or three times longer than that of HEMTs. GaAs/AlGaAs heterojunction bipolar transistors (HBT) should achieve the same high-speed performance as the HEMT. The ulti-

TABLE 4.1 Technological Advantages of HEMTs Compared With Various Competing High-Speed Device Approaches

Device Approach	Performances		Uniformity and Controllability (σV_T/Swing Ratio)	Fabricability	Material Problems	Total Advantages
	Speed	Power				
HEMT $L_G = 0.5–1\ \mu m$	Excellent 10 ps Highly geometry controllable	Very good 0.1 mW	Excellent 10–20 mV (1%)	Excellent Simple MBE and OMVPE Dry etching	Good Defect and trap-free epi High throughput	Excellent
GaAs MESFET $L_G = 0.5–1\ \mu m$	Good 20–30 ps Poor geometry controllable	Good 1 mW	Good 60 mV (10%)	Excellent Simple	Good Defect-free ingot	Good
GaAs/AlGaAs HBT	Excellent 10–30 ps	Good 1 mW	Excellent (<1%)	Complex New process required	Good Defect and trap-free epi High throughput	Unknown
SiMOSFET	Very poor 80 ps	Very good 0.1 mW	Excellent (1%)	Complex	Excellent	Difficult to high speed
Si bipolar	Good 30–60 ps	Poor 1–10 mW	Excellent (<1%)	Complex	Excellent	Difficult to large scale

mate speed capability, limited by cutoff frequency f_T, is over 100 GHz, and the HBT also has the merit of flexible fan-out loading capability. The silicon MOS-FET and bipolar transistor are excellent for both designing due to threshold voltage uniformity and controllability with no material problems, and for ease of fabrication in spite of complex processing steps. Configuration for both high-speed and large-scale integration with low-power performance, however, may be difficult for Si-based technology. HEMTs are very promising devices for high-speed VLSI, but require new technological breakthroughs to achieve the LSI quality of GaAs/AlGaAs material, using MBE and/or organic metal vapor-phase epitaxy (OMVPE) and the self-alignment device fabrication technologies described in the following sections.

To evaluate the high-speed capability of HEMTs in complex logic circuits, a single-clocked divide-by-two circuit based on the master–slave flip–flop, consisting of eight direct coupled FET logic (DCFL) NOR gates, one inverter, and four output buffers was fabricated. The circuit has a fan-out of up to 3 and employs 0.5 mm long interconnects, giving a more meaningful indication of the overall performance of HEMT integrated circuits (ICs) than that obtained with a simple ring oscillator. The basic gate consists of 0.5 × 20 μm gate E-HEMT and saturated resistors as loads. Direct writing electron-beam lithography and lift-off techniques were used throughout the fabrication process. Divide-by-two operation is demonstrated at up to 8.9 GHz at 77 K and up to 5.5 GHz at 300 K.[11] The values of 8.9 and 5.5 GHz, respectively, correspond to internal logic delays of 22 ps/gate with power dissipation of 2.8 mW/gate at 77 K and 36 ps/gate with power dissipation of 2.9 mW/gate at 300 K, with an average fan-out of about 2. A frequency divider circuit composed of dual-gate, selectively doped heterojunction transistors (SDHT) with 0.7 μm gate lengths was also fabricated, showing a maximum clock frequency of 13 GHz at 77 K.[12] Figure 4.3 compares switching delay and power dissipation for a variety of frequency dividers.[11,12,20] The switching speed of HEMT is roughly three times as fast as that of a GaAs MESFET.

4.2.3 HEMT Performances in the Submicron Dimensional Range

HEMT has a performance advantage over conventional devices because of superior electron dynamics of HEMT channels and the unique electrical properties of the HEMT structure. During switching, the speed of the device is limited by both low-field mobility and saturated drift velocity. Low-field mobility routinely obtained is 8,000 cm^2/V · s at 300 K and 40,000 cm^2/V · s at 77 K. Saturated drift velocity measured with HEMT structures at room temperature has been reported to be 1.5 to 1.9 × 10^7 cm/s. These superior transport properties in HEMT channels would result in a high average current-gain cutoff frequency f_T value.

Going to LSI circuits with low-power dissipation per gate, logic voltage swing should be minimized. A high transconductance g_m value with a small logic voltage swing is achieved. The transconductance g_m in gradual channel approximation is given by $g_m = K(V_{GS} - V_T)$, where notations have their usual meanings. K is given by $K = (\varepsilon\mu_n W_G/2dL_G)$, where ε is the dielectric constant, μ_n the electron

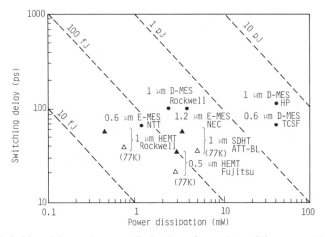

Fig. 4.3 Switching delay and power dissipation of a variety of frequency dividers. The symbol ● denotes GaAs MESFET at 300 K; ▲ and △ HEMTs at 300 K and 77 K, respectively.

mobility, W_G the channel width, d the spacing between the gate and the channel, and L_G the gate length. The K value of a 0.5 μm gate HEMT at 77 K is determined to be 900 mA/V^2 per millimeter of gate width. This K value is about eight times higher than for conventional GaAs MESFETs. The smaller level of logic voltage swing requires more precisely controlled threshold voltages with the smaller standard deviation. State-of-the-art standard deviation of threshold voltages of enhancement mode HEMTs is 6 mV over a 1 cm^2 LSI chip area described in the following section. This value means a controllability of less than 1% to the logic voltage swing of 0.8 V.

In Fig. 4.4, the current-gain cutoff frequency f_T vs. gate length summarizes the typical performances of experimental HEMTs and GaAs MESFETs reported so far.[21-25] At room temperature, the values of f_T were 38 GHz[21] and 80 GHz[12] for HEMTs with gate lengths of 0.5 μm and 0.25 μm, respectively, about twice those for GaAs MESFETs. No significant variation in threshold voltages with gate length was also observed in the range from L_G = 1.4 μm to L_G = 0.28 μm.[9] This horizontal sensitivity indicates that reducing the geometry of HEMTs is an acceptable way to increase performance with no short-channel effect problems.

For application of FET device with the submicron dimensional range to integrated circuits, the short-channel effect is one of the most serious problems. However, the HEMT structure has the inherent advantage of reducing the short-channel effect. This is because the gate-to-channel capacitance can be increased by raising the doping concentration in the n-AlGaAs layer using the modulation doping technique. This shields the drain fields without degrading the semiconductor mobility. This would give easily designable and stable current–voltage characteristics for gates in the submicron range. Figure 4.5 shows how the threshold voltage varies with gate length for both HEMTs and self-aligned gate GaAs MESFETs.[9] It can be seen that the threshold voltage variation of a HEMT whose gate is in the sub-

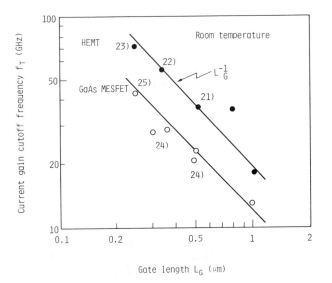

Fig. 4.4   Current-gain cutoff frequency vs. gate length for experimental HEMTs and GaAs MESFETs.

micron range is much smaller than that of a GaAs MESFET. For a HEMT, the variation was less than 30 mV when the gate length varied from 1.4 μm to 0.28 μm. Therefore, existing HEMTs can potentially allow 0.25 μm LSI to be produced. To suppress the short-channel effect in conventional FETs such as GaAs MESFETs, the doping concentration of the channel must be increased to

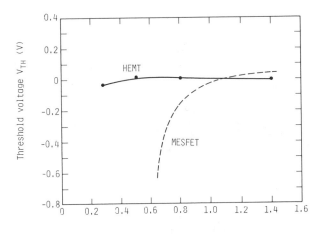

Fig. 4.5   Dependence of threshold voltage V_{TH} on gate length L_G for HEMT and GaAs MESFET.

raise the gate-to-channel capacitance. In this case, the electron mobility eventually decreases with increasing doping concentration in the channel, degrading the high-speed performance.

The dependence of the K factor and transconductance g_m of E-HEMTs on gate length were measured at both 77 and 300 K and are plotted in Fig. 4.6. Dashed lines indicate the L_G^{-1} dependence of the K factor and g_m expected from the gradual channel approximation. Below a 1 μm gate length at 300 K, the K factor and g_m deviate from the L_G^{-1} dependence. A velocity saturation effect and parasitic source resistances probably play a significant role in these results. The 0.5 μm gate E-HEMT at 77 K exhibits a g_m of 500 mS/mm.

It is very important that large enough noise margins are maintained to allow the LSI to operate stably. Figure 4.7 shows the basic propagation delay time t_{pd} vs. the gate length L_G for both high- and low-level noise margins (larger than 200 mV). The supply voltage V_{DD} is 1 V. It can be seen in the figure that t_{pd} increases proportionally with L_G. The average value of t_{pd} for $L_G = 0.5$ μm over a 2 in. wafer was 23 ps, with a standard deviation of 1 ps. The standard deviation at the high level was 12 mV, and that at the low level was 15 mV. If we consider that Si dynamic RAM will be the technology leader, we can expect a gate dimension of 0.5 μm at the production level in the near future. We can anticipate that by then, HEMT 10-kgate gate arrays with a 100 ps propagation delay per gate and HEMT 64 kbit static RAMs with 1 ns access time will be feasible.

4.3   HEMT TECHNOLOGY FOR VLSI

The development of a high-performance VLSI requires new technological breakthroughs. This section describes state-of-the-art HEMT technology, including the material and self-alignment device fabrication technologies involved.

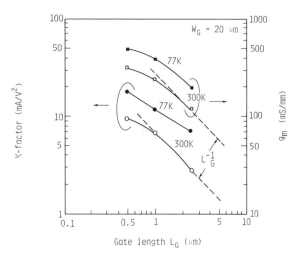

Fig. 4.6 Dependence of K factor and transconductance g_m of E-HEMTs on gate length L_G at 77 and 300 K, respectively.

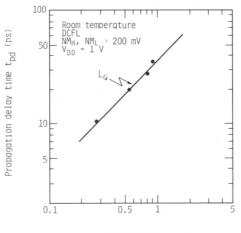

Gate length L_G (μm)

Fig. 4.7 Dependence of propagation delay time t_{pd} on gate length L_G for a HEMT DCFL circuit.

4.3.1   Material Technology for HEMT Fabrication

To realize high-quality material grown by MBE, we optimized the buffer layer between the semi-insulating GaAs substrate and the two-dimensional electron-gas channel layer. The thickness of this layer was 0.6 μm. The electron mobility in this optimized heterostructure was 8×10^3 cm^2/V·s at 300 K and increased to 1.2×10^5 cm^2/V·s at 77 K due to reduced phonon scattering.[26]

The surface-defect problem of MBE is a serious one for fabrication of LSI-level complexity.[27] The surface irregularities are called oval defects. The oval defects typically are from submicrometer to several micrometers, which is comparable to the size of devices in LSI circuits. These oval defects seriously affect the current–voltage characteristics of HEMTs.[28] We have already achieved a density of less than 10 cm^{-2} with a size of over 20 μm^2 by optimizing the growth conditions to make it possible to develop an LSI with complexity 10 kgate logic and 64 kbit static RAM circuits.

An important problem to be solved in fabricating HEMT LSI is highly uniform epitaxial wafer growth technology with high throughput and large size. We optimized both the geometrical configurations between source and substrate in the molecular-beam epitaxy (MBE) system and the growth conditions for highly uniform epitaxial layers on a 3 in. diameter semi-insulating GaAs substrate, making high throughput and high quality possible. Selectively doped GaAs/n-AlGaAs heterostructure were grown on semi-insulating GaAs substrates mounted on a substrate holder with a diameter of 190 mm,[29] as shown in Fig. 4.8. The substrate temperature during growth was held at 660°C. The uniformity of ±1% for both the thickness and the carrier concentration of the AlGaAs layer is required to control the threshold voltage of HEMT characteristics. Figure 4.9 shows the lateral uniformity of the thickness in n-Al$_x$Ga$_{1-x}$As ($x = 0.29$).[29] The variation of the

Fig. 4.8 Three semi-insulating GaAs substrates with 3 in. diameter mounted on a substrate holder with a 190 mm diameter.

thickness over a diameter of 180 mm was less than $\pm 1\%$. The average carrier concentration of the Si-doped n-AlGaAs epitaxial layer was 1.0×10^{18} cm^{-3}, with a variation over the diameter of 180 mm of less than $\pm 1\%$. Based on the above data, we grew a selectively doped GaAs/n-AlGaAs heterostructure which consisted of an 800 nm undoped GaAs buffer layer, a 6 nm undoped AlGaAs spacer layer, a 90 nm Si-doped AlGaAs layer ($N_D = 1 \times 10^{18}$ cm^{-3}). Figure 4.10 shows the lateral uniformity of the electron mobility and the sheet electron concentration of two-dimensional electron gas (2DEG) obtained by Hall measurement at 77 K.

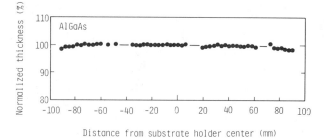

Fig. 4.9 Lateral uniformity of thickness in an n-Al$_x$Ga$_{1-x}$As ($x = 0.29$) epitaxial layer grown over a 190 mm diameter.

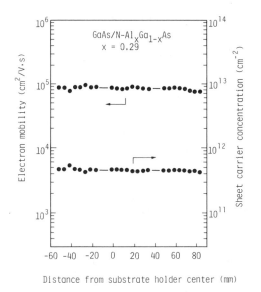

Fig. 4.10 Lateral uniformity of the mobility and the concentration of 2DEG by Hall measurements at 77 K.

It was confirmed that the variations of the mobility and the concentration at 77 K in the selectively doped GaAs/n-AlGaAs heterostructure was less than ±2% over a diameter of 180 mm.

The epitaxial material growth of HEMT LSI quality with AlGaAs/GaAs heterostructure has been also performed by atmospheric pressure OMVPE technology. To achieve highly uniform material characteristics for LSI fabrication, three-wafer (multiwafer) growth is carried out by using an rf-heated graphite suscepter, not only revolving the suscepter itself, but also rotating each wafer simultaneously.[30] The speeds of revolution and rotation were 8 rpm and 20 rpm, respectively. The Cr-doped GaAs substrates were oriented 2.5° off the (100) toward the ⟨110⟩. The source materials used in the hydrogen carrier were trimethylgallium (TMG), trimethylaluminum (TMA), AsH_3, and Si_2H_6.

The growth temperature was 660°C. The growth rates of GaAs and $Al_{0.28}Ga_{0.72}As$ were 58 nm/min and 21 nm/min, respectively. To get abrupt interfaces, the compositions of the source gases were changed instantaneously by a so-called vent/run system.[31] The uniformities of the thickness and the carrier concentration of n-AlGaAs film across a 2 in. wafer were better than ±2.0% and ±1.5%, respectively. The sheet carrier concentration and the electron mobility at 300 K of an AlGaAs/GaAs selectively doped heterostructure with a 2.5 nm spacer were 1.1×10^{12} cm^{-2} and 6,400 cm^2/V · s, respectively, and those at 77 K were 9.3×10^{11} cm^{-2} and 46,000 cm^2/V · s, respectively. Increasing the spacer thickness to 7.5 nm increased the electron mobility at 77 K to 90,000 cm^2/V · s. To check the feasibility of wafers grown by the present system, we have fabricated HEMT inverters with E-HEMTs and D-HEMTs. Standard deviations of the threshold voltages of

23 mV and 35 mV for E-HEMTs and D-HEMTs, respectively, have been obtained across a 2 in. wafer. The transconductance and the current-gain cutoff frequency were 250 mS/mm and 23 GHz, respectively, for an OMVPE-grown HEMT with a gate length of 0.8 μm. These values compare favorably with those for an MBE-grown HEMT.

4.3.2 Self-Alignment Device Fabrication Technology

Figure 4.11 is a cross-sectional view of a typical self-aligned structure of enhancement-mode (E-) and depletion-mode (D-) HEMTs forming an inverter for a DCFL circuit configuration.[32] The basic epilayer structure consists of a 600 nm undoped GaAs layer, a 30 nm $Al_{0.3}Ga_{0.7}As$ layer doped to 2×10^{18} cm^{-3} with Si, and a 70 nm GaAs top layer successively grown on a semi-insulating substrate by MBE. The low-field electron mobility was found from Hall measurements to be 7,200 cm^2/ V $\cdot$ s at 300 K and 38,000 cm^2/V $\cdot$ s at 77 K. The concentration of two-dimensional electron gas (2DEG) was 1.0×10^{12} cm^{-2} at 300 K and 8.4×10^{11} cm^{-2} at 77 K. The AlAs mole fraction tentatively selected was 0.3, although it can be expected that higher AlAs mole fractions would increase the maximum achievable concentration of 2DEG, resulting in an increase in transconductance of HEMTs. $Al_xGa_{1-x}As$ with a high AlAs mole fraction, however, exhibits inferior surface morphology and an increase in deep traps, making device fabrication difficult. A thin $Al_{0.3}Ga_{0.7}As$ layer, which prevents dry etching, is embedded in the top GaAs layer to fabricate E- and D-HEMTs in the same wafer. By adopting this new device structure, we can apply the selective dry etching of GaAs to AlGaAs to achieve precise control of the gate recessing process for E- and D-HEMTs.

Figure 4.12 indicates the process sequence for the self-aligned gate process in the fabrication of HEMT LSIs, including E- and D-HEMTs forming an inverter for a DCFL circuit configuration. First of all, the active region is isolated by implanted oxygen at 130 keV with a dose of 10^{12} cm^{-2} to make planar structure. The

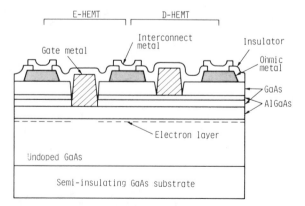

Fig. 4.11 Cross-sectional view of a typical self-aligned structure of E/D-HEMTs forming an inverter for DCFL circuit configuration.

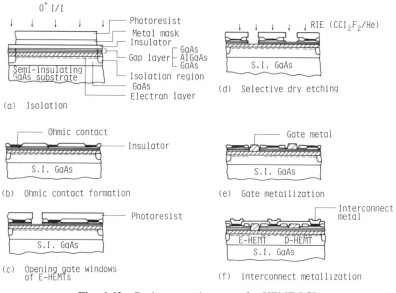

Fig. 4.12   Basic processing steps for HEMT LSI.

source and drain for E- and D-HEMTs are metallized with AuGe eutectic alloy and a Au overlay alloying to form ohmic contacts. Then, fine gate patterns are formed for E-HEMTs, and the top GaAs layer and thin $Al_{0.3}Ga_{0.7}As$ layer are etched off by nonselective chemical etching. Using the same resist after the formation of gate patterns for D-HEMTs, selective dry etching is performed to remove the top GaAs layer for D-HEMTs and also to remove the GaAs layer under the thin $Al_{0.3}Ga_{0.7}As$ layer for E-HEMTs. Next, Schottky contacts for the E- and D-HEMT gates are provided by depositing Al, the Schottky gate contacts and GaAs top layer for ohmic contact being self-aligned to achieve high-speed performance. Finally, Ti/Pt/Au electrical connections running from the interconnecting metal to the device terminals are provided through contact holes etched in a crossover SiON insulator film deposited by plasma-enhanced CVD.

As described above, a unique epistructure in combination with self-terminating selective dry recess etching makes it possible to fabricate superuniform E- and D-HEMTs simultaneously, reflecting the uniformity of MBE-grown epitaxial film. The key technique in achieving stable fabrication of self-aligned gate HEMTs is the selective dry etching of the GaAs/AlGaAs layer. Etching characteristics in CCl_2F_2 + He discharges achieved a high selectivity ratio of more than 260, where the etching rate of $Al_{0.3}Ga_{0.7}As$ is as low as 2 nm/min and that of GaAs is about 520 nm/min. The histograms of threshold uniformities for E- and D-HEMTs are shown in Fig. 4.13. The standard deviations in threshold voltages are 6 mV for E-HEMTs and 11 mV for D-HEMTs, respectively, for an LSI chip size of 10 × 10 mm area. The ratio of the standard deviation (6 mV) of threshold voltage to the logic voltage swing (0.8 V for DCFL) is 0.7%, indicating excellent control-

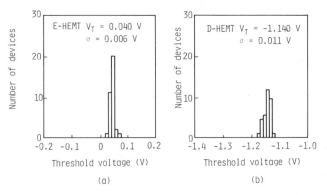

Fig. 4.13 Histograms of threshold voltages for (*a*) E-HEMTs and (*b*) D-HEMTs for an LSI chip size of 10 × 10 mm² area.

lability of MBE growth and the LSI fabrication process. This strongly supports the viability of these technologies for realizing the ICs with LSI/VLSI-level complexities. The vertical threshold sensitivity is calculated to 70 mV/nm[32] at a V_T of 0.13 V at a carrier concentration of 2×10^{18} cm^{-3}. As shown in Fig. 4.13, the deviation in threshold voltage over the wafer for the E-HEMT is 50 mV at a V_T of 0.04 V. This corresponds to a thickness deviation of only 0.7 nm in the LSI chip size, indicating excellent controllability of MBE growth and the device fabrication process. Figure 4.14 is a cross-sectional SEM micrograph of a HEMT with 0.5 μm gate length in the circuit.

4.4 HEMT LSI CIRCUIT IMPLEMENTATIONS

The current implementations and recent advances of HEMT logic and memory LSI circuits are reviewed and discussed in this section.

4.4.1 Logic Circuits

For logic circuits, a HEMT 4.1-kgate gate array with E/D-type DCFL circuits is designed and fabricated.

Figure 4.15 is a microphotograph of a 4.1-kgate gate array.[16] This gate array consists of 156 I/O cells and 4096 basic cells. The basic cell includes one depletion-type (D-) HEMT and 3 enhancement-type (E-) HEMTs with a gate length of 0.8 μm. It can be programmed as a 3-input NOR gate. The cell size is 37.5 × 45 μm. The basic cell array consists of 32 columns with 128 cells each. Between the columns are 15 interconnection tracks, each line being 2 μm wide with 2 μm spacing. The chip of this gate array has 100 pads, including 72 for I/O signals and 28 for power supply. To obtain sufficient noise margin, this chip is designed to minimize the V_{DD} voltage drop and GND voltage rise by careful arrangement of the power supply pads. Therefore, the chip has a relatively large

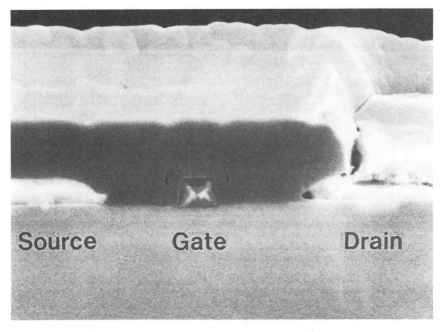

Fig. 4.14 Cross-sectional SEM microphotograph of a HEMT with 0.5 μm gate length in the circuit.

number of power supply pads. The chip measures 6.3 × 4.8 mm and contains 17,692 devices.

The dynamic performance of the basic gate is measured with several ring oscillators arranged on the test element groups. The top data of the delay time of the inverter without load is 22 ps. The average value of the delay time of the inverter and the basic gate is 27 ps and 40 ps, respectively. The difference between 27 and 40 ps results from crossover capacitance between gate electrode and power supply lines. The basic gates are covered with power supply lines in order to make the array more compact. The dependence of the delay time on fan-out and the interconnection lines are 22 ps/fan-out and 12 ps/mm, respectively, at 300 K. The delay time and the power dissipation per gate are 95 ps and 1.6 mW, respectively, at 300 K, under a loaded condition of FI = FO = 3 and interconnecting line length of 1 mm.

A 4.1-kgate gate array uses a 16 × 16 bit parallel multiplier as a test vehicle. A 16 × 16 bit multiplier constructed of 93% of this array consists of registers for a 16 × 16 bit multiplier products, 15 half-adders, 210 full-adders, and a carry look-ahead circuit. Figure 4.16 is the logic diagram of the multiplier. The critical path is that from Y_0 to S_{30}, which consists of 49 stages, including a 5-stage I/O buffer. To evaluate the performance of the multiplier, the multiplication time was measured. When the multiplication X (= 1111111111111111) × Y (= 100000000000000Y_0) is performed, the output signal of each bit is inverted by

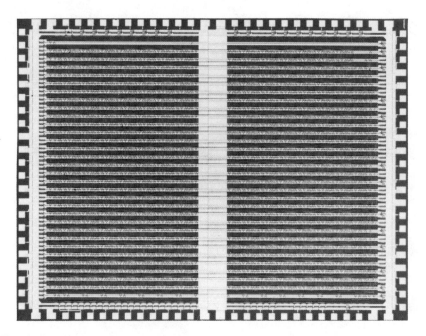

Fig. 4.15 Microphotograph of HEMT 4.1-kgate gate array, which measures 4.8 × 6.3 mm^2 and contains 17,692 E/D-HEMTs.

inputting a low level (0) and high level (1) to Y_0 alternatively. Figure 4.17 shows the Y_0 input and S_{30} output waveforms. A multiplication time of 4.1 ns at 300 K, including a 5-stage I/O buffer delay, was achieved, where the supply voltage V_{DD} was 1.1 V and total chip power dissipation was 6.2 W. This is the fastest multipli-

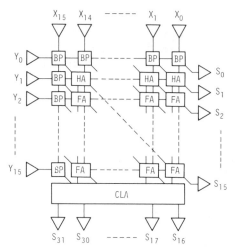

Fig. 4.16 Logic diagram of a 16 × 16 bit parallel mutliplier.

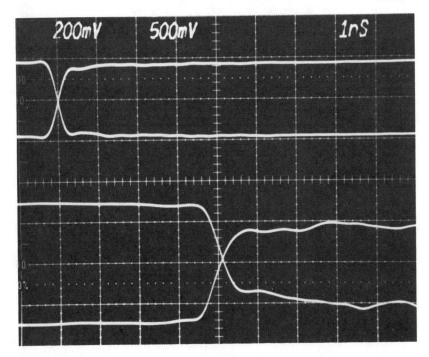

Fig. 4.17 Oscillograph for multiplication operation. The upper signal shows input to Y_0 and the lower signal, output waveforms from S_{30}. The horizontal scale is 1 ns/div.

cation time ever reported for a 16×16 bit parallel multiplier. From a simulation using the SPICE II program, we confirmed that the multiplication time was about 49 times the typical gate delay of 80 ps, with a loading of 2.6 fan-outs and a 363 μm interconnection line. This simulation gave an I/O buffer delay of about 8% of the multiplication time. So, the intrinsic multiplication time was found to be 3.8 ns. As a comparison between state-of-the-art GaAs MESFET[33,34] and HEMT multipliers,[16,35-38] circuit gate delay/power performance is shown in Fig. 4.18 as a function of gate power dissipation.

Recently, a multibit data register circuit using HEMTs with 0.5 μm gate length has been developed.[39] The register is a circuit synchronizing data signals for data transfer in a system with high clock rate. The clock pulse is applied to each latch through the clock chopper, and then 4×9 bit latched input data is transferred to the output ports, synchronized by the clock signal. The input and output buffers were designed to provide signal levels compatible with the ECL interface. The chip contains 1137 gates and measures 6.1×6.2 mm. This chip was designed to minimize the difference in the propagation delay from the clock input to each output. The width and spacing of the interconnecting lines were both 2 μm. The logic circuit is 2.4×2.4 mm and includes 3335 HEMTs. Speed of the multibit data register was measured by the delay time from clock input to data output at room temperature using a coaxial probe card. The delay time was 490 ps at room

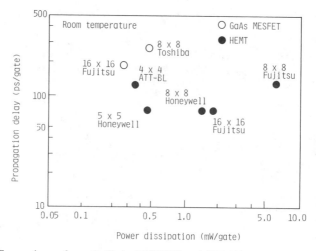

Fig. 4.18 Comparison of recent GaAs MESFET and HEMT multiplier gate propagation delays as a function of gate power dissipation.

temperature and the power dissipation was 4.12 W. The gate delay through the path from clock input to data output was estimated to be 43 ps/gate.

Performance of HEMT VLSI for future high-speed computers is projected and discussed based on the results with the HEMT performance described above.[2,11,40] Chip delay time is the sum of intrinsic gate delay, logic layout delay on fan-out capability, and delay in the wiring on the chip. Chip delays are calculated based on experimental data for HEMTs with a 0.5 μm gate length at 300 and 77 K, respectively. Here, we assume that fan-out is 3, logic swing is 0.8 V, wiring capacitance is 100 fF/mm, average line length is 1 mm in the chip, and heat flux for liquid cooling is 20 W/cm^2. Figure 4.19 shows chip delay as a function of complexity under 0.5 μm design rule HEMT technology. At 10^4 gates, the chip delays are 70 ps at 300 K and 40 ps at 77 K. This sub-100 ps performance is sufficient for future high-speed computer requirements.

4.4.2 Memory Circuits

A 1 kword × 4 bit static RAM using 0.5 μm gate HEMT technology has been designed and fabricated.[14] Figure 4.20 shows a mircophotograph of the RAM. The layout design rule of 1.5 μm line/2.0 μm space was used for interconnections. The minimum via hole is 2 × 2 μm. The memory cell is 24.5 × 23 μm, a very small size for a GaAs memory LSI. The chip measures 2.8 × 3.0 mm.

The block diagram of a 1 kword × 4 bit static RAM is shown in Fig. 4.21. The memory cell array is divided into four 1-kbit memory planes, each with 32 rows and 32 columns. This configuration has a reduced interconnections and access time. The RAM has ECL-compatible I/O interface circuits, and the supply voltages are −2 V and −3.6 V. The E/D-type DCFL was used for the basic logic gate and the memory cell in the RAM circuit design, and source follower buffers

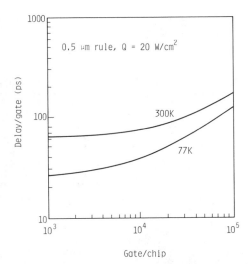

Fig. 4.19 Chip delay calculated as a function of LSI complexity under a 0.5 μm design rule HEMT technology.

were also used for the driver circuits and level shifters. In this RAM circuit, the data line equalization technique was adopted for reduction of the address access time. The address transition is detected by the address transition detector (ATD), and an ATD pulse is generated and used for the data line equalizer (DLE), as shown in Fig. 4.21. In the circuit simulation, an E-HEMT threshold voltage of 0.2 V and a D-HEMT threshold voltage of -0.6 V were used, and the propagation delay time of the basic DCFL gate (FI = FO = 1) was 22 ps. Using this DLE technique, the address access time was reduced from 0.68 ns to 0.54 ns. The design value of the chip power dissipation was 5 W. This power dissipation was rather large, but can be reduced to less than 2 W by using another supply voltage of -1 V.

Performance of the half-micron HEMT DCFL gates was measured by using different types of ring oscillators. The basic propagation delay time t_{pdo} was 22.5 ps, with a power dissipation of 3.9 mW/gate and a supply voltage V_{DD} of 2 V. The standard deviation of t_{pdo} was 1.0 ps in an area of 30 × 30 mm, and the noise margins of the basic inverters were 220 mV (N_{ML}) and 280 mV (N_{MH}). The dependences of t_{pd} is shown in Fig. 4.22, as a function of the supply voltage. The loaded delay time (FI = FO = 3, 1 = 1 mm) is 84 ps. The measurements of the RAM were done on a wafer at room temperature using coaxial probe cards. Figure 4.23 shows the oscilloscope photograph of the basic read/write address-access operation of the RAM. An address access time of 0.5 ns was achieved at room temperature, with a chip power dissipation of 5.7 W. The chip select access time was 0.25 ns. Figure 4.24 shows an oscilloscope photograph of the superimposed 32 address signals and 32 outputs (Dout1). The variation of the address access time in the 1 kbit memory plane was about 0.15 ns.

Fig. 4.20 Microphotograph of HEMT 1 kword × 4 bit static RAM, which measures 2.8 × 3.0 mm^2 and contains 29,994 E/D-HEMTs.

A HEMT 16 kword × 1 bit fully decoded static RAM has been successfully developed with the E/D-type DCFL circuit configuration.[15] Using a D-HEMT for load devices, E/D-type DCFL circuits were employed as the basic circuit. The memory cell is a six-transistor, cross-coupled flip–flop circuit with switching devices having gate lengths of 1.2 μm. As a result of the RAM layout design, the chip size is 4.3 × 5.5 mm and the RAM cell size is 23 × 30 μm (690 μm^2). The RAM has a total device count of 107,519. Figure 4.25 is a microphotograph of the 16 kword × 1 bit static RAM. Dynamic performance such as the address access time of the HEMT 16 kbit static RAM was evaluated both at room temperature and liquid nitrogen temperature. The minimum address access time obtained was 3.4 ns, with a chip dissipation power of 1.34 W at 77 K. The GaAs MESFET

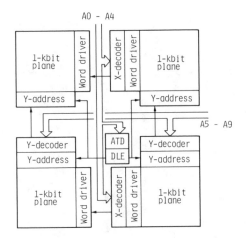

Fig. 4.21   Bock diagram of HEMT 1 kword × 4 bit static RAM.

4 kword × 4 bit static RAM with 1 μm gate length devices has an address access time of 4.1 ns, with a chip dissipation power of 2.52 W at 300 K.[41] Figure 4.26 shows the address access time and power dissipation of the static RAM, compared to Si MOS, bipolar, and GaAs MESFET static RAMs. The plots show the performances of the HEMT 1 kbit, 4 kbit, and 16 kbit static RAMs.[13-15,42,43] The shadowed area shows the projected performances assuming device technologies between 1 and 0.5 μm. By using 0.5 μm gate device technology, subnanosecond address access time can be projected for the 64 kbit static RAM.

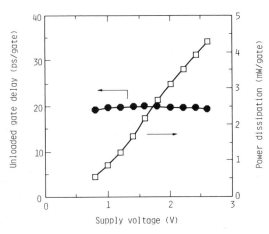

Fig. 4.22   Basic propagation delay time and power dissipation with unloaded condition as a function of supply voltage.

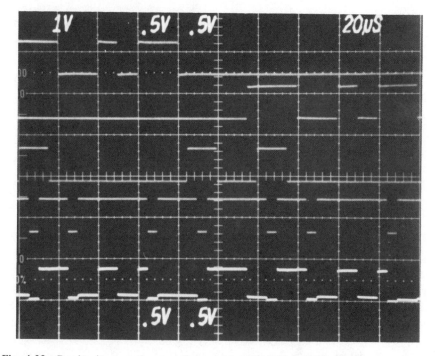

Fig. 4.23 Read/write operation waveforms of the HEMT RAM. The top signal shows the X address input, the second, Y address input, the third, data input, the fourth, write/enable, and the bottom, data output.

4.5 SUMMARY

The current status and recent advances in HEMT technology for high-performance VLSI were reviewed with a focus on material, self-alignment device fabrication, and HEMT LSI implementations.

HEMTs are very promising devices for VLSI because of their ultrahigh speed and low-power dissipation. The projected HEMT performance target suitable for VLSI is a fundamental switching delay below 10 ps. By evaluating the gate length dependence of threshold voltage and the K factor of short-channel HEMTs, short-channel effects were found not to be a problem in microstructures scaled down in the submicron dimensional range.

Since the first HEMT integrated circuit was developed in 1981, device, circuit design, processing, and material technologies have progressed and continue to grow rapidly. The evolution of HEMT IC complexity is fourfold each year,[32] and HEMT has already caught up with GaAs MESFET integration. As the HEMT shifts from the phase of research and development toward that of industry, new development for the material technology or breakthroughs of an alternative growth technique will have to be focused to match not only the LSI requirements of

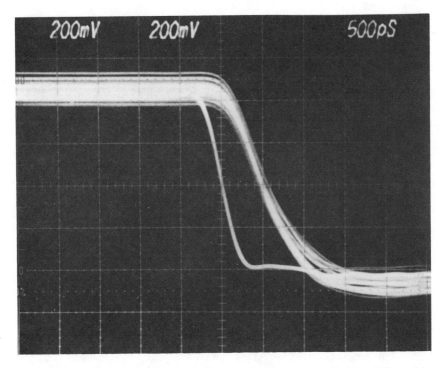

Fig. 4.24 Oscillograph for memory address access operations, showing 500 ps address access time and superimposed signals of address access in 1 kbit memory plane. The horizontal scale is 500 ps/div.

highly uniform and perfect epitaxial materials, but also the wafer supply and production with high throughput and large size.

A HEMT 4.1-kgate gate array with a 16 × 16 bit parallel multiplier has been developed to achieve a multiplication time of 4.1 ns. A HEMT 4 kbit static RAM with an address access time of 500 ps and a 16 kbit static RAM with an address access time of 3.4 ns have been developed to demonstrate the feasibility of high-performance VLSIs. With the 0.5 μm gate device technology, a HEMT 64 kbit static RAM should achieve subnanosecond access operations. Using the experimental data on HEMT logic, we projected an optimized chip delay of 40 ps at 10 kgate integration with HEMT VLSI at liquid nitrogen temperature. This performance will achieve speeds required for future large-scale computers. Based on the results described above, it is a certainty that HEMT technology will grow into one of the most important semiconductor technologies in the twentieth century.

ACKNOWLEDGMENTS

The authors wish to thank Drs. T. Misugi, M. Kobayashi, and M. Fukuta for their encouragement and support. The authors also wish to thank their colleagues, whose many contri-

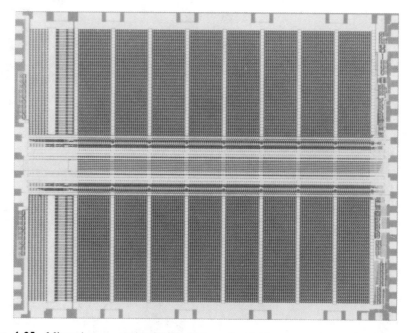

Fig. 4.25 Microphotograph of HEMT 16 kword × 1 bit static RAM, which measures 4.3 × 5.5 m² and contains 107,519 E/D-HEMTs.

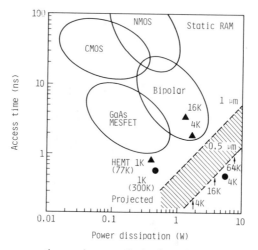

Fig. 4.26 Address access time and power dissipation of the static RAM, compared with Si MOS, bipolar, and GaAs MESFET static RAMs. The symbols ● and ▲ denote HEMT at 300 K and 77 K, respectively.

butions have made possible the results described here. The present research effort is part of the National Research and Development Program on the "Scientific Computing System," conducted under a program set by the Agency of Industrial Science and Technology, Ministry of International Trade and Industry, Japan.

REFERENCES

1. T. Mimura, S. Hiyamizu, T. Fujii, and K. Nanbu, "A New Field-Effect Transistor with Selectively Doped GaAs/n-Al$_x$Ga$_{1-x}$As Heterojunctions," *Jpn. J. Appl. Phys.*, **19**, L225 (1980).

2. M. Abe, T. Mimura, N. Yokoyama, and H. Ishikawa, "New Technology Towards GaAs LSI/VLSI for Computer Applications," *IEEE Trans. Electron Devices*, **ED-29**, 1088 (1982).

3. R. Dingle, H. L. Störmer, A. C. Gossard, and W. Wiegmann, "Electron Mobilities in Modulation-Doped Semiconductor Heterojunciton Superlattice," *Appl. Phys. Lett.*, **33**, 665 (1978).

4. T. Mimura, K. Joshin, and S. Hiyamizu, K. Hikosaka, and M. Abe, "High Electron Mobility Transistor Logic," *Jpn. J. Appl. Phys.*, **20**, L598 (1981).

5. P. N. Tung, P. Delescluse, D. Delagebeaudeuf, M. Laviron, J. Chaplart, and N. T. Linh, "High Speed Low Power DCFL Using Planar Two-Dimensional Electron Gas FET Technology," *Electron. Lett.*, **18**, 517 (1982).

6. J. V. DiLorenzo, R. Dingle, M. Feuer, A. C. Gossard, R. Hendel, J. C. Hwang, A. Kastalsky, V. G. Kerasmidas, R. A. Kiehl, and P. O'Connor, "Material and Device Considerations for Selectively Doped Heterojunction Transistor," *Tech. Dig. — Int. Electron Devices Meet.*, p. 578 (1982).

7. N. C. Cirillo, Jr. and J. K. Abrokwah, "8.5-Picosecond Ring Oscillator Gate Delay with Self-Aligned Gate Modulation-Doped n^+-(Al, Ga)As/GaAs FET's" *43rd Annu. Device Res. Conf.*, Pap. IIA-7 (1985).

8. N. J. Shah, S.-S. Pei, C. W. Tu, and R. C. Tiberio, "Gate-Length Dependence of the Speed of SSI Circuits Using Submicrometer Selectively Doped Heterostructure Transistor Technology," *IEEE Trans. Electron Devices*, **ED-33**, 543 (1986).

9. Y. Awano, M. Kosugi, T. Mimura, and M. Abe, "Performance of a Quarter-Micrometer-Gate Ballistic Electron HEMT," *IEEE Electron Device Lett.*, **EDL-8**, 451 (1987).

10. K. Nishiuchi, T. Mimura, S. Kuroda, S. Hiyamizu, H. Nishi, and M. Abe, "Device Characteristics of Short Channel High-Electron Mobility Transistor (HEMT)," *41st Annu. Device Res. Conf.*, Pap. IIA-8 (1983).

11. M. Abe, T. Mimura, K. Nishiuchi, A. Shibatomi, and M. Kobayashi, "HEMT LSI Technology for High Speed Computers," *Tech. Dig. — IEEE GaAs IC Symp.*, p. 158 (1983).

12. R. H. Hendel, S. S. Pei, C. W. Tu, B. J. Roman, N. J. Shah, and R. Dingle, "Realization of Sub-10 Picosecond Switching Times in Selectively Doped (Al,Ga)As/GaAs Heterostructure Transistors," *Tech. Dig. — Int. Electron Devices Meet.*, p. 857 (1984); see also, *Electronics*, **58**, 22 (1985).

13. S. Kuroda, T. Mimura, M. Suzuki, N. Kobayashi, K. Nishiuchi, A. Shibatomi, and M. Abe, "New Device Structure for 4Kb HEMT SRAM," *Tech. Dig.—IEEE GaAs IC Symp.*, p. 125 (1984).

14. S. Notomi, Y. Awano, M. Kosugi, T. Nagata, K. Kosemura, M. Ono, N. Kobayashi, H. Ishiwari, K. Odani, T. Mimura, and M. Abe, "A High Speed 1K × 4 Bit Static RAM Using 0.5 μm-Gate HEMT," *Tech. Dig.—IEEE GaAs IC Symp.*, p. 177 (1987).

15. M. Abe, T. Mimura, S. Notomi, K. Odani, K. Kondo, and M. Kobayashi, "Ultrahigh Speed High Electron Mobility Transistor Large Scale Integration Technology," *J. Vac. Sci. Technol.*, A [2], **5**, 1387 (1987).

16. K. Kajii, Y. Watanabe, M. Suzuki, I. Hanyu, M. Kosugi, K. Odani, T. Mimura, and M. Abe, "A 40 ps High Electron Mobility Transistor 4.1 K Gate Array," *Custom Integrated Circuits Conf. Dig. Tech. Pap.*, p. 199 (1987).

17. T. Mimura, S. Hiyamizu, K. Joshin, and K. Hikosaka, "Enhancement-Mode High Electron Mobility Transistors for Logic Applications," *Jpn. J. Appl. Phys.*, **20**, L317 (1981).

18. S. Hiyamizu, "Recent Developments in MBE GaAs/n-AlGaAs Heterostructures and HEMTs," *Collect. Pap. Int. Symp. Mol. Beam Epitaxy Relat. Clean Surf. Tech., 2nd, 1982*, Pap. A-7-1, p. 113 (1982).

19. D. V. Lang, R. A. Logan, and M. Jaros, "Trapping Characteristics and a Donor-Complex (DX) Model for the Persistent-Photoconductivity Trapping Center in Te-Doped $Al_xGa_{1-x}As$," *Phys. Rev. (B): Condens. Matter* [3] **19**, 1015 (1979).

20. C. P. Lee, D. Hou, S. J. Lee, D. L. Miller, and R. J. Anderson, "Ultrahigh Speed Digital Integrated Circuits Using GaAs/GaAlAs High Electron Mobility Transistor," *Tech. Dig.—IEEE GaAs IC Symp.*, p. 162 (1983).

21. K. Joshin, T. Mimura, M. Niori, M. Yamashita, K. Kosemura, and J. Saito, "Noise Performance of Microwave HEMT," *IEEE MTT-S Int. Microwave Symp.* Dig. p. 563 (1983).

22. L. H. Caminitz, P. J. Tasker, H. Lee, D. V. D. Merwe, and L. F. Eastman, "Microwave Characterization of Very High Transconductance MODFET," *Tech. Dig.—Int. Electron Devices Meet.*, p. 360 (1984).

23. P. C. Chao, S. C. Palmateer, P. M. Smith, U. K. Mishra, K. H. G. Duh, and J. C. M. Hwang, "Millimeter-Wave Low-Noise High Electron Mobility Transistors," *IEEE Electron Device Lett.*, **EDL-6**, 531 (1985).

24. M. Feng, H. Kanber, V. K. Eu, E. Watkins, and L. R. Hackett, "Ultrahigh Frequency Operation of Ion-Implanted GaAs Metal-Semiconductor Field-Effect Transistors," *Appl. Phys. Lett.*, **44**, 231 (1984).

25. W. Chye and C. Huang, "Quartermicron Low Noise GaAs FET's," *IEEE Electron Device Lett.*, **EDL-3**, 401 (1982).

26. S. Hiyamizu, T. Mimura, and T. Ishikawa, "MBE-Grown GaAs/n-AlGaAs Heterostructures and Their Application to High Electron Mobility Transistors," *Jpn. J. Appl. Phys.*, **21**, Suppl. 21-1, 161 (1982).

27. M. Abe, T. Mimura, K. Nishiuchi, A. Shibatomi, and M. Kobayashi, "Recent Advances in Ultrahigh Speed HEMT Technology," *IEEE J. Quantum Electron.*, **QE-22**, 1870 (1986).

28. T. Mimura, M. Abe, and M. Kobayashi, "High Electron Mobility Transistors," *Fujitsu Sci. Tech. J.*, **21**, 370 (1985).

29. J. Saito, T. Igarashi, T. Nakamura, K. Kondo, and A. Shibatomi, "Growth of Highly Uniform Epitaxial Layers over Multiple Substrates by Molecular Beam Epitaxy," *J. Cryst. Growth,* **81,** 188 (1987).

30. H. Tanaka, H. Itoh, T. O'hori, M. Takikawa, K. Kasai, M. Takechi, M. Suzuki, and J. Komeno, "Multi-Wafer Growth of HEMT LSI Quality AlGaAs/GaAs Heterostructure by MOCVD," *Jpn. J. Appl. Phys.,* **26,** L1456 (1987).

31. E. J. Thrush, G. Wale-Evans, J. E. A. Whiteaway, and B. L. Lamb, D. R. Wight, N. G. Chew, A. G. Cullis, and R. J. M. Griffiths, "Evidence for Transient Composition Variations at GaAs/Ga$_{1-x}$Al$_x$As Heterostructure Interfaces Prepared by Metal-Organic Chemical Vapour Deposition," *J. Electron. Mater.,* **13,** 969 (1984).

32. M. Abe, T. Mimura, K. Nishiuchi, A. Shibatomi, M. Kobayashi, and T. Misugi, "Ultrahigh Speed HEMT Integrated Circuits," in R. Dingle, Ed., *Semiconductors and Semimetals,* Vol. 24, p. 249, Academic, New York, 1987.

33. Y. Nakayama, K. Suyama, H. Shimizu, N. Yokoyama, A. Shibatomi, and H. Ishikawa, "A GaAs 16 × 16b Parallel Multiplier Using Self-Alignment Technology," *Int. Solid-State Circuit Conf., Dig. Tech. Pap.,* p. 48 (1983).

34. N. Toyoda, N. Uchitomi, Y. Kitaura, M. Mochizuki, K. Kanazawa, T. Terada, Y. Ikawa, and A. Hojo, "A 42ps 2K-Gate GaAs Gate Array," *Int. Solid-State Circuit Conf., Dig. Tech. Pap.,* p. 206 (1985).

35. A. R. Schlier, S. S. Pei, N. J. Shah, C. W. Tu, and G. E. Mahoney, "A High Speed 4 × 4 Bit Parallel Multiplier Using Selectively Doped Heterostructures," *Tech. Dig.— IEEE GaAs IC Symp.,* p. 91 (1985).

36. D. K. Arch, B. K. Betz, P. J. Vold, J. K. Abrokwah, and N. C. Cirillo, Jr., "A Self-Aligned Gate Superlattice (Al, Ga)As/n^+-GaAs MODFET 5 × 5-Bit Parallel Multiplier," *IEEE Electron Device Lett.,* **EDL-7,** 700 (1986).

37. Y. Watanabe, K. Kajii, K. Nishiuchi, M. Suzuki, I. Hanyu, M. Kosugi, K. Odani, A. Shibatomi, T. Mimura, M. Abe, and M. Kobayashi, "High Electron Mobility Transistor 1.5K Gate Array," *Int. Solid-State Circuit Conf., Dig. Tech. Pap. 29,* p. 80 (1986).

38. N. C. Cirillo, Jr., D. K. Arch, P. J. Vold, B. K. Betz, I. R. Mactaggart, and B. L. Grung, "8 × 8-Bit Pipelined Parallel Multiplier Utilizing Self-Aligned Gate n^+-(Al, Ga)As/GaAs MODFET IC Technology," *Tech. Dig.—IEEE GaAs IC Symp.,* p. 257 (1987).

39. Y. Watanabe, S. Saito, N. Kobayashi, M. Suzuki, T. Yokoyama, E. Mitani, K. Odani, T. Mimura, and M. Abe, "A HEMT LSI for a Multibit Data Register," *Int. Solid-State Circuit-Conf., Dig. Tech. Pap.,* p. 86 (1988).

40. M. Abe, T. Mimura, K. Nishiuchi, and N. Yokoyama, "GaAs VLSI Technology for High-Speed Computers," in N. G. Einspruch and W. R. Wisseman, Eds., *VLSI Electronics: Microstructure Science,* Vol. 11, p. 333, Academic, New York, 1985.

41. Y. Ishii, M. Ino, M. Idda, M. Hirayama, and M. Ohmori, "Processing Technologies for GaAs Memory LSIs," *Tech. Dig.—IEEE GaAs IC Symp.,* p. 121 (1984).

42. K. Nishiuchi, N. Kobayashi, S. Kuroda, S. Notomi, T. Mimura, M. Abe, and M. Kobayashi, "A Subnanosecond HEMT 1Kb SRAM," *Int. Solid-State Circuit Conf., Dig. Tech. Pap. 27,* p. 48 (1984).

43. N. H. Sheng, H. T. Wang, S. J. Lee, G. L. Sullivan, and D. L. Miller, "A High-Speed 1K-Bit High Electron Mobility Transistor Static RAM," *Tech. Dig.—IEEE GaAs IC Symp.,* p. 97 (1986).

5 Resonant Tunneling Devices and Their Applications

FEDERICO CAPASSO, SUSANTA SEN,[†] FABIO BELTRAM,
and ALFRED Y. CHO
AT&T Bell Laboratories
Murray Hill, New Jersey

5.1 INTRODUCTION

Resonant tunneling (RT) through double barrier quantum well (DBQW) structures was first demonstrated by Chang, Esaki, and Tsu in 1974.[1] Since then, the material quality has improved to the point that negative differential resistance (NDR) can be observed at room temperature.[2] Recently, peak-to-valley ratios in current of nearly 4:1 were reported at room temperature.[3,4] At 77 K, the observed value of peak-to-valley is as high as 15:1.[4] Further improvement in the peak-to-valley ratio (14:1 at room temperature) was obtained by the use of pseudomorphic structures.[5] Mendez et al.[6] have also observed RT of holes. Recently, Reed et al.[7] showed that the replacement of AlGaAs in the barriers with an AlAs/GaAs superlattice with the same average composition considerably improves the current–voltage (I–V) characteristics of RT diodes by making it symmetric. Nakagawa et al.[8,9] have also reported RT of electrons and holes in triple barrier diodes. The RT of electrons from two-dimensional (2D) states in the valence band to 2D states in the conduction band (resonant Zener tunneling) was investigated by Allam et al.[10] Sen et al.[11] have recently observed RT through parabolic quantum wells. As many as 14 resonances were observed in the I–V of one such sample. Unlike rectangular quantum wells, the multiple resonances of the parabolic well were nearly equally spaced.

The first practical application of RT diodes was made by Sollner et al.[12] at the MIT Lincoln Laboratories in 1983. They used the NDR of RT diodes in detectors and mixers at frequencies up to the terahertz range. With improvement of technology, the parasitic series resistance of RT diodes reduced and they could be used for microwave generation[13] as well. Recently, Brown et al. have also reported oscillation frequencies up to 200 GHz[14] using RT diodes.

[†]On leave from the Institute of Radio Physics and Electronics, University of Calcutta, Calcutta 700 009, India.

Apart from these early applications, RT structures have assumed great importance in recent years as functional devices for circuit applications.[15] A variety of three terminal devices (both unipolar and bipolar) have been proposed and implemented.[16-30] The RT approach to circuits is one of the several ones proposed to circumvent the limits imposed by scaling laws to the ever-increasing functional density.[31] In fact, Sen et al.[32] have shown that through the use of RT devices many circuits can be implemented with less devices per function. In particular, RT devices with multiple negative resistance regions are of considerable interest for a variety of potential applications which could be realized with greatly reduced circuit complexity.

This chapter deals with the operation of some of these interesting RT devices and their circuit applications.

5.2 RESONANT TUNNELING DIODES

In this section, we first discuss the physics of RT through double barrier quantum wells followed by some state-of-the-art results on RT diodes. Finally, the last part of the section is devoted to the discussion of some novel applications involving integration of RT diodes.

5.2.1 Physics of Resonant Tunneling

5.2.1.1 The Origin of Negative Differential Resistance. RT through a double barrier occurs when the energy of an incident electron in the emitter matches that of an unoccupied state in the quantum well (QW) corresponding to the same lateral momentum. NDR arises simply from momentum and energy conservation considerations and does not require the coherence of the electron wave function (as is the case for the electronic analogous of the optical Fabry–Perot effect). In fact, NDR is a manifestation of the reduction in the dimensionality of the electronic states; in the present case from the 3D emitter to the 2D quantum well, but 2D to 1D was also considered.[20] This has been clarified by Luryi[33] and is illustrated in Fig. 5.1.

Consider the Fermi sea of electrons in the degenerately doped emitter. Their energy can be expressed as

$$E_{3D} = E_C + \frac{\hbar^2 k_z^2}{2m^*} + \frac{\hbar^2 k_\perp^2}{2m^*}, \tag{1}$$

where E_C is the bottom of the conduction band and $k_\perp^2 = k_x^2 + k_y^2$. In doing so, we are ignoring the quantization at the interface of the injecting electrode. This would simply add further structure to the peaks in the I–V without altering the main finding, i.e., the NDR. The energy in the 2D target states in the QW, on the other hand, is given by

$$E_{2D} = E_n + \frac{\hbar^2 k_\perp^2}{2m^*}, \tag{2}$$

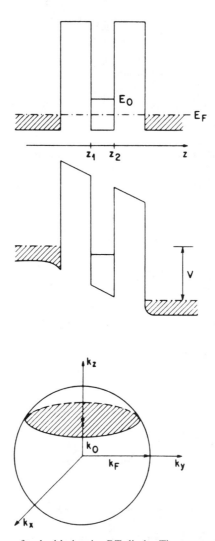

Fig. 5.1 The operation of a double barrier RT diode. The top part shows the electron energy diagram in equilibrium. The middle displays the band diagram for an applied bias V, when the energy of certain electrons in the emitter matches unoccupied levels of the lowest sub-band E_0 in the QW. The bottom illustrates the Fermi surface for a degenerately doped emitter. Assuming conservation of the lateral momentum during tunneling, only those emitter electrons whose momenta lie on a disk $k_z = k_0$ (shaded disk) are resonant. The energy separation between E_0 and the bottom of the conduction band in the emitter is given by $\hbar^2 k_0^2 / 2m^*$. In an ideal double barrier diode at zero temperature, RT occurs in a voltage range, during which the shaded disk moves down from the pole to the equatorial plane of the emitter Fermi sphere. At higher V (when $k_0^2 < 0$), resonant electrons no longer exist.

where E_n is the bottom of the relevant subband in the QW. The RT of electrons into the 2D states requires the conservation of energy and of lateral momentum ($k_\perp$). It is, of course, assumed that the barriers are free from impurities and inhomogeneities. The momentum conservation condition requires that the last term in (1) and (2) be equal. So from energy conservation, we find that tunneling is possible only for electrons whose momenta lie in a disk corresponding to $k_z = k_0$ (shaded disk in the figure in which the case $n = 0$ is illustrated) where

$$k_0^2 = \frac{2m^*}{\hbar^2}(E_0 - E_C).$$ (3)

Only those electrons have isoenergetic states in the quantum well with the same $k_\perp$. This is a general feature of tunneling into a 2D system of states. As the emitter–base potential rises, so does the number of electrons that can tunnel; the shaded disk moves downward to the equatorial plane of the Fermi sphere. For $E_n = E_C$, which corresponds to $k_0 = 0$, the number of tunneling electrons per unit area equals $m^* E_F / \pi \hbar^2$. When E_C rises above E_n, then, at $T = 0$ temperature, there are no electrons in the emitter that can tunnel into the quantum well while conserving their lateral momentum. Therefore, one can expect an abrupt drop in the tunneling current. This has been experimentally observed by Morkoç et al.[34] The same effect was also utilized by Beltram et al.[30] in the operation of the gated quantum well RT transistor presented in Section 5.4.4. Of course, similar arguments of conservation of lateral momentum and energy leading to NDR apply also to systems of lower dimensionality, for example, to tunneling of 2D electrons through a quantum wire and to RT in one dimension.

5.2.1.2 Coherent (Fabry–Perot-Type) Resonant Tunneling.

Let us now consider the Fabry–Perot effect. In the presence of negligible scattering of the electrons in the well, the above NDR effect is accompanied by a coherent enhancement of the transmission analogous to that occurring in an optical Fabry–Perot. In the case of a symmetric structure (such as the one discussed in Section 5.3), after a transient in which the electron wave function builds up in the quantum well, an equilibrium is reached where the portion of the incident wave reflected by the first barrier is exactly canceled by the fraction of the electron wave function leaking from the QW to the left. The net effect is then a total transfer of electrons from the left to the right through the double barrier. This is shown in Fig. 5.2 by the unity transmissivity peaks. In this case of coherent RT, the peak transmission at resonance is given approximately by T_{min}/T_{max}, where T_{min} is the smallest among the transmission coefficients of the two barriers and T_{max} is the largest.[35] It is therefore possible to achieve unity transmission at the resonance peaks as discussed above by making the transmission of the left and right barriers equal. This crucial role of symmetry has been discussed in detail by Ricco and Azbel.[35] Application of an electric field to a symmetric double barrier introduces a difference between the transmissions of the two barriers, thus significantly decreasing below unity the overall transmission at the resonance peaks. Unity transmission can be restored if

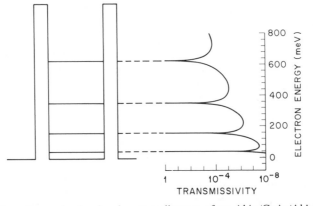

Fig. 5.2 Schematic conduction band energy diagram of an AlAs/GaAs/AlAs RT double barrier with 25 Å barriers and 100 Å well (left) and the calculated transmissivity as a function of incident electron energy (right).

the two barriers have different and appropriately chosen thicknesses; obviously, with this procedure, one can only optimize the transmission of one of the resonance peaks. However, as recently elaborated by Luryi,[36] these arguments do not strictly apply to the DB under the large biases required for RT operation due to the broad energy distribution of the incoming electrons. Under this condition it can be shown that the peak current is proportional to T_{min}.[36] This is because the exit barrier becomes lower and therefore has a higher transmission, under application of an electric field, than the input barrier. Unity transmission can be restored by making the exit barrier thicker.

5.2.1.3 The Role of Scattering: Incoherent (Sequential) Resonant Tunneling.
RT through a double barrier has been investigated experimentally by many researchers.[1–7,11–14,37–42] These investigations assumed that a Fabry–Perot-type enhancement of the transmission was operational in such structures. However, as previously discussed, the observation of NDR does not imply a Fabry–Perot mechanism. Other types of tests are necessary to show the presence of the coherence of the wave function, such as the dependence of the peak current on the thickness of the exit barrier.

The presence of scattering gives rise to another physical picture for RT. Once the electrons are injected into the QW, as discussed in Section 5.2.1.1, scattering events can randomize the phase of the electronic wave function; this considerably weakens the coherent enhancement of the transmission. RT is then a two-step process in which the electrons first tunnel into the well and then out of it through the second barrier. *The first step is the one that gives rise to NDR.* A lucid discussion of this point has recently been given by Stone and Lee[43] in the context of RT through an impurity center. Unfortunately, their work has gone unnoticed among researchers in the area of quantum well structures. Their conclusions can also be applied to RT through quantum wells and we shall discuss them in this context.

To achieve the resonant enhancement of the transmission (Fabry–Perot effect), the electron probability density must be peaked in the well. The time constant for this phenomenon τ_0 is on the order of $\hbar/\Gamma_r$, where Γ_r is the full width at half maximum of the transmission peak. Collisions in the double barrier tend to destroy the coherence of the wave function, and therefore the electronic density in the well will never be able to build up to its full resonant value. If the scattering time τ is much shorter than τ_0, the peak transmission at resonance is expected to be decreased by the ratio τ_0/τ. The principal effects of collisions are to decrease the peak transmission by the ratio $\tau_0/(\tau_0 + \tau)$ and to broaden the resonance. In addition, the ratio of the number of electrons that resonantly tunnel without undergoing collisions to the number that tunnel after undergoing collisions is equal to τ/τ_0.[43] To summarize, coherent RT is observable when the intrinsic resonance width ($\approx \hbar/\tau_0$) exceeds or equals the collision broadening ($\approx \hbar/\tau$). In the other limit, electrons will always tunnel through one of the intermediate states of the well, but they will do it incoherently without resonant enhancement of the transmission. We shall now apply the above criterion to RT through the AlGaAs/GaAs double barrier investigated in many experiments.

Consider a 50 Å thick GaAs well sandwiched between two $Al_{0.30}Ga_{0.70}As$ barriers. Table 5.1 shows the ground state resonance widths Γ_r (full width at half maximum of the transmission curve) calculated for different values of the barrier thicknesses L_B (assumed equal). Note the strong dependence of Γ_r on L_B. This is due to the fact hat Γ_r is proportional to the transmission coefficient of the individual barriers, which decreases exponentially with increasing L_B. The case $L_B = 50$ Å corresponds to the microwave oscillator reported by Sollner et al.[13]

Because of dimensional confinement in the wells and because the wells are undoped, one can obtain a good estimate of the scattering time of electrons in the wells from the mobility of 2D electron gas (in the plane of the layers), measured in selectively doped AlGaAs/GaAs heterojunctions.[44] For state-of-the-art selectively doped AlGaAs/GaAs heterojunctions, the electron mobility at 300 K is ~ 7000 cm^2/V · s. From this value, we can infer an average scattering time $\cong 3 \times 10^{-13}$ s, which corresponds to a broadening of $\cong 2$ meV. In Table 5.1, the ratio of the resonance width Γ_r to the collision broadening Γ_c is also shown. For the 50 and 70 Å barrier case, the resonance width is much smaller than the collision broadening so that, by the previously discussed criterion, there is very little resonant enhancement of the transmission via the Fabry–Perot mechanism at

TABLE 5.1 Resonance and Collision Widths of $Al_{0.30}Ga_{0.70}As$/GaAs RT Diode (at Zero Bias) for Different Barrier Thicknesses

L_W (Å)	L_B (Å)	Γ_r (meV)	Γ_r/Γ_c ($\equiv$ resonance width/collision broadening)		
			300 K	200 K	70 K
50	70	1.28×10^{-2}	6×10^{-3}	1.93×10^{-2}	2.6×10^{-1}
50	50	1.5×10^{-1}	7.5×10^{-2}	2.26×10^{-1}	3.08
50	30	1.76	8.8×10^{-1}	1.32	3.62
20	50	6.03	3.02	4.56	124.02

300 K. However, the latter effect should become visible in structures with thinner barriers (<30 Å), as seen from Table 5.1. Consider now a temperature of 200 K; from the mobility ($\approx 2 \times 10^4$ cm^2/V · s),[44] one deduces $\tau \approx 1$ ps, which corresponds to a broadening of ≈ 0.67 meV. This value is comparable to the resonance width for a barrier width of 50 Å. This implies that in Sollner's microwave oscillators[13] (which operated at 200 K), coherent RT effects were probably present. This is definitely not the case for the mixing and detection experiments performed up to terahertz frequencies[12] in double barrier RT structures with $L_W = L_B = 50$ Å, $x = 0.25$–0.30 at a temperature of 25 K. In this case, the well was intentionally doped to $\approx 10^{17}$ cm^{-3}, which would correspond to a mobility of ≈ 3000 cm^{-2}/V · s, which in turn gives a collision broadening of 4 meV, significantly larger than the resonant width. Thus, in this case, electrons are tunneling incoherently (i.e., sequentially) through the double barriers.

Finally, in Table 5.1, we have estimated Γ_r/Γ_c for a temperature of 77 K. State-of-the-art mobilities in selectively doped interfaces exceed 10^5 cm^2/V · s, so that scattering times are typically longer than 1 ps and the broadenings are less than 0.5 meV. Thus, coherent RT will significantly contribute to the current for barrier widths $\lesssim 70$ Å and dominate for $L_B \lesssim 30$ Å. The values of Γ_r/Γ_c at 70 K in Table 5.1 were obtained using a mobility of 3×10^5 cm^2/V · s.[44]

The situation appears to be different in the case of AlAs/GaAs double barrier with well widths of 50 Å. The confining barriers in this case are much higher (≈ 1.35 eV),[23] and for barrier thicknesses in the 30–70 Å range, the resonance widths are $\lesssim 10^{-2}$ meV. Thus, coherent RT is negligible at room temperature, but is expected to become dominant at 70 K for $L_B \lesssim 70$ Å in high-quality double barriers.

5.2.2   Room Temperature Operation of $Ga_{0.47}In_{0.53}As/Al_{0.48}In_{0.52}As$ Resonant Tunneling Diodes

Many efforts have been directed toward achieving better performance, such as room temperature operation and improved peak-to-valley ratio. So far, RT structures composed of the GaAs/AlGaAs material system have received the major attention. The highest peak-to-valley current ratio at room temperature obtained so far in this material system is $3.9:1$.[3] Recently, Muto et al.[41] have reported a systematic study of $Ga_{0.47}In_{0.53}As/Al_{0.48}In_{0.52}As$ RT diodes at low temperature (77 K). The same material system has been used by Yokoyama et al. in their resonant tunneling hot-electron transistor (RHET)[45] to obtain improved performance at 77 K. Sen et al. have presented data on the room temperature operation[4] of a $Ga_{0.47}In_{0.53}As/Al_{0.48}In_{0.52}As$ RT-diode-grown lattice matched to InP substrate by molecular-beam epitaxy (MBE). The I–V characteristics of these diodes exhibit peak-to-valley ratios of $4:1$ at room temperature and $15:1$ at 80 K. The improved peak-to-valley ratio compared with AlGaAs/GaAs RT diodes is due to the lower electron effective mass in the barriers (0.075 m_0 for $Al_xIn_{1-x}As$ at $x = 0.48$ compared with 0.092 m_0 for $Al_xGa_{1-x}As$ at $x = 0.3$), resulting in higher tunneling current density, and to the large conduction band discontinuity which strongly reduces the thermionic emission current across the barrier.

The structure in Ref. 4 consisted of 1 μm thick n^+ ($\sim 3 \times 10^{17}$/cm^3) Ga$_{0.47}$In$_{0.53}$As buffer layer grown on an n^+ InP substrate. On top of the buffer layer is grown the RT double barrier, consisting of an undoped 50 Å wide Ga$_{0.47}$In$_{0.53}$As quantum well sandwiched between two 50 Å wide undoped Al$_{0.48}$In$_{0.52}$As barriers. The growth ends with a 1 μm thick Ga$_{0.47}$In$_{0.53}$As cap layer doped to $n^+ \sim 3 \times 10^{17}$/cm^3.

The structures were etched into 50 μm diameter mesas using 1H$_2$O$_2$ + 3H$_3$PO$_4$ + 50H$_2$O etchant at room temperature. Ge (60 Å)/Au (135 Å)/Ag (500 Å)/Au (750 Å) deposited in sequence and alloyed at 420°C for 30 s were used for 30 μm diameter ohmic contacts. For the bottom contact, Ni (50 Å)/Au (385 Å)/Ge (215 Å)/Au (750 Å) were deposited on the etched surface of the buffer layer and alloyed at 420°C for 30 s.

The samples were tested at different temperatures in a Helitran dewar equipped with microprobes. Figure 5.3 shows the I–V characteristics of the diodes in both polarities measured at room temperature and 80 K. Positive polarity refers to the top contact being positively biased with respect to the bottom. The room temperature characteristics indicate a peak-to-valley ratio of 4:1 in one polarity and 3.5:1 in the other. Figure 5.4 shows the room temperature I–V in one polarity on a blown-up scale. At low temperature (80 K), the peak-to-valley ratio increases to 15:1. It should be noted that though the peak-to-valley ratio increases dramatically upon cooling down, the peak current remains the same. The peak in the I–V occurs at $\sim$600 mV and does not change with temperature. An electron tunneling transmission calculation shows that the first resonance is at $E_1 = 126$ meV from the bottom of the quantum well. Note that the peak in the I–V appears at a voltage greater than $2 E_1/e = 252$ mV. This can be explained by considering the voltage

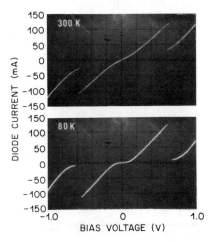

Fig. 5.3 Typical current–voltage characteristics of the Al$_{0.48}$In$_{0.52}$As/Ga$_{0.47}$In$_{0.53}$As resonant tunneling diode at 300 K (top) and 80 K (bottom). Positive polarity refers to the top contact being positively biased with respect to the bottom.

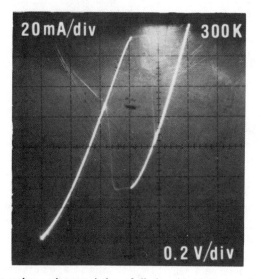

Fig. 5.4 Current–voltage characteristics of diode with highest peak-to-valley ratio obtained at room temperature.

drop in the depletion and accumulation regions in the collector and emitter layers adjacent to the double barrier. Thus, a larger voltage must be applied across the entire structure to line up the first sub-band in the well with the bottom of the conduction band in the emitter to quench RT. A simple calculation, taking the above effects into account, indicates that the peak should occur at $\simeq 580$ mV applied bias, which is in reasonable agreement with the measured value.

The large peak-to-valley ratio observed at room temperature makes this device suitable for many circuit applications. A circuit with a 30 Ω load resistance in series with the device and a 3.0 V supply has two stable operating points which are measured to be 0.47 V and 0.85 V, respectively, at room temperature. The corresponding load line drawn on the room temperature I–V characteristics indicate the stable operating points at 0.46 V and 0.84 V, respectively, which are in close agreement with the measured values. The circuit can therefore be used as a static random access memory (RAM) cell involving only one device. Such a RAM cell is also suitable for integration in a large memory array, as discussed in the next section, in connection with multistate memory.

5.2.3 Resonant Tunneling Through Parabolic Quantum Wells

Parabolic quantum wells have interesting possibilities for device applications[16] because the levels in such a well are equally spaced, unlike rectangular wells. The I–V characteristics of RT structures with parabolic wells therefore are expected to produce equally spaced peaks in voltage. The first experimental observation of such resonances was reported by Sen et al.[11] in 1987.

The samples in Ref. 11 were grown by MBE on silicon-doped (100) GaAs substrates at a substrate temperature of 680°C. The growth was computer-controlled and calibrated by ion-gauge flux measurement at the position of the substrate. Parabolically graded well compositions were produced by growth of short-period (~15 Å), variable duty cycle, GaAs/Al_xGa_{1-x}As superlattices in which the Al content within each period of the superlattice corresponded to the Al content at the same point in a smooth parabolic well.[46] A cross-sectional transmission electron micrograph (TEM) of one such structure is shown in Fig. 5.5. The structure consists of a 439 Å parabolic quantum well of Al_xGa_{1-x}As, with x varying from 0.3 at the edges to 0 at the center, sandwiched between two 35 Å AlAs barriers. The parabolic part of the structure is composed of variable gap superlattice with a period of nearly 10 Å, as discussed above. The brighter lines in the well part of the TEM picture represent $Al_{0.3}Ga_{0.7}$As layers; the darker lines represent GaAs layers. Notice how the relative widths of the bright and the dark lines change from the edges of the well to its center. The electrons, of course, "sense" the local average composition, since their de Broglie wavelength is much greater than the superlattice period.

Two types of structures were grown. In one, sample (A), the 300 Å undoped well is sandwiched between two 20 Å AlAs undoped barriers. The parabolic well composition is effectively graded from $x = 1$ at the edges to $x = 0$ at the center.

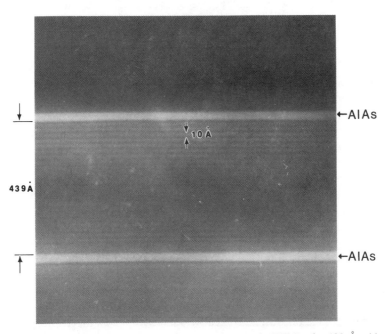

Fig. 5.5 A cross-sectional tranmission electron micrograph (TEM) of a 439 Å wide parabolic quantum well composed of Al_xGa_{1-x}As, with x varying from 0.3 at the edges to 0 at the center, sandwiched between two 35 Å AlAs barriers.

The portion of the well from $x = 0.49$ to $x = 0$ is achieved by means of an $Al_{0.50}Ga_{0.50}As/GaAs$ superlattice, whereas the rest (from $x = 0.49$ to $x = 1$) is achieved using an $AlAs/GaAs$ superlattice. Undoped 20 Å thick GaAs spacer layers were used between the barriers and the Si-doped ($n = 10^{18}/cm^3$) 5000 Å thick GaAs contact layers. Systematic studies have shown that offsetting the doping in the regions adjacent to the barriers significantly improves the I–V of RT diodes.[39]

In sample (B), the 439 Å undoped well is bound by 35 Å AlAs undoped barriers and the composition of the well is graded from $x = 0.30$ at the edges to $x = 0$ at the center using an $Al_{0.325}Ga_{0.675}As/GaAs$ superlattice. $Al_{0.02}Ga_{0.98}As$ 1000 Å thick layers Si-doped to $5 \times 10^{17}/cm^3$ (with a doping offset of 50 Å from the barriers) were used as contact regions to the RT double barrier. The composition of these layers was chosen in such a way that the bottom of the conduction band in the emitter is nearly lined up with (but always below) the first energy level of the well, a technique successfully used in improving the peak-to-valley ratio of RT transistors and diodes.[23–40] These layers are followed by 1000 Å regions compositionally graded from $x = 0.02$ to $x = 0$ Si-doped to $n = 5 \times 10^{17}/cm^3$ and by 4000 Å thick Si-doped ($n = 1 \times 10^{18}/cm^3$) GaAs. The thickness of the layers were determined by cross-sectional transmission electron microscopy.

The energy band diagrams at the Γ point[†] for sample A and B are shown in Fig. 5.6.

The structures were processed into usual 50 μm diameter mesa diodes with top and bottom ohmic contacts (Section 5.2.2). The diodes were tested at temperatures in the range from 7 to 300 K in a Helitran dewar equipped with microprobes. The I–V and the differential conductance were measured with an HP4145 parameter analyzer.

Figure 5.7 shows the I–V and the corresponding differential conductance dI/dV for a representative diode from sample A measured at 7.5 K. Positive and negative polarity refer to the top contact being positively and negatively biased with respect to the bottom contact.

Consider first the positive polarity data. These display five equally spaced inflections in the I–V and five corresponding minima in the conductance. For negative polarity, four of the conductance minima occur at practically identical voltage magnitudes as the corresponding minima for positive polarity with the exception of the resonance at $+0.3$ V, which is not observed experimentally for negative polarity. The resonances for negative polarity are more pronounced, and the one at ≈ 0.72 V actually exhibits negative differential resistance.

Energy levels in the wells under bias were determined by electron tunneling transmission calculations for the grown layer sequence; thus, the effect of the superlattice grading is directly included. The effects of depletion and accumulation in the emitter and collector layers are also taken into account. A conduction band offset of 0.60 times the direct energy gap differences was assumed.[46] Electron effective mass dispersion with Al content was included. From these calculations, the

---

[†]Recent systematic studies in Ref. 42 have shown that electron RT through $GaAs/Al_xGa_{1-x}As$ diodes with thin barriers (30 Å) is dominated by the barrier height at the Γ point also in the indirect gap region ($0.45 \leq x \leq 1$).

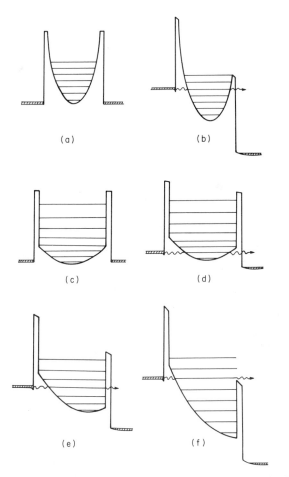

Fig. 5.6 (a)–(b): Band diagram of sample A in equilibrium and under RT conditions. (c–f): Band diagram of sample B in equilibrium and under different bias conditions. The wells are drawn to scale; however, for the sake of clarity, only half the number of levels in an energy interval are shown.

voltage position of the transmission peaks can be directly obtained. These voltages are indicated by the vertical segments at the bottom of Fig. 5.7 and correspond, respectively, to the first six energy levels of the well. The first resonance is not seen experimentally because the corresponding current is below the detection limit of the apparatus (≈ 1 pA). The other calculated positions of the resonances (E_2 through E_6) are in good agreement with the observed ones. The calculations show that not only are the energy levels, for a given bias, nearly equally spaced, but also that the spacing ΔE between the quasi-bound states is little dependent on the electric field as the voltage is varied from 0 to 1 V. For example, at zero bias, $\Delta E \simeq 90$ meV, whereas at 1 V, $\Delta E \simeq 80$ meV. The latter effect is easily under-stood if one considers that the application of a uniform electric field to a parabolic

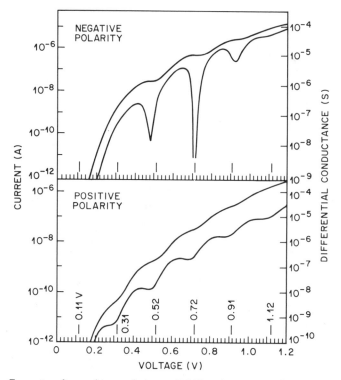

Fig. 5.7   Current–voltage characteristics at 7.4 K and corresponding conductance for a representative diode of sample A under opposite bias polarity conditions. The vertical segments near the horizontal axis indicate the calculated positions of the resonances.

well (Fig. 5.6a) preserves the curvature of the parabola and therefore the spacing ΔE while shifting its origin to the right (Fig. 5.6b).

Figure 5.8 shows the I–V and corresponding conductance for sample B for opposite bias polarities. The band diagrams at different voltages are shown in Figs. 5.6c–f.

It is interesting to note that the group of resonances from the 5th to the 11th are the most pronounced and actually display NDR. A total of 14 resonances are observed in the sample of Fig. 5.8, for positive polarity. In a few diodes, two additional resonances were also observed. The resonances were observed up to temperatures ≈ 100 K, but are considerably less pronounced. The vertical segments near the horizontal axis indicate the calculated positions of the transmission peaks. Overall good agreement with the observed minima in the conductance is found.

The overall features of the I–V can be interpreted physically by means of the band diagrams of Figs. 5.6c–f and of the calculations. At zero bias, the first six energy levels of the well are confined by a parabolic well 225 meV deep, corresponding to the grading from $x = 0$ to $x = 0.30$, and their spacing is ≈ 35 meV. When the bias is increased from 0 to 0.3 V, the first four energy levels probed by

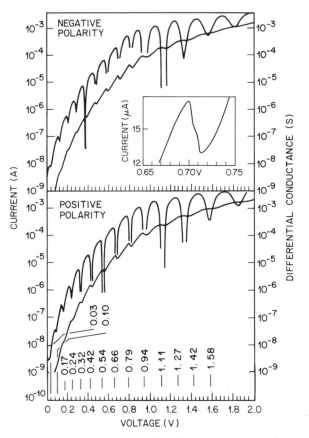

Fig. 5.8 Current–voltage characteristic at 7.1 K and conductance for a representative diode of sample B under opposite bias polarity conditions. The inset show the eighth resonance on a linear scale. The vertical segments near the horizontal axis indicate the calculated positions of the resonances.

RT (Fig. 5.6d) remain confined by the parabolic portions of the well, and their spacing is practically independent of bias for reasons identical to those discussed in the context of sample A. This gives rise to the calculated and observed equal spacing of the first four resonances in the I–V characteristic (Fig. 5.8). Consider now the higher energy levels confined by the rectangular part of the well (>230 meV) at zero bias. When the voltage is raised above 0.3 V, these levels become increasingly confined on the emitter side by the parabolic portion of the well and on the collector side by a rectangular barrier, thus becoming progressively more separated, although retaining the nearly equal spacing (Fig. 5.6e). This leads to the observed gradual increase in the voltage separation of the resonances as the bias is increased from 0.3 to 1.0 V. Above 1 V, the electrons injected from the emitter probe the virtual levels in the quasi-continuum above the collector barrier (Fig. 5.6f). These resonances result from electron interference effects[47] associated with multiple quantum mechanical reflections at the well–barrier interface for energies above the barrier height. It should be noted that these reflections give rise to the existence of quasi-2D states in the well region. The observed NDR is due to tunneling into these states. These interference effects give rise to the four resonances observed above 1 V and must be clearly distinguished from the ones occurring at lower voltages, which are due to RT through the double barrier.

A simple physical explanation of why the resonances above the 4th (5th to 12th) are the most pronounced, leading to negative differential resistance, can be easily given in terms of the calculated voltage dependence of the transmission.

Up to 0.3 V, tunneling out of the well (Fig. 5.6d) occurs through the thick parabolic part of the collector barrier and the resulting widths of the transmission resonances are very small (<1 μeV). As the bias is increased above 0.3 V, not only is the barrier height further reduced but now electrons tunnel out of the well through the thin (20 Å) rectangular part of the barriers. This greatly enhances the barrier transmission and the resonance widths (Fig. 5.6e). This behavior is clearly observed in the calculation of the total transmission vs. bias. For example, the calculated energy width of the level corresponding to the 10th resonance (0.9 V) is 1 meV, which is not negligible compared with the width of the incident energy distribution in the emitter. As the bias voltage is further increased, the competing effect of the decrease of the peak-to-valley ratio takes over, as shown by the calculations. This explains why the highest resonances (above 1 V) become less pronounced.

5.2.4 Integration of Resonant Tunneling Diodes and Their Circuit Applications

A variety of potential applications could be realized with devices exhibiting multiple peaks in the I–V. These include ultrahigh-speed analog-to-digital coverters, parity bit generators, and multiple valued logic.[15,16] RT devices offer interesting possibilities toward realizing this characteristic. One way of obtaining multiple peaks in the I–V, of course, is by using the multiple resonances of a quantum well. This approach however, suffers from the difficulty that the peaks corresponding to the excited states generally carry significantly higher current than that associated with the ground state for a variety of structural and material reasons. On the other hand, the above-mentioned circuit applications require nearly equal peak currents. A novel approach for obtaining multiple peaks in the I–V of RT devices, introduced by Capasso et al.,[48] is the integration of a number of RT diodes. In this method, a single resonance of the quantum well is used to generate the multiple peaks. Hence they occur at almost the same current level and exhibit similar peak-to-valley ratios. The prototype device, fabricated using two RT diodes, exhibits two peaks of nearly equal current at 100 K. The device has been used to demonstrate, for the first time, a number of practical circuits using devices with multiple peaks in the I–V.

5.2.4.1 Structure and Processing. The device structure, grown on a $\langle 100 \rangle$ n^+ Si-doped GaAs substrate, is shown in Fig. 5.9a. An undoped GaAs layer 2500 Å

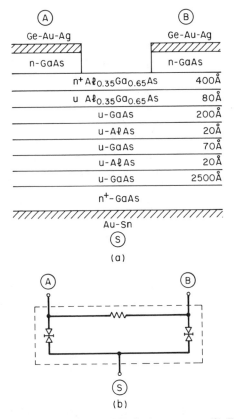

Fig. 5.9 (a) Schematics of the integrated RT diode structure. (b) Equivalent circuit: two RT diodes in parallel connected by the resistance R of the 200 Å GaAs channel between A and B. The resistance of the $Al_{0.35}Ga_{0.65}As$ layer between A and B is much higher than R due to carrier depletion by electron transfer into the GaAs channel. The choice of the circuit symbol for the RT diode (two back-to-back tunnel diodes) is motivated by the symmetry of the current–voltage characteristic of the RT diode.

thick is grown on a 1 μm thick n^+ $(= 5 \times 10^{17}/cm^3)$ GaAs buffer layer and is followed by the RT double barrier. The latter consists of a 70 Å GaAs quantum well sandwiched between two 20 Å AlAs barriers. A modulation-doped $Al_{0.35}Ga_{0.65}As/$ GaAs heterojunction is then grown on top of the double barrier; the 200 Å thick GaAs is undoped, whereas the 480 Å thick $Al_{0.35}Ga_{0.65}As$ layer is doped with Si to $2 \times 10^{18}/cm^3$ except for an 80 Å spacer region adjacent to the GaAs channel. The channel contains a high-density $(\simeq 10^{18}/cm^3)$ high-mobility electron gas spatially separated from the parent donors in the AlGaAs layer. As a result, the AlGaAs layer is completely depleted. The growth ends with an n^+ GaAs 1400 Å contact layer doped to $n \simeq 2 \times 10^{17}/cm^3$.

Two rectangular (240 × 80 μm) contact pads separated by a distance of 6.5 μm, along the long side, were defined by evaporating in succession Ge (120 Å), Au (270 Å), Ag (1000 Å), and Au (1500 Å) and using lift-off techniques. The metallizations were then alloyed at 380°C for 10 s and used as a mask for wet chemical etching. A selective stop etch (H_2O_2 and NH_4OH, pH = 7.2) was used to reveal the $Al_{0.35}Ga_{0.65}As$ barrier.

Note that the thickness of the cap layer and the composition of the two top contacts and the alloying temperature and time were the same used for the fabrication of the charge injection transistor, which is structurally similar to this device.[49] This ensures that the contacts to the electron gas in the GaAs layer beneath the AlGaAs barrier do not penetrate through the RT double barrier.

5.2.4.2 *Equivalent Circuit and the I–V Characteristics.*

This device consists essentially of two monolithically integrated RT diodes in parallel (Fig. 5.9*b*) The resistance shown in Fig. 5.9*b* is that of the GaAs 200 Å channel connecting the two diodes, and its measured value is $\approx 12\ \Omega$. In the two-diode case discussed in this section, this resistance is not essential for the operation of the device. However, the scheme can be obviously extended to more than two RT diodes; in the latter case, the resistance of the channel linking the devices provides the useful function of a monolithically integrated voltage divider. For proper biasing of this voltage divider, the structure should be suitably designed so that the current in the divider network is sufficiently large compared with that through the RT diodes.

The use of the modulation-doped heterojunction allows the formation of a low-resistance ohmic contact to the RT diodes while keeping the dopants away from the double barrier. In addition, the AlGaAs passivates the GaAs channel between the two metallizations.

The substrate current (i.e., the one through terminal S) is measured as a function of positive bias applied between terminals S and A (which is grounded) for different values of the potential difference V_{BA} applied between B and A. The substrate current consists primarily of the sum of the two RT currents flowing through the two RT diodes. For zero potential difference V_{BA}, the structure behaves like a conventional RT diode and the *I–V* displays one peak (Figs. 5.10*a* and *b*). The negative conductance region is, of course, due to the quenching of RT through the two double barriers under terminals A and B, respectively. When terminal B is biased negatively with respect to terminal A (which is grounded) the *I–V* characteristic (see Fig. 5.10*a*) develops an additional peak at lower voltages; the position of one peak remains unchanged, whereas that of the other moves to lower bias as the potential difference V_{BA} between B and A is made more negative. Note that by appropriate choice of the bias between B and A, the two peak currents can be made nearly equal.

This effect is explained as follows. As a result of the bias applied between A and B, the potential differences across the two double barriers are different and for B negatively biased with respect to A, RT through the double barrier under B is obviously quenched at a lower substrate bias than in the double barrier under terminal A, leading to two peaks in the *I–V*.

The peak that does not shift with varying V_{BA} is, of course, associated with quenching of RT through diode A. Note also that, as expected, the separation between the peaks is nearly equal to the bias applied between A and B. Finally, if

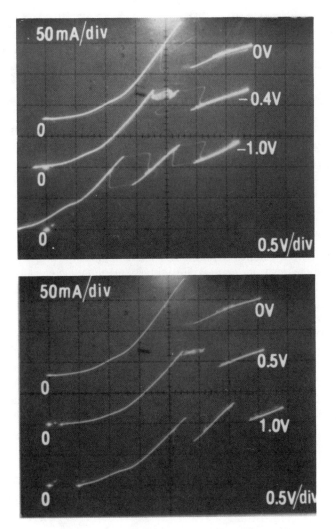

Fig. 5.10 Substrate current vs. positive substrate bias at 100 K with (*a*) negative and (*b*) positive potential difference (V_{BA}) between terminals B and A as the parameter. Terminal A is grounded.

terminal B is positively biased with respect to A, a higher voltage is required to quench RT through the double barrier B, leading to a second peak which shifts to higher voltages as V_{BA} is increased (Fig. 5.10*b*). Similar results are obtained with negative bias applied to S.

The characteristics of Fig. 5.10 were obtained for an operating temperature of 100 K. With improved processing and material quality, it should be possible to operate the device at room temperature. This has already been achieved in RT diodes and transistors.

5.2.4.3. Circuit Applications. The *I–V* of Fig. 5.10 has been used in a variety of circuit applications ranging from frequency multipliers to multiple valued logic elements.[31,48] Such applications for devices with multiple peaks in the *I–V* had been predicted before,[15,16] but could not be implemented for the lack of a practical device exhibiting such characteristics. The circuits constructed using the new device are discussed and the experimental results are presented in this section.

Frequency Multiplier. The circuit of Fig. 5.11 has been used for frequency multiplication. Its operation is understood from the diagrams of Fig. 5.12, which show the *I–V* for a typical bias voltage V_{BA} between the terminals A and B of the device. Figures 5.12a and b show the operation with a sawtooth and a sinewave input, respectively. Let us consider the operation with a sawtooth input first (Fig. 5.12a). The substrate bias V_{SS} is adjusted to select the quiescent operating point at A_2 of the *I–V*. As the sawtooth input voltage increases from A_1 to B_1, the operating points shift from A_2 to B_2 along the *I–V*, with the substrate current I_s increasing almost linearly. The output voltage across the resistance R is proportional to I_s and hence also increases from A_3 to B_3 linearly. As the input increases beyond B_1, the current I_s suddenly drops to the valley point B_2', resulting in a sudden drop in the output voltage from B_3 to B_3'. Between B_3 and C_3, the output continues to rise again, followed by a second drop at C_2 and rise thereafter as the input continues to rise up to D_1. At D_1, the input returns to zero to start a new cycle and the operating point also shifts back to A_2 with a drop in the output as well. Thus, the frequency of the sawtooth input signal has been multiplied by a factor of 3. It should be noted here that the multiplier circuit described is independent of the input signal frequency. Conventionally, a phase-lock loop in conjunction with digital frequency divider is used to construct a frequency independent multiplier.

The operation of the circuit with a sinewave input is shown in Fig. 5.12b. The output waveform in this case can also be explained following similar arguments and is found to be rich in the fifth harmonic of the input. Figures 5.13a and b show the experimental results for a sawtooth and a sinewave input, respectively, with the device biased at $V_{BA} = 1$ V and $V_{SS} = 2.3$ V. The efficiency of this device in generating the fifth harmonic is thus found to be much better than conven-

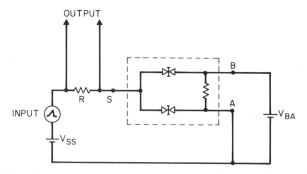

Fig. 5.11   The frequency multiplier circuit using the integrated RT diodes.

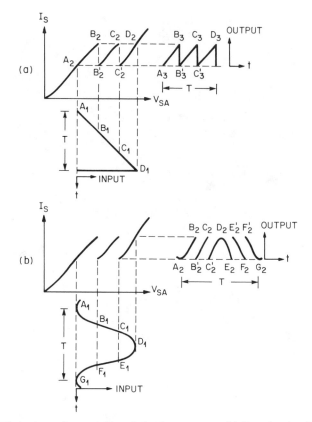

Fig. 5.12 The schematic operation of the frequency multiplier circuit of Fig. 5.11 for (*a*) sawtooth and (*b*) sinewave inputs.

tional devices, like a step-recovery diode, used in frequency multiplier circuits. It should also be noted here that if V_{BA} is adjusted to produce a single peak in the $I-V$ instead of two, the sawtooth will be multiplied by a factor of 2 instead of 3 and the output for sinewave will be rich in the third harmonic. Sinewave multiplication by a factor of 2 using a single peak in the $I-V$ of a RT hot-electron transistor has been demonstrated before.[17]

Multiple-Valued Logic. The circuit shown in the inset of Fig. 5.14 can be used as a memory element in a three-state logic system. The bias voltage V_{BA} between the terminals A and B is again adjusted to produce the $I-V$ as in Fig. 5.14, with two nearly equal peaks at the same current level. For a suitable supply voltage V_{SS} and load resistance R_L, the load line intersects the $I-V$ at five different points, of which three (Q_1, Q_2, and Q_3) are in the positive slope parts of the curve and are hence stable operating points. The output voltage of the circuit corresponding to the three operating points, Q_1, Q_2, and Q_3, are V_1, V_2, and V_3, respectively, as shown in Fig. 5.14. The circuit can stay indefinitely on any one of the three

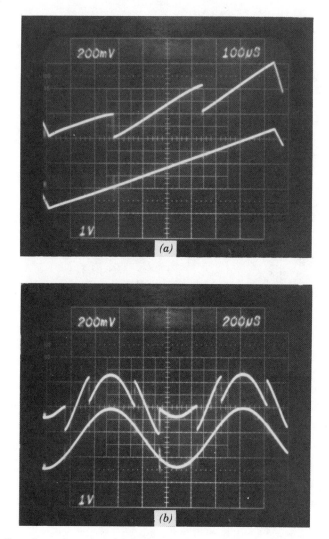

Fig. 5.13 Experimental results of frequency multiplication: (*a*) sawtooth input, (*b*) sine-wave input.

points, thus retaining the last voltage information impressed on it. It can therefore be used as a memory element in a three-state logic circuit, with V_1, V_2, and V_3 being the voltages corresponding to the three logic states. This is a significant component reduction over the existing three-state logic circuits, which require four conventional transistors and six resistors to construct a memory cell.[50] The circuit can be switched from one stable state to another by applying a short voltage pulse. In the experimental studies,[31,48] the operating point was shifted from one state to another by momentarily changing the supply voltage V_{SS}, which has the same effect as applying a short voltage pulse. With a supply voltage $V_{SS} = 16$ V, load re-

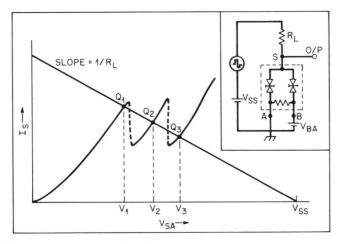

Fig. 5.14 The current–voltage characteristic and schematic of a three-state memory cell using the integrated RT diodes. The associated load line shows three stable operating points, Q_1, Q_2, and Q_3.

sistance $R_L = 215\ \Omega$, and the device biased to $V_{BA} = 0.7$ V, the three stable states were measured to be at 3.0 V, 3.6 V, and 4.3 V. The corresponding load line drawn on the measured I–V characteristic of the device at $V_{BA} = 0.7$ V intersects at 2.8 V, 3.4 V, and 4.1 V, respectively, which are in close agreement with the measured values of the three stable operating points.

The three-state memory cell discussed above is also suitable for integration in memory ICs with read/write and decoding network laid out as shown in Fig. 5.15. The memory cells are placed in a matrix array and a particular element in the array is addressed by activating the corresponding row and column select lines. A row select connects each device in that row to the corresponding column lines. The column select finally connects the selected column to the data bus. Consider the element (i, j) of the memory matrix shown in Fig. 5.15. When the row select line is activated, it turns the driving switch Q1 on. It also turns on the switches for every element in the ith row. The column select logic now connects the jth column only to the data bus. The ternary identity cell T^{51} acts as the buffer between the memory element and the external circuit for reading data. For reading data from the memory, the identity cell is activated with the read/enable line and data from the element no. (i, j) in the matrix goes, via the data bus, to the in/out pin of the IC. When the Write/enable line is activated, data from external circuit is connected to the data bus and subsequently forced on the (k, j)th element in the array and is written there.

Parity Generator. A 4 bit parity generator circuit using the new device is shown in Fig. 5.16a. The operation of the circuit can be understood from Fig. 5.16b, which shows the *I–V* of the device with the bias voltage V_{BA} properly adjusted and the resultant voltage waveforms at various points in the circuit for different input conditions. The four digital inputs are added in the inverting summing amplifier

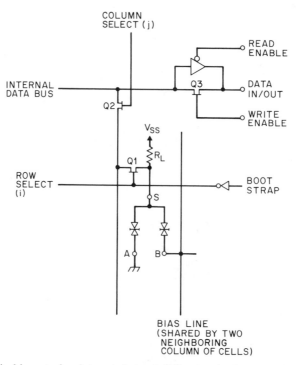

Fig. 5.15 Typical layout of an integrated circuit (IC) using the three-state memory cells of Fig. 5.14.

A_1 to produce five distinct voltage steps at its output corresponding to the number of digital bits being in the high state. Normally, the output of A_1 would be negative for positive input voltages. The addition of a suitable negative offset voltage V_{OFF} at the input shifts the whole waveform up to produce the A_1 output, as shown at the bottom of Fig. 5.16b. The substrate bias voltage V_{SS} is adjusted to select the operating points of the device at the five dots shown in the $I-V$ curve, corresponding respectively to the five different voltage levels at the A_1 output. The substrate current of the device generates a voltage across the 7.5 Ω resistor, which is picked up by the buffer amplifier A_2. Note that the output is high when the number of input bits set high is odd and vice versa. The circuit can thus be used as a 4 bit parity generator. The two difference amplifiers in the circuit can be constructed using three transistors each. There is considerable reduction in the number of components compared with a conventional circuit, which needs three exclusive-OR gates, each requiring eight transistors.

Typical experimental output from the circuit of Fig. 5.16a is shown in Fig. 5.17. The four digital inputs are driven by the outputs of a 4 bit binary counter. The top trace of Fig. 5.17 shows the output of amplifier A_1, and the bottom trace, that of A_2. Considering the dotted line as the reference level of a logic circuit, we find that the output is high for the second and the fourth voltage levels of

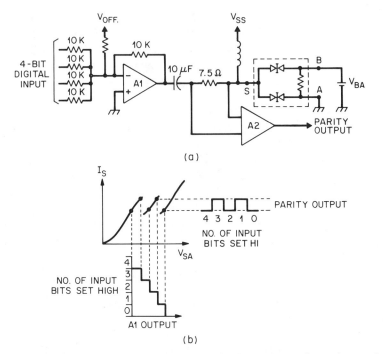

Fig. 5.16 (*a*) The 4 bit parity generator circuit. (*b*) The current–voltage characteristic of the device and the waveforms at various points in the circuit for different input conditions.

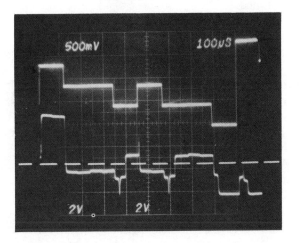

Fig. 5.17 Experimental results of the parity generator circuit. The top trace shows the output of the amplifier A₁, and the bottom trace, the parity output. The dotted line is the threshold voltage level of the output logic.

A_1; it is low at the first, third, and fifth levels. It should be noted however, that there are considerable decoding spikes in the output whenever the operating point of the device is shifted across a negative resistance region. This is believed to be due to the inherent oscillations of a circuit involving a negative resistance device and could be taken care of by proper circuit design.

5.2.4.4 Summary. The structure also lends itself to the realization of multiple peaks using the same operating principle and a series of metallizations. The resistance of the channel in this case performs the useful function of a monolithically integrated voltage divider, with the bias V_{BA} being applied between the pads at the two extreme ends. Multiple peaks in the *I–V* characteristic could, of course, also be realized by connecting in parallel a series of tunnel diodes with resistors in between, which basically amounts to the circuit of Fig. 5.9*b* in the case of two diodes. The approach presented here has, however, two clear potential advantages: (1) it is a monolithic integration of RT devices and the voltage divider with resulting reduced parasitic resistances and capacitances (note that the monolithic integration of tunnel diodes would be a much more demanding task); and (2) reproducibility of the *I–V*s. The *I–V* of tunnel diodes is notoriously not sufficiently reproducible for logic applications due to the extremely high doping levels and the sensitive dependence of the *i–V* on the latter. On the other hand, RT diodes, although their technology is still in its infancy in comparison with that of tunnel diodes, have potentially more reproducible characteristics, since the peak current and negative resistance regions depend primarily on the well width and alloy composition of the barrier, factors which can be controlled to a great precision (a few percent) using MBE.

Following the above demonstration, Soderstrom and Andersson[52] have come up with a new idea of combining RT diodes to obtain multiple-valued charcteristics. Their structure does not require the additional bias supply to separate the peaks due to the different diodes.[52]

5.3 RESONANT TUNNELING BIPOLAR TRANSISTOR (RTBT)

The negative differential resistance of RT double barriers showed enough potential to be included in three terminal devices, to take advantage of the NDR as well as the transistor action. The first of these kinds of structures was the RT bipolar transistor (RTBT), proposed by Capasso and Kiehl at Bell Laboratories in 1985.[16] The structures initially proposed are shown in Fig. 5.18 and 5.19. They consisted of AlGaAs/GaAs heterojunction bipolar transistors with a quantum well in the *p*-type base layer. In order to satisfy consistently the condition of tunneling through symmetric double barriers discussed in Section 5.2.1, these structures employed high-energy or ballistic injection of minority carriers into the base to achieve RT through the double barrier, rather than applying a field across the latter. This method does not alter the transmission of the two barriers and therefore should lead to near-unity transmission at all resonance peaks and to larger negative conductance and peak-to-valley ratios than conventional RT structures.

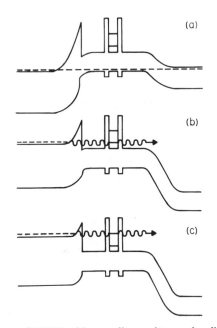

Fig. 5.18 Band diagram of RTBT with tunneling emitter under different bias conditions: (*a*) in equilibrium, (*b*) RT through the first level in the well, (*c*) RT through the second level. (Not to scale.)

Figure 5.18 shows the band diagram of one of these devices. The structure is a heterojunction bipolar transistor with a degenerately doped tunneling emitter and a symmetric double barrier in the base. The collector current as a function of base–emitter voltage V_{BE} should exhibit a series of peaks corresponding to RT through the various quasi-stationary states of the well. Multiple negative conductance in the collector circuit can therefore be achieved.

An alternative injection method is the abrupt or nearly abrupt emitter, which can be used to launch electrons ballistically into the quasi-eigenstates with high momentum coherence. As V_{BE} is increased, the top of the launching ramp eventually reaches the energy of the quasi-eigenstates so that electrons can be ballistically launched into the resonant states (Fig. 5.19*a*).

To achieve equally spaced resonances in the collector current, the rectangular quantum well in the base should be replaced by a parabolic one (Fig. 5.19*b*). RT through parabolic quantum wells was discussed in Section 5.2.3. Assuming the depth of the parabolic well in the conduction band to be 0.34 eV (corresponding to grading from $Al_{0.45}Ga_{0.55}As$ to GaAs) and its width to be 200 Å, one finds that the first state is at an energy of 32 meV from the bottom of the well and that the resonant states are separated by ≈ 64 meV. This gives a total of five states in the well. In a recent experiment, as many as 16 resonances through a parabolic quantum well were observed as discussed in Section 5.2.3.

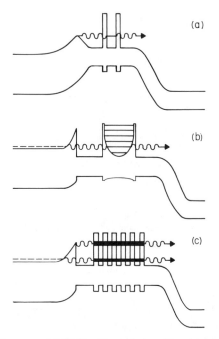

Fig. 5.19 (*a*) Band diagram of RTBT with graded emitter (at resonance). Electrons are ballistically launched into the first quasi-eigenstate of the well. (*b*) RTBT with a parabolic quantum well in the base and tunneling emitter. A ballistic emitter can also be used. (*c*) RTBT with superlattice base. (Not to scale.)

Finally, in Fig. 5.19*c*, we illustrate another method, that of high-energy injection and transport in the minibands of a superlattice, using ballistic launching or tunnel injection.

5.3.1   Circuit Applications of RTBT

These new functional devices, because of their multiple resonant characteristic, can have many potential applications, leading to tremendous reduction in circuit complexity and size. These are discussed in this section.

Multiple-Valued Logic.   Consider the common-emitter circuit shown in Fig. 5.20a. For an input voltage V_i in the base for which the electrons undergo RT, the transistor strongly conducts and the output voltage V_o is low. Off resonance, instead, the device basically does not conduct and the output voltage is high. This results in the multiple-valued voltage transfer characteristic of Fig. 5.20*b* having as many peaks as the number of resonances in the well. The output voltage V_o takes on one of the two values in accordance with the level of the input voltage V_i. Thus, the device provides a binary digital output for an analog input, or a multiple-valued digital input.

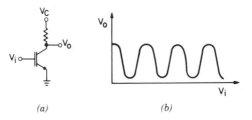

(a)   *(b)*

Fig. 5.20 Common emitter amplifier circuit using the RTBT and the corresponding multiple-valued voltage transfer characteristics.

Parity Generator. The multiple-valued characteristics of Fig. 5.20b can be used to design the parity generator circuit shown in Fig. 5.21. In this circuit, the binary bits of a digital word are added in the resistive network at the input of the RTBT. With proper weighting of the resistors R_o and R_B, the operating point would be placed either on a peak or a valley of the I–V depending on whether the total number of 1's at the input is even or odd, respectively. The advantage of this approach over conventional circuits is that the RTBT implementation, apart from being smaller in size and simpler, should also be extremely fast, since it uses a single high-speed switching device. Conventional transistor implementation requires complex circuitry involving many logic gates with a consequent reduction in speed.

Compare this implementation of the parity generator with that discussed in Section 5.2.4.3. The advantage of using a transistor structure, as shown in this section, is further simplification of the circuit. The function of adding the voltages corresponding to the digital input bits, as performed by the amplifier A_1 in Fig. 5.1, is performed by the transistor in this case. Also, since the input and the output in the transistor are isolated, the amplified A_2 is not necessary, as the output automatically comes as ground referenced.

Analog-to-Digital Converter. The circuit shown in Fig. 5.22 can be used as an analog-to-digital converter. In this application, the analog input is simultaneously applied to an array of RTBT circuits having different voltage scaling networks. To understand the operation of the circuit, consider the simplest system comprised of only the two transistors Q_1 and Q_2. The voltages at different points of this circuit are shown in Fig. 5.23a for various input voltages V_i. Consider that the resistances R_0, R_1, and R_2 are so chosen that the base voltages of the transistors Q_1 and

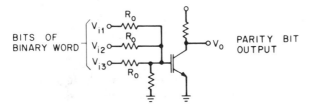

Fig. 5.21 The parity generator circuit using RTBT.

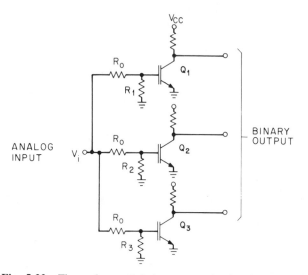

Fig. 5.22 The analog-to-digital converter circuit using RTBT.

Q_2 vary according to the curves V_{B1} and V_{B2} respectively with V_i. With the input voltage at V_1, the output of both the transistors will be at the operating point P_1 (Hi-state). With the input changing to V_2, the output of Q_1 will become low (P_2), whereas that of Q_2 will still remain high (closer to P_1). Applying this logic to the input voltages V_3 and V_4, it can be easily shown that circuit indeed follows the truth table of Fig. 5.23b. The outputs of the RTBT array thus constitute a binary code representing the quantized analog input level. The system can be extended to more bits with larger number of peaks in the I–V. Again, the circuitry involved in this approach is simple and should be very fast.

Multiple-State Memory. This application takes advantage of the ability to achieve a multiple-valued negative differential resistance characteristic. This type of characteristic is achieved at the emitter–collector terminals by holding the base–collector junction at fixed bias V_{BC}, as shown in the inset of Fig. 5.24. With V_{BC} fixed, variations in V_{CE} produce variations in V_{EB} which cause the collector current to peak as V_{EB} crosses a tunneling resonance (Fig. 5.24). When connected to a resistive laod R_L and voltage supply V_{CC}, as shown in the circuit of Fig. 5.24, the resulting load line intersects the I–V at N stable points, where N is the number of resonant peaks. The circuit thus acts as an N-state memory element, providing the possibility of extremely high-density data storage. Such an element can be latched onto any one of he stable states by momentarily applying a voltage close to the desired state. It can therefore be integrated in a large memory array, as discussed in Section 5.2.4.3.

A N-state latch such as this can also be used to build a counter in the N-state logic system, as the circuit can be switched from one stable state to the immediately adjacent one by a voltage pulse of height V_{LS} to the circuit, forcing the oper-

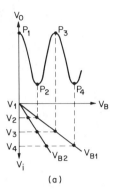

(a)

TRUTH TABLE

INPUT	OUTPUT	
V_i	Q_2	Q_1
V_1	1	1
V_2	1	0
V_3	0	1
V_4	0	0

(b)

Fig. 5.23 The schematic operation of the analog-to-digital converter circuit of Fig. 5.22, involving only 2 bits: (a) the voltages at different points of the circuit at various input voltages, (b) the truth table.

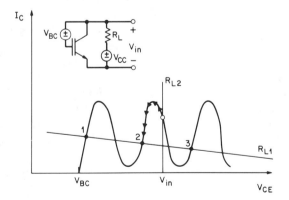

Fig. 5.24 The current–voltage characteristics with multiple-valued negative differential resistance when the base–emitter voltage of the RTBT is held fixed, as shown in the circuit in the inset of the figure. The load line corresponding to R_{L1} demonstrates its operation as a multiple-valued memory element. The solid circles denote stable states. The load line R_{L2} and the path indicated by the arrows show how such a memory can be pulsed from one stable state to another.

ating point to that of the open circle on the unstable part of the characteristics in Fig. 5.24. It can be easily shown that as the input pulse is removed the operating point will move along the indicated trajectory, finally latching onto the State 2. Circuits such as these and others, such as multipliers and dividers, in the N-state logic system had been of considerable interest for some time.[50] However, since no physical device indicating multiple-valued negative resistance previously existed, such circuits were possible only with combinations of binary devices. To achieve N state, N two-state devices were connected, resulting in a complex configuration with reduced density and speed.

5.3.2   Resonant Tunneling Spectroscopy

The RT bipolar transistor structures initially proposed and discussed so far relied heavily on the ability to ballistically transfer the electrons through the p-type base region up to the double barrier structure. Recent experiments with resonant tunneling spectroscopy[53–55] of hot electrons injected at a high energy into a p-type GaAs well indicate that it may be extremely difficult to achieve that goal.

Figure 5.25 illustrates the energy band diagram of the structure used for resonant tunneling electron spectroscopy. It consists basically of a reverse biased *pin* heterojunction and can be used to investigate hot minority-carrier transport. Low-intensity incident light is strongly absorbed in the wide-gap p^+ layer. Photogenerated minority-carrier electrons diffuse to an adjacent low-gap layer. Upon entering this region, electrons are ballistically accelerated by the abrupt potential step and gain a kinetic energy $\cong \Delta E_c$ and a forward momentum $p_\perp \simeq \sqrt{2m_e^* \Delta E_c}$. Collisions in the low-gap layer tend to randomize the injected, nearly mono-energetic distribution. Hot electrons subsequently impinge on the double barrier in the collector. From simple considerations of energy and lateral momentum $p_\parallel$ conservation in the tunneling process, it can be shown that only those electrons with a perpendicular energy $E_\perp$ ($p_\perp^2/2m_e^*$ for a parabolic band) equal (within the resonance width) to the energy of the bottom of one of the sub-bands of the quantum well, resonantly tunnel through the quantum well and give rise to a current. Thus, by varying the applied bias (i.e., changing the energy difference between the resonance of the quantum well and the bottom of the conduction band in the low-gap p^+ layer) and measuring the current, one directly probes the electron energy distribution $n(E_\perp)$. One has therefore

$$E_\perp = E_n - \frac{e(V + V_{bi})(L_B + L_w/2 + L_{sp})}{L_c} \qquad (4)$$

where V is the reverse bias voltage, V_{bi} the built-in potential of the *pin* diode, L_c the total collector layer thickness, L_B and L_w the barrier and well layer thicknesses, respectively, and L_{sp} the thickness of the undoped spacer layer (20 Å) between the p-type region and the double barrier (see later in text). E_n is the energy of the bottom of the nth sub-band measured with respect to the bottom of the center of the well. (Note that E_n is assumed to be independent of the electric field F, which is a

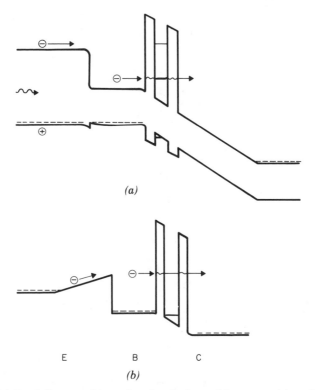

E   B   C

(b)

Fig. 5.25 (*a*) Band diagram of heterojunction diode used for resonant tunneling spectroscopy of hot minority-carrier electrons. By measuring the photocurrent as a function of the
reverse bias, the hot-electron energy distribution $n(E_\perp)$ can be directly probed. (*b*) Unipolar
transistor structure for resonant tunneling spectroscopy of hot majority-carrier electrons in
the n^+ base layer.

good approximation as long as E_n is significantly greater than eFL_w.) Identical
arguments apply to the case of the unipolar transistor structure of Fig. 5.25*b*,
which can be used to analyze the electron distribution in the layer by measuring
the collector current as a function of the collector–base voltage. In the above
arguments, we have assumed that thermionic currents over the double barrier can
be minimized. This can be done by operating the structure at sufficiently low temperature and by suitably designing the double barrier.[2] To obtain the actual energy
distribution $n(E_\perp)$ from the current, the latter must be properly normalized by taking into account the field dependence of the resonant tunneling probability (integrated over the resonance width). This procedure does not alter, of course, the
position of the peaks in the *I–V* characteristic (see Fig. 5.26), since the above
probability varies monotonically with the electric field, irrespective of whether
electrons resonantly tunnel sequentially or coherently through the double barrier.[33]
Thus, the main features of the electronic transport can still be obtained directly
from the current, without normalization.

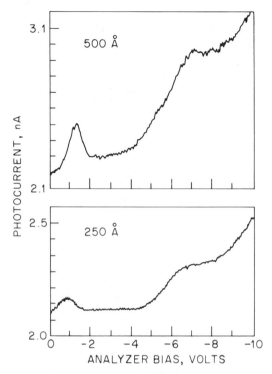

Fig. 5.26 Photocurrent as a function of reverse bias for the structure of Fig. 5.25a, with GaAs p^+ layer thickness of 500 Å (top) and 250 Å (bottom).

The structures were grown by MBE on a $\langle 100 \rangle p^+$ GaAs substrate and consist of *pin* heterojunction diodes. Their band diagram is shown in Fig. 5.25a at a given reverse bias. The growth starts with a 2000 Å thick $n^+ = 2 \times 10^{17}/\text{cm}^3$ buffer layer followed by an undoped ($|N_D - N_A| \simeq 10^{14}/\text{cm}^3$) 5000 Å GaAs layer and an AlAs/GaAs/AlAs double barrier, with barrier and well thicknesses of 20 Å and 80 Å, respectively. A 20 Å undoped GaAs spacer layer separates the double barrier from the p^+ ($= 3 \times 10^{18}/\text{cm}^3$) GaAs layer, in which electrons are launched. Different thicknesses were used for this region (250 Å, 500 Å, 1800 Å) while keeping everything else the same. The last layer consists of 2 μm thick $\text{Al}_{0.3}\text{Ga}_{0.7}\text{As}$ doped to $p = 3 \times 10^{18}/\text{cm}^3$. This provides a launching energy of $\cong 225$ meV, that is, the conduction band discontinuity between GaAs and $\text{Al}_{0.3}\text{Ga}_{0.70}\text{As}$ (obtained from $\Delta E_c = 0.6 \, \Delta E_g$). The depletion width on the p^+ side of the junction is negligible with respect to the p^+ GaAs well thickness up to the highest applied bias (-10 V), due to the high doping. The parameters of the double barrier were chosen in such a way that over the applied voltage range (0–10 V) the electron energy distribution is probed essentially by one resonance at a time. For this double barrier, the first resonance is at $E_1 = 60$ meV from the well bottom and the second at $E_2 = 260$ meV, with full widths at half maximum of ≈ 0.1 meV and ≈ 1 meV, respectively.[2] It is easily shown from Eq. (4) that over the range of applied bias (Fig. 5.26), the first resonance samples the $E_\perp$ energy range from 37 meV to 0 meV, whereas the second resonance samples the energy range from 225 meV to 80 meV. The thickness of the collector layer L_c was made much greater than that of the double barrier to enhance the energy resolution of the spectrometer.

The samples were processed into mesa devices, using standard photolithographic, wet etching, and metallization techniques. The photosensitive area of these detectors is 10^{-4} cm^2. Light from a He–Ne laser ($\lambda = 6328$ Å) heavily absorbed in the $Al_{0.3}Ga_{0.7}As$ region (absorption length $\simeq 5000$ Å) was used to achieve pure electron (minority-carrier) injection and the dc photocurrent was measured with an HP4145 parameter analyzer as a function of reverse bias at low temperature in a Helitran dewar. Figure 5.26 illustrates the measured photocurrent at 9.2 K for the structure with a 500 Å thick GaAs p^+ layer. At these current levels and higher (up to 10 μA), space charge effects are negligible, as shown by varying the light intensity and monitoring the photocurrent–voltage characteristic. The dark current was completely negligible ($\lesssim 10^{-13}$ A) in the same voltage range. Two distinct features are present at 1.3 V and 7 V, respectively. Using Eq. (4), one can easily see that the first peak corresponds to electrons with perpendicular energy of a few tens of meV ($\simeq 17$ meV) that have resonantly tunneled through the first resonance of the quantum well. The second peak is much broader and corresponds to incident electrons with energy $E_\perp \simeq 130$ meV which have resonantly tunneled through the second resonance of the well. It is therefore clear that the energy distribution of the electrons in the p^+ GaAs layer, following high-energy injection, consists of two parts. One has relaxed close to bottom of the conduction band, whereas the other has considerably higher perpendicular kinetic energy $E_\perp$. Therefore, the distribution is strongly non-Maxwellian, similar to what has been found in the case of majority-carrier electrons in the base of hot-electron planar doped barrier transistors.[56] Note that the peaks in the photocurrent were observed at temperatures as high as 70 K and did not appreciably shift with temperature.

Similar results are found by decreasing the GaAs p^+ layer thickness from 500 Å to 250 Å (Fig. 5.26). The peaks are located at somewhat lower voltages (corresponding to 10–20% higher energies), implying that the relaxation of carriers is somewhat less, due to the thinner layer, as expected. Overall, however, the shape of the energy distribution has not changed significantly, which implies that already over a length of a few hundred Å the near-ballistic injected distribution has been strongly randomized by scattering and reached a quasi-steady state. Additional manifestation of strong scattering comes from the fact that no evidence is found in the data of the quantized sub-bands of the 250 Å p^+ well into which electrons are injected. Since this sub-band structure should be reflected on the electron distribution, one would expect to observe peaks in the photocurrent at such voltages that the resonances of the double barrier coincide in energy with the resonances of the 250 Å thick layers. The fact that this is not observed implies that the collisional broadening $\hbar/\tau$ must be comparable to or greater than the typical energy separation between the resonances of the p^+ layer. The latter varies roughly from 30 meV (between the first two resonances of the p^+ well) to 80 meV (between the highest two quasi-bound resonances of the p^+ well). This implies that the scattering time (averaged over the hot-electron distribution) is $\lesssim 10^{-14}$ s. This estimate is consistent with recent studies of electron dynamics in p-type GaAs. Hopfel et al.[57] have investigated minority electron transport in GaAs quantum wells following picosecond photoexcitation of small carrier densities ($n \ll p$). Their results show that the high-density hole plasma in the wells ($p = 1.5 \times 10^{11}/cm^2$) induces strong electron-hole scattering. At very low electric fields (a few tens of V/cm) and lattice temperatures of $\simeq 15$ K, their measured minority electron mobilities give a total momentum relaxation time of 4×10^{-14} s. Such time is expected to be $\leq 10^{-14}$ s at higher hole densities ($\approx 5 \times 10^{12}/cm^2$), comparable to that of the p^+ well in these samples. Previous work by Hopfel, Shah, and Gossard[58] had also shown that the energy relaxation rate of minority electrons in quantum wells is considerably higher than that of majority electrons. Although their results were obtained at a lattice temperature of 300 K, this is expected to be valid at low temperatures as well.

These results clearly show that quasi-ballistic or ballistic transport of a significant fraction of the electrons injected in p^+ GaAs does not occur even over distances ≤ 500 Å. Indeed, recent electron spectroscopy measurements in heterojunction bipolar transistors with base thicknesses as short as 400 Å have also demonstrated that electrons undergo strong relaxation in the p^+ ($= 2 \times 10^{18}/cm^3$) base.[59]

On the other hand, previous work on hot-electron transistors[60,61] has shown that a significant fraction of the electrons injected at similar energies ($\simeq 0.2$ eV) in n^+ GaAs of comparable doping density ($\approx 10^{18}/cm^3$) traverses the base quasi-ballistically or ballistically for base thicknesses in the 300–500 Å range. In fact, recent experimental determinations of the scattering rates of nonequilibrium injected electrons in $n^+ = 1 \times 10^{18}/cm^3$ GaAs give scattering times in the 3×10^{-14}–5×10^{-14} s range.[56] These values are considerably longer than the ones estimated in the present experiment for p^+ GaAs at comparable injection energies and doping levels.

Measurements have also been performed in similar structures with a thicker p^+ region (1800 Å). From the photocurrent–voltage characteristics, it is found that electrons undergo RT through the double barrier starting from a few meV energy from the bottom of the conduction band in the p^+ GaAs layer, and that there is not a hot-electron distribution at higher energy. This is to be expected, since by making the p^+ layer much thicker, electrons have had time to thermalize at the bottom of the band before undergoing RT.

In summary, the results show that nonequilibrium energetic electrons injected in p^+ GaAs undergo very strong scattering. This implies that the implementation of ballistic heterojunction bipolar transistor[62] and ballistic RT transistors[16] will most likely not be possible.

5.3.3 Resonant Tunneling Bipolar Transistor (RTBT) With Thermal Injection

The first operating RTBT, demonstrated by Capasso et al.[23] in 1986, was designed to have the minority electrons *thermally* injected into and tranported through the base, rather than *hot-electron* or *quasi-ballistic* transport. This made the operation of the device much less critical, and the structure implemented in the AlGaAs/ GaAs material system showed resonance peaks even at *room temperature.*

The band diagram of this transistor is shown in Fig. 5.27. Thermal injection is achieved by adjusting the alloy composition of the portion of the base adjacent to the emitter in such a way that the conduction band in this region lines up with or is slightly below the bottom of the ground state sub-band of the quantum well (Fig. 5.27a). For a 74 Å well and 21.5 Å AlAs barriers, the first quantized energy level is $E_1 = 65$ meV.[2] Thus, the Al mole fraction was chosen to be $x = 0.07$ (corresponding to $E_g = 1.521$ eV), so that $\Delta E_c \simeq E_1$. This equality need not be rigorously satisfied for the device to operate in the desired mode as long as E_1 does not exceed ΔE_c by more than a few kT. The quantum well is undoped; nevertheless, it is easy to show that there is a high concentration of ($\cong 7 \times 10^{11}/\text{cm}^2$) 2D hole gas in the well. These holes have transferred from the nearby $Al_{0.07}Ga_{0.93}As$ region by tunneling through the AlAs barrier in order to achieve Fermi level line-up in the base.

The structures were grown by MBE on an n^+ Si doped GaAs substrate. A 2100 Å $n = 3 \times 10^{17}/\text{cm}^3$ GaAs buffer layer is followed by a 1.6 μm thick GaAs n-type ($= 1 \times 10^{16}/\text{cm}^3$) collector. The base layer starts with a Be doped p^+ ($= 1 \times 10^{18}/\text{cm}^3$) 1900 Å GaAs region adjacent to the collector, followed by a 210 Å undoped GaAs set-back layer. The double barrier is then grown. It consists of a 74 Å undoped GaAs quantum well sandwiched between two undoped 21.5 Å

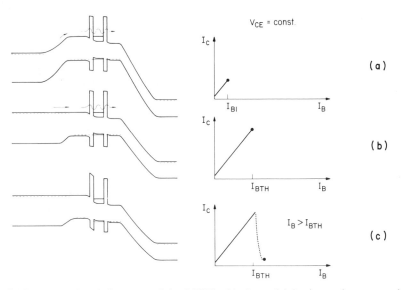

Fig. 5.27 Energy band diagrams of the RTBT with thermal injection and corresponding schematics of collector current I_C for different base currents I_B at a fixed collector emitter voltage V_{CE} (not to scale). As I_B is increased, the device first behaves as (a) a conventional bipolar transistor with current gain until (b) near flat-band conditions in the emitter are achieved. For $I_B > I_{BTH}$, a potential difference develops across the AlAs barrier between the contacted and uncontacted regions of the base. This raises the conduction band edge in the emitter above the first resonance of the well, thus quenching resonant tunneling and the collector current (c).

AlAs barriers. The last portion of the base is 530 Å thick $Al_{0.07}Ga_{0.93}As$, of which 105 Å adjacent to the double barrier is undoped and the rest is p-type doped to $\approx 1 \times 10^{18}/cm^3$. The purpose of the two set-back layers is to offset Be diffusion into the double barrier during the high-temperature growth ($T = 680°C$) of the AlGaAs graded emitter.[63] The latter consists of a 530 Å $n^+ \approx 3 \times 10^{17}/cm^3$ region linearly graded between $x = 0.07$ and $x = 0.24$, adjacent to the base, and of 3200 Å thick $Al_{0.25}Ga_{0.75}As$ doped to $\cong 3 \times 10^{17}/cm^3$. The growth ends with a 1000 Å $n^+ = 3 \times 10^{18}/cm^3$ GaAs contact layer separated from the emitter by a 530 Å $n^+ = 3 \times 10^{18}/cm^3$ region linearly graded from $x = 0.24$ to $x = 0$.

Test structures with 7.5×10^{-5} cm^2 emitter area were fabricated using photolithography and wet and anodic etching techniques. The base layer was revealed by anodic etching in H_3PO_4/H_2O. The portion of the base ($Al_{0.07}Ga_{0.93}As$) adjacent to the emitter was also anodically etched off at 12 Å/V, while the mesa height was continually monitored with a Dektak depth profiler. The rest of the base was contacted using AuBe (1% Be by weight, 400 Å)/Au (1100 Å) alloyed at 400°C in a H_2 flow for 2 s. Au (500) Å/Sn (250 Å)/Au (2000 Å) alloyed at 450°C for 1 s was used as the n-type contact to the emitter and collector.

The present structure therefore consists of a HBT with a double barrier *in the base region*. In an alternative RTBT design, the double barrier can be placed *between a wide-gap graded emitter and a GaAs base*, giving rise to RT transistor action similar to that discussed here.

To understand the operation of this device, consider a common emitter bias configuration. Initially, the collector–emitter voltage V_{CE} and the base current I_B are chosen in such a way that the base–emitter and the base–collector junctions are respectively forward and reverse biased. If V_{CE} is kept constant and the base current I_B is increased, the base–emitter potential also increases until flat-band condition in the emitter region is reached [Fig. 5.27b (left)]. In going from the band configuration of Fig. 5.27a to that of Fig. 5.27b, the device behaves like a conventional transistor with the collector current linearly increasing with the base current (Figs. 5.27a–5.27b (right)]. The slope of this curve is, of course, the current gain β of the device. In this region of operation, electrons in the emitter overcome, by thermionic injection, the barrier of the base–emitter junction and undergo RT through the double barrier. If now the base current is further increased above the value I_{BTH} corresponding to the flat-band condition, the additional potential difference drops primarily across the first semi-insulating AlAs barrier (Fig. 5.27c), between the contacted and uncontacted portions of the base, since the highly doped emitter is now fully conducting. This pushes the conduction band edge in the $Al_{0.07}Ga_{0.93}As$ above the first energy level of the well, thus quenching the RT. The net effect is that the base transport factor and the current gain are greatly reduced. This causes an abrupt drop of the collector current as the base current exceeds a certain threshold value I_{BTH} [Fig. 5.27c (right)]. This is the most important manifestation of the inherent negative transconductance of this device. It should be noted that although the base metallization penetrates into the AlAs barrier, the latter can still sustain the potential drop in the region under the emitter, as discussed above, since the base contact is placed away from the emitter mesa and the barrier is undoped (semi-insulating). Nevertheless, it should be noted that RT transistor action can also be achieved if the *p*-type ohmic contact is made *only* to the portion of the base adjacent to the collector and not to the quantum well. Obviously, the band diagram configuration in the base region under operating conditions will be somewhat different. In this case, although there is no direct ohmic contact to the quantum well, electrical connection between the 2D hole gas in the well and the part of the base adjacent to the collector is established via tunneling of holes through the AlAs barrier.

The devices were biased in a common emitter configuration at 300 K and the *I–V* characteristics were displayed on a curve tracer. For base currents ≤2.5 mA, the transistor exhibits normal characteristics; for $I_B \geq 2.5$ mA, the behavior previously discussed was observed. Figure 5.28 shows the collector current vs. base current at $V_{CE} = 12$ V, as obtained from the common-emitter characteristics. The collector current increases with the base current and there is clear evidence of current gain ($\beta = 7$ for $I_C > 4$ mA). As the base current exceeds 2.5 mA, there is a drop in I_C because the current gain mechanism is quenched by the suppression of RT. The transistor characteristics were also measured in a pulsed mode using

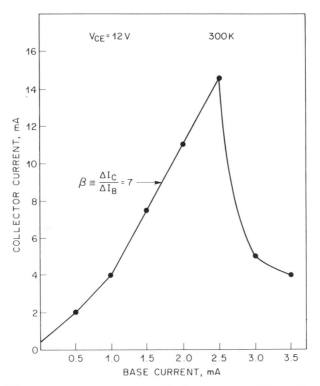

Fig. 5.28 Collector current vs. base current in the common-emitter configuration of the RTBT of Fig. 5.27 at room temperature with the collector–emitter voltage held constant. The line connecting the data points is drawn only to guide the eye.

300 μs pulses. No changes were detected and a behavior identical to that of Fig. 5.28 was observed, thus ruling out heating effects.

The devices exhibited similar behavior in the investigated temperature range between 100 and 300 K, although the negative conductance effects are more pronounced at lower temperatures because the quenching of RT is more abrupt than at higher temperatures. Figure 5.29 illustrates the common-emitter characteristics at 100 K. At relatively low V_{CE}, both the emitter–base and the collector–base junctions are forward biased (saturation region) and the collector current is negative, corresponding to a net flow of electrons from the collector contact into the collector layer. This gives rise to a collector–emitter offset voltage, which is typical of asymmetric heterojunction bipolar transistors.[64] In the present case, this offset is large due to the relatively small α of this structure at low temperatures and the relatively large emitter resistance and ideality factor ($n \approx 2$) of the base–emitter pn junction.[64] This offset is greatly reduced at room temperature. As V_{CE} is further increased, the collector–base junction becomes sufficiently reverse biased, and for base currents $I_B \leq 4$ mA, the characteristics are similar to that of a conventional bipolar (Fig. 5.29). For $I_B > 4$ mA, at sufficiently high V_{CE}, the collector current instead decreases with increasing base current. It is apparent from Fig. 5.29 that in addition to this behavior, which was previously discussed, there is also a large negative conductance in the I_C vs. V_{CE} curve for base currents in excess of the threshold value (=4 mA). This is easy to understand by noting that in order to reach the band configuration of Fig. 5.27c and quench RT (at a fixed $I_B > I_{BTH}$) the collector–emitter voltage V_{CE} should be large enough for the collector–base junction to be reverse biased and draw a significant collector current. In the present structure, the higher I_{BTH} at lower temperatures (4 mA at 100 K compared with 2.5 mA at 300 K) is a consequence of the larger collector–emitter offset voltage.

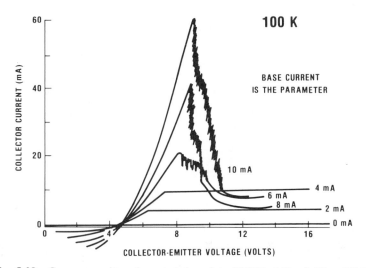

Fig. 5.29 Common-emitter characteristics of the RTBT of Fig. 5.27 at 100 K.

These bipolars in the common-emitter configuration act as oscillators when biased in the negative conductance region of the characteristic. The current oscillation in the collector circuit was picked up by a loop and displayed. Figure 5.30 shows the oscillation in the time domain at room temperature and in the frequency domain at both 300 K and 100 K at an operating point $V_{CE} = 12$ V, $I_B = 6$ mA. Note the high spectral purity (near-single-frequency response), particularly at low temperature. The oscillation frequency ($\simeq 20$ MHz) is limited at present by the probe stage used and the collector-bias circuit. Much higher oscillation frequencies (>10 GHz) may be ultimately expected.

By changing the bias conditions (I_B, V_{CE}), the oscillation frequency can be tuned over a few MHz range. At room temperature, the device ceases to oscillate for base currents $\lesssim 2.5$ mA, since at such base currents the device is out of the negative conductance region (RT is not quenched; see Fig. 5.28). A simple, physical picture of the oscillations can be given. As the dc base bias current is increased above the threshold value, RT is suppressed. Thus, the collector current is reduced, which implies a reduction of the emitter current since the bias current I_B is kept fixed. This, in turn, implies a decrease of the voltage applied across the emitter, followed by a restoration of the flat-band conditions (since $I_B > I_{BTH}$), and the cycle is repeated.

5.3.4 Alternate Designs of RTBT

An alternative RTBT design, with the double barrier in the base–emitter junction rather than in the base region, was reported shortly after by Futatsugi et al.[24] of Fujitsu Laboratories. The operation of this device is very similar to the one discussed in the previous section. With the emitter–base junction forward biased, resonance peak could be observed at 77 K in the emitter and, hence, in the collector current also as a function of base–emitter voltage V_{BE} at a fixed collector–emitter voltage V_{CE}. However, in this design, the tunneling probability becomes a strong function of the base–emitter voltage V_{BE}. This makes it difficult to observe NDR at room temperature. Recently, the Fujitsu group has implemented an RTBT in the AlInAs/GaInAs material system, grown lattice matched to an InP substrate, with the double barrier in the emitter layer.[65] This structure also suffers from the drawback that the tunneling probability is a strong function of the base–emitter voltage. However, the parameter advantages of this material system, discussed earlier, made it possible to observe NDR at room temperature.[65]

5.4 RESONANT TUNNELING UNIPOLAR TRANSISTORS

Following the announcement of the RTBT from Bell Laboratories in 1984, several unipolar three-terminal devices were proposed and implemented that utilized the RT structure as electron injectors to generate voltage tunable NDR and negative transconductance characteristics. The first of these unipolar structures was the RHET[17] developed at the Fujitsu Laboratories in 1985. The same year, Luryi and Capasso proposed the quantum wire transistor,[20] a novel device in which the quan-

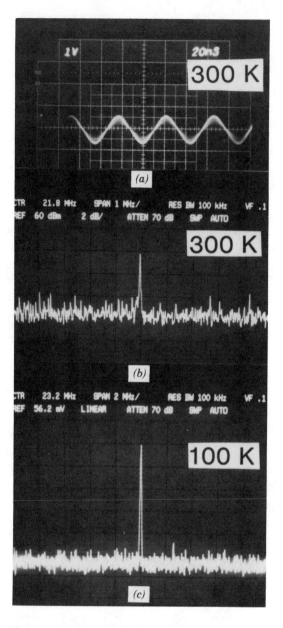

Fig. 5.30 (*a*) Oscillations and (*b*) spectral characteristics for a resonant tunneling transistor operating as an oscillator at room temperature (center frequency 21.8 MHz; span 1 MHz). (*c*) Frequency response at 100 K of the same device (center frequency 23.2 MHz; span 2 MHz). The oscillation frequency is limited by the external circuit.

tum well is linear rather than planar and the tunneling is of 2D electrons into a 1D density of states. In 1987, the resonant tunneling gate field-effect transistor (RT-FET)[25,27,28] was developed at Bell Laboratories. Recently, a device exhibiting the Stark effect[18] and the quantum capacitance effect[66] has been demonstrated at Bell Laboratories.[30] Among other proposals of unipolar three-terminal devices, the one involving integration of RT diodes and FETs, developed jointly by California Institute of Technology and Xerox PARC[19,26,67] and their circuit applications,[68] deserve mention. These devices are discussed in this section.

5.4.1 Resonant Tunneling Hot-Electron Transistor (RHET)

The schematic band diagram of the RHET, demonstrated by Yokoyama et al. of Fujitsu Laboratories,[17] is shown in Fig. 5.31. The structure consisted of a RT double barrier placed between GaAs base and emitter layers. The RT structure was made of a 56 Å thick GaAs quantum well sandwiched between two 50 Å thick $Al_{0.33}Ga_{0.67}As$ barriers. The RT double barrier between the base and the emitter simply served the purpose of injecting, through RT, high-energy electrons into the base region. The high-energy electrons are transported ballistically through the 1000 Å thick n^+ base region before being collected at the 3000 Å thick $Al_{0.20}Ga_{0.80}As$ collector barrier. The barriers and the quantum well were undoped

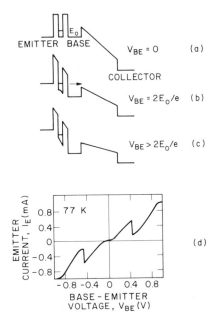

Fig. 5.31 The band diagrams of the RHET at (a) $V_{BE} = 0$, (b) $V_{BE} = 2E_0/e$ (RT established), and (c) $V_{BE} > 2E_0/e$ (RT quenched), illustrating the operating principle of the device, and (d) base–emitter current–voltage characteristics measured at 77 K.

whereas the emitter, base, and the collector layers were n-type doped to $1 \times 10^{18}/cm^3$.

The operation of the device in the common-emitter configuration with a fixed collector–emitter voltage V_{CE} is schematically shown in the band diagrams of Fig. 5.31. When the base–emitter voltage V_{BE} is zero (Fig. 5.31a), there is no electron injection and, hence, the emitter as well as the collector currents are zero even with a positive V_{CE}. A peak in the emitter and the collector current occurs when V_{BE} is nearly equal to $2E_0/e$ (where E_0 is the energy of the first resonant state in the quantum well; Fig. 5.31b). With further increase in V_{BE}, RT is quenched (Fig. 5.31c) and, hence, there is a sudden drop in the emitter current (Fig. 5.31d) and a corresponding drop in the collector current as well.[17]

This device could in principle be used for the same applications discussed in Section 5.3.1. The Fujitsu group has demonstrated its application as an exclusive-NOR gate,[17] which is essentially the parity generator circuit of Fig. 5.21 with two inputs. The base emitter characteristics were also used to design a flip–flop,[69] as shown in the inset of Fig. 5.32. The circuit consists of a RHET with a 51 kΩ collector resistance and a 1.6 kΩ resistance in series with the base. The collector and the base supply voltages are 4.0 V and 1.0 V, respectively. It should be noted that the base current vs. base–emitter voltage also exhibits a resonance peak. The load line corresponding to the 1.6 kΩ base resistance and 1.0 V base supply voltage, drawn on the I_B vs. V_{BE} characteristic, has two stable points P_1 and P_2 (Fig. 5.32). The circuit can thus stay in either of these two points indefinitely and may also be shifted from one to the other by applying suitable pulses. At P_1, the transistor is heavily conducting and therefore the output at the collector node will be low; at P_2, since RT has been quenched, the transistor is essentially in the poorly conducting state, resulting in a high state at the output.

The RHET is, however, operable at a low temperature (77 K) only and possesses very little current gain, since it is an unipolar device. A large fraction of the hot electrons injected into the base cannot surmount the collector barrier because of large phonon scattering in the n^+ base region. Recently, the RHET has been

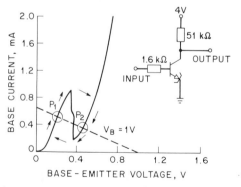

Fig. 5.32 Schematic operation of the flip–flop circuit (shown in the inset) using the RHET.

implemented in the $Al_{0.48}In_{0.52}As/Ga_{0.47}In_{0.53}As$ material system[45] with improved characteristics, as is expected.

5.4.2   Resonant Tunneling Gate Field-Effect Transistor (RT-FFET).

5.4.2.1   *Structure and Processing.*   Figure 5.33a shows the schematics of the device grown by MBE on an LEC semi-insulating substrate. The structure consists of an n-type GaAs channel, 1 μm thick, and doped to $4 \times 10^{17}/cm^3$. The double barrier, consisting of a 70 Å undoped GaAs well layer sandwiched between two 25 Å undoped AlAs barriers, was grown on top of the channel. The gate contact layer is 0.4 μm thick GaAs and doped to $5 \times 10^{17}/cm^3$. The undoped GaAs spacer layers (~25 Å) are left on the two sides of the undoped double barrier to offset the effect of Si diffusion from the adjacent n-type layers during the high-temperature growth process.

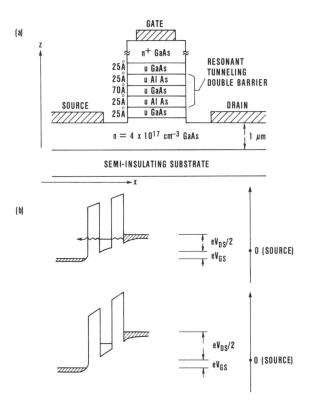

Fig. 5.33   (a) Schematic cross section of the resonant tunneling gate field-effect transistor RT-FET. (b) Band diagram showing RT through double barrier in the gate, electrons injected from the channel, and quenching of RT. Note the combined action of the gate and drain voltages in biasing the double barrier.

The devices have a 5.65 μm gate length, 308 μm gate width, 2.15 μm source-gate spacing, and 9.8 μm source-drain spacing. The asymmetry between the source-gate and drain-gate spacing was unintentionally introduced during processing.

The devices were processed by evaporating the gate electrode followed by mesa etching. The gate mesa was formed by wet etching using, alternately, the stop etches H_2O_2 + NH_4OH (pH = 7.2) for the GaAs and 1:1 HCl + H_2O for the AlAs layers. The height of the gate mesa was 5500 Å. The source and drain electrodes were then deposited on the exposed channel. The gate contact was composed of Ge (60 Å)/Au (135 Å)/Ag (500 Å)/Au (750 Å) alloy. Ni (50 Å)/Au (385 Å)/Ge (215 Å)/Au (750 Å) alloy was used as the source and drain contacts. All the contacts were alloyed together at 400°C for 10 s in a hydrogen flow.

5.4.2.2 Principle of Operation.

The operation of the device is based on the quenching of RT through the double barrier between the gate electrode and the channel. The biasing of the double barrier when a gate-to-source voltage (V_{GS}) as well as a drain-to-source voltage (V_{DS}) is applied is shown in Fig. 5.33b. Note the combined action of the gate voltage and half the drain voltage in controlling RT. The voltage appearing across the double barrier is, to a first approximation, given by

$$V_{DB} = V_{GS} - r \cdot V_{DS}, \tag{5}$$

where r is a factor determined by the ratio of the gate-to-source and the drain-to-gate spacings. Ideally, r should be 0.5. In the devices tested, however, this factor was 0.42 due to the structural asymmetry unintentionally introduced during processing. Thus, RT through the double barrier may be controlled by either the gate or the drain voltage to produce peaks in the gate-current characteristics. This, in turn, generates structures in the drain current as well, since the gate current adds to the drain-to-source current in the channel. Thus, in addition to negative differential resistance, negative transconductance was obtained as well, by controlling structures in the drain current with the variation of the gate voltage.

I_D and I_G vs. V_{GS}. To understand the $I-V$ characteristics quantitatively, it is necessary to discuss the potential distribution inside the device. Let us first consider the gate I_G and the drain I_D current variation with V_{GS} when V_{DS} = 0 V. Under this condition, the electrons tunneling from the gate to the channel, and vice versa, form two equal (in the ideal case, when r = 0.5) and opposite current flows in the channel between its center and the source and drain electrodes. Half the gate current thus forms the drain current. The resulting potential distribution is shown in Figs. 5.34a and b for positive and negative gate voltages, respectively. The conduction band energy diagram across the double barrier is shown on the left of Fig. 5.34; the electron potential energy distribution due to ohmic drop along the channel in the region immediately adjacent to the double barrier is shown on the right. The interesting points to note from these diagrams are the following:

1. For positive gate bias, an electron accumulation layer is formed at the heterointerface between the channel and the bottom AlAs barrier while the gate

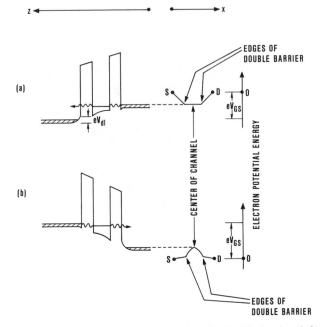

Fig. 5.34 Conduction band energy diagram across the double barrier (left) and electron potential energy distribution along the channel in the region immediately adjacent to the double barrier (right) at $V_{DS} = 0$ V for (a) $V_{GS} > 0$ V and (b) $V_{GS} < 0$ V.

side of the double barrier is depleted of carriers. The density of electrons in the thin (100–200 Å) accumulation layer substantially exceeds the average carrier concentration in the channel, thus strongly reducing, by screening, the source-to-drain electric field in the immediate vicinity of the double-barrier–channel interface, compared with the rest of the channel, as shown at the right of Fig. 5.34a. The near equipotential at the double barrier channel interface allows RT to be quenched everywhere at the same gate bias, resulting in an abrupt drop of the current with voltage, as observed in the experimental data (Fig. 5.35a). The important role of the accumulation and depletion layers adjacent to the double barrier on the I–V characteristics of RT diodes has been discussed by Goldman et al.[70] These data provide the first direct evidence of such effect.

2. For negative gate bias, on the other hand, the accumulation layer is formed on the gate side while the region of the channel adjacent to the double barrier is depleted of carriers. As a result, the flow of electrons causes a near-triangular potential variation in the channel with the center being at the highest energy and the edges of the double barrier being at the lowest (Fig. 5.34b). Thus, quenching of RT is initiated at the edges of the double barrier and gradually proceeds to the center with increasing magnitude of negative gate voltage. The I–V therefore exhibits a gradual drop, as opposed to the abrupt drop with positive gate bias (Fig. 5.35a). The same effect is

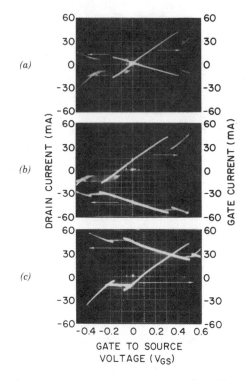

Fig. 5.35 Drain and gate currents vs. gate-to-source voltage V_{GS} measured at 100 K for (a) $V_{DS} = 0$ V, (b) $V_{DS} = -0.2$ V, and (c) $V_{DS} = 0.2$ V.

also observed for finite voltages applied externally between the drain and the source (Figs. 5.35b and c), since the accumulation layer in the channel also screens the applied drain-to-source field in the immediate vicinity of the double barrier, rendering it near equipotential. For negative gate bias in this latter case, the electron potential energy diagram in the channel (which is left to the reader as an exercise) implies that RT is quenched initially at either the drain or the source end (depending on the polarity of V_{DS}) of the double barrier and gradually proceeds to the other end with increasing magnitude of gate voltage.

3. For positive gate voltage (Fig. 5.34a), there is a significant potential difference between the equipotential region under the double barrier and the source and drain electrodes (which are at the same potential, since $V_{DS} = 0$). This is due to the flow of electrons in the channel that come from the gate by RT. Because of this ohmic drop, only a part of the applied V_{GS} appears across the double barrier. Also, there is a potential drop V_{dl} (~ 0.1 V) in the depletion layer adjacent to the double barrier on the gate side (Fig. 5.34a, left). Consequently, the voltage appearing across the double barrier is substantially less than that applied between the gate and source electrodes externally. Therefore, a larger V_{GS} (0.4 V) is required to quench RT than what is expected from a simple tunneling calculation ($2E_1 \simeq 0.2$ V).

For negative V_{GS}, on the other hand, the portion of the channel under the gate immediately adjacent to the double barrier is depleted. Hence, the ohmic potential drop is larger in this part of the channel than in the rest, bringing the center of the channel to the highest energy point (Fig. 5.34b, right) Also, as there is no depletion layer in the gate side of the barrier, the potential drop V_{dl} is absent. The voltage appearing across the double barrier at its edges is almost equal to the applied voltage V_{GS}. Quenching of RT is therefore initiated at the edges of the double barrier and at a much lower $|V_{GS}|$ ($\simeq 0.2$ V, as expected from tunneling calculations) than with positive gate bias. With increasing negative V_{GS}, this propagates inward until RT is finally quenched at the center. The absence of the drop V_{dl} also helps it to quench RT totally at a lower V_{GS} (0.3 V), as opposed to 0.4 V required in the case of positive V_{GS}.

When the drain is negatively biased with respect to the source, the negative conductance region and the overall $I-V$ characteristics shift to a lower V_{GS} (Fig. 5.35b). This is a result of the increase of the electron potential energy in the region of the channel under the gate, so that less positive bias is required on the gate to quench RT. For negative gate bias, however, a larger voltage is necessary to obtain the same effect. Note that, as RT is quenched, there is a change not only in I_G but also in I_D, as in the case with $V_{DS} = 0$. This can be explained as follows. With $V_{DS} < 0$ and $V_{GS} > 0$, the drain current consists of electrons flowing from the drain to the source and the drain to the gate. When RT is quenched, the latter flow is reduced and I_D drops.

For $V_{DS} < 0$ and $V_{GS} < 0$, however, the picture is somewhat different. With $V_{GS} \simeq 0$, the electron potential energy in the center of the channel is higher than that in the gate due to the drain-to-source current flow. Electrons thus flow from the channel to the gate by RT, resulting in a positive gate current for small negative gate voltages. With increasing negative bias on the gate, this flow decreases and finally changes direction, giving rise to a negative gate current at a sufficiently large V_{GS}. Before RT is quenched, electrons flowing from the gate to the channel by RT add up to those coming from the drain by ohmic conduction and flow out in the source electrode, causing an ohmic potential drop in the source-gate part of the channel. The electron potential energy in the region under the gate is thus higher than what would have been expected in the absence of the gate current. When RT is quenched, the potential drop in the source-gate part, therefore, is decreased. But as the drain-to-source voltage is fixed, the potential drop in the drain-gate part of the channel has to increase, resulting in an increase in the drain current flow. This is observed as a sudden increase in the magnitude of the negative drain current.

For $V_{DS} > 0$ (Fig. 5.35c), a negative gate current is produced even at $V_{GS} = 0$ by the electrons resonantly tunneling from the gate to the channel, since the positive V_{DS} has brought the electron potential energy in the middle of the channel below that in the gate. These electrons also add to the drain current. With V_{GS} becoming positive, the RT gate current is first reduced to zero and then changes direction with a corresponding reduction in the drain current, as some of the electrons coming from the source are directed to the gate. When RT is finally quenched with a drop in the gate current, there is a corresponding abrupt rise in the drain current because all the electrons coming from the source are now directed to the drain. For negative V_{GS}, the resonantly tunneling electrons from the gate always add to the drain current and, hence, the latter increases with increasing $|V_{GS}|$. A drop in the gate as well as the drain current is observed when RT is quenched.

I_D and I_G vs. V_{DS} for Different V_{GS}.   The foregoing discussions indicate that a negative drain bias has the same effect as a positive gate bias and vice versa. It therefore is easy to understand that the drain I_D and the gate I_G current variation with V_{DS} at different V_{GS} will show similar NDR characteristics, with the suppression of RT being abrupt at negative V_{DS} and gradual with positive V_{DS}. Also, for a change ΔV_{GS} in the gate bias, the NDR region will shift by $\Delta V_{DS} \simeq 2 \Delta V_{GS}$.

Figure 5.36 shows I_G and I_D as a function of the drain-to-source bias V_{DS} for $V_{GS} = 0$ (top) and $V_{GS} = +0.2$ V (bottom). Let us consider the curves for $V_{GS} = 0$ first. Note that for positive V_{DS}, the region of the channel adjacent to the double barrier is depleted and electrons resonantly tunnel from the gate contact layer into the channel. For negative V_{DS}, on the other hand, electrons resonantly tunnel from the accumulation layer at the double-barrier–channel interface into the gate contact layer. In fact, from Fig. 5.33b, we find that the application of a positive gate voltage has the same effect, as far as biasing the double barrier and controlling RT, as

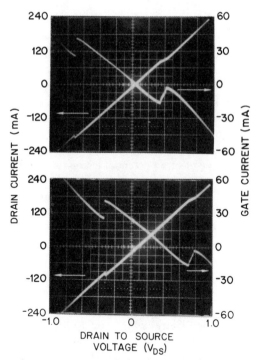

Fig. 5.36 Drain and gate currents vs. drain-to-source voltage V_{DS} measured at 100 K for $V_{GS} = 0$ V (top) and $V_{GS} = 0.2$ V (bottom).

that of a negative drain bias. It is then easy to explain the differences in the steepness of the I–V (Fig. 5.36) in the NDR regions for different bias polarities in a manner similar to those in Fig. 5.35. Following similar arguments, it is also understood why the NDR region occurs at a significantly lower $|V_{DS}|$ when the drain is positively biased than when it is negatively biased. Application of a positive gate bias ($V_{GS} = +0.2$ V, Fig. 5.36, bottom) shifts the NDR region to higher V_{DS} by nearly 0.4 V, as expected. Also, at $V_{DS} = 0$, electrons now tunnel from the channel into the gate, giving rise to positive gate current. A positive V_{DS} is then required to suppress this tunneling first ($I_G = 0$ at $V_{DS} \approx 0.4$ V for $V_{GS} = 0.2$ V) and then reverses the direction of resonant tunneling electrons.

An even more interesting situation occurs when the applied gate bias is large enough to quench RT even at $V_{DS} = 0$. The conduction band energy diagram across the double barrier under this condition (with a positive V_{GS}) is shown in Fig. 5.37a (left). Even if RT is suppressed, there will be some flow of electrons from the channel to the gate by thermionic emission over the barrier and inelastic

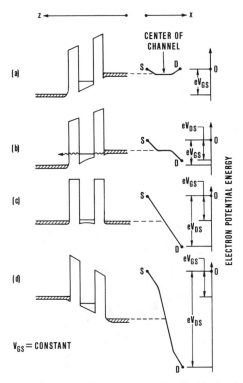

Fig. 5.37 Conduction band energy diagram across the double barrier (left) and electron potential energy distribution along the channel in the region immediately adjacent to the double barrier (right) at a large positive V_{GS} (constant): (a) $V_{DS} = 0$ V, RT quenched; (b) small positive V_{DS}, RT established; (c) larger V_{DS}, gate current is zero; and (d) large positive V_{DS}, RT quenched again.

tunneling, resulting in a positive gate current (Fig. 5.38). These electrons are supplied by the drain and the source contacts. The resulting ohmic drop will place the drain and the source points at a higher energy than the region under the double barrier, as shown in the right half of Fig. 5.37a. As a positive drain voltage is applied, the electron flow from the drain end is reduced, resulting in a reduction in the negative drain and positive gate currents until a condition is reached when the drain is at the same potential as the middle of the channel and the drain current is reduced to zero. As the drain voltage is further increased, a positive drain current starts flowing. The positive drain voltage also pulls down the middle of the channel in the energy diagram, and the gate current continues to decrease until RT is restored at some V_{DS} (Fig. 5.37b). At this point, a large fraction of the electrons are again transferred to the gate by resonant tunneling, resulting in a sudden increase in the gate current and corresponding decrease in the drain current, as observed experimentally (Fig. 5.38). Note that the onset of RT is quite abrupt in this case because of the accumulation layer in the channel (Fig. 5.37b). Beyond this point, the gate current continues to decrease with a decreasing RT flow of electrons and the drain current increases.

At larger V_{DS}, the drain current is large compared with the gate current, and hence the potential in the channel can be determined solely by V_{DS}. So, at $V_{DS} \simeq 2V_{GS}$, the gate is at the same potential as the center of the channel (Fig. 5.37c) and no gate current flows. The experimental data taken at $V_{GS} = 0.5$ V (Fig. 5.38) in fact indicate that $I_G = 0$ at $V_{DS} \simeq 1.0$ V. In this situation, the region of the channel adjacent to the double barrier cannot be considered equipotential, as there cannot be any accumulation of electrons. The part of the region nearer the source is at a higher energy than the gate while the other half is at a lower energy. Electrons therefore transfer from the channel to the gate at the source end while in the opposite direction at the other end. When these two flows are equal, gate current is zero.

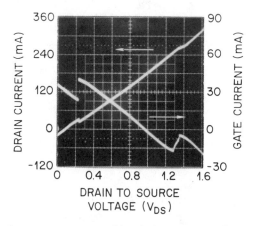

Fig. 5.38 Drain and gate currents vs. positive drain-to-source voltage at $V_{GS} = 0.5$ V, exhibiting two peaks as illustrated in Fig. 5.37.

With a further increase in the drain voltage, electrons start flowing from the gate to the channel by RT, giving rise to a negative gate current. These electrons also add up in the channel with those coming from the source to increase the positive drain current further (Fig. 5.38). When RT is quenched again at a sufficiently large V_{DS} (Fig. 5.37d), there is a drop in the gate as well as the drain currents (Fig. 5.38). It should be noted that the quenching of RT is gradual in this case, as is expected, because of the depletion of the channel adjacent to the gate (Fig. 5.37d).

A similar situation can also be obtained with large negative bias applied to the gate and the characteristics taken against negative V_{DS}. Figure 5.39 shows the experimental data taken against $-V_{DS}$ at $V_{GS} = -0.3$ V. These curves can be explained quantitatively, following similar arguments as before. It may also be noted that, unlike in the previous figure, the first NDR region ($V_{DS} = -0.2$ V) in this case is broad, whereas the second one ($V_{DS} = -1.3$ V) is abrupt. With $V_{GS} = -0.3$ V, the channel is at a lower energy than the gate when $V_{DS} = -0.2$ V, and hence its depletion region adjacent to the double barrier leads to a gradual suppression of RT. For $V_{DS} = -1.3$ V, on the other hand, an accumulation layer is formed in the channel and thus the quenching is abrupt.

This part of the characteristics may be important in applications involving multiple peaks,[16,32,48] since the two peaks are obtained by a suitable manipulation of only the ground state resonance of the quantum well.

All these qualitative and quantitative agreements of the experimental data for various bias configurations verify the proposed model of the device.

In another attempt to integrate RTDB with field-effect transistors (FET), developed jointly by California Institute of Technology and Xerox PARC,[19,26,67] the double barrier is placed in the channel, adjacent to the source electrode. The operation of this class of structures, called DB/FET by its developers, is quite different from the resonant tunneling gate field-effect transistor (RT-FET) discussed

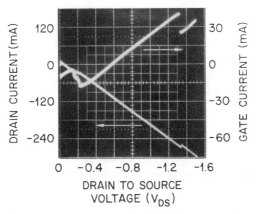

Fig. 5.39 Drain and gate currents vs. negative drain-to-source voltage at $V_{GS} = -0.3$ V, exhibiting two peaks.

above. In principle, the DB/FET is electrically equivalent to a RT diode with a resistance (that of the channel) in series with it. The role of the resistance is to push the NDR region in the I–V characteristics to higher voltages, since a significant fraction of the applied voltage drops across the same. In the DB/FET, since the gate voltage changes the channel resistance, it effectively controls the position of the peak. Thus, a voltage-controllable negative resistance characteristic is obtained.[19] The DB/FETs have been implemented both in the vertical FET[26] and the MESFET[67] geometry. The I–V characteristics in both the geometries illustrate the same basic nature, namely, gate-controllable negative resistance, although there are differences in details arising out of the differences in the field distributions in the two geometries. The DB/FETs have been used to construct frequency multipliers and flip–flops.[68]

5.4.3   Quantum Wire Transistor

This novel device, proposed by Luryi and Capasso,[20] uses a linear rather than a planar quantum well as the active region. In this device, electrons resonantly tunnel from 2D to 1D (quantum wire) states. The properties of zero- and one-dimensional systems are receiving increasing attention as new techniques are developed to realize them. Sakaki[71] discussed the possibility of obtaining an enhanced mobility along quantum wires because of the suppression of the ionized-impurity scattering. He proposed a V-groove etch of a planar heterojunction quantum well as a means of achieving the one-dimensional confinement. Chang et al.[72] proposed a technique involving epitaxial overgrowth of a vertical $\langle 110 \rangle$ edge of pre-grown $\langle 100 \rangle$ heterostructure as a means of obtaining the quantum wire. Petroff et al.[73] reported experimental attempts of implementing one-dimensional confinement by etching techniques. Cibert et al.[74] experimentally demonstrated carrier confinement in one and zero dimensions. They used a novel technique involving electron-beam lithography and laterally confined interdiffusion of aluminum in a GaAs/AlGaAs heterostructure grown by MBE. Recently, Reed et al.[75] reported dimenionsal quantization effects in GaAs/AlGaAs "quantum dots."

The idea of the quantum wire transistor is illustrated in Fig. 5.40, assuming a GaAs/AlGaAs heterostructure implementation. The device consists of an epitaxially grown undoped planar quantum well and a double AlGaAs barrier sandwiched between two undoped GaAs layers and heavily doped GaAs contact layers. The working surface, defined by a V-groove etching, is subsequently overgrown epitaxially with a thin AlGaAs layer and gated. The thickness of the gate barrier layer ($d \gtrsim 100$ Å) and the Al content in this layer ($x \gtrsim 0.5$) should be chosen so as to minimize gate leakage. The thicknesses of the quantum well barrier layers are chosen so that their projection on the slanted surface should be $\lesssim 50$ Å each. The Al content in these layers should be typically $x \lesssim 0.45$. Application of a positive gate voltage V_G induces 2D electron gases at the two interfaces with the edges of undoped GaAs layers outside the quantum well. These gases will act as the source (S) and drain (D) electrodes. At the same time, there is a range of V_G in which electrons are not yet induced in the quantum wire region (which is the edge of the quantum well layer) because of the additional dimensional quantization.

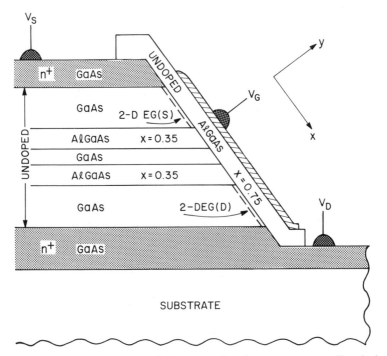

Fig. 5.40 Schematic cross section of the proposed surface resonant tunneling device, the quantum wire transistor structure. A V-groove implementation of the quantum wire is assumed. Thicknesses of the two undoped GaAs layers outside the double barrier region should be sufficiently large ($\gtrsim 1000$ Å) to prevent the creation of a parallel conduction path by the conventional (bulk) RT.

To understand the operation of the device, consider first the band diagram in the absence of a source to drain voltage, $V_{DS} = 0$ (Fig. 5.41a). The diagram is drawn along the x direction (from S to D parallel to the surface channel). The y direction is defined as the one normal to the gate and z, that along the quantum wire. Dimensional quantization induced by the gate results in a zero-point energy of electronic motion in the y direction, represented by the bottom E_0 of a 2D subband corresponding to the free motion in the x and z directions. The thicknesses of the undoped S and D layers are assumed to be large enough ($\gtrsim 1000$ Å) so that the electronic motion in the x direction in these layers can be considered free. On the other hand, in the quantum well region of the surface channel, there is an additional dimensional quantization — along the x direction — which defines the quantum wire.[71] If t is the x projection of the quantum well layer thickness, then the additional zero-point energy is approximately given by

$$E_0' - E_0 = \frac{\pi^2 \hbar^2}{2m^* t^2}.$$
(6)

This approximation is, of course, good only when the barrier heights substantially exceed E_0'.

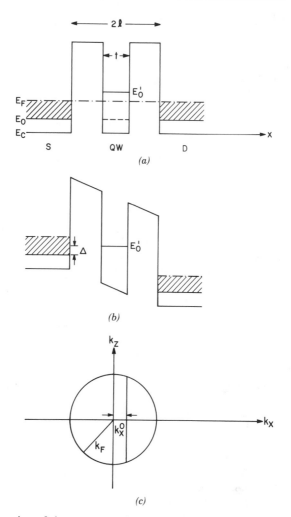

Fig. 5.41 Illustration of the quantum wire transistor operation. (a) Band diagram along the channel in "equilibrium," that is, in the absence of a drain bias. (b) Band diagram for an applied bias V_{DS} when the energy of certain electrons in the source (S) matches unoccupied levels of the lowest 1D subband E_0' in the quantum wire. (c) Fermi disk coresponding to the 2D degenerate electron gas in the source electrode. Vertical chord at $k_x = k_x^0$ indicates the momenta of electrons which can tunnel into the quantum wire while conserving their momentum k_z along the wire.

Application of a gate voltage V_{GS} moves the 2D sub-band E_0 with respect to the (classical) bottom of the conduction band E_C and the Fermi level E_F. The operating regime of this device with respect to V_{DS} at $V_{GS} = 0$ corresponds to the situation when E_F lies in the gap $E_0'-E_0$. Resonant tunneling condition is set in by the application of a positive V_{DS}, as illustrated in Fig. 5.41b. In this situation, the energy of certain electrons in S matches unoccupied levels in the quantum wire

(Fig. 5.41c). Compare this with the tunneling of 3D electrons into 2D density of states discussed in Section 5.2.1.1. In the present case, the dimensionality of both the emitter and the base is reduced by one. Hence, the emitter Fermi sea of Fig. 5.1 has become a disk in this case and the Fermi disk of the previous case is replaced by a resonant segment, as shown in Fig. 5.41c. Since both k_x and k_y are quantized in the quantum wire, resonant tunneling requires conservation of energy and the lateral momentum k_z. This is true only for those electrons whose momenta lie in the segment $k_x = k_x^0$ (Fig. 5.41), where

$$\hbar^2(k_x^0)^2 2m^* = \Delta .\tag{7}$$

It should be noted that the energies of all electrons in this segment ($k_x = k_x^0$) lie in the band $E_0 + \Delta \leq E \leq E_F$. However, only those electrons in this energy band that satisfy the momentum conservation condition are resonant. As V_{DS} is increased, the resonant segment moves to the left (Fig. 5.41c), toward the vertical diameter $k_x = 0$ of the Fermi disk, and the number of tunneling electrons grows, reaching a maximum $[2m^*(E_F - E_0)]^{1/2}/\pi\hbar$ per unit length in the z direction when $\Delta = 0$. At higher V_{DS}, when $\Delta < 0$, there are no electrons in the source that can tunnel into the quantum wire while conserving their lateral momentum. This gives rise to the NDR in the drain circuit.

In the present device, apart from obtaining high electron mobility, as discussed by Sakaki,[71] additional flexibility is achieved through the gate electrode. The gate voltage in this structure not only determines the number of electrons available for conduction but also controls the position of the E_0' level in the quantum wire with respect to E_0 in the source. This latter control is effected by the fringing electric fields and gives rise to the interesting possibility of negative transconductance, as in the RT-FET. Luryi and Capasso[20] have solved the corresponding electrostatic problem by suitable conformal mappings and have shown that in the operating regime of the device, an increasing $V_{GS} > 0$ *lowers* the electrostatic potential energy in the base (quantum wire) with respect to the emitter (source) nearly as effectively as does an increasing V_{DS}.

5.4.4 The Gated Quantum Well Resonant Tunneling Transistor

The device discussed in this section is similar to the one proposed by Bonnefoi et al.[18] under the name of Stark effect transistor (SET). The key idea of that transistor was the use of a quantum well collector and the inverted sequence of layers in which the controlling electrode (here referred to as the gate[†]) was placed "behind" the collector layer. It was predicted[18] that the gate field would modify the positions of the collector sub-bands with respect to the emitter Fermi level and thus modulate the tunneling current. The SET structure was reexamined by Luryi,[76] revealing

---

[†]It appears that the nomenclature "emitter, collector, and gate" is best suited for a transistor of this kind. Like all potential-effect transistors, this one has an emitter, a collector, and controlled injection process, but the control is effected by an insulated gate through which no dc current is flowing.

its additional possibilities, in particular the existence of NDR at a fixed gate bias and negative transconductance at a fixed emitter–collector bias. A practical device demonstrating these effects was demonstrated at Bell Laboratories in 1988[30] and is discussed here.

Figure 5.42 shows the schematics of the device grown by MBE in the AlGaAs material system. It consists of an undoped quantum well collector separated from a 5000 Å thick n^+-doped emitter layer by a thin undoped barrier. A thicker undoped barrier on the other side of the well is followed by a n^+-doped gate. The layer thicknesses indicated were measured by transmission electron microscopy; the doping level in the n^+ GaAs layers was nominally $2 \times 10^{18}/cm^3$. The devices were defined by standard photolithographic techniques and wet-etched with a $H_3PO_4/H_2O_2/H_2O$ solution. The evaporated contacts were provided by Ge/Au/Ag/Au for the emitter and collector and Ni/Au/Ge/Au for the gate.

The emitter–collector I–V characteristics of the device are expected to peak at biases which maximize the RT of the emitter electrons into the 2D collector subbands. The expected NDR is similar to that in double barrier RT structures and results solely from the tunneling from a 3D into a 2D system. The presence of two tunnel barriers is not essential as discussed in Section 5.2.1.1.

Transistor action in the structure is obtained via the influence of the gate field on the alignment of the 2D electron gas (2DEG) energy levels relative to the emitter Fermi level. This occurs for two distinct reasons. One, which in accordance with Ref. 18 can be called a generalized Stark effect, is associated with the field penetration into the quantum well.[†] The other effect is the *quantum capacitance* of a 2DEG,[66] because of which the gate field partially penetrates beyond the 2D metal in the quantum well and induces charges on the collector electrode.

In Fig. 5.43, the band diagram of the device is shown in the common-collector configuration with applied biases $V_G > 0$ and $V_E < 0$ such that the bottom of the conduction band in the emitter is in resonance with the second collector sub-band; this corresponds to a peak in the current. The RT current can be subsequently

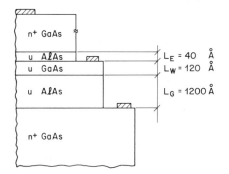

Fig. 5.42   Schematic cross section of the gated quantum well resonant tunneling transistor.

[†]Quantum-confined Stark effect (shift of the quantum well levels by a transverse electric field has been considered in a different context by D. A. B. Miller et al)[77].

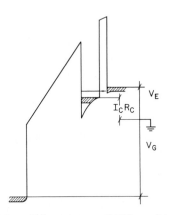

Fig. 5.43 Band diagram of the gated quantum well RT transistor, with the collector at reference and the biases $V_G > 0$ and $V_E < 0$ corresponding to peak resonant tunneling of emitter electrons into the second sub-band of the well.

quenched by increasing V_G; this leads to negative transconductance. As will become clear later, the observed peaks correspond to the RT into the second sub-band rather than the first, because at the experimental bias conditions corresponding to the peaks, the ground sub-band bottom is below the conduction band edge in the emitter. Moreover, the ground sub-band wave function in the triangular part of the well is displaced away from the emitter barrier, which further suppresses tunneling into it.

Figure 5.44 shows the experimental data at 7 K. The expected features present in the I–V were observed (though not as pronounced) up to liquid nitrogen tem-

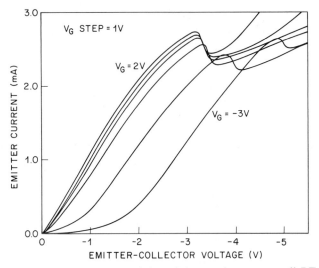

Fig. 5.44 Common-collector characteristics of the gated quantum well RT transistor at various gate voltages $(2, 1, 0, -1, -2, -3 \text{ V})$. The measurements were performed at 7 K.

perature. Both NDR in the emitter current and the control of the I–V by the gate are observed. In particular, if we examine I_E at a fixed V_E, we see that the device can be brought into and out of resonance by varying V_G. This control is electrostatic in nature, as evidenced by the fact that the gate current is always several orders of magnitude smaller than the emitter current. At the resonances, for instance, I_G varies from ~ 1 pA with $V_G = -3$ V up to ~ 10 nA with $V_G = 2$ V.

In order to understand quantitatively the operation of the device in the range $V_G > 0$, let us first analyze the shift of the I-V peaks with varying positive V_G. As discussed above, to estimate these shifts $(\Delta V_E)_p$, we need to calculate the shift of the second sub-band relative to $E_{F,E}$. This will be performed in two steps, first by calculating with first-order perturbation theory the "Stark" shift due to the penetration of the gate field into the well, and then by taking into account the quantum capacitance effect and the corresponding additional shift due to the gate field penetration into the emitter barrier.

Neglecting the quantum capacitance effect, the first-order variation in E_2 in response to a variation in the gate field is given by

$$\delta E_2 = \langle \Psi_2 | e\, \delta\phi | \Psi_2 \rangle \tag{8}$$

where $\delta\phi(z)$ is the variation in the electrostatic potential in the well, which can be calculated from the Poisson equation

$$\frac{d^2 \delta\phi}{dz^2} = \frac{e\, \delta n}{\varepsilon} |\Psi_1|^2. \tag{9}$$

Here, δn is the variation in surface electron concentration and ε the dielectric constant of the well layer. The ground sub-band, which is in the triangular part of the well (Fig. 5.43), can be approximated by the variational function of Stern and Howard,[78] $|\Psi_1|^2 = (b^3/2)z^2 e^{-bz}$, where $b = 3/\langle z \rangle_1$, $\langle z \rangle_1$ being the average penetration of the wave function in the well given by

$$\langle z \rangle_1 = \left(\frac{72\varepsilon\hbar^2}{11 m e^2 n} \right)^{1/3} = 81 \text{ Å} \times \left(\frac{10^{12}}{n} \right)^{1/3} \tag{10}$$

Equation (9) can then be integrated analytically. For the second state, which lies in the rectangular part of the energy diagram, on the other hand, one can use the second sine solution of the corresponding square well, that is, $\Psi_2 = (2/L_W)^{1/2} \sin(2\pi z/L_W)$. Equation (8) then gives

$$\delta E_2 = \frac{6e^2 \delta n}{\varepsilon b^2 L_W} \left(1 - \frac{\alpha^4 + 3\alpha^2 + 6}{6(1 + \alpha^2)^3} \right), \tag{11}$$

where $\alpha \equiv 4\pi/bL_W$. In this case of a rather thin well, the second term in the brackets in (11) is small. Defining the *gate leverage* factor λ as the variation in E_2

per unit variation in the gate voltage, $\lambda \equiv \delta E_2 / e \, \delta V_G$, one obtains the Stark effect contribution to this factor in the form

$$\lambda_S = \frac{2\langle z \rangle_1^2}{3 L_G L_W} \approx \frac{1}{51}, \qquad (12)$$

where $\langle z \rangle_1$ was evaluated from (10) with $n \approx 2 \times 10^{12}/\text{cm}^2$.

The second contribution to λ results from the gate field penetration into the emitter barrier. One can easily estimate the additional shift from the expression[66] for the quantum capacitance ideality factor ($\equiv \lambda^{-1}$), which gives

$$\lambda_Q = \frac{C_1}{C_1 + C_2 + C_Q} \approx \frac{1}{58}, \qquad (13)$$

where $C_1 \approx \varepsilon / L_G$ and $C_2 \approx \varepsilon / (L_E + L_W)$ are the gate–collector and emitter–collector geometric capacitances and C_Q is the quantum capacitance of a GaAs 2DEG,

$$C_Q = \frac{me^2}{\pi \hbar^2} \approx 4.5 \ \mu\text{F}/\text{cm}^2. \qquad (14)$$

The total leverage factor $\lambda = \lambda_S + \lambda_Q = 1/27$ is directly measured in the I–V as a shift $(\Delta V_E / \Delta V_G)_p$. The predicted value of $\lambda (\approx 37$ mV per 1 V variation in V_G) is in good agreement with the experimentally measured ~ 40 mV per 1 V variation in V_G.

The second question to be considered is why the peaks in the I–V occur at such a high applied bias V_E. This is believed to be associated with the series resistance introduced by the exposed part of the collector layer between the ring contact and the emitter mesa (in the present devices, this separation is $\approx 10 \ \mu$m). The Fermi level pinning at the surface depletes the exposed collector channel much like the gate of a field-effect transistor.† It is easily shown that, for the whole range of V_G examined, the portion of the collector layer between the mesa and the ring contact is always pinched off by the surface potential (≈ 0.8 V). It will then present a constant resistance R_C characteristic of a space-charge-limited conduction with a constant saturation velocity. The potential drop across this resistance is in series with the internal emitter-to-collector bias. Referring to Fig. 5.43, we can write (at resonance)

$$e I_E R_C + (E_2 - E_1) - E_{F,W} = e V_E - E_{F,E}, \qquad (15)$$

where $E_{F,E}$ and $E_{F,W} = \pi \hbar^2 n / m$ are the quasi Fermi levels in the emitter and collector, respectively, and E_1 and E_2 are respectively, the bottoms of the first and the

---

†It should be noted that while the exposed portion of the channel is depleted, the portion under the emitter is not. This follows unambiguously from the identification of the current peaks with resonant tunneling. At the bias conditions corresponding to the resonances in the range $V_G > 0$, the field drop associated with the charge in the quantum well is always $\gtrsim 2 \times 10^5$ V/cm, corresponding to $n \gtrsim 1 \times 10^{12}$ cm^2.

second sub-bands in the quantum well. Taking a variation of Eq. (15), using (11) and a similar relation easily derived for δE_1, and substituting the experimental variations of V_E and I_E at resonance, one obtains a constant $R_C \cong 1$ kΩ in a wide range of V_G. This constancy of R_C for the different characteristics indicates that the operation of the structure cannot be understood as resulting from the series combination of the RT emitter-to-collector diode and a parasitic FET.[26] It should be noted that the existence of a series resistance is indispensable for the understanding of the operation of this structure. The nonvanishing R_C ensures the charge accumulation and the resulting sharp field drop inside the well. If R_C were to vanish, then the quasi Fermi level in the well would be anchored to the ground and the well would be depleted by the negative emitter bias.

The operation of the device for negative V_G — at least for large magnitudes — is very different, because in that limit no charge is induced in the collector layer. The field is then constant across the structure and can be calculated simply as $(V_G - V_E)/(L_G + L_E + L_W)$. To reestablish the resonant condition after a variation of V_G, one has to vary V_E by the same amount. This is observed experimentally for V_G varying from -2 V to -3 V. For small negative V_G, it is difficult to estimate the amount of charge induced in the well, but, qualitatively, a smooth transition to the "dry-collector" regime can be expected. This is experimentally observed for V_G ranging from 0 V to -2 V where steps of 1 V determine (at resonance) increasing steps in V_E, ranging between the 40 mV and the 1 V limits discussed above.

ACKNOWLEDGMENTS

It is a pleasure to acknowledge the following colleagues for collaboration and technical discussions: A. C. Gossard, W. T. Tsang, R. J. Malik, S. Luryi, A. L. Hutchinson, R. A. Kiehl, R. C. Miller, D. L. Sivco, S. N. G. Chu, and K. Mohammed.

REFERENCES

1. L. L. Chang, L. Esaki, and R. Tsu, "Resonant Tunneling in Semiconductor Double Barriers," *Appl. Phys. Lett.*, **24**, 593 (1974).

2. M. Tsuchiya, H. Sakaki, and J. Yoshino, "Room Temperature Observation of Differential Negative Resistance in an AlAs/GaAs/AlAs Resonant Tunneling Diode," *Jpn. J. Appl. Phys.*, **24**, L466 (1985).

3. C. I. Huang, M. J. Paulus, C. A. Bozada, S. C. Dudley, K. R. Evans, C. E. Stutz, R. L. Jones, and M. E. Cheney, "AlGaAs/GaAs Double Barrier Diodes with High Peak-to-Valley Current Ratio," *Appl. Phys. Lett.*, **51**, 121 (1987).

4. S. Sen, F. Capasso, A. L. Hutchinson, and A. Y. Cho, "Room Temperature Operation of $Ga_{0.47}In_{0.53}As/Al_{0.48}In_{0.52}As$ Resonant Tunneling Diodes," *Electron. Lett.*, **23**, 1229 (1987).

5. T. Inata, S. Muto, Y. Nakata, S. Sasa, T. Fujii, and S. Hiyamizu, "A Pseudomorphic $In_{0.53}Ga_{0.47}As/AlAs$ Resonant Tunneling Barrier with a Peak-to-Valley Current Ratio of 14 at Room Temperature," *Jpn. J. Appl. Phys.*, **26**, L1332 (1987).

6. E. E. Mendez, W. I. Wang, B. Ricco, and L. Esaki, "Resonant Tunneling of Holes in AlAs-GaAs-AlAs Heterostructures," *Appl. Phys. Lett.*, **47**, 415 (1985).

7. M. A. Reed, J. W. Lee, and H.-L. Tsai, "Resonant Tunneling through a Double GaAs/AlAs Superlattice Barrier, Single Quantum Well Heterostructure," *Appl. Phys. Lett.*, **49**, 158 (1986).

8. T. Nakagawa, H. Imamoto, T. Kojima, and K. Ohta, "Observation of Resonant Tunneling in AlGaAs/GaAs Triple Barrier Diodes," *Appl. Phys. Lett.*, **49**, 73 (1986).

9. T. Nakagawa, T. Fujita, Y. Matsumoto, T. Kojima, and K. Ohta, "Resonant Tunneling of Holes in AlAs/GaAs Triple Barrier Diodes," *Appl. Phys. Lett.*, **50**, 974 (1987).

10. J. Allam, F. Beltram, F. Capasso, and A. Y. Cho, "Resonant Zener Tunneling of Electrons Between Valence-Band and Conduction-Band Quantum Wells," *Appl. Phys. Lett.*, **51**, 575 (1987).

11. S. Sen, F. Capasso, A. C. Gossard, R. A. Spah, A. L. Hutchinson, and S. N. G. Chu, "Observation of Resonant Tunneling Through a Compositionally Graded Parabolic Quantum Well," *Appl. Phys. Lett.*, **51**, 1428 (1987).

12. T. C. L. G. Sollner, W. D. Goodhue, P. E. Tannenwald, C. D. Parker, and D. D. Peck, "Resonant Tunneling Through Quantum Wells at Frequencies up to 2.5 THz," *Appl. Phys. Lett.*, **43**, 588 (1983).

13. T. C. L. G. Sollner, P. E. Tannenwald, D. D. Peck, and W. D. Goodhue, "Quantum Well Oscillators," *Appl. Phys. Lett.*, **45**, 1319 (1984).

14. E. R. Brown, T. C. L. G. Sollner, W. D. Goodhue, and C. L. Chen, "High-Speed Resonant-Tunneling Diodes," *Proc. Soc. Photo-Opt. Instrum. Eng.*, (to be published).

15. F. Capasso, "New High Speed Quantum Well and Variable Gap Superlattice Devices," in G. A. Mourou, D. M. Bloom, and C. H. Lee, Eds., *Picosecond Electronics and Optoelectronics*, p. 112, Springer-Verlag, Berlin and New York, 1985.

16. F. Capasso and R. A. Kiehl, "Resonant Tunneling Transistor with Quantum Well Base and High-Energy Injection: A New Negative Differential Resistance Device," *J. Appl. Phys.*, **58**, 1366 (1985).

17. N. Yokoyama, K. Imamura, S. Muto, S. Hiyamizu, and H. Nishi, "A New Functional Resonant Tunneling Hot Electron Transistor (RHET)," *Jpn. J. Appl. Phys.*, **24**, L853 (1985).

18. A. R. Bonnefoi, D. H. Chow, and T. C. McGill, "Inverted Base-Collector Tunnel Transistors," *Appl. Phys. Lett.*, **47**, 888 (1985).

19. A. R. Bonnefoi, T. C. McGill, and R. D. Burnham, "Resonant Tunneling Transistors with Controllable Negative Differential Resistance," *IEEE Electron Device Lett.* **EDL-6**, 636 (1985).

20. S. Luryi and F. Capasso, "Resonant Tunneling of Two Dimensional Electrons Through a Quantum Wire: A Negative Transconductance Device," *Appl. Phys. Lett.*, **47**, 1347 (1985); erratum: *ibid.*, **48**, 1693 (1986).

21. Y. Nakata, M. Asada, and Y. Suematsu, "Novel Triode Device using Metal Insulator Superlattice Proposed for High Speed Response," *Electron. Lett.*, **22**, 58 (1986).

22. F. Capasso, K. Mohammed, and A. Y. Cho, "Resonant Tunneling Through Double Barriers, Perpendicular Quantum Transport Phenomena in Superlattices, and their Device Applications," *IEEE J. Quantum Electron.*, **QE-22**, 1853 (1986).

23. F. Capasso, S. Sen, A. C. Gossard, A. L. Hutchinson, and J. H. English, "Quantum Well Resonant Tunneling Bipolar Transistor Operating at Room Temperature," *IEEE Electron Device Lett.*, **EDL-7,** 573 (1986).

24. T. Futatsugi, Y. Yamaguchi, K. Ishii, K. Imamura, S. Muto, N. Yokoyama, and A. Shibatomi, "A Resonant Tunneling Bipolar Transistor (RBT): A New Functional Device with High Current Gain," *Jpn. J. Appl. Phys.*, **26**, L131 (1987).

25. F. Capasso, S. Sen, F. Beltram, and A. Y. Cho, "Resonant Tunneling Gate Field-Effect Transistor," *Electron. Lett.*, **23**, 225 (1987).

26. T. K. Woodward, T. C. McGill, and R. D. Burnham, "Experimental Realization of a Resonant Tunneling Transistor," *Appl. Phys. Lett.*, **50**, 451 (1987).

27. S. Sen, F. Capasso, F. Beltram, and A. Y. Cho, "The Resonant Tunneling Field-Effect Transistor: A New Negative Transconductance Device," *IEEE Trans. Electron Devices*, **ED-34**, 1768 (1987).

28. F. Capasso, S. Sen, and A. Y. Cho, "Negative Transconductance Resonant Tunneling Field Effect Transistor," *Appl. Phys. Lett.*, **51**, 526 (1987).

29. F. Capasso, S. Sen, and A. Y. Cho, "Resonant Tunneling: Physics, New Transistors and Superlattice Devices," *Quantum Well Superlattice Phys.*, **SPIE-792**, 10 (1987).

30. F. Beltram, F. Capasso, S. Luryi, S. N. G. Chu, and A. Y. Cho, "Negative Transconductance Via Gating of the Quantum Well Subbands in a Resonant Tunneling Transistor," *Appl. Phys. Lett.*, **53**, 219 (1988).

31. G. H. Heilmeir, "Microelectronics: End of the Beginning or Beginning of the End?," *Tech. Dig. — Int. Electron. Devices Meet.*, **IEDM-84**, 2 (1984).

32. S. Sen, F. Capasso, A. Y. Cho, and D. Sivco, "Resonant Tunneling Device with Multiple Negative Differential Resistance: Digital and Signal Processing Applications with Reduced Circuit Complexity," *IEEE Trans. Electron Devices*, **ED-34**, 2185 (1987).

33. S. Luryi, "Frequency Limit of Double-Barrier Resonant-Tunneling Oscillators," *Appl. Phys. Lett.*, **47**, 490 (1985).

34. H. Morkoç, J. Chen, U. K. Reddy, T. Henderson, and S. Luryi, "Observation of a Negative Differential Resistance due to Tunneling Through a Single Barrier into a Quantum Well," *Appl. Phys. Lett.*, **49**, 70 (1986).

35. B. Ricco and M. Ya. Azbel, "Physics of Resonant Tunneling. The One Dimensional Double-Barrier Case," *Phys. Rev.*, **B29**, 1970 (1984).

36. S. Luryi, "Coherent versus Incoherent Resonant Tunneling and Implications for Fast Devices," to be published in *Superlattices and Microstructures*.

37. M. Tsuchiya and H. Sakaki, "Precise Control of Resonant Tunneling Current in AlAs/GaAs/AlAs Double Barrier Diodes with Atomically-Controlled Barrier Widths," *Jpn. J. Appl. Phys.*, **25**, L185 (1986).

38. M. Tsuchiya and H. Sakaki, "Dependence of Resonant Tunneling Current on Well Widths in AlAs/GaAs/AlAs Double Barrier Diode Structures," *Appl. Phys. Lett.*, **49**, 88 (1986).

39. S. Muto, T. Inata, H. Ohnishi, N. Yokoyama, and S. Hiyamizu, "Effect of Silicon Doping Profile on $I–V$ Characteristics of an AlGaAs/GaAs Resonant Tunneling Barrier Structure Grown by MBE," *Jpn. J. Appl. Phys.*, **25**, L577 (1986).

40. H. Toyoshima, Y. Ando, A. Okamoto, and T. Itoh, "New Resonant Tunneling Diode with a Deep Quantum Well," *Jpn. J. Appl. Phys.*, **25**, L786 (1986).

41. S. Muto, T. Inata Y. Sugiyama, Y. Nakata T. Fujii, H. Ohnishi, and S. Hiyamizu, "Quantum Well Width Dependence of Negative Differential Resistance of $In_{0.52}Al_{0.48}As/In_{0.53}Ga_{0.47}As$ Resonant Tunneling Barriers Grown by MBE," *Jpn. J. Appl. Phys.*, **26**, L220 (1987).

42. M. Tsuchiya and H. Sakaki, "Dependence of Resonant Tunneling Current on Al Mole Fractions in $Al_xGa_{1-x}As$-GaAs-$Al_xGa_{1-x}As$ Double Barrier Structures," *Appl. Phys. Lett.*, **50**, 1503 (1987).

43. A. D. Stone and P. A. Lee, "Effect of Inelastic Processes on Resonant Tunneling in One Dimension," *Phys. Rev. Lett.*, **54**, 1196 (1985).

44. C. W. Tu, R. Hendel, and R. Dingle, "Molecular Beam Epitaxy and the Technology of Selectively Doped Heterostructure Transistors," in D. K. Ferry, Ed., *Gallium Arsenide Technology,* Howard & Sams, Indianapolis, IN; p. 107, 1985.

45. N. Yokoyama, K. Imamura, H. Ohnishi, T. Mori, S. Muto, and A. Shibatomi, "Resonant Tunneling Hot Electron Transistor (RHET)," *Solid-State Electron.*, **31**, 577 (1988).

46. A. C. Gossard, R. C. Miller, and W. Wiegmann, "MBE Growth and Energy Levels of Quantum Wells with Special Shapes," *Surf. Sci.*, **174**, 131 (1986).

47. M. Heiblum, M. V. Fischetti, W. P. Dumke, D. J. Frank, I. M. Anderson, C. M. Knoedler, and L. Osterling, "Electron Interference Effects in Quantum Wells: Observation of Bound and Resonant States," *Phys. Rev. Lett.*, **58**, 816 (1987).

48. F. Capasso, S. Sen, A. Y. Cho, and D. Sivco, "Resonant Tunneling Devices with Multiple Negative Differential Resistance and Demonstration of a Three-State Memory Cell for Mutliple-Valued Logic Applications," *IEEE Electron Device Lett.*, **EDL-8,** 297 (1987).

49. S. Luryi, A. Kastalsky, A. C. Gossard, and R. H. Hendel, "Charge Injection Transistor Based on Real-Space Hot-Electron Transfer," *IEEE Trans. Electron Devices,* **ED-31,** 832 (1984).

50. C. Rine, Ed., *Computer Science and Multiple Valued Logic,* North-Holland, Amsterdam; 1977.

51. A. Heung and H. T. Mouftah, "An All-CMOS Ternary Identity Cell for VLSI Implementation," *Electron. Lett.*, **20**, 221 (1984).

52. J. Söderström and T. G. Andersson, "A Multiple-State Memory Cell Based on the Resonant Tunneling Diode," *IEEE Electron Device Lett.*, **EDL-9,** 200 (1988).

53. F. Capasso, S. Sen, A. Y. Cho, and A. L. Hutchinson, "Hot Electron Resonant Tunneling Through a Quantum Well: A New Electron Spectroscopy," in W. T. Lindley, Ed., *Gallium Arsenide and Related Compounds 1986,* p. 539, Institute of Physics, Bristol, UK, 1986.

54. F. Capasso, S. Sen, A. Y. Cho, and A. L. Hutchinson, "Resonant Tunneling Electron Spectroscopy," *Electron. Lett.*, **23**, 28 (1987).

55. F. Capasso, S. Sen, A. Y. Cho, and A. L. Hutchinson, "Resonant Tunneling Spectroscopy of Hot Minority Electrons Injected in Gallium Arsenide Quantum Wells," *Appl. Phys. Lett.*, **50**, 930 (1987).

56. J. R. Hayes and A. F. J. Levi, "Dynamics of Extreme Nonequilibrium Electron Transport in GaAs," *IEEE J. Quantum Electron.*, **QE-22,** 1744 (1986).

57. R. A. Hopfel, J. Shah, P. A. Wolff, and A. C. Gossard, "Negative Absolute Mobility of Minority Electrons in GaAs Quantum Wells," *Phys. Rev. Lett.*, **56**, 2736 (1986).

58. R. A. Hopfel, J. Shah, and A. C. Gossard, "Nonequilibrium Electron-Hole Plasma in GaAs Quantum Wells," *Phys. Rev. Lett.*, **56**, 765 (1986).

59. J. R. Hayes, A. F. J. Levi, A. C. Gossard, and J. H. English, "Base Transport Dynamics in a Heterojunction Bipolar Transistor," *Appl. Phys. Lett.*, **49**, 1481 (1986).

60. A. F. J. Levi, J. R. Hayes, P. M. Platzman, and W. Wiegmann, "Injected-Hot-Electron Transport in GaAs," *Phys. Rev. Lett.*, **55**, 2071 (1985).

61. M. Heiblum, M. Nathan, D. C. Thomas, and C. N. Knoedler, "Direct Observation of Ballistic Electron Transport in GaAs," *Phys. Rev. Lett.*, **55**, 2200 (1985).

62. D. Ankri and L. F. Eastman, "GaAlAs-GaAs Ballistic Heterojunction Bipolar Transistor," *Electron Lett.*, **18**, 750 (1982).

63. R. J. Malik, F. Capasso, R. A. Stall, R. A. Kiehl, R. W. Ryan, R. Wunder, and C. G. Bethea, "High Gain, High Frequency AlGaAs/GaAs Graded Band-Gap Base Bipolar Transistors with a Be Diffusion Setback Layer in the Base," *Appl. Phys. Lett.*, **46**, 600 (1985).

64. D. Ankri, R. A. Zoulay, E. Caquot, J. Dangla, C. Dubon, and J. Palmier, "Analysis of D.C. Characteristics of GaAlAs/GaAs double Heterojunction Bipolar Transistors," *Solid-State Electron.*, **29**, 141 (1986).

65. T. Futatsugi, Y. Yamaguchi, S. Muto, N. Yokoyama, and A. Shibatomi, "InAlAs/InGaAs Resonant Tunneling Bipolar Transistor (RBTs) Operating at Room Temperature with High Current Gains," *Tech. Dig.—Int. Electron Devices Meet.*, **IEDM-87**, 877, (1987).

66. S. Luryi, "Quantum Capacitance Devices," *Appl. Phys. Lett.*, **52**, 501 (1988).

67. T. K. Woodward, T. C. McGill, H. F Chung, and R. D. Burnham, "Integration of a Resonant-Tunneling Structure with a Metal-Semiconductor Field-Effect Transistor," *Appl. Phys. Lett.*, **51**, 1542 (1987).

68. T. K. Woodward, T. C. McGill, H. F. Chung, and R. D. Burnham, "Applications of Resonant-Tunneling Field-Effect Transistors," *IEEE Electron Device Lett.*, **EDL-9**, 122 (1988).

69. N. Yokoyama and K. Imamura, "Flip-Flop Circuit Using a Resonant-Tunneling Hot Electron Transistor (RHET)," *Electron. Lett.*, **22**, 1228 (1986).

70. V. J. Goldman, D. C. Tsui, and J. E. Cunningham, "Resonant Tunneling in a Magnetic Field: Evidence for Space-Charge Build-Up," *Phys. Rev.*, **B35**, 9387 (1987).

71. H. Sakaki, "Scattering Suppression and High-Mobility Effect of Size-Quantized Electrons in Ultrafine Semiconductor Wire Structures," *Jpn. J. Appl. Phys.*, **19**, L735 (1980).

72. Y. C. Chang, L. L. Chang, and L. Esaki, "A New One-Dimensional Quantum Well Structure," *Appl. Phys. Lett.*, **47**, 1324 (1985).

73. P. M. Petroff, A. C. Gossard, R. A. Logan, and W. Wiegmann, "Toward Quantum Well Wires: Fabrication and Optical Properties," *Appl. Phys. Lett.*, **41**, 636 (1982).

74. J. Cibert, P. M. Petroff, G. J. Dolan, S. J. Pearton, A. C. Gossard, and J. H. English, "Optically Detected Carrier Confinement to One and Zero Dimensions in GaAs Quantum Well Wires and Boxes," *Appl. Phys. Lett.*, **49**, 1275 (1986).

75. M. A. Reed, R. T. Bate, K. Bradshaw, W. M. Duncan, W. R. Frensley, J. W. Lee, and H. D. Shih, "Spatial Organization in GaAs-AlGaAs Multiple Quantum Dots," *J. Vac. Sci. Technol.*, **B4**, 358 (1986).

76. S. Luryi in F. Capasso and G. Margaritondo, Eds., *Heterojunction Band Discontinuities: Physics and Device Applications*, North-Holland, Amsterdam, 1987.

77. D. A. B. Miller, D. S. Chemla, T. C. Damen, A. C. Gossard, W. Wiegmann, T. H. Wood, and C. A. Burrus, "Band-Edge Electroabsorption in Quantum Well Structures: The Quantum-Confined Stark Effects," *Phys. Rev. Lett.*, **53**, 2173 (1984).

78. F. Stern and W. E. Howard, "Properties of Semiconductor Surface Inversion Layers in the Electric Quantum Limit," *Phys. Rev.*, **163**, 816 (1967).

6 Electrical Modeling of Interconnections

J. R. BREWS

AT&T Bell Laboratories
Murray Hill, New Jersey

6.1 INTRODUCTION

For chip design in the VLSI era, interconnections have new importance. Today, a chip no longer contains only a circuit, but may contain a good part of an entire system, for example, a computer on a chip. For such a complex chip, flexibility unavailable in small circuits means that interconnections are important at more than the nuisance level of added delay, power consumption, and noise. More profoundly, interconnections change the entire architecture of a VLSI system from that appropriate for a traditional design.

Traditionally, interconnection was a key factor in design because interconnections formed an hierarchy based on packages (modules), circuit-packs (cards or printed wiring boards), and backplanes (panels).[1,2] Optimal allocation of components among these assemblies minimized the impact of the great difference in speed of communication at different levels. For example, Russo and Wolff[3] and Kernighan and Lin[4] considered how to assign gates to modules and how to connect the resulting modules so that minimal delay resulted. With VLSI, some of the same delay strategies have moved onto the chip itself (for reviews, see Hong and Nair[5] and Brady and Blanks[6]). One might think that the resulting reductions of delay and dissipation would be the main advantages of consolidation of multiple packages into one VLSI package.

Besides such straightforward progress, more fundamental redesigns of systems have come about. Easy replication of circuits or subcircuits in VLSI cuts interconnect penalties simply by adding computational structures where needed, instead of sending the work out of the neighborhood. This easy replication alters the allocation problem, as shown by Lawler et al.[7] Partition into large intercommunicating blocks can be replaced by nucleation into smaller, more self-sufficient local entities, trading costs of replication against costs of communication. These issues have been reviewed by Seitz,[8] and examples are provided by Mead and Conway[9] (Chapter 8, p. 263).

269

To optimize replication–interconnection trade-offs, interconnection should be designed in conjunction with the algorithms that are to be executed (see, e.g., Allen[10] or Flynn and Hennessy[11]). For specific computations, the notion of optimal interconnection has been refined. By analyzing the flow of information through a system, bounds can be placed on the minimum area and maximum speed of execution of a computation, providing a criterion for optimality, as proposed in detail by Thompson,[12] and further developed by Leiserson,[13] Leighton,[14] Savage,[15] and Chazelle and Monier.[16] An overview has been written by Ullman.[17] To avoid restriction to a special purpose, programmable interconnection is being explored.[18-20] At an extreme are the educable neural networks of Hopfield,[21] and attempts at their implementation as described by Graf et al.[22]

The pervasive and even dominating role of interconnection in VLSI forms a backdrop for this chapter, which examines electrical models of interconnections. Good models for interconnections help to grasp the trade-offs between subsystem autonomy and interconnect topology or hierarchy. Also, good models can suggest innovations to alter the dependence of interconnect delay, noise, and dissipation upon scaling rules and line lengths, modifying the strategy for nucleation as technology develops.

To illustrate the possible influence of interconnection in future systems, consider two scenarios that could break up a system on a chip into component islands. These scenarios stem from very basic behavior. The first scenario is based on high-frequency performance, the second on heat dissipation.

6.1.1   An Interconnect Scenario Based on Performance

As the size of a system increases, so does its average interconnect length. For instance, Sorkin[23] suggests that experience shows the wiring length for an average net in a good placement is roughly one-fifth the semiperimeter of the circuit. (A net is a set of electrically equivalent nodes to be wired together.) That is, the average interconnect length in a circuit $\bar{\ell}$ is related to the circuit area A as

$$\bar{\ell} \approx \frac{\sqrt{\pi}}{5}\sqrt{A} \approx \frac{\sqrt{A}}{3}. \tag{1}$$

There are several consequences of any relation that allows $\bar{\ell}$ to increase with A, even if the dependence is not that of (1). For the electrical parameters of a line, resistance, capacitance, and inductance all are proportional to line length. Thus, an increase in $\bar{\ell}$ with A implies that lower-frequency performance is expected from larger-area circuits. To perform as rapidly as a simple circuit, a complex circuit must be made with smaller design rules.

At high complexity and high speeds, the performance-allowed area can be insufficient to include the whole system. To build the system, one solution is nucleation into separated, smaller subsystems. The whole system remains on one chip, but it is organized in pieces that can be constructed inside subareas consistent with speed requirements. The smaller subsystems can perform rapidly, but they also must communicate. If communication is too slow, the overall combined performance will be too slow. Hence, the design of the intersystem interconnection becomes crucial. If this interconnection is fast, much faster than the component systems, then it becomes immaterial how the subsystems are designed. If the interconnection is slower or comparable in speed to the component systems, then the chip must be nucleated so that the requisite communication between subsystems is less demanding than that within a subsystem. Such organization may not be easy, or may be impossible. Clearly, it is desirable that the intersystem interconnection have as high a performance rate as possible.

How can the intersystem interconnection be made rapid? Various strategies are in use. First, the interconnection itself can be improved. To do this, many layers of interconnections are desirable. In this way, interconnections can be shortened, with immediate benefit to performance. Different levels can be isolated from one another, reducing crosstalk problems. Different design rules can be used on different levels, allowing larger cross sections for interconnections on levels with longer lengths (thus reducing resistance) and wider spacings (thus reducing capacitance).

Second, long interconnections can be broken up with repeaters. The advantages of repeaters have been described by Glasser and Dobberpuhl,[24] (pp. 260–261) and by Bakoglu and Meindl.[25] An intuitive grasp of the advantage stems from the dependence of the delay of a long distributed RC line on the square of its length ℓ. (This point is discussed in more detail later.) This delay therefore exceeds the sum of the delays of segments of the same line divided into parts. If the line is segmented in n equal parts and if repeaters are inserted that introduce an output impedance R_R and an input capacitance C_R, then the delay is

$$\tau = n\left[\left(R_R + \frac{0.5R\ell}{n}\right)\frac{C\ell}{n} + \left(R_R + \frac{R\ell}{n}\right)C_R\right],\qquad(2)$$

where R and C are the line resistance and capacitance per unit length, respectively, and we assume the driver impedance is R_R and the load capacitance C_R. Minimizing with respect to n, to find the optimal number of repeaters for least delay,

$$n \sim \left(\frac{\tau_L}{\tau_R}\right)^{1/2},\qquad(3)$$

where τ_L is the uninterrupted line delay, $\tau_L = 0.5RC\ell^2$, and τ_R is the "repeater delay" $\tau_R \sim R_R C_R$. Then, the least delay is given by

$$\tau = 2\tau_L\left[\left(\frac{\tau_R}{\tau_L}\right)^{1/2} + \frac{R_R}{R\ell} + \frac{C_R}{C\ell}\right],\qquad(4)$$

which can be less than τ_L provided the total line resistance $R\ell \gg R_R$ and the total line capacitance $C\ell \gg C_R$. Although oversimplified, this argument does show the opportunity for improvement, and suggests that the best number of stages should be based on delay considerations. Thus, this use of repeaters differs from that in submarine cables, for example, where the repeater spacing is based not on minimizing delay, but on reconstruction of signals degraded by attenuation.

Finally, one can design drivers and loads to match the interconnect properties better.[25]

6.1.2   An Interconnect Scenario Based on Power Dissipation

Clearly, the faster a circuit works, the more power it will dissipate. There is a maximum power density that can be tolerated in a given heat-removal technology. As performance goals are raised, this limit will be reached and the chip design must be restructured.

Kanai[26] has considered the case where a complex system is constrained by an upper bound on temperature. The temperature T reached when the power P is being dissipated is determined by the thermal resistance of the package Θ as follows:

$$T = \Theta P. \tag{5}$$

Naturally, the thermal resistance depends in part on the spatial arrangement of the heat sources. In particular, if the heat source is subdivided and the component parts are spread out over an area larger than the original single heat source, then Θ is reduced.

This reduction of thermal resistance is illustrated in Fig. 6.1 (from Bar-Cohen[27]). Here, thermal resistance is plotted against the number of chips in a multichip module per cm^2. As the number of chips/cm^2 is reduced, Θ is also reduced. Three curves result, depending on the cooling ambient (air, water, or immersion cooling). A more effective cooling results in less heat spreading beneath the chip, so the slope of a curve is lower for a better cooling method. The data points indicate a number of different modules.

Similar reduction of thermal resistance should apply to a single chip if its heat production is spread out by dividing the circuitry into islands, so that the chip itself houses a number of interconnected component circuits. Reducing Θ by lower-

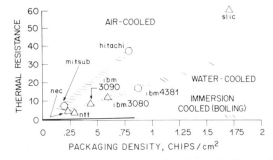

Fig. 6.1   Thermal resistance vs. packaging density for multichip modules. (From Bar-Cohen.[27] Reprinted with permission, copyright 1987, IEEE).

ing the overall circuit density allows the total power to be increased without raising T, and more power can be given to the component circuits. To the extent that power is (switching energy)/delay, an increased allotment for component circuit power allows a faster operation.

Thus, where speed is not limited by interconnect delay but by heating, allowing the on-chip circuitry to nucleate into separated islands can allow faster operation. Here again, success depends on the quality of intercircuit interconnection, including the power dissipated by this interconnection that serves as overhead on the entire separation philosophy. Included in the overhead are special drivers, loads, or repeaters needed to make this interconnection perform.

6.1.3 Discussion

Both these scenarios lead to a nucleation of a complex circuit into subcircuits and quality intercircuit interconnections. To this degree, nucleation of a chip design parallels partitioning of older systems into multichip packages, circuit-packs, and so forth. However, nucleation of a system differs from traditional partitioning because replication is used deliberately to increase subsystem autonomy. A major aim of nucleation is to lighten demands on interconnection, to relieve communication by simplifying or reducing the variety of messages exchanged between islands. Such simplification is accomplished in part by distributing memory. More than it resembles older partitioning, nucleation resembles the division of a GaAs system among chips in a multichip package. Although yield limits the nucleation that is possible in these GaAs designs, replication has implications that are discussed by Multinovic et al.,[28] such as single-chip processor design [which can result in a RISC (reduced instruction set computer) design], a maximal use of registers, and *in situ* execution of some operations.

Can we use the two scenarios just discussed quantitatively? For example, can the area allowed for a given speed be predicted? If we try to derive (1), some basic difficulties emerge: for example, what is the distribution of wire lengths? Attempts to answer these questions tend to begin with Rent's rule, and this subject is discussed next.

6.1.4 Rent's Rule

Rent's rule is an empirical observation that the number of input/output (I/O) ports to a circuit is related to the complexity of the circuit, as measured by the number of gates in the circuit. The relation is

$$P = KG^p, \tag{6}$$

where K is the average number of I/Os per gate, G the number of gates, p the Rent exponent, and P is the number of I/Os for the circuit. The exponent p is thought to lie in the range $0.5 < p < 1.0$, and its value is thought to reflect the circuit organization. Landman and Russo[29] experimentally verified this formula for a number of logic circuits and blocks in early computer circuits. More recently, Ferry et al.[30]

have suggested that the exponent $p = 0.2$ for chips other than gate arrays, such as microprocessors and memory. Striny[31] has assembled data that shows DRAM complexity has almost no effect on I/O count, but that microprocessor I/O, although lower than for gate arrays, increases about as rapidly with complexity.

Rent's rule has implications about the distribution of line lengths in a circuit. The basic idea relating line lengths to Rent's rule is that if line lengths are short, then only devices in an annulus near the circuit periphery can contribute to I/O, whereas if they are long, devices anywhere inside may be connected to I/O. Hence, for an approximately uniform density of gates, one might expect in the first case $P \propto G^{1/2}$ and in the second case $P \propto G$, because $G^{1/2} \propto$ perimeter while $G \propto$ area.

These ideas were formulated by Feuer[32] for the following model of a circuit:

(a) The gates are uniformly distributed.
(b) There is a distribution of wire lengths ℓ given by $q(\ell)$.
(c) The wire length between two gates is the least Manhattan distance between the gates. That is, the wire length ℓ between points (x_1, y_1) and (x_2, y_2) is

$$\ell = |x_1 - x_2| + |y_1 - y_2| .$$

(d) The circuit is square and oriented at 45° to the $x–y$ axes. This condition is a mathematical convenience, but the same model can be worked out for a rectangle of any orientation.

Feuer found that Rent's rule was satisfied for such a model with $0.5 < p < 1.0$ provided the length distribution was given by

$$q(\ell) = \frac{\Gamma}{\ell^{4-2p}} , \tag{7}$$

where Γ is some constant. This length distribution drops off rather slowly with length, so that an appreciable number of rather long interconnections exists. That is, a randomly interconnected circuit appears to have its performance compromised by long interconnections.

Another approach to the wire length distribution was made by Donath in a series of papers based on the idea that subdivisions of a circuit separately obey the same Rent's rule. This subdivision method is described in most detail in Heller et al.[33] and has been generalized by Masaki and Yamada,[34] who estimate the advantages of three-dimensional wiring. The derived wire length distribution is quite similar to Feuer's Eq. (7). However, as Rent's rule is itself empirical, no insight into Rent's rule comes from this approach, regardless of its practical value. Applicability to nucleated systems has yet to be explored.

A more basic attack is the work of Vuillemin,[35] who shows any planar circuit performing a certain class of computations requires an area A bounded by

$$A > a_G N + a_{I/O} D + a_W D^2, \tag{8}$$

where N is the number of bits permuted, D is the data rate (number of input bits/time between successive problem submittals), and the three terms represent the areas devoted to gates, I/O, and wiring. The basic sizing of gates, wire spacings, etc., satisfies a Mead–Conway[9] type of scaling behavior based on a minimum dimension. According to (8), the data rate helps decide the relative importance of the various contributions to area. This idea was pursued by Kuhn,[36] who showed that distribution of memory could reduce the required data rate in several specific cases, causing the Rent exponent to decrease, supporting the empirical argument of Ferry et al.[30] More generally, the decreased Rent exponent is a result of an increased self-sufficiency of neighborhoods not restricted to distribution of memory alone.

For optimally divided systems, probably the distribution of wire lengths breaks up into a bimodal (or even multimodal) distribution, as shown in Fig. 6.2 (from Kang[37]). A similar distribution was suggested by Phillips et al.[38] The wire length distribution of the entire system is a superposition of the inner subsystem (local) wire length distribution and the external intersystem (network) wire length distribution. The two distributions very probably have quite different dependencies on complexity because of the different local communication demands inside subsystems compared with the between-subsystem network demands. The behavior of the average wire length for the overall system is then a mix of the local average and the network average, and its dependence on system complexity depends on how local autonomy and network complexity are traded off. The individual $\bar{\ell}$-values for each component distribution would be more illuminating than an overall $\bar{\ell}$.

To sum up, fundamental results concerning the relation between I/O and circuit complexity are within reach. These relations are tied closely to the distribution of

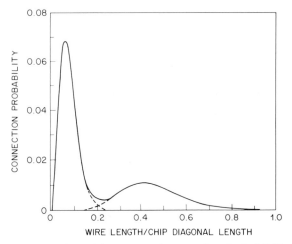

Fig. 6.2 Bimodal distribution of interconnection wavelengths in VLSI. (From Kang.[37] Reprinted with permission, copyright 1987, IEEE).

wirelengths. The Rent exponent generally decreases as the organization of the system changes to increase local self-sufficiency.

Here ends our introduction to the impact interconnection can have in future systems as speed and complexity build up. The trade-offs involved in designing interconnection, its drivers, loads, and repeaters affect the entire architecture of such circuits. As a result, good models for interconnect design have greater significance than that expected from a straightforward extrapolation of the past.

We begin detailed discussion of electrical modeling of interconnections with the very common *RC* line model.

6.2   *RC* LINE MODELS

The earliest and easiest model of an interconnect line, other than a simple short circuit, is a capacitor or a capacitor and resistor. Thus, the simplest model is as shown in Fig. 6.3, where R and C are the line resistance and capacitance per unit length, respectively, and ℓ is the line length. A more sophisticated version is the distributed *RC* line, which can be represented as a cascade of T-sections, each T like that of Fig. 6.3 (see Fig. 6.4). As the number of subcircuits tends to infinity while the total line length remains fixed, the cascade becomes the transmission line governed by the equations

$$\frac{\partial I}{\partial z} = -C\frac{\partial V}{\partial t}\,;\qquad \frac{\partial V}{\partial z} = -RI\,, \tag{9}$$

which combine to produce the diffusion equation

$$\frac{\partial^2 V}{\partial z^2} = RC\frac{\partial V}{\partial t}\,. \tag{10}$$

In (9) and (10), I and V are current and voltage, respectively, and the coordinate z measures distance along the line. Intuitively, the transmission line appears closer to reality than the lumped circuit of Fig. 6.3. Solutions to the diffusion equation are well known from the theory of heat conduction, so the behavior of such a circuit is well understood so long as R and C are voltage-independent and the terminal conditions are not too unusual (this matter is examined in Section 6.2.2).

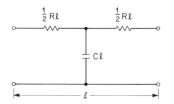

Fig. 6.3   A T-section lumped *RC* model for interconnection.

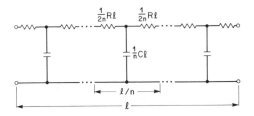

Fig. 6.4   A distributed *RC* line model for interconnection approximated by a cascade of T-sections.

In a VLSI context, the analysis of the *RC* line frequently has been restricted to a single parameter describing the overall response of the system driver + line + load, namely, the Elmore delay time, given by

$$\tau_d = \frac{1}{2} RC\ell^2 + RC_L \ell + R_D C\ell + R_D C_L, \tag{11}$$

where R_D is the driver resistance, C_L the load capacitance, and ℓ the line length. Using an expression like (11) for the delay, a host of interesting questions can be explored. For example, if it is known how R, C, R_D, and C_L scale, one can discuss how delay scales. If one knows how these parameters depend on the choice of technology, for example, GaAs vs. Si, bipolar vs. MOSFET, one can make technology comparisons. If one knows the distribution of line lengths, one can estimate the impact on delay of single vs. multilevel interconnection, and so forth.

These matters are of interest in searching for the best path to follow in future chip design, quite apart from the utility of the *RC* line for quantitative design questions. In what follows, some published conclusions based on (11) are presented. However, lest these conclusions be overvalued, let us first examine (11) to see how well it actually describes the *RC* line and also ask how well the *RC* line describes interconnections.

6.2.1 *RC* Line Response

The transient response of damped networks was discussed by Elmore.[39] He introduced the use of a delay time and a rise time, as shown in Fig. 6.5, as determined by the output response to a step function input. If the output response is $v_0(t)$ and its derivative is $v_0'(t)$, then

$$\tau_d \equiv \int_0^\infty dt\, v_0'(t)t, \tag{12}$$

$$\tau_r \equiv (2\pi)^{1/2} \left\{ \int_0^\infty dt\, [v_0'(t)t^2] - \tau_d^2 \right\}^{1/2}, \tag{13}$$

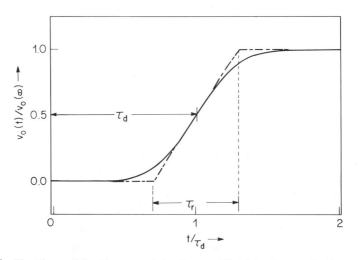

Fig. 6.5 The Elmore delay time τ_d and rise time τ_r. The delay time marks the centroid of the derivative of the waveform. The rise time measures the moment of the derivative about τ_d. The waveform shown here has a Gaussian derivative.

where the amplitude of the step input is adjusted to make

$$\int_0^\infty dt\, v_o'(t) = 1.$$ (14)

Alternatively, (12) and (13) can be divided by the integral in (14). The big advantage of the Elmore transient response times is that they are easy to calculate. For example, if the response is monotonic, as for an RC line, and we Laplace transform, $v_o(t) \rightarrow V_o(s)$, one finds (see Ghausi and Kelly,[40] p. 83)

$$\tau_d = -V_o'(0),$$ (15)

$$\tau_r = (2\pi)^{1/2}[V_o''(0) - \tau_d^2]^{1/2},$$ (16)

where $V_o'(0)$ and $V_o''(0)$ represent the first and second derivatives, respectively, of the Laplace transform of $v_o(t)$ evaluated at $s = 0$ (i.e., evaluated at the zero-frequency limit).

Applying (12) to the distributed RC line, one obtains (11) for the circuit of Fig. 6.6a, which happens to be identical to τ_d for the circuit of Fig. 6.6b. However, Fig. 6.6b is a lumped approximation to Fig. 6.6a valid only at low frequencies. The implication is that the many analyses of interconnection based on the Elmore delay could be based just as well on the simple lumped approximation of Fig. 6.6b.

Sakurai[41] made extensive studies of distributed RC lines driven by a driver with resistive input impedance and terminated with a capacitive load. He determined

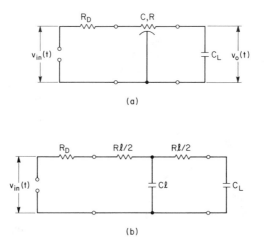

Fig. 6.6 (a) The distributed *RC* line with a driver of impedance R_D and a capacitive load C_L. (b) The lumped T-section approximation to the circuit in (a).

that to within 4% the time for the output voltage to reach 90% of final value $t_{0.9}$, defined by

$$v_o(t_{0.9}) \equiv 0.9 v_o(\infty), \tag{17}$$

was given by

$$t_{0.9} = 1.02 RC\ell^2 + 2.21(RC_L\ell + R_D C\ell + R_D C_L). \tag{18}$$

This expression is very similar to (11). This similarity is increased by Sakurai's observation that to the same accuracy the output voltage is described by

$$v_o(t) = a - e^{-t/\tau}, \tag{19}$$

where $v_o(t)$ is normalized according to (14). From (19),

$$t_{0.9} = \tau \ln \frac{10}{a}. \tag{20}$$

Using (19) in the Elmore definition (12), one finds $\tau = \tau_d$, and taking an approximate value of $a \approx 0.91$ from Sakurai's Fig. 4, one finds

$$\tau = \frac{t_{0.9}}{\ln(10/a)}$$

$$\sim 0.43 RC\ell^2 + 0.92(RC_L\ell + R_D C\ell + R_D C_L). \tag{21}$$

It is clear that for general discussion either (11) or (21) will lead to the same conclusions, although the differences may be significant for quantitative design purposes (see, e.g., Rubinstein et al.,[42] Lin and Mead,[43] and Zukowski[44]).

The conclusion drawn from the similarity between Sakurai's (21) and Elmore's (11) is that discussion of RC line response based on the distributed RC line will not differ significantly from that based on a simple lumped T-section, at least for the general considerations described in this review. Moreover, both the lumped and distributed RC models fail at high frequencies, as will become clear later.

6.2.2   Terminal Conditions

The response for a capacitative load and a resistive driver may not be very realistic, considering that the load and driver are actually time varying in practice. Bilardi et al.[45] have considered modeling a MOSFET driver as follows:

$$I = \begin{cases} \dfrac{V}{R_D}, & 0 < V < V_{sat} \\ I_{sat}, & V_{sat} \le V \le V_{dd}. \end{cases} \qquad (22)$$

Assuming the supply voltage $V_{dd} > V_{sat}$, the driving conditions change as V passes V_{sat}. The constant driver impedance of Sakurai's study does not apply, and the results are somewhat different.

Using either approach, we can examine different contributions to the system delay, as is done in Fig. 6.7. The driver- and load-limited region is the region in which system delay is dominated by R_D and C_L, and its boundary is arbitrarily taken as

$$R_D C_L = RC_L \ell + R_D C \ell \qquad (23)$$

because the $R_D C_L$ delay exceeds the delay added by the line inside this region. The line-limited region is the region governed by the delay term quadratic in the line length, $1/2RC\ell^2$, and we can estimate the importance of this term using the ratio based on (11):

$$\varepsilon_c = \frac{0.5RC\ell^2}{R_D C_L + RC_L \ell + R_D C \ell}. \qquad (24)$$

For the variable driver impedance model of Bilardi et al.,[45] the corresponding ratio ε_v is related to the ratio ε_c for a fixed driver impedance by

$$\varepsilon_v = 2\frac{\varepsilon_c}{\mu^2} - 1, \qquad (25)$$

where μ is the smallest positive solution of

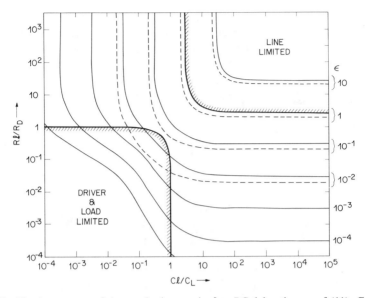

Fig. 6.7 The importance of the quadratic term in ℓ to *RC* delay time τ_d of (11). Contours of constant ε_c [Eq. (24)] and ε_v [Eq. (25)] are shown, as well as the device- and load-limiting boundary [Eq. (23)]. In the line-limited region, τ_d is dominated by $\frac{1}{2}RC\ell^2$, whereas in the load- and driver-limited region it is dominated by $R_D C_L$. The region outside the line-limited and termination-limited regimes is affected by both the line properties *and* its terminations, as expressed by the terms in τ_d linear in ℓ [Eq. (11)]. A similar plot was introduced by Bilardi et al.[45]

$$\frac{\tan(\mu)}{\mu} = \frac{(RC\ell^2/\mu^2 - 1)}{(R_D C\ell + RC_L \ell)}.$$ (26)

Oddly, there is no dependence of ε_v on the ratio V_{sat}/V_{dd} that decides the extent of exploration of the nonlinear region. This independence cannot be general because, for example, $V_{\text{sat}}/V_{dd} > 1$ implies a fixed impedance model according to (22). Bilardi et al. do not comment on this point, although they mention that for their example V_{sat}/V_{dd} is 0.8.

The contours of constant ε_v are shown in Fig. 6.7 as solid lines for the variable impedance model and the contours of constant ε_c as dashed lines for the fixed impedance model. The contour $\varepsilon_v = 1$ is taken arbitrarily as the boundary of the line-dominated region because the nonlinear line delay that is independent of R_D and C_L exceeds the linear delay inside this region. From Fig. 6.7, it is clear that both models predict much the same boundary. Thus, Fig. 6.7 indicates that the complication of time-varying driver impedance can be ignored for our purposes, and the simple Elmore delay time (11) can be used.

Between the two limiting corner regions of Fig. 6.7 is the regime in which both the line and the terminations affect the delay, and in this region the delay is not simply the sum of a line delay $1/2RC\ell^2$ and a $R_D C_L$ delay, as sometimes assumed intuitively.

6.2.3   Applications of the *RC* Model

The main applications of the *RC* delay expression (11) are found in Mead and Rem,[46] who recognized the implications of this expression for multilevel interconnect and driver design, Carter and Guise,[47] who used it to compare GaAs and Si technologies, and Bakoglu and Meindl,[25] who provided numerous graphs comparing different interconnect materials and drivers. Driver staging has been discussed extensively by Lewis,[48] Lee and Soukup,[49] and Hedenstierna and Jeppson.[50]

All these papers require expressions relating the various resistances and capacitances to the design rules for drivers, loads, and lines. With such expressions, one knows how the electrical parameters change with design rules and with changes from Si to GaAs or from CMOS to bipolar technology. The expressions used by Mead and Rem were primitive, ignoring several terms in the delay expression itself, and ignoring such matters as fringing capacitance and self-loading of drivers by their own output capacitance. Carter and Guise used computer-calculated values. Bakoglu and Meindl used the fitted formulas of Sakurai and Tamaru.[51] Their results for interconnection delay for aluminum lines of several widths are shown in Fig 6.8. For lines of various C and R and several types of drivers, their results shown in Fig. 6.9 indicate that operation in the gigahertz region will require the use of cascaded drivers or optimal repeaters (optimal repeaters are sized and spaced

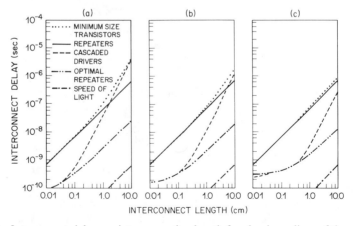

Fig. 6.8   Interconnect delay vs. interconnection length for aluminum lines of three widths: (*a*) 0.25 μm, (*b*) 0.5 μm, and (*c*) 1.0 μm. The minimally sized inverter is assumed to have an output resistance of $R_D = 10$ kΩ and an input capacitance of $C_L = 1.17$ fF/μm $\times$ width. Several driving schemes are considered. (After Bakoglu and Meindl.[25] Reprinted with permission, copyright 1985, IEEE).

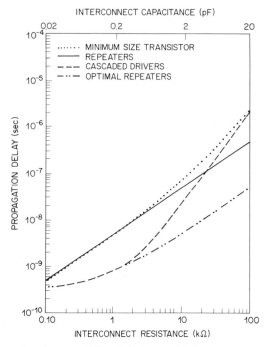

INTERCONNECT CAPACITANCE (pF)

PROPAGATION DELAY (sec)

INTERCONNECT RESISTANCE (kΩ)

.... MINIMUM SIZE TRANSISTOR
——— REPEATERS
– – – CASCADED DRIVERS
– ·· — OPTIMAL REPEATERS

Fig. 6.9 Delay vs. interconnect $C\ell$ or $R\ell$. $R_D = 10$ kΩ and $C_L = 2$ F. (After Bakoglu and Meindl.[25] Reprinted with permission, copyright 1985, IEEE.)

for minimal delay). However, developments using combined bipolar and CMOS technology could alter these conclusions.[52]

The *RC* line model ignores inductance. If frequencies are pushed into the gigahertz region, this neglect may not be justified, introducing our next topic.

6.3 *RLC* LINE MODELS

Inductance has two roles. As a circuit element, it can introduce ringing or overshoot, phenomena not found in *RC* circuits. In addition, the introduction of inductance allows wave propagation and transmission line effects, quite different from the diffusive propagation of distributed *RC* lines. To simplify presentation, let us begin with inductance in a lumped circuit.

6.3.1 Lumped *RLC* Circuits

The T-section equivalent circuit for interconnection analogous to the *RC* T-section is shown in Fig. 6.10. Because inductance appears everywhere in series with the line resistance, the transfer function for this circuit incorporates L, always in the combination $R + j\omega L$. As a result, a sufficient condition for a valid *RC* approximation is

$$\omega L \ll R . \tag{27}$$

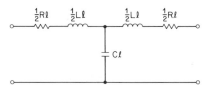

Fig. 6.10 A lumped T-section approximation to interconnect that includes inductance L as well as R and C.

Should (27) be violated, for example, for a lossless line, another condition validating the use of an RC circuit is derived later, namely,

$$\omega L\ell \ll R_D . \tag{28}$$

These inequalities show that at low-enough frequencies an RC circuit will be a good approximation, but as frequency increases this approximation will fail. At sufficiently high frequencies, a distributed LC line approximation will suffice, to which the resistance contributes only an attenuation constant, $\alpha \sim R/(2\sqrt{L/C}\,)$.

An advantage of the right combination of resistance and inductance can be found using the simple circuit of Fig. 6.11. Such an L-section approximation is less accurate than a T-section approximation, but for qualitative purposes it is more transparent. Many texts contain a discussion of such circuits; for example, see Kuo,[53] pp. 314–326, pp. 590–596 [there is an error in the sign of the second term in Kuo's Eq. (6-84) which carries over to a wrong sign for his (6-88)]. The analysis of this circuit shows that the circuit is underdamped or oscillatory if

$$(R\ell + R_D)(C\ell + C_L) < \frac{4L\ell}{R\ell + R_D} , \tag{29}$$

and it is overdamped or nonoscillatory if the inequality is reversed. We define Q by

$$Q^2 = \frac{L\ell/(R\ell + R_D)}{(R\ell + R_D)(C\ell + C_L)} , \tag{30}$$

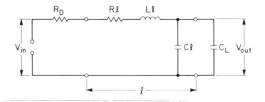

Fig. 6.11 A lumped L-section approximation to interconnect.

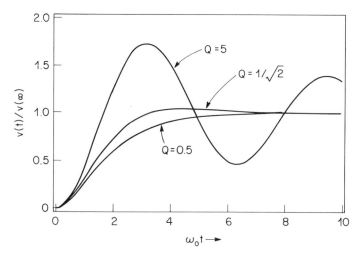

Fig. 6.12 Response to a step input of the ladder circuit of Fig. 6.11 according to (37). The plotted response is critically damped for $Q = 0.5$.

where the right side is the ratio of the inductive to the capacitive time constant. The step response of the circuit of Fig. 6.11 is then as shown in Fig. 6.12 for several values of Q. The critically damped case corresponds to $Q = 0.5$, but not much overshoot ($\sim 4\%$) occurs for the case $Q = 1/\sqrt{2}$, which is discussed later. Equation (29) is a condition that natural oscillations of the *RLC* circuit can occur as decided by its characteristic equation and it does not depend on the frequencies in the applied voltage pulse.

The response to a general input voltage can be deduced from the transfer function of the circuit

$$H(s) = \frac{1}{s^2 L\ell(C\ell + C_L) + s(R\ell + R_D)(C\ell + C_L) + 1} \tag{31}$$

as the Laplace transform of $V_{out}(s)$, given by

$$V_{out}(s) = V_{in}(s)H(s). \tag{32}$$

Setting $s = j\omega$, the magnitude of the transfer function can be written as

$$|H(\omega)| = \frac{1}{\sqrt{[1 - (\omega/\omega_0)^2]^2 + (\omega/\omega_0)^2/Q^2}}, \tag{33}$$

where ω_0 is defined as

$$\omega_0^2 \equiv \frac{1}{L\ell(C\ell + C_L)}. \tag{34}$$

The magnitude of the transfer function is shown for several values of Q in Fig. 6.13. It can be seen that the circuit is a low-pass circuit with a cutoff frequency ω_c given by

$$
\begin{aligned}
\left(\frac{\omega_c}{\omega_0}\right)^2 &= \sqrt{\left(1 - \frac{1}{2Q^2}\right)^2 + 1} - \frac{1}{2Q^2} + 1 \\
&\sim Q^2 + 2Q^4, \qquad\qquad\qquad\text{low } Q, \\
&\sim (1 + \sqrt{2})\left(1 - \frac{1}{2\sqrt{2}\,Q^2}\right), \qquad \text{high } Q.
\end{aligned}
\tag{35}
$$

(Here, ω_c is defined as the 3 dB bandwidth, i.e., the value of ω making $|H|^2 = 1/2$). The transfer function peaks at a frequency ω_p given by

$$
\omega_p = \omega_0 \sqrt{1 - \frac{1}{2Q^2}},
\tag{36}
$$

provided $Q > 1/\sqrt{2}$, which is not the same condition as (29), $Q > 0.5$.

For $Q > 1/\sqrt{2}$, the peak in the transfer function at ω_p means that higher frequencies are accentuated in the output more than for a more heavily damped cir-

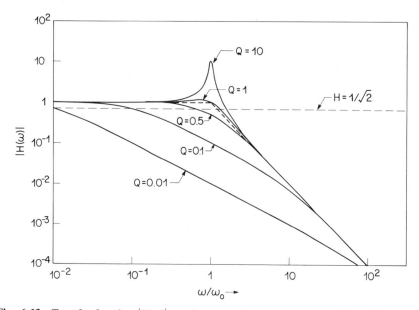

Fig. 6.13 Transfer function $|H(\omega)|$ vs. frequency for the L-section of Fig. 6.11. Quality factor Q and corner frequency ω_0 are defined by (30) and (34).

cuit. Hence, the rise time is shorter, as can be seen in Fig. 6.12. The step response shown in Fig. 6.12 is given by[53]

$$\frac{v(t)}{v(\infty)} = 1 - \frac{\exp(-\zeta\omega_0 t)}{\sqrt{1 - \zeta^2}} \sin[\omega_0 t \sqrt{1 - \zeta^2} + \Theta], \tag{37}$$

with $\Theta = \tan^{-1}(\sqrt{1 - \zeta^2}/\zeta)$ and $\zeta = 1/(2Q)$. If the damping is too low (as for $Q = 5$ in Fig. 6.12), this rapid response is accompanied by ringing, which is undesirable because it leads to delays (settling times) or false triggering. Overdamping prevents ringing, but slows the response. A compromise is the case $\omega_p = 0$, or $Q = 1/\sqrt{2}$. For this case and a step input, the response reaches 50% of its final value in a time t_d, given by

$$t_d \simeq \left(\frac{3}{4}\right)(R\ell + R_D)(C\ell + C_L),$$

which is about the same as for the *RC* version of the same circuit with $L = 0$. Thus, ringing can be avoided without sacrifice of speed by matching the load and driver C_L and R_D to the line $R\ell$ and $L\ell$ (or the reverse) so that $Q \approx 1/\sqrt{2}$ results.

Although ringing can be avoided this way, other drawbacks to inductance exist. These drawbacks are well known in packaging, where the inductance of package leads causes unacceptable voltage fluctuations due to $L\,dI/dt$ (Ho et al.[54]) and crosstalk problems due to inductive coupling between different leads.[55] These topics are considered later.

6.3.2 Distributed *RLC* Lines

A distributed *RLC* line is a better approximation than the simple lumped circuit of the previous section. How does the transfer function of a distributed *RLC* line compare with that of the lumped T-section or L-section? No comparison comparable in detail to that for *RC* lines exists. Here, some limited comparisons are made.

The simplest starting point for the distributed *RLC* line is the transmission line equations relating current and voltage along the line at any position z for any time t namely,

$$\frac{\partial V}{\partial z} = -RI - L\frac{\partial I}{\partial t}; \qquad \frac{\partial I}{\partial z} = -C\frac{\partial V}{\partial t}.$$

Combining these equations, one finds

$$\frac{\partial^2 V}{\partial z^2} = RC\frac{\partial V}{\partial t} + LC\frac{\partial^2 V}{\partial t^2}. \tag{38}$$

The appearance of the second derivative in time means that the RLC line, unlike the RC line, supports wave propagation and not simple diffusion. The coefficient of the second derivative determines the velocity of propagation for the lossless line to be

$$v = \frac{1}{\sqrt{LC}}. \tag{39}$$

Using these equations and the boundary conditions at the driver ($z = 0$) and at the load ($z = \ell$) and some initial condition stating the condition of the line at $t = 0$, the transmission line equations can be solved. If we are interested in the transfer function particularly, we can assume a time dependence $\exp(j\omega t)$ for V_{in}, V, and I. One then finds the general solution to the transmission line equations is

$$V(z) = A \, \exp(-\gamma z) + B \, \exp(\gamma z), \tag{40}$$

$$I(z) = \frac{[A \, \exp(-\gamma z) - B \, \exp(\gamma z)]}{Z_0}, \tag{41}$$

where

$$Z_0 = \frac{R + j\omega L}{\gamma} = \frac{\gamma}{j\omega C} \tag{42}$$

is called the characteristic impedance of the line and γ is the complex propagation constant. The ratio B/A is found using the boundary condition at $z = \ell$:

$$\frac{V(\ell)}{I(\ell)} = Z_L; \qquad \frac{B}{A} = \exp(-2\gamma\ell)\left(\frac{Z_L - Z_0}{Z_L + Z_0}\right), \tag{43}$$

where the load impedance $Z_L = 1/(j\omega C_L)$. The transfer function then is found as

$$H(\omega) \equiv \frac{V_{out}}{V_{in}} = \frac{V(\ell)}{V((0)} \frac{V(0)}{V_{in}}, \tag{44}$$

where the boundary condition at the driver provides

$$\frac{V_{in}}{V(0)} = 1 + \frac{I(0)}{V(0)} R_D. \tag{45}$$

Evaluating (40) and (41) at both ends of the line to find the various ratios and substituting for B/A from (43), one finds

$$H(\omega) = \frac{Z_L Z_0}{Z_0(Z_L + R_D) \cosh(\gamma\ell) + (Z_0^2 + R_D Z_L) \sinh(\gamma\ell)}. \tag{46}$$

For the case of on-chip interconnections, we do not expect lines with lengths of many wavelengths. Hence, we look at (46) in the case $\gamma\ell \ll 1$. Expanding the *cosh* and *sinh,* we find (46) becomes

$$H(s) = \frac{1}{1 + \tau_d s + [C_L L\ell + 0.5C\ell^2(L + RC_L R_D)]s^2},\qquad(47)$$

which agrees in form with (31). Here, τ_d is the Elmore delay (11), and terms higher than quadratic in s or ℓ have been dropped. In Fig. 6.14, the results for the distributed line, the above approximation (labeled briefly as the ladder approximation), and the *RC* approximation [i.e., (47) without the term quadratic in s] are

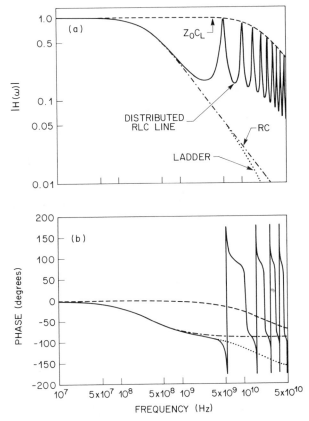

Fig. 6.14 (*a*) Magnitude of transfer function vs. frequency for a lossless interconnection line: $R = 0$, $L = 15$ nH/cm, $C = 0.5$ pF/cm, $R_D = 1$ kΩ, $C_L = 50$ fF, and length $\ell = 1$ cm. The distributed line shows rapid oscillations at larger frequencies, with an envelop (labeled $Z_0 C_L$) given by (57). Several approximations to this response also are shown. The curve labeled "ladder" refers to (47) and is close to the response of the T-section in Fig. 6.10. (*b*) Phase of transfer function in (*a*).

compared. As is evident from (47), the quadratic term in s and, hence, the line inductance are important only if

$$s \gg \frac{R_D C_L + (RC_L + R_D C)\ell + 0.5RC\ell^2}{LC_L\ell + 0.5C\ell^2(L + RC_L R_D)}, \tag{48}$$

which, for a low loss line such that $C_L R_D < L/R$, implies (setting $s = j\omega$)

$$\omega L\ell \gg R_D. \tag{49}$$

Thus, if the driver impedance is high, as shown in Fig. 6.14, where $R_D = 1\ \mathrm{K}\Omega$, then (49) is not satisfied except at high ω and the inductance has little effect; the RC approximation works as well as the ladder approximation. However, for a low driver impedance, as shown in Fig. 6.15, where $R_D = 50\ \Omega$, the ladder approximation of (47) is much better than the RC model out to the first phase reversal.

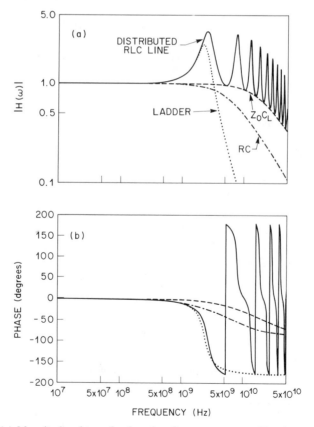

Fig. 6.15 (a) Magnitude of transfer function for same case as Fig. 6.14, but for $R_D = 50\ \Omega$. (b) Phase of transfer function in (a).

The ladder of (47) is not that of (31); although both have denominators quadratic in s, the coefficients are different. Incidentally, the results in these figures for the ladder approximation (47) are found to agree very closely with the approximation using the T-section shown in Fig. 6.10. Thus, these two approximations are not significantly different. From Figs. 6.14 and 6.15, it appears that the simple approximations using the ladder of (47) or the T-section of Fig. 6.10 are quite good out to the first phase reversal of the distributed line, that is, for the initial range of phase between 0° and $-180°$. However, if the frequencies of operation extend higher than this point, the models must be improved.

For the lossless line, where $\gamma = j\beta = j\omega\sqrt{LC}$, and the characteristic impedance is $Z_0 = \sqrt{L/C}$, the transfer function can be written simply as

$$|H(\omega)|^{-2} = [1 + (\omega Z_0 C_L)^2]\left\{\frac{1}{2}\left[1 + \left(\frac{R_D}{Z_0}\right)^2\right] + \frac{1}{2}\left[1 - \left(\frac{R_D}{Z_0}\right)^2\right]\right.$$
$$\left. \times \left[\frac{1 - (\omega Z_0 C_L)^2}{1 + (\omega Z_0 C_L)^2}\cos 2\beta\ell - \frac{2\omega_0 Z_0 C_L}{1 + (\omega Z_D C_L)^2}\sin 2\beta\ell\right]\right\}. \quad (50)$$

According to the result (50), oscillations in amplitude occur with the periodicity of $\cos(2\beta\ell)$ or $\sin(2\beta\ell)$. Using $\beta = \omega\sqrt{LC}$, a phase of $-180°$ occurs for

$$\ell\omega\sqrt{LC} \simeq \pi/2\,, \quad (51)$$

which corresponds to a frequency of 3 GHz for the example of Figs. 6.14 and 6.15. If we recognize that the velocity of propagation down the line [see (39)] is $1/\sqrt{LC}$, the condition (51) implies simply that for these lumped models of interconnect to work we require either

$$\ell \lesssim \frac{\lambda}{4} \quad \text{or} \quad \tau_{\text{prop}} \lesssim \frac{T}{4}\,, \quad (52)$$

where λ is the wavelength at frequency ω and T is the period. That is, these lumped models are good for lengths of interconnect less than a quarter wavelength, or lengths with a propagation time τ_{prop} less than a quarter period.

How high a frequency must be considered for a given bit rate? The answer depends in part on the pulse shape. If the pulse has very sharp corners, then the frequency spectrum extends to quite high frequencies. For example, for a ramp input of rise time t_r, a common estimate of the maximum frequency is $f_{\text{max}} = 0.35/t_r$. However, differently shaped pulses with the same rise times can have different frequency contents. Real pulses are somewhat rounded. To provide an optimistic appraisal, we could suppose the pulse shape resembles the Tukey pulse described in the time domain by

$$v(t) = \begin{cases} \dfrac{1}{2}\left(1 + \cos\left(\dfrac{\pi t}{\tau}\right)\right), & |t| < \tau \\ 0, & |t| \geq \tau. \end{cases} \quad (53)$$

Half the pulse is shown in Fig. 6.16. This pulse shape has one of the most tightly confined frequency spectrums of any pulse shape localized in time to the interval $|t| < \tau$, as also is shown in Fig. 6.16. The frequency spectrum is

$$V(f) = \tau\left(\frac{\sin(2\pi f\tau)}{2\pi f\tau}\right)\left(\frac{1}{1 - (2f\tau)^2}\right). \tag{54}$$

From Fig. 6.16, the spectrum is almost null for frequencies greater than $f_{max} = 1/\tau$. The pulse width is 2τ, and if the pulses are spaced at least 2τ apart, the bit rate is $B = 1/(4\tau)$ pulses/s. Thus,

$$f_{max} \approx \frac{1}{\tau} = 4B \tag{55}$$

is an approximate lower bound on the maximum frequency that must be considered. More sharply shaped pulses than the Tukey pulse will require larger values of f_{max}. With this value of f_{max}, the condition (52) for applicability of a lumped circuit model becomes

$$\ell \leq \frac{1}{\sqrt{LC}}\left(\frac{1}{16B}\right) \qquad \text{or} \qquad \tau_{prop} \leq \frac{1}{16B}, \tag{56}$$

or, for the example of Figs. 6.14 and 6.15 where $1/\sqrt{LC} = 1.15 \times 10^{10}$ cm/s, $\ell \leq 3/4$ cm at 1 Gbit/s.

For a distributed line, the ringing phenomena of lumped RLC circuits is replaced by repeated reflections of pulses between driver and load. For lossless (LC) lines,

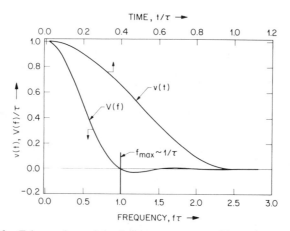

Fig. 6.16 Half a Tukey pulse and its frequency spectrum. The pulse extends symmetrically to $t = -\tau$. This pulse has a spectrum confined to lower frequencies, approximately below $1/\tau$, where τ is half the pulse width.

reflections are avoided by use of matched terminations and terminations in the characteristic impedance $Z_0 = \sqrt{L/C}$, which is frequency independent. Thus, the same characteristic impedance suffices for all the component frequencies of a pulse. For a driver impedance matched to the line $R_D = Z_0$, the expression (50) simplifies immediately to the nonoscillatory form

$$|H(\omega)| = \frac{1}{\sqrt{1 + (\omega Z_0 C_L)^2}}, \tag{57}$$

which is the curve labeled $Z_0 C_L$ in Figs. 6.14 and 6.15. This curve bounds the oscillations above if $R_D > Z_0$ and bounds them below if $R_D < Z_0$.

Avoidance of reflections is difficult for a lossy line because the characteristic impedance of a lossy line varies with frequency. A suggestion to avoid ringing was made by Ho et al.[54] based on an analysis by Cases and Quinn.[56] Assuming capacitive loads negligible compared with the line capacitance, and also negligible driver impedance, they suggest that line resistance and inductance be tailored so

$$\frac{2}{3} \sqrt{\frac{L}{C}} < R\ell < 2 \sqrt{\frac{L}{C}}. \tag{58}$$

Under these conditions, the pulse reflected at the nonmatching capacitive load attenuates to a negligible level before returning to the driver. This suggestion was followed up by Yuan et al.[57] and by Frye,[58] who included the role of driver impedance, and searched for values of R, L, and C that would provide a critically damped response. A very similar result is found using the ladder approximation of (47) and the condition for no peaking of the magnitude of the transfer function from (36), namely,

$$Q^2 = \frac{1}{(\omega_0 \tau_d)^2} \leq \frac{1}{2}, \tag{59}$$

where ω_0^{-2} now is the coefficient of s^2 in (47). Explicitly, (59) is

$$R_D + \frac{1}{2}R\ell + (R_D + R\ell)\left(\frac{C_L}{C\ell}\right) \geq \sqrt{\frac{L}{C}}\left[1 + 2\frac{C_L}{C\ell} + \left(\frac{R}{L}\right)C_L R_D\right]^{1/2}, \tag{60}$$

which agrees with Yuan et al. for the case they considered when C_L can be neglected, and can be compared with the slightly different criterion of Frye, also for C_L negligible, namely,

$$R_D + 0.567R\ell \geq \sqrt{\frac{L}{C}}. \tag{61}$$

According to (60), C_L is negligible provided $C_L/(C\ell) \ll 1$ and $C_L R_D \ll L/R$. Because our derivation is based on a ladder approximation valid at frequencies obeying (52), it would seem (60) needs further examination for higher frequencies.

A problem for all types of lines is that load impedances may be voltage dependent and may be attached at many positions along the line, each position becoming a possible source of reflections. In addition, lines are branched. In an early review, Jarvis[59] recognized all these problems, and they still remain.

6.4 *RLCG* LINE MODELS

The general transmission line includes a conductance G well as R, L, and C. For many applications, this conductance represents the dielectric loss in the line, and is negligible for most VLSI analysis. However, when an interconnection is laid over a semiconductor surface, the conductance of the semiconductor makes it necessary to add a transmission line conductance G to the interconnect model.

The simplest way to understand this need is to consider a very wide line (but, of course, not so wide as a wavelength). In cross section, as shown in Fig. 6.17, the line resembles an MOS capacitor. If the semiconductor is silicon, ordinary MOS capacitor theory (see, e.g., Nicollian and Brews,[60] pp. 73–75) suggests the line capacitance is the series combination of the oxide and substrate capacitance. That is, we might expect the C of the RLC line to be

$$C = \frac{C_{\mathrm{ox}} C_s}{C_{\mathrm{ox}} + C_s}. \tag{62}$$

Here, for a lightly doped silicon,

$$C_s = \frac{\kappa_s \varepsilon_0}{t_s}. \tag{63}$$

Here, t_s is the thickness of the silicon substrate, $\kappa_s = 11.7$ is the dielectric constant of silicon, and $\varepsilon_0 = 8.85 \times 10^{-14}$ F/cm is the permittivity of empty space. This case corresponds to a dielectric composed of two layers. On the other hand, for a heavily doped silicon, $C_s \rightarrow \infty$, so

$$C \rightarrow C_{\mathrm{ox}}. \tag{64}$$

That is, C is given by a single dielectric, the oxide layer.

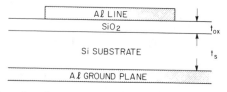

Fig. 6.17 Cross section of an interconnect line over an oxide layer of thickness t_{ox} on a silicon substrate of thickness t_s.

At intermediate doping levels, ignoring any depletion layer at the silicon surface due to applied voltages, charge flow is not purely capacitive, but contains a conductive component due simply to the nonzero conductivity of the silicon. Thus, the equivalent circuit of the cross section becomes that of Fig. 6.18. As first pointed out by Guckel et al.,[61] and pursued by Hasegawa et al.[62] and Owens and Jaggers,[63] the resulting behavior of the line is quite complicated. Three regimes can be identified according to the resistivity of the substrate and the frequencies of interest. These regimes are illustrated in Fig. 6.19.

The boundaries between the various regimes in Fig. 6.19 are not really sharp, but are determined approximately by the various time constants in the equivalent circuit. For example, the "dielectric" regime is the regime where the silicon behaves like a dielectric, namely, for frequencies above the dielectric relaxation frequency, $\omega > G_s/C_s = 1/(\kappa_s \varepsilon_0 \rho)$. The mobile carriers in the silicon cannot respond at these frequencies, and the return current flows in the ground plane. The inductance is governed by the dimension $t_{ox} + t_s$.

The "skin-effect" regime applies for frequencies below the relaxation frequency but for frequencies large enough that the skin depth given by $\delta = \sqrt{[2\rho/(\omega\mu_0)]}$ is exceeded by the substrate thickness. The electromagnetic fields are limited to the skin-effect layer, and the return current flows within this part of the silicon. The silicon thus acts as a ground plane, and the inductance is governed (approximately) by the dimension $t_{ox} + \delta$.

Finally, at lower frequencies and for a wide range of resistivities, one has the "slow-wave" regime. In this regime, the skin depth exceeds the substrate thickness, but the silicon conductivity is great enough to short out the silicon capacitance. As a result, the capacitance is C_{ox}, but the return current is primarily in the ground plane so the inductance is governed by the metal-to-ground plane separation $t_{ox} + t_s$, just as in the dielectric regime. Hence, the velocity of propagation $v = 1/\sqrt{(LC)}$ is less than for a line in the dielectric regime, thus the name "slow-wave." The slow-wave regime has been studied extensively as a possible delay line, particularly by Jager et al.[64,65] Recently, this regime has become a target for

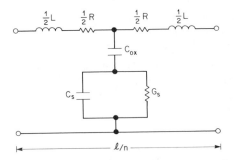

Fig. 6.18 An equivalent T-section network for a line over a semiconductor substrate. A cascade of such T-sections forms a line of length ℓ. All parameters are per unit length and should be multiplied by ℓ/n for a section of length ℓ/n, n = number of subsections. An infinitely wide line is assumed.

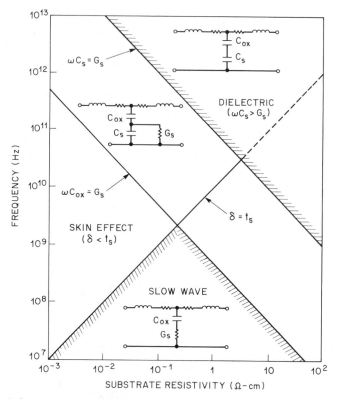

Fig. 6.19 A frequency vs. resistivity plot for an infinitely wide line over 0.5 μm of oxide on a 500 μm silicon substrate. Various regimes are indicated. The skin depth parameter is given by $\delta = [2\rho/(\mu_0\omega)]^{1/2}$ where μ_0 = permeability of empty space, $4\pi \times 10^{-9}$ H/cm.

very detailed numerical studies by Krowne,[66] Aubourg et al.,[67] Sorrentino et al.,[68] and Tzuang and Itoh.[69,70]

Such numerical analysis is necessary, particularly for realistic (i.e., finite-width) lines, because two-dimensional effects are difficult to treat. For example, the frequency-dependent skin depth makes the fringing effect frequency dependent, as follows. The skin-depth region is enclosed in a cylinder located a skin depth away from the top conductor. Depending on the size of the skin depth, this cylinder may enclose part of the ground plane directly under the line, but it always lies in the silicon farther to the sides of the line. As a result, part of the return current may flow in the ground plane immediately under the line, and part always flows in the silicon at the sides of the line. Thus, we end up with a frequency-dependent, two-dimensional combination of slow-wave and skin-effect regimes.

Attempts have been made to avoid numerical work by a complex merging of one- and two-dimensional approximations and by approximate mapping techniques that cannot be justified except for solving the Laplace equation. Conductance is modeled in terms of vertical current flow from the top conductor to the ground plane, and resistance in terms of lateral current flow parallel to the top conductor. These attempts have led to a bewildering variety of approximations for the inductance and resistance in the equivalent circuits for both the skin-effect and slow-wave regimes. For example, the work of Hasegawa and Seki[71] is of this sort. While some such approximations may eventually prove accurate, numerical support would prove reassuring. Unfortunately, little effort has been made to summarize the numerical studies in terms of equivalent circuits.

In fact, the losses introduced by a semiconductor substrate are due to currents flowing in such involved patterns as to make one wonder whether a division of the loss into "lateral" R and "vertical" G components makes any sense. Also, one often is cautioned that the usual approaches for evaluating inductance and capacitance for transmission lines are based on Laplace's equation (see e.g., Ramo et al.,[72] pp. 398, 399) and do not apply when losses are large or when axial field components exist. Nonetheless, for the case where one mode propagates, the author has shown that the usual transmission line equation still hold and has provided formulas for determining R, L, C, and G.[73]

It should be noted that the general transmission line equivalent circuit is a cascade of T-sections such as that in Fig. 6.20, which is not of the form of the circuit in Fig. 6.18, for example. When the more complex circuit of Fig. 6.18 is cast in terms of equivalent elements in a circuit like that of transmission line of Fig. 6.20, in place of physically meaningful, easily interpreted elements, one has complex frequency dependent expressions for R, L, C, G. Thus, the frequency independent R, L, and C that were used in the transmission lines for the simple situations discussed earlier usually do not exist.

6.5 FINDING INTERCONNECTION PARAMETERS

Assuming some circuit model for interconnect, how are the parameters of the circuit to be found? Many calculations can be found in the literature, most of them

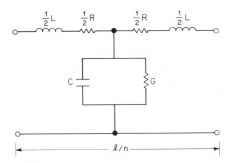

Fig. 6.20 The T-section for the general transmission line. The circuit connecting the top and bottom rails of this T-section is not the same as for Fig. 6.18, which means the circuit in Fig. 6.18 must be transformed to an equivalent C and G for use in the transmission line equations.

based on dc evaluations of capacitance and inductance. We will begin with the formulation for a single line, and then introduce more complicated situations.

6.5.1 Formulation for a Single Line

The simplest case is a single line, for example, a microstrip or coplanar line. Let us assume a two-dimensional case, with line properties independent of position in the direction of propagation (the z-direction).

For simplicity, we consider a linear situation, where there is no dependence of the material parameters on field. Then we can consider a single frequency and build up a general time dependence by superposition. To treat a single line, one must characterize the materials forming the line. For example, we need the dielectric constants and conductivities of the materials. Let us suppose the material is described in terms of the complex permittivity and permeability given by

$$\varepsilon = \kappa(x, y)\varepsilon_0 + \frac{\sigma(x, y)}{j\omega}, \tag{65a}$$

$$\mu = \mu_1(x, y) - j\mu_2(x, y). \tag{65b}$$

(We depart from Ref. 73 in the sign of μ_2, which usually is positive if chosen as done here.) The permittivity and permeability in (65) are allowed to be frequency dependent.

Because of the assumed uniformity along the third dimension, the z-dependence of the solution is simple. We can write the fields in the form

$$\mathbf{E}(x, y, z) = \mathbf{E}_n(x, y, z) + \mathbf{E}_z(x, y, z) \tag{66a}$$

$$= \mathbf{E}_t(x, y)V(z) + \eta\mathbf{E}_l(x, y)I(z), \tag{66b}$$

$$\mathbf{H}(x, y, z) = \mathbf{H}_n(x, y, z) + \mathbf{H}_z(x, y, z) \tag{67a}$$

$$= \mathbf{H}_t(x, y)I(z) + \left(\frac{1}{\eta}\right)\mathbf{H}_l(x, y)V(z), \tag{67b}$$

where subscripts t and l refer to transverse and longitudinal field components with respect to the z direction, and where η is the intrinsic impedance of empty space.

Substituting in Maxwell's equations (following the discussion of Ref. 73) using the separation of variables technique, one introduces two constants of separation taken conveniently in the form

$$c_1 = -\gamma Z_0, \qquad c_2 = -\gamma/Z_0, \tag{68}$$

where γ and Z_0 later turn out to be the propagation constant and characteristic impedance of a mode of propagation, respectively. Going back to Maxwell's equations, one finds

$$\frac{dI(z)}{dz} = -\frac{\gamma}{Z_0}V(z), \qquad \frac{dV(z)}{dz} = -\gamma Z_0 I(z), \tag{69}$$

which are the usual transmission line equations. From the solution to these equations, γ can be identified as the propagation constant. Besides (69), Maxwell's equations provide the transverse and longitudinal field components; for the transverse electric field, for example,

$$\mu\nabla\mathbf{x}\left[\frac{1}{\mu}\nabla\mathbf{x}\mathbf{E}_t\right] - \left\{\nabla\left[\frac{1}{\varepsilon}\nabla\cdot(\varepsilon\mathbf{E}_t)\right] + (\gamma^2 + \omega^2\varepsilon\mu)\mathbf{E}_t\right\} = 0, \qquad (70\text{a})$$

and for the transverse magnetic field,

$$\varepsilon\nabla\mathbf{x}\left[\frac{1}{\varepsilon}\nabla\mathbf{x}\mathbf{H}_t\right] - \left\{\nabla\left[\frac{1}{\mu}\nabla\cdot(\mu\mathbf{H}_t)\right] + (\gamma^2 + \omega^2\varepsilon\mu)\mathbf{H}_t\right\} = 0. \qquad (70\text{b})$$

Notice the symmetry of these equations under the exchange of ε and μ. Equations (70a) and (70b) have multiple solutions, with eigenfunctions corresponding to the same complex eigenvalues for γ^2. Thus, there are numerous modes possible.

For microstrip, usually a particular mode is considered, called the "quasi-static" or "quasi-TEM" mode, which is the mode approximated best by the static values of inductance and capacitance found from Laplace's equation for the potential generated by a static voltage on the various conductors of the line. For inhomogeneous dielectrics (e.g., layered dielectrics) and for lossy metals or insulators, or for semiconductor layers, longitudinal field components exist, causing some departure from this static formulation. To proceed, let us suppose that only one mode, say this quasi-static mode, is important to propagation, perhaps because the higher-order modes can propagate only at higher frequencies. Then, a characteristic impedance for the line can be found for this mode, namely,

$$Z_0 = \sqrt{\frac{R + j\omega L}{G + j\omega C}}, \qquad (71)$$

where[73]

$$R = \frac{1}{|I(z)|^2}\int dx \int dy\{\sigma|\mathbf{E}_z(x,y,z)|^2 + \omega\mu_2|\mathbf{H}_n(x,y,z)|^2\}, \qquad (72\text{a})$$

$$L = \frac{1}{|I(z)|^2}\int dx \int dy\{\mu_1|\mathbf{H}_n(x,y,z)|^2 - \kappa\varepsilon_0|\mathbf{E}_z(x,y,z)|^2\}, \qquad (72\text{b})$$

$$C = \frac{1}{|V(z)|^2}\int dx \int dy\{\kappa\varepsilon_0|\mathbf{E}_n(x,y,z)|^2 - \mu_1|\mathbf{H}_z(x,y,z)|^2\}, \qquad (72\text{c})$$

$$G = \frac{1}{|V(z)|^2}\int dx \int dy\{\sigma|\mathbf{E}_n(x,y,z)|^2 + \omega\mu_2|\mathbf{H}_z(x,y,z)|^2\}. \qquad (72\text{d})$$

From (72c) and (72d), the usually assumed static symmetry of G and C under exchange of σ and $\kappa\varepsilon_0$ fails when there is a z-component of H.

It is worth noting that (72b) and (72c) provide a quantitative estimate of the importance of the z-component of the fields in estimating L and C, namely, the z-components are negligible if

$$\int dx \int dy \, \kappa\varepsilon_0 |\mathbf{E}_z|^2 \ll \int dx \int dy \, \mu_1 |\mathbf{H}_n|^2 \tag{73a}$$

and

$$\int dx \int dy \, \mu_1 |\mathbf{H}_z|^2 \ll \int dx \int dy \, \kappa\varepsilon_0 |\mathbf{E}_n|^2. \tag{73b}$$

These terms are often neglected without any estimate of the possible errors involved.

To obtain (72), it is necessary to assume that only one mode propagates. Thus, the field functions in (72) refer to this mode. The resulting impedance (71) also is for only this mode. In addition, the values of R, L, C, and G from (72) force the complex power given by Poynting's vector to agree with that from the simple circuit expression

$$P = \frac{I^*(z)V(z)}{2}, \tag{74}$$

where * denotes complex conjugation. Frequently, this condition has not been imposed or only the real part of the power has been forced to agree. However, reactive power must be considered in design, as the interconnection can store and release energy as well as transfer it. Gucket et al.[61] made no condition on the power in deriving their equivalent circuits, but instead used an intuitive factorization of γ. As a result, there is no guarantee that their circuits lead to correct power flow.

If multiple modes can propagate and the line is ideal (i.e., has no losses or discontinuities), then the above analysis can be generalized by introducing separately the power transported by each mode, as each mode is independent. However, when the line is not ideal, the modes are not orthogonal. One then has at least two choices: (1) expand the modes in terms of a closely approximating set of orthogonal modes and introduce some (weak) coupling terms (see Johnson,[74] pp. 149ff.) or (2) introduce the adjoint waveguide.[75] Multiple modes are excited in the immediate neighborhood of line discontinuities introducing three-dimensional considerations.

A fundamental ambiguity in the characteristic impedance should be noted.[76-78] Because current and voltage are not field concepts, some definition of these quantities is needed to make the transition from the field world to the circuit world. When longitudinal field components exist, the usual path integrals used to make this transition are no longer independent of the path used for their evaluation — that is, these path integral evaluations of I and V are not unique. Hence, an alternative approach seems necessary. One possible approach is to force the agreement of circuit power with field power, as was done above. Ambiguity then is reduced to ambiguity to within a frequency-dependent proportionality factor. Depending on this factor, the equivalent circuit for the three-dimensional transition from driver to line or from line to load can be simplified. Hence, this ambiguity may be best resolved in conjunction with the modeling of these terminations.

A complication for semiconductors not included in this analysis and yet of interest is that the material constants of (65) incorporate free-carrier contributions. Under ac excitation, this free-carrier contribution implies an ac variation of the material parameters, in turn implying ac screening variations of the materials neglected in (65). Inclusion of such effects cannot be done within the complex permittivity approximation, and require a solution of Poisson's equation, and even of the carrier transport equations for complicated situations.

6.5.2   Multiple Lines

The majority of work on interconnect has concerned multiple lines and has treated losses and dielectric inhomogeneities as causing little departure from electrostatic values of capacitance and inductance. These assumptions exclude the treatment of semiconductors for most resistivities.

From the circuit standpoint, the effects of multiple lines differ from single lines because of mutual inductance between lines and interline capacitances. Denoting the vector of line voltages by $[V]$, of current by $[I]$, and the inductance and capacitance matrices by $[L]$, and $[C]$, respectively, Amemiya[79] and Marx[80] considered the transmission line equations

$$-\frac{d}{dz}[V] = j\omega[L][I], \tag{75a}$$

$$-\frac{d}{dz}[I] = j\omega[C][V]. \tag{75b}$$

The general case with resistance and conductance and some niceties of solution were considered by Paul.[81] A useful overview has been written by Frankel.[82] Equations (75a) and (75b) can be combined to provide

$$\frac{d^2}{dz^2}[V] = -\omega^2[L][C][V]. \tag{76}$$

As a consequence of (76), it can be seen that in this case where $[R]$ and $[G]$ are negligible, it is the eigenvalues of $[L][C]$ that decide the propagation constants (velocities) of the modes of the set of multiple conductors. Also, the symmetry of this matrix will decide the symmetry of the propagating modes and their degeneracies. Moreover, the number of modes will be the same as the dimension of this matrix, which is at least N for N lines plus a ground plane, and could be larger if some lines propagate several modes.

The above formulation shows that the computation of inductance as the reciprocal of the capacitance matrix as

$$[L] = \frac{1}{v^2}[C]^{-1} \tag{77}$$

assumes that all the modes have the same velocity of propagation v. Such an assumption is not realistic for the usual case of layered dielectric media. For example, when the conductors are located at the interfaces between stacked dielectric layers, each mode has its own characteristic field distribution, which samples the dielectric constants of the various layers in its own way. This different sampling leads to different "effective" dielectric constants and different velocities of propagation. The simplest example is the even and odd symmetry modes for two lines, which have the field lines shown in Fig. 6.21.

Although the reciprocal relation (77) is inadequate for multiple dielectric layers, an alternative procedure, very commonly used (e.g., Green[83]), and recently rejustified by Wei et al.,[84] is to find the inductance matrix as the reciprocal of a different capacitance matrix. In place of the capacitance for the layered dielectric, one removes all the dielectric layers and then finds the capacitance matrix of the conductors in empty space $[C_0]$. Then the inductance matrix for the real system of dielectrics and conductors is estimated as

$$[L] = \frac{1}{c^2}[C_0]^{-1} \tag{78}$$

where c is the speed of light in empty space.

This procedure can be justified for two-dimensional cases with layered (piecewise homogeneous) media and where there is no magnetic field penetration into

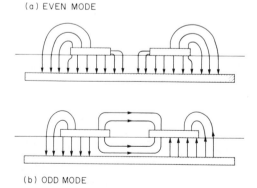

(a) EVEN MODE

(b) ODD MODE

Fig. 6.21 The even and odd mode field distributions for two parallel lines. Because the field pattern weights the two dielectrics differently for the two modes, the two modes propagate at different speeds.

the conductors by noting (see Landau and Lifshitz,[85] footnote, p. 121; Mattick,[86] Appendix 8A, p. 344) that for currents flowing only in the z direction the magnetic field can be expressed in terms of a scalar potential which satisfies exactly the same Poisson equation as an electrostatic problem, but currents replace charges and $(1/\mu)$ replaces ε. For our case where μ is uniform, therefore, the electrical analog for our magnetic field determination employs a uniform dielectric medium. It also is necessary that the cross-sectional dimensions be small enough that wave propagation in the cross-sectional plane can be ignored.

Another derivation of (78) which provides some idea of its limitations can be obtained from (72b, c). If there are no losses in the system, the analysis leading to (72b, c) then applies to each mode individually, regardless of the number of conductors carrying the mode currents. If, in addition, the z-components of the fields contribute a negligible amount to the inductance from (72b) and to the capacitance of (72c), then from (72b, c) for each mode *separately* we have

$$LC = \frac{\int dx \int dy\, \mu |\mathbf{H}|^2}{|I|^2} \cdot \frac{\int dx \int dy\, \kappa\varepsilon_0 |\mathbf{E}|^2}{|V|^2} = \frac{1}{v^2}, \tag{79}$$

where we have used the fact that the numerator is the square of twice the energy density in the mode, while the denominator is the square of twice the power transported by the mode. This power is the stored energy density times the speed of propagation. On a mode by mode basis, (79) is more generally true than the global result of (78). That is, to obtain (78) from (79), we need extra restrictions.

Equation (79) applies for L and C corresponding to a particular mode. In such a mode representation, $[L][C]$ is diagonal,[79–81,87] but each diagonal entry is different unless the different modes propagate with the same speed. To make this speed the same for all modes, we can follow the prescription leading to (78). That is, we can substitute a homogeneous dielectric for the original layered dielectric. In addition, we must exclude all fields from the interior of the conductors. [For lossy conductors, the skin effect causes this exclusion to occur at high frequencies, so this assumption can be viewed as introducing a high-frequency approximation to the inductance (see Bertin[88] and Ruehli[89]).] Then the fields for all modes see the same dielectric constant, and all have the same speed of propagation. Now $\mathbf{E}$ becomes $\mathbf{E}_0$, the electric field for a uniform dielectric, and C from (72) becomes C_0 for our particular mode. It remains to show that this replacement of dielectrics leaves L unaffected.

From (72b), L will be the same for both systems if $\mathbf{H}$ is the same for the layered and homogeneous dielectrics. We already have seen that $\mathbf{H}$ is indifferent to the layered dielectrics for a system of perfect conductors because the electric analogy for $\mathbf{H}$ depends on the substitution of $1/\mu$ for ε, and μ is not layered. However, the derivations of (78) do not apply if the magnetic permeability varies spatially, an uncommon occurrence for ICs. Second, in our derivation we are forced to assume small z-components of the fields, the conditions (73a, b). For lossless layered media, such z-components always exist, although they usually are not large.

However, for low-conductivity conductors and for semiconductors, evaluation of $[L]$ using (78) is definitely a high-frequency (lower) limit, because field penetration into these materials always occurs and commonly is accompanied by large z-components of the fields. As is evident from (72b), field penetration adds to the volume over which the integrations are performed, and generally will increase L.

6.5.3   Numerical Determination of L and C in the Static Approximation

The calculation of inductance and capacitance matrices for multiconductor interconnection systems can be pursued by many methods. Of these methods, the oldest and most popular are those based on the use of Green's functions. It is easiest for intuition to grasp the determination of the capacitance matrix. This matrix, $[C]$, is defined electrostatically by the relation

$$[Q] = [C][V] \tag{80}$$

where $[Q]$ is the vector of surface charges on the various conductors and $[V]$ is the vector of voltages. Actually, voltage *differences* matter, so one conductor can be chosen as a reference, leaving a matrix $[C]$ that is of dimensions $N \times N$ for N conductors plus a ground. For any given voltages, the charges can be found by solving Laplace's equation, finding the potential, taking its normal derivative at the surface of the conductors, and integrating over the conductor surface. This normal field is proportional to the charge density on the conductor surface according to Gauss's law.

To find $[C]$, one must solve this potential problem for N independent voltage vectors $[V]$, assuming $[C]$ is $N \times N$. Usually, the choice is made to place 1 V on each conductor in turn, and 0 V on all the rest. For example, for two lines above a ground plane, the result is

$$[C] = \begin{bmatrix} C_1 + C_m & -C_m \\ -C_m & C_2 + C_m \end{bmatrix}, \tag{81}$$

where C_m is the mutual capacitance between lines in the presence of the ground plane and C_1 and C_2 are the direct capacitances between lines 1 and 2, respectively, and the ground plane. For instance, if the two lines are shorted together, their combined capacitance to ground is $C_1 + C_2$. Also, $C_1 + C_m$ is the capacitance between line 1 and the combination of the grounded line 2 and the ground plane, as is consistent with (81) using the voltage vector $[V] = [1, 0]$. Similarly, for the same voltage vector, the charge on line 2 due to 1 V on line 1 is $Q_2 = -C_m$, a negative charge as it is induced by the positive unit voltage on line 1.

Equivalent to solving Laplace's equation is the integral form of this equation, which finds the potential as the integral of the Green's function times the surface charge density on the conductors, integrated over all the conductors. The various Green's function methods differ in (a) the choice of Green's function which can be a free space Green's function or one that builds in one or more boundary conditions, and (b) the choice of functions used for expansion of the charge distributions, potentials, or fields.

Among the first to use this method were Chestnut,[90] Weeks,[91] Patel,[92] and Ruehli and Brennan.[93,94] A comparison of these methods is contained in a paper by Benedek,[95] and a survey article was written by Ruehli.[96] The big advantage of this approach is that it approximates the solution for the potential at a given point in terms only of the boundary conditions, leading to small (though dense) matrices to manipulate. Hence, for large problems, computation still is tractable. Recent proponents of these methods include Wei et al.,[84] Harrington and Wei,[97] Chilo and Arnaud,[98] Kamikawai et al.,[99] Taylor et al.,[100] and Ning et al.[101] Detractors of the method claim that the matrices obtained are dense and poorly conditioned and require direct solution methods.[102] Proponents have discussed weighting and basis functions, choice of which is critical.[103,104] A recent comparison of three approaches was made by Scheinfein et al.[105]

Other authors solve the Laplace equation using finite elements and sparse matrix techniques[102] or finite difference methods.[106,107] Some interesting comments on effects of discretization at finite element boundaries are contained in Hammond and Tsiboukis.[108] A discussion of hybrid methods which involve both boundary and volume finite elements was made by Salon.[109]

6.5.4 Beyond the Static Approximation

We have seen how a high-frequency inductance calculation can be based on a capacitance calculation. As frequency varies from dc to high frequencies, the current distribution in conductors changes due to the skin effect, first in the width direction as skin depth and conductor width become comparable, and then in the depth direction, as skin depth and conductor thickness become comparable. For IC lines, width and thickness may be comparable, causing both effects to occur together. This nonuniform current distribution affects the inductance and the resistance of the lines. Also, when lines are near each other, each line perturbs the fields and current distributions of the others, causing proximity effects in the inductance and resistance. Related effects occur in semiconductors. One result is that $[R]$ (and $[G]$ for semiconductors) are not diagonal matrices.

For metals, the simplification can be adopted that displacement current is negligible. Within this approximation, the skin effect for conductors has been calculated numerically by Brennan et al.[110] and by Weeks et al.[111] A time-domain analysis was presented by Yen et al.[112]

For semiconductors, as discussed in connection with Fig. 6.19, displacement current cannot be neglected, and the approaches useful for metals are not valid. Tripathi and Bucolo[106] have used a network analog based on a finite difference approximation to Laplace's equation. These authors include both conduction and displacement currents. However, $\mathbf{E}$ is assumed to be the gradient of a scalar potential, so the skin effect, which depends essentially on $curl\,\mathbf{E} = -j\omega\mu\mathbf{H}$, is not included. Harrington and Wei[97] use a perturbation treatment of the skin effect appropriate for metals that ignores displacement current and that cannot treat large effects. An alternative to perturbation theory that includes both currents is Eq. (70) and (72), but no evaluation has been attempted to date.

For more complete solutions, we are then left with the numerical methods applied in the main to individual lines. We have already mentioned the work on slow-wave structures by Aubourg et al.,[67] Sorrentino et al.,[68] and Tzuang and Itoh.[69,70]

A very popular approach is the spectral domain method, which is restricted to infinitesimal conductor thicknesses but which can treat semiconductor substrates. Applications have been made by Kennis and Faucon[113] to microstrip, and by Mu et al.[114] and by Farr et al.[115] to multiconductor, multilayer substrate microstrip lines. The restriction to thin metallization is unrealistic for ICs, and Chilo et al.[116] approximately extended the spectral domain method to thicker lines by assuming only the width dependence of the current in the metal varied, while it remained the same throughout the strip thickness. The correct expansion functions are most important in this method and are discussed by Farr et al.,[115] Jansen,[117] and Fukuoka et al.[118]

Fukuoka et al.,[118] Sorrentino et al.,[68] and Tzuang and Itoh[70] applied the mode-matching method, which allows nonzero metallization thicknesses. This method divides the cross section of the line into multiple regions and makes a series expansion in each region. Matching across the boundaries provides self-consistency conditions that lead to an eigenvalue equation. This equation in turn determines the propagation constants of the allowed modes. Various versions can be found in Kowalski and Pregla,[119] Mittra et al.,[120] Crombach,[121] Kuznetsov and Lerer,[122] Solbach and Wolff,[123] Vahldieck,[124] and Bornemann.[125]

Direct solution in the time domain has been improved by Choi and Hoefer,[126] Yoshida and Fukai,[127] and Koike et al.[128]

6.5.5 Ground Planes

Pulses on a signal line above a ground plane induce currents in the ground plane. In a dc situation, the current in the ground plane is simply the return current, which spreads out over the ground plane. As a result, the return path resistance is very low, far less than that of the signal line. Under ac conditions, especially at higher frequencies, the return current becomes confined tightly below the signal line and the return path resistance increases. If the current on the signal line is abruptly stopped, the current in the ground plane can circulate in loops until killed by resistance. Discussion of this behavior can be found in Bertin,[88] Ruehli et al.,[129] and in Brennan et al.[110] The ground plane problem discussed in this work neglects displacement current in the ground plane and is a special case of the eddy current problem (e.g., see Rodger,[130] and Weis and Garg[131]). A formulation in the time domain was presented by Fang et al.[132] However, because of the neglect of displacement current, these approaches do not apply to semiconductor ground planes. The possibility of surface wave propagation away from the signal line over the ground plane hardly has been mentioned (see Mosig and Sarkar[133] and references therein).

6.6   PULSE PROPAGATION

With this formalism behind us, let us examine the results obtained so far for pulse propagation along interconnection lines. Some interesting experimental work in this area has begun using optoelectronics. Following some initial work by Auston,[134] a number of workers, particularly Cooper,[135] Eisenstadt et al.,[136] Bowman et al.,[137] and Ketchen et al.,[138] measured pulses as they propagated down microstrip or coplanar lines. The advantage of such measurements over conventional methods is that they have better time resolution, can probe circuits internally, and are free from the need to subtract a contribution from probes, jigs, and shielding. A more detailed comparison of various methods was made by Weingarten et al.[139]

Pulses are launched by several methods. One is to discharge a capacitor onto the line using an optically activated semiconductor switch (see Fig. 6.22). Switch operation has been analyzed by Auston,[134] Weiner et al.,[140] and Iverson and Smith.[141] The switch consists of a block of high-resistivity silicon or polysilicon upon which metal electrodes have been evaporated. Normally, no current is drawn between the electrodes, but if a pulse of light from a laser creates electron-hole pairs, the semiconductor becomes conducting and a current can flow. Hence, a charged capacitor at the end of a transmission line can be optically discharged onto the line, launching a pulse. Additional switches located along the line can monitor the pulse as it propagates. At times controlled precisely using an optical delay line, the monitoring switches along the line are illuminated, allowing the voltage on the line to charge monitoring capacitors. With a series of pulses and various delays, one can assemble a sequence of voltages that allows the pulse waveform to be reconstructed at each monitoring position.

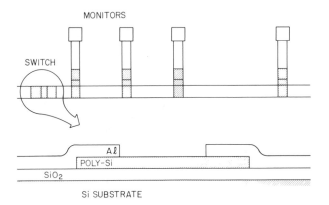

Fig. 6.22   A transmission line with an optically controlled switch at the left end (see inserted blowup at bottom of figure). This switch allows discharge of a capacitor onto the line, launching the pulse. Along the side of the line are monitoring capacitors that can be connected to the line by other optically controlled switches. (After Bowman et al.,[137] Reprinted with permission, copyright 1985, IEEE.)

Some example pulses are shown in Fig. 6.23. In Fig. 6.23*a*, the pulse sampled opposite the launch point is shown. The trailing edge of the pulse (later times) shows some structure not fitted by the Gaussian curve. This structure usually is attributed to interference from the laser pulse bouncing back, reflecting off the ground plane. In Fig. 6.23*b*, which shows the pulse 1.3 cm further down the line, the structure at the trailing edge of the pulse is due to the dispersive nature of the line, which allows lower-frequency components of the pulse to propagate faster than the rest.

Of course, the switch-on and decay to off-state of the optically activated switch is not instantaneous. Neither is the charging or discharging of the capacitors through the optically controlled resistance. The overall transient is convolved with the pulse shape one wishes to study. Because we cannot model the transient quantitatively, pulse shape information observed over times of the order of the switching transient is not meaningful. To date, the duration of the traveling pulse has been comparable to the charging time of the monitor. Hence, experiment has been analyzed mainly to find delay times and attenuation. Approximate extraction of these two parameters is based on a simple exponential decay model for the switch-

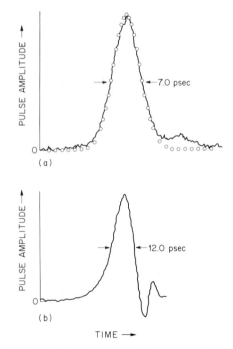

Fig. 6.23   Optically monitored pulses on a microstrip line. The line parameters are $Z_0 = 50\ \Omega$, width 180 μm, substrate 1.0 μm epilayer of Si on 180 μm sapphire, metallization is Au, 150 nm thick. (*a*) Pulse sampled directly opposite launch site following 3 ps optical pulse. (*b*) Pulse shape 1.3 cm down the line. (After Cooper.[135] Reprinted with permission, copyright 1985, *Applied Physics Letters*.)

ing transient. Further development of the method is needed to allow measurement of more detailed line parameters.

For GaAs, a different optical monitoring is possible, because GaAs exhibits the Pockels or electro-optic effect. An electric field in GaAs changes the polarization of light that passes through it. Observation of light intensity through crossed polarizers thus monitors electric field, and hence can monitor the passage of a pulse down a transmission line. This subject has been reviewed by Valdmanis and Mourou[142] and more recently by Weingarten et al.[139]

Another measurement method uses ejected electrons. In one such approach, a scanning electron microscope (SEM) is modified to produce a pulse of electrons (generated by a laser and a photocathode). The pulse of electrons hits the circuit at a node selected using the SEM in its normal mode of operation. The striking electrons eject secondary electrons from the selected node that have an energy distribution shifted by the voltage on the node at the time of ejection. Measuring this shift in distribution produces a signal proportional to the circuit node voltage. Such a system was described by May et al.[143] with a spatial resolution of 0.1 μm and a temporal resolution of 5 ps. Direct use of a laser beam to eject the electrons from a circuit node was described by Bokor et al.[144] and by Blacha et al.[145]

There are at least two interesting but incompletely discussed aspects of these ejected electron measurements. (1) In a picosecond time domain, circuit interconnection takes on a waveguide aspect where voltage and current are ambiguously related to the more fundamental notions of field.[77,78] The energy gained or lost by an electron during transit through a nonconservative field is a function of its trajectory, not just the end points of its path. What does "voltage" mean for these measurements under these circumstances? (2) As the collected electrons travel from the probed point of ejection at the interconnect line to the electron collector, they are electrically coupled to the circuit environment. For example, they expend energy in inducing image currents in nearby conductors and ground planes. What effect do these nonlocal interactions have on the spatial and temporal resolution of these measurements? In short, how accurately do the collected electrons reflect conditions at the point of ejection? Response time errors of several picoseconds were estimated for conservative fields by Fujioka et al.[146] and Clauberg.[147]

Let us return to pulse propagation. Assuming a single mode propagates, the transmission line equations relate the transfer function of the line at any position z to the load and driver impedances, Z_L and Z_D, via the transmission line γ and Z_0. Thus, at any point along the line the pulse shape can be found by the inverse Laplace transform of the product of the input pulse transform times the transfer function, $H(s, z)$, given by [compare (76)]

$$H(s, z) = \frac{Z_P}{(Z_P + Z_D) \cosh \gamma z + (Z_0 + Z_P Z_D / Z_0) \sinh \gamma z}, \tag{82}$$

where the line input impedance seen at point z, Z_P, is given by

$$Z_P = \frac{Z_L \cosh \gamma(\ell - z) + Z_0 \sinh \gamma(\ell - z)}{\cosh \gamma(\ell - z) + (Z_L / Z_0) \sinh \gamma(\ell - z)}. \tag{83}$$

In optoelectronic measurements, the pulse usually is measured before any reflections occur, that is, the line appears to be terminated in Z_0 so that (83) yields $Z_P = Z_0$ and from (82)

$$H(s, z) = \frac{Z_0}{Z_0 + Z_D} \exp(-\gamma z) . \tag{84}$$

In the case of low losses,

$$\gamma \approx j\omega \sqrt{\kappa_{\text{eff}}(\omega)\varepsilon_0\mu_0} , \tag{85}$$

where the effective dielectric constant is given approximately by the empirical Pramanick–Bhartia formula

$$\kappa_{\text{eff}}(\omega) = \kappa_s - \frac{\kappa_s - \kappa_{\text{eff}}}{1 + (\kappa_{\text{eff}}/\kappa_s)(\omega/\omega_t)^2} , \tag{86}$$

where $\omega_t = \pi Z_0/(\mu_0 h)$, h = height of the microstrip above the ground plane, and κ_{eff} can be found, for example, from formulas in Edwards[148] (pp. 41–47). The sensitivity of predicted pulse shapes to various choices for the approximate dispersion formula (88) was examined by Veghte and Balanis,[149] who found that several formulas were adequate and convenience could dictate the choice. Direct computation of $\kappa_{\text{eff}}(\omega)$ based on (85) with γ from a Fourier transformed time-domain analysis was made by Zhang et al.[150] Their work showed the Pramanick–Bhartia formula to be preferable. Presumably, Refs. 149 and 150 are compatible because pulse shapes in the low-loss case are less sensitive to $\kappa_{\text{eff}}(\omega)$ than ratios of pulse shapes at different positions.

Hasnain et al.[151] compared dispersion for Gaussian input pulses of widths ranging from 0.5 to 2 ps. Qualitative agreement with experiment was obtained using an empirical fit to the dispersion of the effective dielectric constant, such as (86). Cooper[135] has inverted the procedure of Hasnain et al. and has derived $\kappa_{\text{eff}}(\omega)$ by Fourier transform of pulse shapes observed on microstrip on silicon-on-sapphire substrates. However, in view of the approximations and sensitivity issues discussed earlier, before accepting these results an error analysis would be advisable.

The interpretation of these short-duration experiments on low-loss substrates is based on (84) and (85). For results applicable to circuits, however, the effects of load and driver impedances are important and the more general form for $H(s, z)$ of (82) is needed. Also, (85) applies only to some low-loss situations, and for other cases $\gamma(\omega)$ must be calculated from the characteristic equation. On the other hand, Z_0 is not determined by solving the field problem, as already discussed. Different choices for Z_0 are tantamount to different definitions relating the fields in the microstrip to currents and voltages in the line.[76–78] Tzuang and Itoh[70] have found little sensitivity of Z_0 to the choice of definition for their structures up to 100 GHz.

Tzuang and Itoh[70] used the mode-matching method to compute the propagation properties of some specific coplanar lines over insulating and semiconductor layers. The coplanar structure is shown in Fig. 6.24. They determined the propaga-

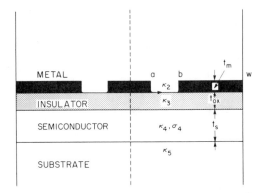

Fig. 6.24 A coplanar waveguide structure. (After Tzuang and Itoh.[70] Reprinted with permission, copyright 1987, IEEE.)

tion constant as shown in Fig. 6.25. The complex power/current definition of Z_0 leads to the characteristic impedance seen in Fig. 6.26. The real part is constant, but there is an approximately linearly increasing imaginary part. Sorrentino et al.[68] found that the inductive or capacitive nature of the reactive part of Z_0 depended on the geometry of the structure, particularly the ratio of top insulator thickness to the thickness of the high-conductivity semiconductor layer.

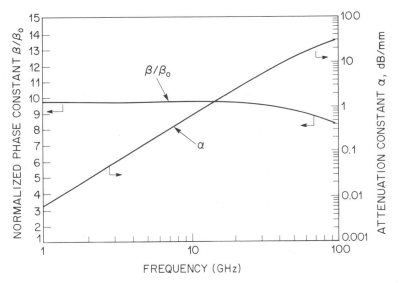

Fig. 6.25 The propagation constant for a coplanar waveguide structure as computed by the mode-matching method. Here, $\gamma = \alpha + j\beta$, and β is normalized to empty space, $\beta/\beta_0 = \beta c/\omega$. Parameters of the waveguide (notation of Fig. 6.24) are $a = 3\ \mu$m, $b = 8\ \mu$m, $w = 29.9\ \mu$m, $t_m = 0.3\ \mu$m, $t_{ox} = 0.3\ \mu$m, $t_s = 0.4\ \mu$m, $\kappa_1 = \kappa_2 = 1$, $\kappa_3 = \kappa_4 = \kappa_5 = 13$, and $\sigma = 2 \times 10^4$ S/m. (After Tzuang and Itoh.[70] Reprinted with permission, copyright 1987, IEEE.)

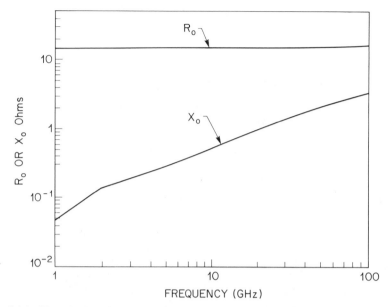

Fig. 6.26 The characteristics impedance $Z_0 = R_0 + jX_0$ for the same structure as in Fig. 6.24. (After Tzuang and Itoh.[70] Reprinted with permission, copyright 1987, IEEE.)

Tzuang and Itoh found broadband power matching of their line was possible using an impedance complex conjugate to Z_0 made up of a parallel RC combination with

$$R = Re(Z_0), \qquad C = \frac{Im(Z_0)}{\omega R^2}. \qquad (87)$$

For a 0.5 mm line of the dimensions given in the caption, their results for the pulse shapes for various driver resistances and a 10 fF load are shown in Fig. 6.27. The ringing is caused by the intrinsic inductive behavior of the characteristic impedance. In Fig. 6.28, the pulse shapes are shown for a matched source of 14.5 Ω in parallel with a 35 fF capacitance and various capacitive loads. These results assume a good conductivity GaAs substrate of 5×10^{-3} Ω · cm. The input pulse has a 12 ps rise and fall time.

For a low-loss coplanar waveguide (resistivity 10 Ω · cm), a decreasing imaginary part of the characteristic impedance is found. The pulse behavior, shown in Fig. 6.29, resembles the results of other workers on low-loss structures.

In passing, some more conventional methods of measuring line parameters should be mentioned.[152–155]

6.7 CROSSTALK

Aside from delay and distortion, the most important aspect of interconnect is its crosstalk properties. Crosstalk is caused by both capacitive and inductive cou-

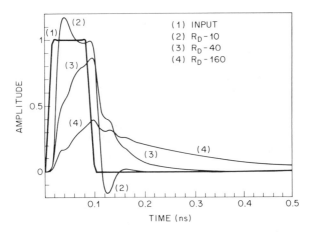

Fig. 6.27 Output pulse shapes for a 0.5 mm long line with driver resistance as parameter. Pulse rise and fall times 12 ps, width 20 ps, $C_L = 10$ fF. (After Tzuang and Itoh.[70] Reprinted with permission, copyright 1987, IEEE.)

pling. An additional source of crosstalk is coupling between circuits due to common impedances in ground loops, and ground planes are an important subject. Finally, proximity effects can make the $[R]$ and $[G]$ matrices of the transmission line model nondiagonal, so coupling can occur through the off-diagonal elements of these matrices as well as those of $[C]$ and $[L]$.

The best introduction to crosstalk is Chapter 9 in H. J. Gray's text,[156] which covers the basic issues very clearly. His treatment of distributed lines is extended

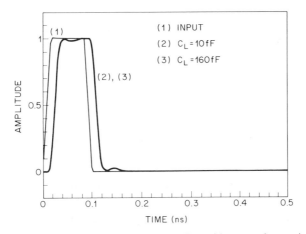

Fig. 6.28 Output pulse shapes for a 0.5 mm long line with a complex-conjugate matched driver and capacitive loads from 10 fF to 160 fF, $Z_D = R_D - j/(\omega C_D)$, with $R_D = 14.5$ Ω, $C_D = 35$ fF. Same input pulse as Fig. 6.27. (After Tzuang and Itoh.[70] Reprinted with permission, copyright 1987, IEEE.)

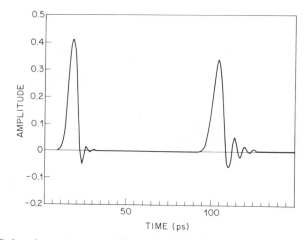

Fig. 6.29 Pulse shapes 1 cm and 2 cm down the line for a low-loss coplanar line, $\sigma_4 = 10$ S/m. The initial pulse was Gaussian with a full width at half-maximum of 5 ps. Other line parameters are as for Fig. 6.25. (After Tzuang and Itoh.[70] Reprinted with permission, copyright 1987, IEEE.)

in a paper by Feller et al.,[157] which, unfortunately, might take the prize for the most typographical errors on record — a study in noise by itself. These authors use the terms *backward* and *forward* crosstalk for the crosstalk at the near and far ends of a quiet line, respectively, and provide an intuitive as well as a quantitative analysis of the case when a single pulse is traveling down the driven line. Another useful discussion of various types of crosstalk for long parallel lines was presented by Catt[158] who introduces the terms fast, slow, and differential crosstalk. *Fast* and *slow* crosstalk correspond to two cases of backward crosstalk, depending on whether the pulse rise time is fast compared with the propagation time down the full line length, or slow. Slow crosstalk overlaps the lumped circuit limit. *Differential* crosstalk arises when the various modes (e.g., the even and odd mode for a pair of lines) propagate at different speeds. We know the amplitude of a pulse on a given line is a sum of various mode amplitudes (e.g., the voltage on the quiet line of a pair is $V_2 = V_{even} - V_{odd}$). Thus, if synchronism among the modes is lost because of different speeds of propagation, then cancellation cannot occur between the components and differential noise is produced. Finally, among the introductory literature we mention the tutorial by DeFalco[159] and the excellent overall discussion of Jarvis.[59] Unfortunately, the introductory work often assumes termination in a characteristic impedance, whereas most IC lines are likely to terminate in a small capacitance, such as a MOSFET gate. Thus, the principles apply, but some modification of the details is expected.

As before the treatment of noise and crosstalk can be based on a lumped circuit or on a distributed circuit model. For simplicity, let us begin with lumped circuits.

6.7.1 Lumped Coupling Analysis

As a beginning, let us consider two parallel lines, coupled both by a mutual inductance L_m and a mutual capacitance C_m. For simplicity, each line is divided into two sections, and the capacitive coupling occurs at the midpoint of the lines. By its nature, the inductive coupling is between current loops, and in this simple two-section model we suppose the near-end loops are inductively coupled only to each other, and likewise for the far-end loops. The general idea is shown in Fig. 6.30a, although the L_m coupling is indicated rather schematically. The resulting circuit can be rearranged, as shown in Fig. 6.30b. In Fig. 6.30b, the impedances Z_{1a} and Z_{2a} include the line model up to midpoint in series with the driver impedance, whereas Z_{1b} and Z_{2b} include the other half of the line model and the load impedances. That is,

$$Z_{1a} = Z_{2a} = Z_D + Z_n , \qquad (88a)$$

$$Z_{1b} = Z_{2b} = Z_f + \frac{1}{sC_L} , \qquad (88b)$$

where Z_D = driver impedance, C_L = load capacitance, Z_n = near-end line impedance, and Z_f = far-end line impedance. The loop currents I_{21}, I_m, and I_{22} can be found by circuit theory, and hence the noise induced in the quiet line can be found, as well as the disturbance of the driven line. The case of greatest interest occurs when the coupling is weak. Keeping only the terms of lowest order in L_m and C_m, a simple result can be found by treating the driven line as uncoupled, so its loop currents are

$$I_{11} \approx \frac{V_{\text{in}}(1 + Z_{1b}Y)}{\Delta_1} ,$$

$$I_{12} \approx \frac{V_{\text{in}}}{\Delta_1} ,$$

with

$$\Delta_1 = Z_{1a} + Z_{1b} + Z_{1a}Z_{1b}Y . \qquad (89)$$

Using superposition, the capacitive and inductive couplings can be treated separately. The current flowing through the coupling capacitance to the quiet line is then

$$I = sC_m \ell V_m ,$$

where V_m is the midpoint voltage of the driven line and $C_m \ell$ the coupling capacitance. This coupled current is split between the three impedances Z_{2a}, Z_{2b}, and

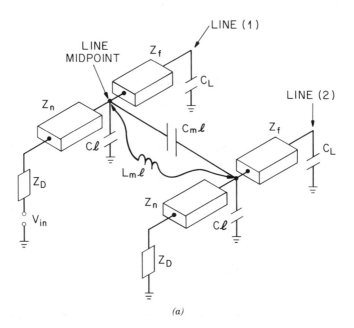

(a)

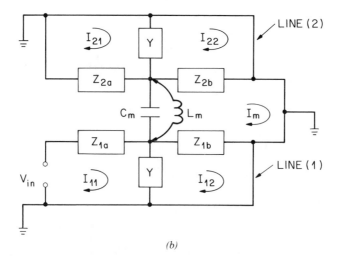

(b)

Fig. 6.30 Two interconnect lines of length ℓ coupled by a mutual inductance L_m and mutual capacitance C_m per unit length. The inductive coupling is indicated rather crudely and is assumed in the text to couple the two near-end current loops to each other using $L_m \ell/2$ and to couple the two far-end current loops using the remaining $L_m \ell/2$. (a) Artist's view. (b) Equivalent circuit showing loop currents.

Y^{-1} in Fig. 6.30b, leading to the following capacitive noise contributions at the near and far ends of the quiet line,

$$V_{\text{ne}}(\text{cap}) = \frac{Z_{2b}}{\Delta_2} I Z_D$$

$$= \frac{s V_{\text{in}} C_m \ell Z_D Z_{1b} Z_{2b}}{\Delta_1 \Delta_2}, \tag{90a}$$

$$V_{\text{fe}}(\text{cap}) = \frac{Z_{2a}}{\Delta_2} I Z_L$$

$$= \frac{s V_{\text{in}} C_m \ell Z_L Z_{1b} Z_{2a}}{\Delta_1 \Delta_2}, \tag{90b}$$

where Δ_2 is given by (89) with the exchange $1 \leftrightarrow 2$.

The inductive contribution to the noise is approximated by supposing that half the mutual inductance couples the near-end loops, and the other half the far-end loops. Then the voltage induced in the near-end loop of the quiet line is

$$-\frac{L_m \ell}{2} s I_{11} = (Z_{2a} + Y^{-1}) I_{21} - Y^{-1} I_{22}, \tag{91a}$$

and that in the far-end loop is

$$-\frac{L_m \ell}{2} s I_{12} = -Y^{-1} I_{21} + (Z_{2b} + Y^{-1}) I_{22}. \tag{91b}$$

Thus, the inductive noise contributions are

$$V_{\text{ne}}(\text{ind}) = -I_{21} Z_D$$

$$= s V_{\text{in}} \frac{L_m \ell Z_D}{\Delta_1 \Delta_2} \left[1 + \frac{1}{2}(Z_{1b} + Z_{2b}) Y + \frac{1}{2} Z_{1b} Z_{2b} Y^2 \right], \tag{92a}$$

and

$$V_{\text{fe}}(\text{ind}) = I_{22} Z_L$$

$$= -s V_{\text{in}} \frac{L_m \ell Z_L}{\Delta_1 \Delta_2} \left[1 + \frac{1}{2}(Z_{2a} + Z_{1b}) Y \right]. \tag{92b}$$

The total noise is the sum of the capacitive contributions (90a, b) and the inductive contributions (92a, b). The same expressions can be obtained from an exact solution of the mesh equations by expanding the exact results in powers of L_m and C_m and keeping only the linear terms. However, the approach just outlined is easier and more intuitive.

The noise voltages from (90) and (92) are useful qualitatively for assessing the importance of capacitive vs. inductive coupling. These equations show that the capacitive noise is proportional to $C_m dV_m/dt$, whereas the inductive noise is related roughly to $L_m dI_m/dt$, where V_m is the voltage at midpoint of line 1 when there is no coupling and I_m is the current at midpoint. This result seems intuitively clear because of the single-point coupling approximation, even if complicated impedances are introduced into the expressions. Approximating the derivatives as

$$\frac{dV_m}{dt} \approx \frac{V_{DD}}{\tau},$$

$$\frac{dI_m}{dt} \approx \frac{I_{sat}}{\tau},$$

where τ is a pulse rise or fall time at the midpoint of the line, we find that large voltages are likely to lead to capacitive noise, whereas large currents are likely to emphasize inductive noise. For small capacitive loads, as are usual in ICs, currents are small and capacitive coupling tends to dominate. A crosstalk analysis based on lumped capacitive coupling was made by Smith and Snyder.[160] Limited comparisons of the lumped capacitive and distributed models were made by Jarvis[59] and by Seki and Hasegawa.[161]

6.7.2   Distributed Coupling Analysis

The majority of qualitative crosstalk analysis is based on a distributed coupling model. Using the transmission line model for a coupled, lossless pair of lines in the weak coupling limit, the usual procedure is to follow Feller et al.[157] or Jarvis[59] to arrive at the backward crosstalk estimate (usually the largest crosstalk component) for a ramp input on the drive line, namely,

$$\frac{V_B}{V_1} = 2\frac{V_B}{V_{in}} = \frac{1}{4}(k_C + k_L)\frac{2\tau_{prop}}{\tau_{rise}}, \qquad 2\tau_{prop} \leq \tau_{rise}, \qquad (93a)$$

$$= \frac{1}{4}(k_C + k_L), \qquad 2\tau_{prop} > \tau_{rise}, \qquad (93b)$$

where V_B is the maximum backward noise voltage, V_{in} the maximum input voltage, $V_1 = V_{in}/2$ the voltage at the near end of the drive line, τ_{prop} the propagation time down the line, τ_{rise} the rise time of the ramp input, $k_C = C_m/C$, and $k_L = L_m/L$. The analysis leading to (93) assumes termination in the characteristic impedance, so no reflections occur.

The difference in noise voltage for short and long lines can be understood as follows. For a long line, the length of drive line over which the voltage is time varying (the "rise length" region) increases from zero length at $t = 0$ to a maximum length $\ell_{rise} = v\tau_{rise}$ (v = velocity) for $t \geq \tau_{rise}$. This rise length region propagates down the drive line, generating a noise pulse that travels backward toward the near end of the quiet line. At $t = \tau_{prop}$, the rise length region reaches the end of the drive line and begins to be swallowed in the Z_0 termination. Hence, the rise length region on the drive line disappears during the time $\tau_{prop} < t < \tau_{prop} + \tau_{rise}$. On the quiet line, this event does not arrive until τ_{prop} later. Hence, the long line noise looks like Fig. 6.31a.

For a short line, there is at first no difference from the long line until the drive ramp reaches the end of the drive line, at time $t = \tau_{prop}$. Once this happens, the entire drive line (length $\ell < v\tau_{rise}$) is subject to time-varying voltage, and the entire line contributes to the backward noise voltage. This maximum noise arrives at the near end of the quiet line τ_{prop} later, at $t = 2\tau_{prop}$. The entire drive line continues to generate the same noise (same time derivatives throughout the drive line) until the pulse rise time is reached. Once $t = \tau_{rise}$, the drive end of the drive line no longer is transient as the voltage has reached its final value, with instant effect on the noise as it is the near ends of both lines that are involved. At time $t = \tau_{rise} + \tau_{prop}$, the drive line has reached its final voltage throughout its length (all time derivatives vanish on the drive line), and this event is signaled at the near end of the quiet line at $t = \tau_{rise} + 2\tau_{prop}$. Thus, the noise voltage pulse on the short line is as shown in Fig. 6.31b.

It is physically evident that the slope of the noise pulse during its rise and fall is the same for both the long and short lines. Hence, the maximum noise values are in the same ratio as the times at which they are reached:

$$\frac{V_{short}}{V_{long}} = \frac{2\tau_{prop}(short)}{\tau_{rise}} = \frac{2\ell(short)}{\ell_{rise}},$$

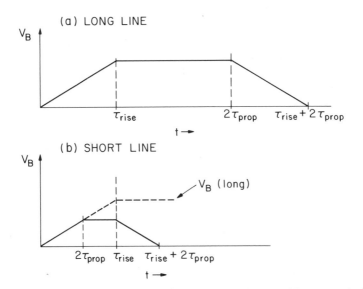

(a) LONG LINE

(b) SHORT LINE

Fig. 6.31 Crosstalk pulses at the near end of a quiet line generated by a ramp input voltage on the drive line: (a) long line, $2\tau_{prop} > \tau_{rise}$; ($b$) short line, $2\tau_{prop} < \tau_{rise}$.

in agreement with (93a), where ℓ(short) is the length of the short line and ℓ_{rise} the maximum rise length, the length of line sufficient to include the entire rising edge of the voltage ramp.

While (93) does not include reflections, it suffices for preliminary assessment of crosstalk. Lewis[162] has outlined how k_C and k_L can be approximated using a capacitance model due to Garg and Bahl[163] evaluating inductance using (78). Carter[164] has used (93) to estimate the maximum coupled line length allowed in ICs. He suggests this length is about 0.1 cm for 100 ps rise times and 1 μm design rules, a short enough length that one might ask how the estimate based on (93) might compare with the lumped circuit estimate from (90) and (92).

To compare, suppose the line is modeled as a simple capacitor, $Z_n = Z_f = 0$, so $Z_{1a} = Z_{1b} = Z_{2a} = Z_{2b} = Z_0$, assuming termination in the characteristic impedance as was done to derive (93). Then the sum of (90a) and (92a) becomes

$$V_{\text{ne}} = \frac{sV_{\text{in}}\ell}{Z_0} \frac{Z_0^2 C_m + (1 + Z_0 Y + Z_0^2 Y^2/2)L_m}{(2 + Z_0 Y)^2}$$

$$\simeq sV_{\text{in}}\frac{1}{4}(k_C + k_L)\tau_{\text{prop}}, \tag{94}$$

where the approximation $Z_0 Y = sC\,\ell Z_0 \ll 1$ has been made and the time of propagation is given by $\ell/v = \sqrt{LC}\,\ell$. Because sV_{in} is the Laplace transform of dV_{in}/dt, (94) predicts a rectangular noise pulse of duration τ_{rise}, rather than the pulse of Fig. 6.31b and with a magnitude

$$\frac{V_{\text{ne}}}{V_{\text{in}}} = \frac{1}{4}(k_C + k_L)\left(\frac{\tau_{\text{prop}}}{\tau_{\text{rise}}}\right), \tag{95}$$

which is the same value predicted by the distributed model (93).

In view of the extreme simplification of the line model necessary to reduce the sum of (90) and (92) to (93), namely, a reduction of the lines to short circuits coupled by $L_m \ell$ and $C_m \ell$, probably (90) and (92) allow a more realistic model of the line for simple noise estimates and a simpler determination of the effects of different terminations. However, the comparisons between lumped and distributed responses by Jarvis[59] and Seki and Hasegawa[161] show the lumped circuit estimates for a coupled pair of lines may be larger or smaller than the actual noise response, with accuracies to perhaps a factor of 2.

For more quantitative analysis of crosstalk, one can use an equivalent circuit model in a circuit analysis program such as SPICE. For two lines, the circuit of Fig. 6.30 could be used, with more mutual coupling points for greater accuracy. For multiple lines, such an approach has been presented by Tripathi and Bucolo[165] and Tripathi and Hill.[166]

Adaptation of such models to treat nonlinear terminations was considered by Djordjevic et al.[167] Another method to treat nonlinear terminations using time-domain scattering parameters of the transmission line was presented by Schutt-

Aine and Mittra.[168] A rather broad review of nonlinear methods was written by Rizzoli and Neri.[169] These numerical works have focused on development of CAD tools, without emphasis on general guidelines for overall planning, trade-offs, or design rules.

Assessment of techniques to reduce crosstalk have been made by Seki and Hasegawa,[161] Chilo and Arnaud,[98] Chilo et al.,[116] and Chilo and Razban.[170] The main suggestions for crosstalk improvement are (a) introduction of grounded screen lines or (b) introduction of ground planes above the signal lines. The drawbacks of screen lines are (a) they reduce the space available for wiring, (b) they should be grounded internally, not just at their end points, and (c) they introduce some degradation in the response of the driven line. The addition of extra ground planes above the signal lines is effective, but adds to the capacitance of the lines, resulting in a lower impedance and speed of propagation. The reduction of crosstalk thus is coupled to other circuit problems and general guidelines for these trade-offs are lacking.

6.8   CLOSING REMARKS

A tutorial and brief summary of the literature describing electrical modeling of interconnections has been presented, directed particularly toward VLSI circuits. Beginning with two interconnect scenarios intended to stress the importance of interconnection in deciding on-chip architecture of large, fast systems, the discussion moved to Rent's rule and the complication of multimodal wire length distributions. This general background led to a discussion of RC, RLC, and $RLCG$ line models, including a generalization of Frye and Yuan's criterion for a suitably damped RLC interconnection that includes both load capacitance and driver impedance.

The design of interconnect systems up to the present has focused largely on the properties of parallel arrays of nearly perfect conductors embedded in dielectric layers. The limitations of this picture are already apparent. Among them are the need for three-dimensional treatments of vias, of transitions from lines to drivers and lines to loads, of bends, and of crossovers. Equivalent circuits for these structures are under development that can be incorporated in simulators and used to develop guidelines for best design of these structures. Power loss and crosstalk due to these discontinuities need assessment.

Because ICs are pulse-operated, circuits valid from dc up to several times the bit rate must be developed. Reflections from discontinuities along the lines cannot be eliminated by a simple resistance because the lines are dispersive, and alternative approaches based on critical damping are being explored. Devices present nonlinear load and driver impedances, which greatly complicate crosstalk and noise analysis. Work in these areas also is just beginning.

For semiconductor substrates, models should include both conduction and displacement currents. Even for the case of a single microstrip line over a semiconductor substrate, we do not have a definitive equivalent transmission line. Semiconductor ground planes have not been examined. The characterization of semiconductor materials where the free-carrier ac response can play a role has hardly been explored, and may lead to new ideas for network control, or optoelectronic integration.

In the area of measurement, much remains to be done as well. S-parameter measurements are too indirect to be very useful in determining the many parameters involved in describing interconnect. Optoelectronic measurements seem promising, but as yet are of only qualitative value.

Our discussion of interconnection began by suggesting that it directly impacts the design of a system, even for a system all on one chip. Perhaps the study of computational VLSI will provide the guidelines needed to trade local autonomy for certain interconnect characteristics. Perhaps figures of merit will emerge for interconnection to replace the present emphasis on simple delay. As these metrics emerge, interconnect design must evolve to deal with them. The host of papers on interconnections centered around the simple Elmore delay expression shows how much we appreciate some simple guidance in evaluating drivers, repeaters, loads, etc., in overall performance. And the Elmore delay does not include power, information, area, or even more than one line.

At the basic level as well as at the computer-aided design level, work on interconnection is only beginning, and there is every possibility for future surprises.

ACKNOWLEDGMENTS

The author is pleased to thank his colleagues for their comments, particularly A. J. Rainal, D. L. Carter, R. C. Frye, S. M. Sze, R. Sharma, and S. K. Tewksbury.

REFERENCES

1. D. P. Seraphim and I. Feinberg, "Electronic Packaging Evolution in IBM," *IBM J. Res. Dev.*, **24**(5), 615 (1981).

2. M. Hatamian, L. A. Hornak, T. E. Little, S. K. Tewksbury, and P. Franzon, "Fundamental Interconnection Issues," *AT&T Tech. J.*, **66**(4), 13 (1987).

3. R. L. Russo and P. K. Wolff, Sr., "A Computer-Based Design Approach to Partitioning and Mapping of Computer Logic Graphs," *Proc. IEEE*, **60**(1), 28 (1972).

4. B. W. Kernighan and S. Lin, "An Efficient Heuristic Procedure for Partioning Graphs," *Bell Syst. Tech. J.*, **49**(2), 291 (1970).

5. S. J. Hong and R. Nair, "Wire-Routing Machines—New Tools for VLSI Physical Design," *Proc. IEEE*, **71**(1), 57 (1983).

6. H. N. Brady and J. Blanks, "Automatic Placement and Routing Techniques for Gate Array and Standard Cell Designs," *Proc. IEEE*, **75**(6), 797 (1987).

7. E. L. Lawler, K. N. Levitt, and J. Turner, "Module Clustering to Minimize Delay in Digital Networks," *IEEE Trans. Comput.*, **C-18**(1), 47 (1969).

8. C. L. Seitz, "Concurrent VLSI Architectures," *IEEE Trans. Comput.*, **C-33**(12), 1247 (1984).

9. C. Mead and L. Conway, *Introduction to VLSI Systems,* Addison-Wesley, Reading, MA, 1980.

10. J. Allen, "VLSI Architectures for Signal Processing," in B. Randall and P. C. Treleaven, Eds., *VLSI Architecture,* Prentice-Hall, Englewood Cliffs, NJ, 1983.

11. M. J. Flynn and J. L. Hennessy, "Parallelism and Representation Problems in Distributed Systems," *IEEE Trans. Comput.,* **C-29**(12), 1081 (1980).

12. C. D. Thompson, *A Complexity Theory for VLSI,* Ph.D. thesis, Carnegie-Mellon University, Computer Sciences Department, Pittsburgh, PA (1980) (University Microfilms International, Ann Arbor, MI).

13. C. E. Leiserson, *Area-Efficient VLSI Computation,* MIT Press, Cambridge, MA, 1983.

14. F. T. Leighton, *Complexity Issues in VLSI: Optimal Layouts for the Shuffle-Exchange Graph and Other Networks,* MIT Press, Cambridge, MA, 1983.

15. J. E. Savage, "Area-Time Tradeoffs for Matrix Multiplication and Related Problems in VLSI Models," *J. Compu. Syst. Sci.,* **22,** 230 (1981).

16. B. Chazelle and L. Monier, "A Model of Computation for VLSI with Related Complexity Results," *J. Assoc. Comput. Mach.,* **32**(3), 573 (1985).

17. J. D. Ullman, *Computational Aspects of VLSI,* Computer Science Press, Rockville, MD, 1985.

18. W. D. Hillis, *The Connection Machine,* MIT Press, Cambridge, MA, 1985.

19. L. Snyder, "Introduction to the Configurable, Highly Parallel Computer," *Computer,* **15**(1), 47 (1982).

20. S. E. Fahlman and G. E. Hinton, "Connectionist Architectures for Artificial Intelligence," *Computer,* **20**(1), 100 (1987).

21. J. J. Hopfield, "Neural Networks and Physical Systems with Emergent Collective Computational Abilities," *Proc. Nat. Acad. Sci. U.S.A.,* **79,** 2554 (1982).

22. H. P. Graf, L. D. Jackel, and W. E. Hubbard, "VLSI Implementation of a Neural Network Model," *Computer,* **21**(3), 41 (1988).

23. G. B. Sorkin, "Asymptotically Perfect Trivial Global Routing: A Stochastic Analysis," *IEEE Trans. Comput.-Aided Des.,* **CAD-6**(9), 820 (1987).

24. L. A. Glasser and D. W. Dobberpuhl, *The Design and Analysis of VLSI Circuits,* Addison-Wesley, Reading, MA, 1985.

25. H. B. Bakoglu and J. D. Meindl, "Optimal Interconnections for VLSI," *IEEE Trans. Electron Devices,* **ED-32**(5), 903 (1985).

26. H. Kanai, "Low Energy LSI and Packaging for System Performance," *IEEE Trans. Components, Hybrids, Manuf. Technol.,* **CHMT-4**(2), 173 (1981).

27. A. Bar-Cohen, "Thermal Management of Air- and Liquid-Cooled Multichip Modules," *IEEE Trans. Components, Hybrids, Manuf. Technol.,* **CHMT-10**(2), 159 (1987).

28. V. Multinovic, D. Fura, and W. Helbig, "An Introduction to GaAs Microprocessor Architecture for VLSI," *Computer,* **19**(3), 30 (1986).

29. B. S. Landman and R. L. Russo, "On a Pin Versus Block Relationship for Partitions of Logic Graphs," *IEEE Trans. Comput.,* **C-20**(12), 1469 (1971).

30. D. K. Ferry, R. O. Grondin, and W. Porod, "Interconnections, Dissipation, and Computation," in N. G. Einspruch, S. S. Cohen, and G. Sh. Gildenblat, Eds., *VLSI Electronics: Microstructure Science,* Vol. 15, Chapter 10, p. 451, Academic, New York, 1987.

31. K. M. Striny, "Assembly Techniques and Packaging of VLSI Devices", in S. M. Sze, Ed., *VLSI Technology*, 2nd ed., Chapter 13, p. 566, McGraw-Hill, New York, 1988.

32. M. Feuer, "Connectivity of Random Logic," *IEEE Trans. Comput.*, **C-31**(1), 29 (1982).

33. W. R. Heller, W. F. Mikhail, and W. E. Donath, "Prediction of Wiring Space Requirements for LSI," *J. Des. Autom. Fault Tolerant Comput.*, **2**(2), 117 (1978).

34. A. Masaki and M. Yamada, "Equations for Estimating Wire Length in Various Types of 2-D and 3-D System Packaging Structures," *IEEE Trans. Components, Hybrids, Manuf. Technol.*, **CHMT-10**(2), 190 (1987).

35. J. Vuillemin, "A Combinatorial Limit to the Computing Power of VLSI Circuits," *IEEE Trans. Compu.*, **C-32**(3), 294 (1983).

36. R. H. Kuhn, "Chip Bandwidth Bounds by Logic-Memory Tradeoffs," in J. P. Gray, Ed., *VLSI 81: Very Large Scale Integration*, p. 279, Academic, New York, 1981.

37. S. M. Kang, "Metal-Metal Matrix (M^3) for High-Speed MOS VLSI Layout," *IEEE Trans. Comput.-Aided Des.*, **CAD-6**(5), 886 (1987).

38. C. D. Phillips, W. C. Selbach, A. J. Narud, and K. K. Lynn, "Complex Monolithic Arrays: Some Aspects of Design and Fabrication," *IEEE J. Solid-State Circuits* **SC-2**(12), 156 (1967).

39. W. C. Elmore, "The Transient Response of Damped Linear Networks with Particular Regard to Wideband Amplifiers," *J. Appl. Phys.*, **19**(1), 55 (1948).

40. M. S. Ghausi and J. J. Kelly, *Introduction to Distributed Parameter Networks*, Chapter 3, Holt, Rinehart & Winston, New York, 1968.

41. T. Sakurai, "Approximation of Wiring Delay in MOSFET LSI," *IEEE J. Solid-State Circuits*, **SC-18**(4), 418 (1983).

42. J. Rubinstein, P. Penfield, Jr., and M. A. Horowitz, "Signal Delay in RC Tree Networks," *IEEE Trans. Comput.-Aided Des.*, **CAD-2**(3), 202 (1983).

43. T.-M. Lin and C. A. Mead, "Signal Delay in General RC Networks," *IEEE Trans. Comput.-Aided Des.*, **CAD-3**(4), 331 (1984).

44. C. A. Zukowski, "Relaxing Bounds for Linear RC Mesh Circuits," *IEEE Trans. Comput.-Aided Des.*, **CAD-5**(2), 305 (1986).

45. G. Bilardi, M. Pracchi, and F. P. Preparata, "A Critique of Network Speed in VLSI Models of Computation," *IEEE J. Solid-State Circuits*, **SC-17**(4), 696 (1982).

46. C. Mead and M. Rem, "Minimum Propagation Delays in VLSI," *IEEE J. Solid-State Circuits*, **SC-17**(4), 773 (1982).

47. D. L. Carter and D. F. Guise, "Analysis of Signal Propagation Delays and Chip Level Performance Due to On-Chip Interconnections," *Proc. IEEE Int. Conf. Comput. Des. VLSI, Comput., 1983*, Port Chester, NY p. 218 (1983).

48. E. T. Lewis, "Optimization of Device Area and Overall Delay for CMOS VLSI Designs," *Proc. IEEE*, **72**(6), 670 (1984).

49. C. M. Lee and H. Soukup, "An Algorithm for CMOS Timing and Area Optimization," *IEEE J. Solid-State Circuits*, **SC-19**(5), 781 (1984).

50. N. Hedenstierna and K. O. Jeppson, "CMOS Circuit Speed and Buffer Optimization," *IEEE Trans. Compu.-Aided Des.*, **CAD-6**(2), 270 (1987).

51. T. Sakurai and K. Tamaru, "Simple Formulas for Two and Three Dimensional Capacitances," *IEEE Trans. Electron Devices*, **ED-30**(2), 183 (1983).

52. H. J. de Los Santos and B. Hoefflinger, "Optimization and Scaling of CMOS Bipolar Drivers for VLSI Interconnects," *IEEE Trans. Electron Devices*, **ED-33**(11), 1722 (1986).

53. B. C. Kuo, *Automatic Control Systems*, 5th ed., Prentice-Hall, Englewood Cliffs, NJ, 1987.

54. C. W. Ho, D. A. Chance, C. H. Bajorek, and R. E. Acosta, "The Thin-Film Module as a High-Performance Semiconductor Package," *IBM J. Res. Dev.*, **26**(3), 286 (1982).

55. A. J. Rainal, "Computing Inductive Noise of Chip Packages," *AT&T Bell Labs. Tech. J.*, **63**(1), 177 (1984).

56. M. Cases and D. M. Quinn, "Transient Response of Uniformly Distributed RLC Transmission Lines," *IEEE Trans. Circuits Syst.*, **CAS-27**(3), 200 (1980).

57. H.-T. Yuan, Y.-T. Lin, and S.-Y. Chiang, "Properties of Interconnection on Silicon, Sapphire and Semi-Insulating Gallium Arsenide Substrates," *IEEE J. Solid-State Circuits*, **SC-17**(2), 269 (1982).

58. R. C. Frye, "Properties and Optimization of Lossy Transmission Lines," *AT&T Bell Labs. Tech. Memo.*, 11528-850277-06TM (1985).

59. D. B. Jarvis, "The Effects of Interconnections on High-Speed Logic Circuits," *IEEE Trans. Electron. Comput.*, **EC-12**(10), 476 (1963).

60. E. H. Nicollian and J. R. Brews, *MOS Physics and Technology*, Wiley, New York, 1982.

61. H. Guckel, P. A. Brennan, and I. Palocz, "A Parallel-Plate Waveguide Approach to Microminiaturized, Planar Transmission Lines for Integrated Cricuits," *IEEE Trans. Microwave Theory Tech.*, **MTT-15**(8), 468 (1967).

62. H. Hasegawa, M. Furukawa, and H. Yanai, "Properties of Microstrip Line on Si-SiO$_2$ System," *IEEE Trans. Microwave Theory Tech.*, **MTT-19**(11), 869 (1971).

63. A. R. Owens and K. A. Jaggers, "Pulse Response of Interconnections in Silicon Integrated Circuits," *Proc. Inst. Electr. Eng.*, **121**(7), 541 (1974).

64. D. Jager, W. Rabus, and W. Eickhoff, "Bias-Dependent Small-Signal Parameters of Schottky Contact Microstrip Lines," *Solid-State Electron.*, **17**, 777 (1974).

65. D. Jager, "Slow-Wave Propagation Along Variable Schottky-Contact Microstrip Line," *IEEE Trans. Microwave Theory Tech.*, **MTT-24**(9), 566 (1976).

66. C. M. Krowne, "Slow Wave Propagation in Waveguides Loaded with a Semiconductor," *Int. J. Electron.*, **58**(2), 249 (1985).

67. M. Aubourg, J.-P. Villotte, F. Godon, and Y. Garault, "Finite-Element Analysis of Lossy Waveguides — Application to Microstrip Lines on Semiconductor Substrate," *IEEE Trans. Microwave Theory Tech.*, **MTT-31**(4), 326 (1983).

68. R. Sorrentino, G. Leuzzi, and A. Silbermann, "Characteristics of Metal-Insulator-Semiconductor Coplanar Waveguides for Monolithic Microwave Circuits," *IEEE Trans. Microwave Theory Tech.*, **MTT-32**(4), 410 (1984).

69. C.-K. Tzuang and T. Itoh, "Finite Element Analysis of Slow-Wave Schottky Contact Printed Lines," *IEEE Trans. Microwave Theory Tech.*, **MTT-34**(12), 1483 (1986).

70. C.-K. C. Tzuang and T. Itoh, "High-Speed Pulse Transmission Along a Slow-Wave Coplanar Waveguide for Monolithic Microwave Integrated Circuits," *IEEE Trans. Microwave Theory Tech.*, **MTT-35**(8), 697 (1987).

71. H. Hasegawa and S. Seki, "Analysis of Interconnection Delay on Very High-Speed LSI/VLSI Chips Using an MIS Microstrip Model," *IEEE Trans. Electron Devices,* **ED-31**(12), 1954 (1984).

72. S. Ramo, J. R. Whinnery, and T. Van Duzer, *Fields and Waves in Communication Electronics,* Wiley, New York, 1965.

73. J. R. Brews, "Transmission Line Models for Lossy Waveguide Interconnections in VLSI," *IEEE Trans. Electron Devices,* **ED-33**(9), 1356 (1986).

74. C. C. Johnson, *Field and Wave Electrodynamics,* McGraw-Hill, New York, 1965.

75. A. D. Bresler, G. H. Joshi, and N. Marcuvitz, "Orthogonality Properties for Modes in Passive and Active Uniform Waveguides," *J. Appl. Phys.,* **29**(5), 794 (1958).

76. B. Bianco, L. Panini, M. Parodi, and S. Ridella, "Some Considerations About the Frequency Dependence of the Characteristic Impedance of Uniform Microstrips," *IEEE Trans. Microwave Theory Tech.,* **MTT-26**(3), 182 (1978).

77. W. J. Getsinger, "Measurement and Modeling of the Apparent Characteristic Impedance of Microstrip," *IEEE Trans. Microwave Theory Tech.,* **MTT-31**(8), 624 (1983).

78. J. R. Brews, "Characteristic Impedance of Microstrip Lines," *IEEE Trans. Microwave Theory Tech.,* **MTT-35**(1), 30 (1987).

79. H. Amemiya, "Time-Domain Analysis of Multiple Parallel Transmission Lines," *RCA Rev.,* **28**(6), 241 (1967).

80. K. D. Marx, "Propagation Modes, Equivalent Circuits, and Characteristic Terminations for Multiconductor Transmission Lines with Inhomogeneous Dielectrics," *IEEE Trans. Microwave Theory Tech.,* **MMT-21**(7), 450 (1973).

81. C. R. Paul, "Useful Matrix Chain Parameter Identities for the Analysis of Multiconductor Transmission Lines," *IEEE Trans. Microwave Theory Tech.,* **MTT-23,** 756 (1975).

82. S. Frankel, *Multiconductor Transmission Line Analysis,* Artech House, Dedham, MA, 1977.

83. H. E. Green, "The Numerical Solution of Some Important Transmission-Line Problems," *IEEE Trans. Microwave Theory Tech.,* **MTT-13**(5), 676 (1965).

84. C. Wei, R. F. Harrington, J. R. Mautz, and T. K. Sarkar, "Multiconductor Transmission Lines in Multilayered Dielectric Media," *IEEE Trans. Microwave Theory Tech.,* **MTT-32**(4), 439 (1984).

85. L. D. Landau and E. M. Lifshitz, *Electrodynamics of Continuous Media,* Pergamon, Oxford, UK, 1960.

86. R. E. Mattick, *Transmission Lines for Digital and Communication Networks,* McGraw-Hill, New York, 1969.

87. J. E. Schutt-Aine and R. Mittra, "Analysis of Pulse Propagation in Coupled Transmission Lines," *IEEE Trans. Circuits Syst.,* **CAS-32**(12), 1214 (1985).

88. C. L. Bertin, "Transmission Line Response Using Frequency Techniques," *IBM J. Res. Dev.,* **8**(1), 52 (1964).

89. A. E. Ruehli, "Inductance Calculations in a Complex Integrated Circuit Environment," *IBM J. Res. Dev.,* **16**(9), 470 (1972).

90. P. C. Chestnut, "On Determining the Capacitances of Shielded Multiconductor Transmission Lines," *IEEE Trans. Microwave Theory Tech.,* **MTT-17**(10), 734 (1969).

91. W. T. Weeks, "Calculation of Coefficients of Capacitance of Multiconductor Transmission Lines in the Presence of a Dielectric Interface," *IEEE Trans. Microwave Theory Tech.*, **MTT-18**(1), 35 (1970).

92. P. D. Patel, "Calculation of Capacitance Coefficients for a System of Irregular Finite Conductors on a Dielectric Sheet," *IEEE Trans. Microwave Theory Tech.*, **MTT-21**(2), 76 (1973).

93. A. E. Ruehli and P. A. Brennan, "Efficient Capacitance Calculations for Three-Dimensional Multiconductor Systems," *IEEE Trans. Microwave Theory Tech.*, **MTT-21**(2), 76 (1973).

94. A. E. Ruehli and P. A. Brennan, "Capacitance Models for Integrated Circuit Metallization Wires," *IEEE J. Solid-State Circuits*, **SC-10**(6), 530 (1975).

95. P. Benedek, "Capacitances of a Planar Multiconductor Configuration on a Dielectric Substrate by a Finite Element Method," *IEEE Trans. Circuits Sys.*, **CAS-23**(5), 279 (1976).

96. A. E. Ruehli, "Survey of Computer-Aided Electrical Analysis of Integrated Circuit Interconnections," *IBM J. Res. Dev.*, **23**(6), 626 (1979).

97. R. F. Harrington and C. Wei, "Losses on Mutliconductor Transmission Lines in Multilayered Dielectric Media," *IEEE Trans. Microwave Theory Tech.*, **MTT-32**(7), 705 (1984).

98. J. Chilo and T. Arnaud, "Coupling Effects in the Time Domain for an Interconnecting Bus in High Speed GaAs Circuits," *IEEE Trans. Electron Devices*, **ED-31**(3), 347 (1984).

99. R. Kamikawai, M. Nishi, K. Nakanishi, and A. Masaki, "Electrical Parameter Analysis from Three-Dimensional Interconnection Geometry," *IEEE Trans. Components, Hybrids, Manuf. Technol.*, **CHMT-8**(2), 269 (1985).

100. C. D. Taylor, G. N. Elkhouri, and T. E. Wade, "On the Parasitic Capacitances of Multilevel Skewed Metallization Lines," *IEEE Trans. Electron Devices*, **ED-33**(1), 41 (1986).

101. Z.-Q. Ning, P. M. Dewilde, and F. L. Neerhoff, "Capacitance Coefficients for VLSI Multilevel Metallization Lines," *IEEE Trans. Electron. Devices*, **ED-34**(3), 644 (1987).

102. P. E. Cottrell and E. M. Buturla, "VLSI Wiring Capacitance," *IBM J. Res. Dev.*, **29**(3), 277 (1985).

103. T. K. Sarkar, "A Note on the Choice of Weighting Functions in the Method of Moments," *IEEE Trans. Antennas Propag.*, **AP-33**(4), 436 (1985).

104. T. K. Sarkar, A. R. Djordjevic, and E. Arvas, "On the Choice of Expansion and Weighting Functions in the Numerical Solution of Operator Equations," *IEEE Trans. Antennas Propag.*, **AP-33**(9), 988 (1985).

105. M. R. Scheinfein, J. C. Liano, O. A. Palusinski, and J. L. Prince, "Electrical Performance of High-Speed Interconnect Systems," *IEEE Trans. Components, Hybrids, Manuf. Technol.*, **CHMT-10**(3), 303 (1987).

106. V. K. Tripathi and R. J. Bucolo, "A Simple Network Analog Approach for the Quasi-Static Characteristics of General, Lossy, Anisotropic, Layered Structures," *IEEE Trans. Microwave Theory Tech.*, **MTT-33**(12), 1458 (1985).

107. C. D. Taylor, G. N. Elkhouri, and T. E. Wade, "On the Parastic Capacitances of Multilevel Parallel Metallization Lines," *IEEE Trans. Electron Devices*, **ED-32**(11), 2408 (1985).

108. P. Hammond and T. D. Tsiboukis, "Dual Finite Element Calculations for Static Electric and Magnetic Fields," *Proc. Inst. Electr. Eng.*, **130-A**(3), 105 (1983).

109. S. J. Salon, "The Hybrid Finite Element-Boundary Element Method in Electromagnetics," *IEEE Trans. Magn.*, **MAG-21**(5), 1829 (1985).

110. P. A. Brennan, N. Raver, and A. E. Ruehli, "Three Dimensional Inductance Computations with Partial Element Equivalent Circuits," *IBM J. Res. Dev.*, **23**(6), 661 (1979).

111. W. T. Weeks, L. L. Wu, M. F. McAllister, and A. Singh, "Resistive and Inductive Skin Effect in Rectangular Conductors," *IBM J. Res. Dev.*, **23**(6), 652 (1979).

112. C.-S. Yen, Z. Fazarinc, and R. L. Wheeler, "Time-Domain Skin-Effect Model for Transient Analysis of Lossy Transmission Lines," *Proc. IEEE*, **70**(7), 750 (1982).

113. P. Kennis and L. Faucon, "Rigorous Analysis of Planar MIS Transmission Lines," *Electron. Lett.*, **17**(13), 454 (1981).

114. T.-C. Mu, H. Ogawa, and T. Itoh, "Characteristics of Multiconductor Asymmetric, Slow-Wave Microstrip Transmission Lines," *IEEE Trans. Microwave Theory Tech.*, **MTT-34**(12), 1471 (1986).

115. E. G. Farr, C. H. Chan, and R. Mittra, "A Frequency-Dependent Coupled-Mode Analysis of Multiconductor Microstrip Lines with Application to VLSI Interconnection Problems," *IEEE Trans. Microwave Theory Tech.*, **MTT-34**(2), 307 (1986).

116. J. Chilo, C. Monllor, and M. Bouthinon, "Interconnection Effects in Fast Logic Integrated GaAs Circuits," *Int. J. Electron.*, **58**(4), 671 (1985).

117. R. H. Jansen, "Unified User-Oriented Computation of Shielded, Covered and Open Planar Microwave and Millimeter-Wave Transmission-Line Characteristics," *IEEE J. Microwaves, Opt. Acoust.*, **3**(1), 14 (1979).

118. Y. Fukuoka, Y.-C. Shih, and T. Itoh, "Analysis of Slow-Wave Coplanar Waveguide for Monolithic Integrated Circuits," *IEEE Trans. Microwave Theory Tech.*, **MTT-31**(7), 567 (1983).

119. G. Kowalski and R. Pregla, "Dispersion Characteristics of Shielded Microstrips with Finite Thickness," *Arch. Elektr. Ubertragungstech.*, **25**(4), 193 (1971).

120. R. Mittra, Y.-L. Hou, and V. Jamnejad, "Analysis of Open Dielectric Waveguides Using Mode-Matching Technique and Variational Methods," *IEEE Trans. Microwave Theory Tech.*, **MTT-28**(1), 36 (1980).

121. U. Crombach, "Analysis of Single and Coupled Rectangular Dielectric Waveguides," *IEEE Trans. Microwave Theory Tech.*, **MTT-29**(9), 870 (1981).

122. V. A. Kuznetsov and A. M. Lerer, "Dispersion Characteristics of Dielectric Waveguides on Substrates," *Radio Eng. Electron Phys. (Engl. Transl.)*, **29**(10), 53 (1984).

123. K. Solbach and I. Wolff, "The Electromagnetic Fields and the Phase Constants of Dielectric Image Lines," *IEEE Trans. Microwave Theory Tech.*, **MTT-26**(4), 266 (1978).

124. R. Vahldieck, "Accurate Hybrid-Mode Analysis of Various Fineline Configurations Including Multilayered Dielectrics, Finite Metallization Thickness, and Substrate Holding Grooves," *IEEE Trans. Microwave Theory Tech.*, **MTT-32**(11), 1454 (1984).

125. J. Bornemann, "Rigorous Field Theory Analysis of Quasiplanar Waveguides," *Proc. Inst. Electr. Eng.*, **132-H**(1), 1 (1985).

126. D. K. Choi and W. J. R. Hoefer, "The Finite-Difference Time-Domain Method and its Application to Eigenvalue Problems," *IEEE Trans. Microwave Theory Tech.*, **MTT-34**(12), 1464 (1986).

127. N. Yoshida and I. Fukai, "Transient Analysis of a Stripline Having a Corner in Three-Dimensional Space," *IEEE Trans. Microwave Theory Tech.*, **MTT-32**(5), 491 (1984).

128. S. Koike, N. Yoshida, and I. Fukai, "Transient Analysis of a Directional Coupler Using a Coupled Microstrip Slot Line in Three-Dimensional Space," *IEEE Trans. Microwave Theory Tech.*, **MTT-34**(3)., 353 (1986).

129. A. E. Ruehli, N. Kulasza, and J. Pivnichny, "Inductance of Nonstraight Conductors Close to a Ground Return Plane," *IEEE Trans. Microwave Theory Tech.*, **MTT-23**(8), 706 (1975).

130. D. Rodger, "Finite Element Method for Calculating Power Frequency 3-Dimensional Electromagnetic Field Distributions," *Proc. Inst. Electr. Eng.*, **130-A**(3), 233 (1983).

131. J. Weis and V. K. Garg, "One Step Finite Element Formulation of Skin Effect Problems in Multiconductor Systems with Rotational Symmetry," *IEEE Trans. Magn.*, **MAG-21**(6), 2313 (1985).

132. J. Fang, X. Zhang, and K. K. Mei, "Dispersion Characteristics of Microstrip Lines in the Vicinity of a Coplanar Ground," *Electron. Lett.*, **23**(21), 1142 (1987).

133. J. R. Mosig and T. K. Sarkar, "Comparison of Quasi-Static and Exact Electromagnetic Fields from a Horizontal Electric Dipole Above a Lossy Dielectric Backed by an Imperfect Ground Plane," *IEEE Trans. Microwave Theory Tech.*, **MTT-34**(4), 379 (1986).

134. D. H. Auston, "Impulse Response of Photoconductors in Transmission Lines," *IEEE J. Quantum Electron.*, **QE-19**(4), 639 (1983).

135. D. E. Cooper, "Picosecond Optoelectronic Measurement of Microstrip Dispersion," *Appl. Phys. Lett.*, **47**(1), 33 (1985).

136. W. R. Eisenstadt, R. B. Hammond, and R. W. Dutton, "On-Chip Picosecond Time-Domain Measurements for VLSI and Interconnect Testing Using Photoconductors," *IEEE Trans. Electron Devices*, **ED-32**(2), 364 (1985).

137. D. R. Bowman, R. B. Hammond, and R. W. Dutton, "Polycrystalline-Silicon Integrated Photoconductors for Picosecond Pulsing and Gating," *IEEE Electron Device Lett.*, **EDL-6**(10), 502 (1985).

138. M. B. Ketchen, D. Grischkowsky, T. C. Chen, C.-C. Chi, I. N. Duling, III, N. J. Halas, J. M. Halbout, J. A. Kash, and G. P. Li, "Generation of Subpicosecond Electrical Pulses on Coplanar Transmission Lines," *Appl. Phys. Lett.*, **48**(12), 751 (1986).

139. K. J. Weingarten, M. J. Rodwell, and D. M. Bloom, "Picosecond Optical Sampling of GaAs Integrated Circuits," *IEEE Quantum Electron.*, **QE-24**(2), 198 (1988).

140. M. Weiner, L. Bovino, R. Youmans, and T. Burke, "Modeling of the Optically Controlled Semiconductor Switch," *J. Appl. Phys.*, **60**(2), 823 (1986).

141. A. E. Iverson and D. L. Smith, "Mathematical Modeling of Photoconductor Transient Response," *IEEE Trans. Electron Devices*, **ED-34**(10), 2098 (1987).

142. J. A. Valdmanis and G. Mourou, "Subpicosecond Electro-Optic Sampling: Principles and Applications," *IEEE J. Quantum Electron.*, **QE-22**(1), 69 (1986).

143. P. G. May, J.-M. Halbout, and G. L.-T. Chiu, "Noncontact High-Speed Waveform Measurements with the Picosecond Photoelectron Scanning Electron Microscope," *IEEE J. Quantum Electron.*, **QE-24**(2), 234 (1988).

144. J. Bokor, A. M. Johnson, R. H. Storz, and W. M. Simpson, "High Speed Circuit Measurements Using Photoemission Sampling," *Appl. Phys. Lett.*, **49**, 226 (1986).

145. A. Blacha, R. Clauberg, and H. K. Seitz, "Photemission Sampling Measurements of a Dispersing Voltage Pulse Traveling on a Transmission Line," *J. Appl. Phys.*, **62**(2), 713 (1987).

146. H. Fujioka, K. Nakamae, and K. Ura, "Analysis of the Transit Time Effect on the Stroboscopic Voltage Contrast in the Scanning Electron Microscope," *J. Phys. D*, **18**, 1019 (1985).

147. R. Clauberg, "Microfields in Stroboscopic Voltage Measurements via Electron Emission. (i) Response Function of the Potential Energy," *J. Appl. Phys.*, **62**(5), 1553 (1987); "(ii) Effects on Electron Dynamics," *ibid.*, **62**(10), 4017 (1987).

148. T. C. Edwards, *Foundations for Microstrip Circuit Design*, Wiley, New York, 1981.

149. R. L. Veghte and C. A. Balanis, "Dispersion of Transient Signals in Microstrip Transmission Lines," *IEEE Trans. Microwave Theory Tech.*, **MTT-34**(12), 1427 (1986).

150. X. Zhang, J. Fang, K. K. Mei, and Y. Liu, "Calculations of the Dispersive Characteristics of Microstrips by the Time-Domain Finite Difference Method," *IEEE Trans. Microwave Theory Tech.*, **MTT-36**(2), 263 (1988).

151. G. Hasnain, A. Dienes, and J. R. Winnery, "Dispersion of Picosecond Pulses in Coplanar Transmission Lines," *IEEE Trans. Microwave Theory Tech.*, **MTT-34**(6), 738 (1986).

152. A. K. Agrawal, K.-M. Lee, L. D. Scott, and H. M. Fowles, "Experimental Characterization of Multiconductor Transmission Lines in the Frequency Domain," *IEEE Trans. Electromagn. Compat.*, **EMC-21**(1), 20 (1979).

153. F.-Y. Chang, "Computer-Aided Characterization of Coupled TEM Transmission Lines," *IEEE Trans. Circuits Syst.*, **CAS-27**(12), 1194 (1980).

154. A. J. Groudis and C. S. Chang, "Coupled Lossy Transmission Line Characterization and Simulation," *IBM J. Res. Dev.*, **25**(1), 25 (1981).

155. P. R. Shepherd and R. D. Pollard, "Direct Calibration and Measurement of Microstrip Structures on GaAs," *IEEE Trans. Microwave Theory Tech.*, **MTT-34**(12), 1421 (1986).

156. H. J. Gray, *Digital Computer Engineering*, Prentice-Hall, Englewood Cliffs, NJ, 1963.

157. A. Feller, H. R. Kaupp, and J. J. Digiacomo, "Crosstalk and Reflections in High-Speed Digital Systems," *Am. Fed. Inf. Process. Soc. (AFIPS), Conf. Proc. Fall Jt. Comput. Conf.*, **27**(1), 511-525 (1965).

158. I. Catt, "Crosstalk (Noise) in Digital Systems," *IEEE Trans. Electron. Comput.*, **EC-16**(6), 743 (1967).

159. J. A. DeFalco, "Reflection and Crosstalk in Logic Circuit Interconnections," *IEEE Spectrum*, **7**(7), 44 (1970).

160. W. R. Smith and D. E. Snyder, "Circuit Loading and Crosstalk Signals from Capacitance in SOS and Bulk Silicon Interconnect Channels," *Proc. IEEE VLSI Multilevel Interconnect. Conf., 1984*, New Orleans, LA, p. 218 (1984).

161. S. Seki and H. Hasegawa, "Analysis of Crosstalk in Very High-Speed LSI/VLSIs Using a Coupled Multiconductor Microstrip Line Model," *IEEE Trans. Microwave Theory Tech.*, **MTT-32**(12), 1715 (1984).

162. E. T. Lewis, "An Analysis of Interconnect Line Capacitance and Coupling for VLSI Circuits," *Solid-State Electron.*, **27**(8/9), 741 (1984).

163. R. Garg and I. J. Bahl, "Characteristics of Coupled Microstrips," *IEEE Trans. Microwave Theory Tech.*, **MTT-27**(7), 700 (1979).

164. D. L. Carter, "An Estimate of On-Chip Crosstalk Signals at High Switching Speeds and High Levels of Integration," *AT&T Bell Labs. Tech. Memo.*, 80-2451-1 (1980).

165. V. K. Tripathi and R. J. Bucolo, "Analysis and Modeling of Multilevel Parallel and Crossing Interconnection Lines," *IEEE Trans. Electron Devices*, **ED-34**(3), 650 (1987).

166. V. K. Tripathi and A. Hill, "Equivalent Circuit Modeling of Losses and Dispersion in Single and Coupled Lines for Microwave and Millimeter-Wave Integrated Circuits," *IEEE Trans. Microwave Theory Tech.*, **MTT-36**(2), 256 (1988).

167. A. R. Djordjevic, T. K. Sarkar, and R. F. Harrington, "Analysis of Lossy Transmission Lines with Arbitrary Nonlinear Terminal Networks," *IEEE Trans. Microwave Theory Tech.*, **MTT-34**(6), 660 (1986).

168. J. E. Schutt-Aine and R. Mittra, "Scattering Parameter Transient Analysis of Transmission Lines Loaded with Nonlinear Terminations," *IEEE Trans. Microwave Theory Tech.*, **MTT-36**(3), 529 (1988).

169. V. Rizzoli and A. Neri, "State of the Art and Present Trends in Nonlinear Microwave CAD Techniques," *IEEE Trans. Microwave Theory Tech.*, **MTT-36**(2), 343 (1988).

170. J. Chilo and T. Razban, "Effets de propagation et de brouillage dans les interconnexions de circuits hybrides," *Ann. Telecommun.*, **42**(1/2), 20 (1987).

7 Impact of VLSI Technology Scaling on Computer Architectures

RAMAUTAR SHARMA

AT&T Bell Laboratories
Murry Hill, New Jersey

7.1 INTRODUCTION

Computers and the underlying fabrication technologies are intimately related. This mutual interdependence has never been more for any other technology than VLSI, and it is increasing further and further. In this chapter, we explore the impact of VLSI technology on computer organization and architectures. In fact, their co-evolution seems to follow a diverging helical path, influencing and enriching each other. Let $A_p(t)$ be the proposed computer architecture and $T_a(t)$ the available technology at a time instant t; then realized architecture $A_r(t)$ can be expressed as

$$A_r(t) = f_{\text{pa}\to r}[A_p(t), T_a(t)]. \tag{1}$$

In Eq. (1) parameter t has been used to emphasize the time dependence of the mutual influence and function $f_{pa\to r}()$ maps the proposed computer architecture to its realization in the technology $T_a(t)$. The realized architecture $A_r(t)$ in terms of a physical machine is available to users and is used to model the newer architectures and technologies thus helping to evolve better technologies and architectures. These relationships are given by

$$T_a(t + 1) = f_{r\to a}[A_r(t), T_a(t), \cdots], \tag{2}$$

and

$$A_p(t + 1) = f_{r\to p}[A_r(t), \cdots]. \tag{3}$$

Note that these relationships are highly complex; they are shown schematically in Figs. 7.1a and b. The improved technology not only influences the newer archi-

332

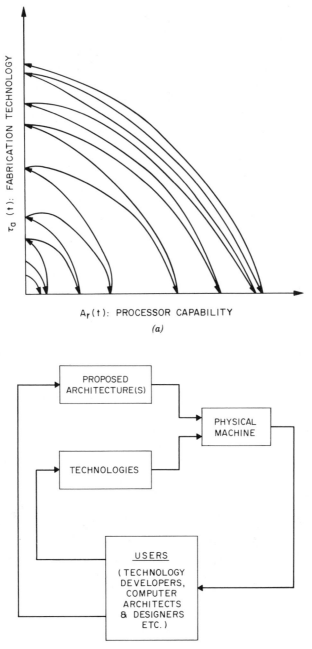

(a)

(b)

Fig. 7.1 (*a*) Interdependence of computer architectures and fabrication technologies. (*b*) Relationship among computer architectures, technologies, and users.

tectures, but also helps in improving the performance of the currently available architectures when these are implemented in the improved technology. We shall explore this relationship in more detail in the following sections.

In order to provide a common ground for our discussions on the impact of VLSI technology on the computer architectures, we begin by looking at the evolution of computers in Section 7.2. In Section 7.3, we classify computers into different types to bring out the essential and common features that are affected by VLSI technology. Requirements for high-performance computers and the technological constraints are treated in Section 7.4. In Section 7.5, we discuss the architectural imbalance caused by VLSI scaling. Besides silicon-based VLSI technology, other materials are also being explored for future technologies, GaAs being the most important one. In Section 7.6, we briefly discuss a combined Si–GaAs VLSI technology. Some areas of application are affected more than the others; these areas are described in Section 7.7. Finally, in Section 7.8, we summarize the symbiosis between computers and VLSI technology and make some conjectures about the future.

7.2 HISTORICAL PERSPECTIVE

The history of mechanical aids in computation goes back many hundreds of years. The development of the abacus predates recorded history. However, the first device that could be called a computer in the modern sense was proposed by Charles Babbage in about 1830. However, it took more than a century to realize those ideas, as the technology was not available for so long, thus going to show how much impact technology can have on computing machines. To study this interrelationship between technology and the computer architecture, we first review the developments in computer design and define some of the terms used throughout the chapter.

In fact, to define what is meant by computer architecture is not a simple task. According to the American College Dictionary, the term "architecture" means "art or science of building, including plan, design, construction, and decorative treatment." In this context, computer architecture includes structure, organization, implementation, and performance evaluation. Thus, we shall use the term "computer architecture" in this broad sense. The traditional view of a computer architecture is that of a stored program computer as proposed by von Neumann. He was the first to recognize and formalize the requirements and structure of such a machine,[1] as shown in Fig. 7.2. In this machine,

1. The INPUT transmits data and instructions from the outside world to memory.
2. The MEMORY stores instructions and data and intermediate and final results.
3. The arithmetic–logic unit (ALU) performs arithmetical and logical operations.

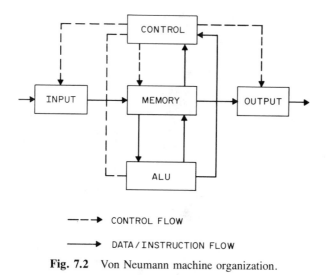

Fig. 7.2   Von Neumann machine organization.

4. The CONTROL interprets the instructions and causes them to be executed.
5. The OUTPUT transmits final results and other information to the outside world.

This machine organization is generally known as von Neumann or Princeton architecture, as it was first proposed at Princeton University's Institute of Advanced Studies.

The memory in this organization has three types of information, namely, data, instructions or procedures, and results. By physically separating the data and instructions into two memories, another machine organization (shown in Fig. 7.3), known as the Harvard machine, results. This organization, which has the potential for operating on both the memories in parallel and the organization, is now becoming common in high-performance machines based on VLSI circuits.

Although the performance gains in computers have been remarkable, the rate of improvement has been declining. Qualitatively, the improvements in performance are shown in Fig. 7.4, where we see that it has been increasing by almost a factor of 10 for every seven years of progress in technology. The performance gains in the computer systems have been due to the combined efforts of new and imaginative architectures and silicon technologies.

Irrespective of the computer organizations, their history can be divided into the following four phases, which are shown in Fig. 7.5:

1. Data processing.
2. Information processing.
3. Knowledge processing.
4. Intelligence processing.

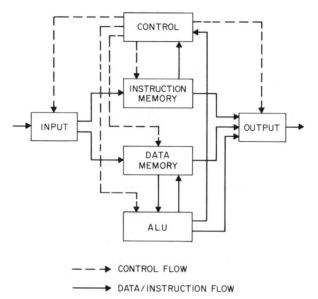

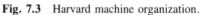

Fig. 7.3   Harvard machine organization.

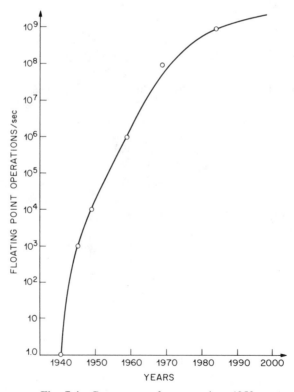

Fig. 7.4   Computer performance since 1950.

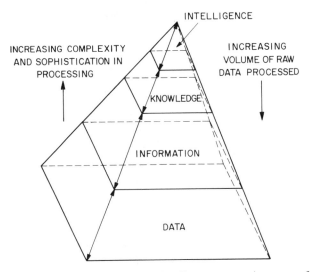

Fig. 7.5 Data, information, knowledge, and intelligence processing spaces from the viewpoint of computer applications.

In the early days of computing, the objective was to have the machine do the computations once the procedure was developed for carrying out the computations automatically. The data space is the largest and includes numeric numbers, various symbols, etc. The data objects are considered mutually unrelated in the space and have no structure. The main characteristic of this space is the combination of the data and program. Data processing is still the major task of today's computers.

In the late 1960s and the early 1970s, the role of the information, that is, data structures, as an entity in its own right was recognized. The machines and program were built around these structures. This gave birth to data-base machines which have essentially a von Neumann type of architecture but make use of the information in the data structures. Most of today's computing is still confined to these two levels, that is, data and information processing. The performance of the computer system at these two levels has been improved significantly through a high degree of parallelism.

As the accumulated experience and knowledge bases about the computers expanded rapidly in the last decade (late 1970s and early 1980s), the use of computers in knowledge processing grew. For example, many expert computer systems with problem-solving capabilities and performance reaching those of human experts have been developed. Some of these expert systems are listed in Table 7.1, and many computer scientists believe that knowledge processing will be the main thrust of computer usage in 1990s.

Today's computers have been successfully used in data, information, and knowledge processing. As the usage moves from data to knowledge and lastly to intelligence processing, the complexity increases manyfold and the performance requirements are far beyond those supported by the most advanced and the fastest machines currently available. Technology alone cannot provide the raw speed

TABLE 7.1   Some Examples of Expert Computer Systems

Computer System	Expert Knowledge Processed
Prospector	Mineral exploration
Dipmeter, advisor	Oil exploration
SPERIL	Earthquake damage estimation
Intermist	Medical consulting

needed to make computers communicate in natural languages, see and recognize patterns and pictures, or hear and respond intelligently like human beings. Computers are far from being satisfactory in performing logical inferences and creating thinking. For this type of intelligence processing, the conventional computer architectures have to be properly blended with signal-processing and artificial intelligence techniques. More parallelism at both the hardware and software level is required. These efforts will be the main focal points of major research efforts well into the twenty-first century, and advanced VLSI technology will significantly influence our ability to design intelligence-processing computers. This chapter will explore these interactions in more detail.

Qualitatively, the influence of silicon VLSI technology and the computer organizations and architectures on the system's performance is shown in Fig. 7.6. A

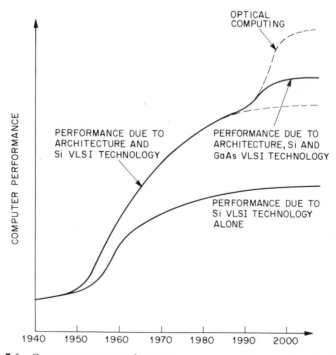

Fig. 7.6   Computer system performance = system architecture + technology.

similar qualitative picture for various phases of processing, shown in Fig. 7.7, can be generated to illustrate the apparent departure of the computer architecture from the conventional von Neumann architectures.[2] Notice that as we progress from standard data-processing applications toward intelligence-processing applications, the departure increases rather rapidly due to performance requirements and the added complexity of the processing itself. From Figs. 7.6 and 7.7, it seems that intelligence processing will be feasible sometime toward the end of the first quarter of the twenty-first century, although many systems will spring up which will mimic human behavior in a limited sense earlier, probably in the first decade of the next century.

7.3 CLASSIFICATION OF COMPUTER ARCHITECTURES

Since the mid-1940s, the von Neumann organization of computer systems has been a dominating factor. This organization is abstracted again in Fig. 7.8 to bring out the problems associated with it. The main problem in this architecture is the common bus through which all the information (data, instructions, intermediate results, etc.) must be exchanged. Therefore, no matter how fast the other components in the organization are, the communications through the bus will be the limiting factor, Backus[3] calls this von Neumann bottleneck.

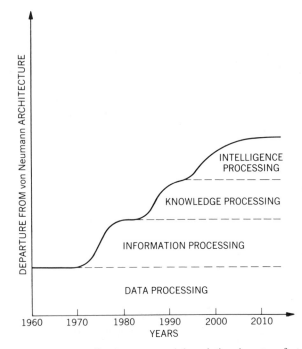

Fig. 7.7 Computer system application areas and the relative departure from von Neumann architecture.

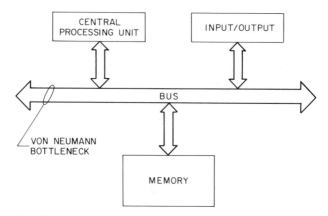

Fig. 7.8 Hardware organization of von Neumann architecture.

To overcome this bottleneck, many computer architectures based on parallel or/ and overlapped machine organizations have been proposed. The parallelism can be at any level: processing, control, etc. Flynn[4] defines the following four types of machine organizations based on the control and processing parallelism.

1. SISD: single instruction stream, single data stream (conventional von Neumann machine).
2. SIMD: single instruction stream, multiple data stream.
3. MISD: multiple instruction stream, single data stream.
4. MIMD: multiple instruction stream, multiple data stream.

However, depending on the following four criteria, we get a generalized picture of machine architectures that include the classifications due to Flynn:

1. Kind of parallelism.
2. The control structures due to Flynn.
3. The processor structure.
4. The interconnection structure.

The resulting multitude of architectures is shown in Fig. 7.9 and is briefly described below.

In pipelined computers,[5] the parallelism is achieved through three types of pipelines:

1. *Instruction.* This includes instruction fetch, instruction decode, and operand fetch and execution. This type of pipeline is common in almost all the high-performance machines, from microprocessors to supercomputers.

2. *Arithmetic.* The entire arithmetic operation is divided into several pipeline stages.

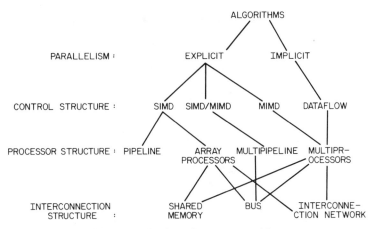

Fig. 7.9   Classification of computer architectures.

3. *Processor.* This type of pipeline involves a cascade of processors in which the output of one processor is an input to another one, and so on.

In the case of array processors, several processing elements synchronized by a special control unit execute the same operation at the same time on a spatial stream of data. The exchange of data is done by an interconnection network or a common memory, as shown in Fig. 7.10.

The multiple pipeline structure could be used for both SIMD and MIMD organizations. In this case, instead of a single pipeline stage between various operations, many pipeline stages can be used, as shown in Fig. 7.11. These types of computer organizations are very effective for vector computations involved in image processing, numerical computations, etc.

The shared memory multiprocessor has n processors and m memory modules that are shared by all the n processors through a switching network (Fig. 7.12). In the case of shared buses, the processors and memories are connected through buses and the communication is through buses, as shown in Fig. 7.13. In this figure, there are two models: one with single bus, another with multiple buses.[6]

In data flow architecture, the instructions are enabled if and only if all the required input data are available (Fig. 7.14). Here, the sequence of instruction exececution depends on the data only and can thus be executed in any order.

7.4   CPU REQUIREMENTS AND VLSI TECHNOLOGY

In all the architectures described in Section 7.3, there are a few common factors that provide high performance:

1. Multiple processors or processing elements which provide simultaneous computation of many partial results that can be combined to get the final answer.

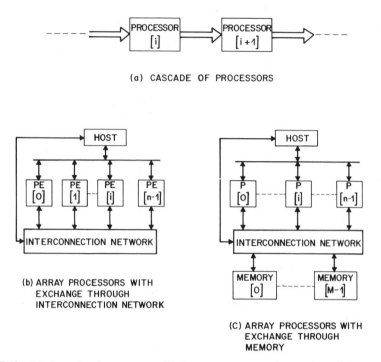

Fig. 7.10 (*a*) Cascade of processors. (*b*) Array processor exchange through interconnection network. (*c*) Array exchange through memory.

2. Some sort of interconnection or switching network between the processors, memories, and other resources that helps in moving data and other information among these elements.

3. Some local and much shared (global) memory. The local memory is useful in reducing traffic in the interconnection network.

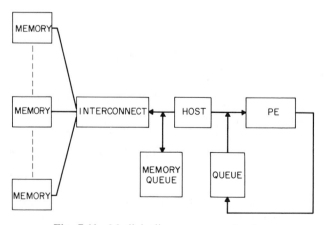

Fig. 7.11 Mutlipipelines processor structure.

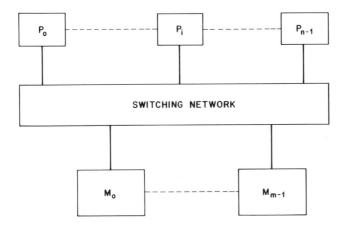

P_i : i^{th} PROCESSOR
M_j : j^{th} MEMORY MODULE

Fig. 7.12   Shared memory multiprocessor organization.

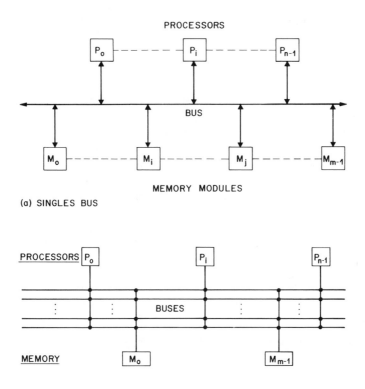

(a) SINGLES BUS

(b) MULTIPLE BUSES

Fig. 7.13   Shared bus multiprocessor.

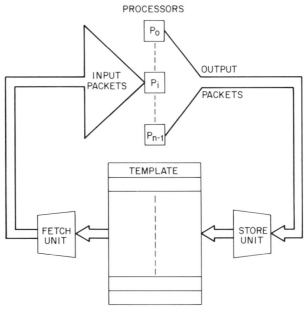

Fig. 7.14   A model of data flow machine.

A careful evaluation of architectures also suggests that the performance of a computer system generally improves by: (a) adding more processing elements, provided these can be kept busy meaningfully; (b) increasing the speed of operation, whether this involves computing, information exchange, or any other type of communication; and (c) adding more high-speed memory. Note that the above discussion gives the impression that a lavish use of VLSI technology to increase hardware contents of a computer system would lead to higher performances. However, this is not true; system and application software are other important components of the overall performance which are not discussed here. However, we would rather concentrate on the hardware aspects as related to central processing units and memories. These are also the areas where the scaling of VLSI technology has the most significant effect.

There are many benefits to using VLSI technology in designing computer systems. For large computers, potential benefits are, in order of importance, (1) high reliability, (2) small size, (3) higher speeds, (4) low parts count due to higher levels of integration, and (5) lower cost due to scale of economy.

High reliability is brought about mainly by the reduction in the interconnections among the chips. The mean time between failures (MTBF) for on-chip connections is more that those for off-chip, and therefore less off-chip interconnections result in more reliable system overall.

As transistor size is reduced further and further, inherent delay in the device becomes negligible and delay associated with the wiring becomes more important.

Thus, the reduction in the interconnection lengths also reduces the delay in the system and higher speeds of operation are achieved. The reduced feature sizes allow more functions to be packed and hence leads to smaller sizes and lower parts count.

Although there is a cost reduction of the VLSI chips that go into a computer system, it is a small fraction of the overall cost and therefore of lesser importance than the other factors. However to use VLSI technology to its full potential, we must understand its limitations and constraints. Below, we explore the consequences of these constraints. The important ones are:

1. *Regularity.* A large computer may contain 100,000 or more components, a significant proportion of which will be VLSI chips. Clearly, the design cost is amortized over a large number of fewer types of chips. This imposes the constraint of regularity on architecture too. To reduce the chip types, one should use a general design that can be particularized to a variety of uses. The computer should also be built by repeated substructures to increase the architectural regularity. Such machines are sometimes called homogeneous; some examples include the massively parallel processor,[7] systolic arrays,[8] CHIP,[9] Cosmic Cube,[10] etc.

2. *Planarity.* VLSI technology is a two-dimensional technology, a constraint further magnified as chips are placed on a two-dimensional board. Although wires on the chips are allowed to cross over each other and boards have considerable backplane wiring, these still are not three dimensional. However, the two-dimensional constraint is not the same as the "planarity" constraint that requires the wires not to cross over each other, but for our general discussion of constraints we shall not distinguish between the two. The planarity influences the arthitecture through the "perimeter problem." It refers to the fact that the number of wires leading off a two-dimensional or planar medium is proportional to the perimeter.

For single chips, an aspect ratio of $1:1$ is preferred; thus, the number of external connections is proportional to the square root of the chip area. It implies that if n bits can be transferred onto the chips in parallel, the number of gates available to process the data is proportional to n^2. If the amount of processing performed is quadratic to the amount of data, then there is no perimeter problem. However, in reality we may not be able to satisfy this constraint, especially with VLSI scaling. For example, if we have a potential of putting 50,000 gates on an 1.00 cm^2 chip and only 50% of these are used, then we need at least 160 pins to transfer the data in parallel to use all the gates effectively, that is, to make the perimeter problem vanish. In conventional technologies, the connections to the outside world are made at the peripheries. A typical pad size is $100 \times 100 \ \mu\text{m}$ and is generally placed with 200 μm center to center spacing, as shown in Fig. 7.15. Therefore, for an 1×1 cm chip, we can put only

$$4 + \left[4 \times \frac{10,000 - (2 \times 200)}{200} \right] = 196 \text{ pads.}$$

However, if the pins are distributed over the entire area of the chip or a grid-array style is used, we can have more pins devoted to the inputs and outputs (Fig. 7.16).

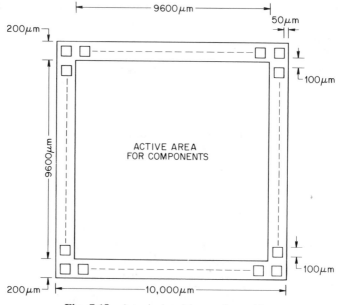

Fig. 7.15 A typical pad layout for a chip.

In this case, we have a potential of putting

$$\left[\frac{10,000 \times 10,000}{(100 + 50) \times (100 + 50)} \right] \approx 4,444 \text{ pads}$$

of the same size as in the previous example. However, not all of these pads can be used, since some space would be required for wiring these to the external world. However, our example just goes to show that the perimeter problem can be resolved by distributing pads all over the chip area. Note that this style of pin distri-

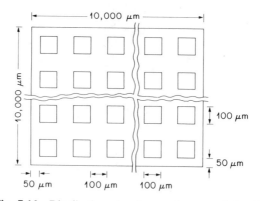

Fig. 7.16 Distribution of pads on active area of the chip.

bution complicates the signal routing on the chip. Another option is to reduce the pad size, but for conventional bonding technologies this is not a viable solution. Advanced packaging technology in which chips are mounted on a silicon wafer,[11] has a potential for providing very high performance systems. Here, the pads on the entire surface of a chip can be connected to the substrate and wired properly. However, problems, such as power and ground distribution through resistive lines, heat removal, testability, maintainability, mechanical and thermal stressing of devices due to pads on active area, and reliability, etc., must be solved before it can be used in a cost-effective manner.

Empirically it has been found that the pin requirements increase at a slower rate than the increase in the gates per chip. The heuristic relationship between the number of pins located on the periphery of a chip and number of gates on it is given by

$$N = C \times G^p, \tag{4}$$

where N is the number of pins (pads), G the number of gates on the chip, and p the exponent, and C a coefficient. The relationship of Eq. (4) is also known as Rent's rule. The values of C and p depend on the level of integration and the performance requirements. This dependence and empirical nature of Rent's rule are demonstrated in Table 7.2. For many chips, the number of pins is plotted as a function of number of gates (see Fig. 7.17) to obtain these values. The multiplicative constant C and the exponent are found to be 6.0 and 0.4, respectively. The number of pins that can be accommodated on a chip are also tabulated alongside in Table 7.2.[12-25] This column has been obtained by using the previous pad size and spacing. Note that the number of pads (pins) predicted by Rent's rule and area consideration to avoid perimeter bottleneck are close enough to what is provided by the manufacturers. Thus, Rent's rule provides a way of evaluating the relation-

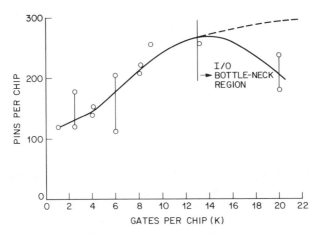

Fig. 7.17 Relationship between number of pins and gates on a chip.

TABLE 7.2   Required and Available Pins vs. Gates per Chip[a]

Number of Gates Per Chip	Pins Per Chip		
	Rent's Rule $c = 6.0, p = 0.4$	Actual Data	$20\sqrt{\text{Area}}$
600	78	120^{12}	157^{12}
2,500	137	$120,^{13}\ 176^{14}$	$134,^{13}\ 100^{14}$
4,032	166	140^{15}	143^{15}
4,368	172	156^{16}	194^{16}
6,000	194	112^{17}	165^{17}
6,032	195	204^{18}	160^{18}
8,064	219	209^{19}	197^{19}
8,370	222	220^{20}	198^{20}
9,000	229	256^{21}	226^{21}
13,313	268	256^{22}	239^{22}
20,000	315	$180,^{23}\ 236^{24}$	$200,^{23}\ 248^{24}$
29,234	366	220^{25}	220^{25}

[a]Superscript numbers are references.

ship between the chip area, number of gates, and pins available for signal transfer. Figure 7.17 also shows this graphically and points to the regions where the size of the chip is big enough for more logic but cannot be incorporated due to the input/output bottleneck.

With further reduction in the feature size, more gates can be put on the same size chip. Unless the pins are put right on the active area through multilevel metallization, it is unlikely that the number of pads and thus available pins would increase. This poses a problem. What should be added on the chip so that it is still useful and does not have perimeter bottleneck? DeRuyck et al. have analyzed this problem for the systolic array type of architectures,[26] but the idea can be extended further to other architectures. One way to reduce the data traffic is to include more local memory and that is what seems to be taking place in most of the high-end single-chip computers in terms of instruction/data caches[27–29] and register windows.[30]

3. *Part-Type Requirements.* The CPU of a large computer normally contains more than 100 K gates. Partitioning these many gates into a VLSI chip poses a problem. Although it is somewhat related to the perimeter problem, it has some unique characteristics. As the level of integration increases from one gate per chip (say a two-input NOR or NAND gate) to a few tens of thousand gates, the number of unique parts to implement the system exhibits the characteristics shown in Fig. 7.18. In principle, the entire system can be implemented using just one gate type, so the number of unique part types is one. On the other hand, when the entire system can be implemented on a single chip, again the number of unique part types is just one. However, in between the unique part types increase up to a certain level and then decrease. The exact shape of the curve in the Fig. 7.18 is fabrication technology and actual system design dependent; however, the general shape is somewhat similar in all cases. This indicates that the VLSI scaling has a signifi-

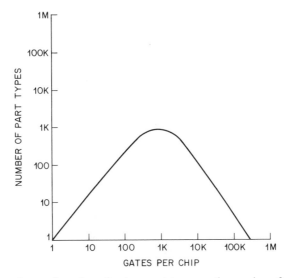

Fig. 7.18   Dependence of number of unique part types on the number of gates per chip.

cant effect and puts a constraint on the unique part types that are needed to implement an architecture. Another effect of scaling is the decrease in the part utilization and an increase in the complexity of the design. Therefore, to minimize design time and cost of a chip, the computer architecture must be partitioned such that we get fast and low-cost design of each part type. The standard gate arrays generally tend to satisfy this requirement. Large-volume standard chips produced by many semiconductor manufacturers also satisfy the low-cost and fast design requirement, but now the system architecture must be modified to take advantage of the second approach.

4. *Interconnection Requirements*. With an increasing density of chips, the number of interconnections between chips tends to decrease as more and more functions are put on the same chip. However, this shifts the problem of interchip connections to intrachip connections. In general, the total length of interconnections on chip increases faster than the number of available gates; thus, the size of the chip is mostly dominated by the wiring rather than devices.[31] Therefore, with increasing number of gates, the chip area will increase much faster to a point where the yield and the production cost are uneconomical. As the number of gates on a chip increases, the average interconnection length between the gates increases. This average length can be related to the number of ports in the chip requiring interconnection. Empirically, it is found that the average length of interconnection L_{av} is given by

$$L_{av} = \alpha P_{av}[n_{port}]^{\beta} \tag{5}$$

where P_{av} is the average pitch of the cell or gate, α the proportionality constant, typically between 0.1 and 0.2, n_{port} the number of ports in the chip to be connected, and β the exponent, typically 0.5. All these parameters generally depend on the technology and design style.

For example, in a typical CMOS technology where the layout has been done in gate-matrix style,[32] metal and polysilicon pitches are 6λ and 9λ, respectively, as shown in Fig. 7.19. Thus, the value of P_{av} is 7.5λ. Assuming $\alpha = 0.2$ and $\beta = 0.5$, the average interconnection length is given by

$$L_{av} = 1.5\lambda\sqrt{n_{port}},\tag{6}$$

and a plot is shown in Fig. 7.20. From this figure, it is clear that for an average interconnection length of about 50 μm we can have more than 16,000 internal ports for λ = 0.25 μm (i.e., 0.5 μm technology), whereas less than 2,000 ports can be permitted for the same average interconnection length in the case of 1.0 μm technology (i.e., λ = 0.5 μm). Assuming that on the average there is one port for every logic gate such as two-input NAND, the average interconnection length can also be related to the number of devices. Empirically, for CMOS technology, it is given by

$$L_{av} = \alpha P_{av}\left(\frac{n_{dev}}{4}\right)^{\beta}.\tag{7}$$

For the same values of α, β, and P_{av}, we get

$$L_{av} = 0.75\lambda\sqrt{n_{dev}}.\tag{8}$$

Thus, for a chip with 160,000 devices and λ = 1.25 μm, we get $L_{av} = 375$ μm. This average length will reduce to 140 μm and 70 μm for λ = 0.5 μm and λ = 0.25 μm technologies, respectively.

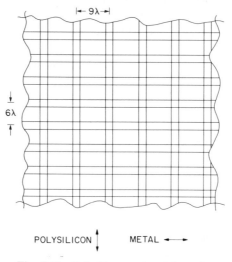

Fig. 7.19 Polysilicon and metal matrix.

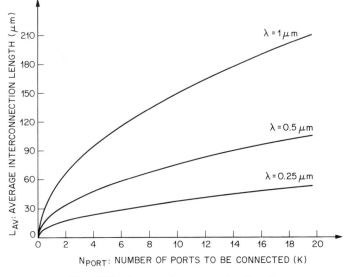

Fig. 7.20   Average interconnection length.

Prediction of average interconnect length is important for determining the dc and high-speed performance of the chip. This is also used to determine the routability of the chip, as the routing should not occupy more than 50% of the available tracks for complete routing.[31] Generally, the routability increases with an increase in the chip size.

5. *System Performance Requirements.* The system performance depends on both on- and off-chip delays and these are functions of the level of integration. The off-chip delays decrease with increased level of integration as the length of interconnection between chips decreases. Conversely, on-chip delays increase with the level of integration for two main reasons. First, the increased interconnect length contributes to this delay. Second, the power dissipation per gate reduces, further increasing the delay.

As on-chip delay increases and off-chip delay decreases with level of integration, there is some optimum level of integration that will give highest performance of the system; this is shown schematically in Fig. 7.21. From the figure, we see that there is not a very well defined optimum point and also that its value depends on many factors: cooling capability, chip and packaging technology, and system architecture, etc. Although such an analysis is essential, it cannot be done here for lack of space; the reader is referred to the literature.[34–35]

7.5 SCALING UPSETS ARCHITECTURAL BALANCE

As the feature size decreases, the speed of processor increases, but the input/output (I/O) rate of the processor does not increase at the same rate. Thus, an architecture that has been balanced for internal processing and I/O rates initially

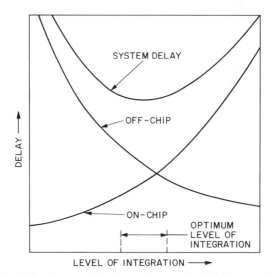

Fig. 7.21   System-delay dependence on level of integration.

becomes unbalanced, that is, the processing power cannot be matched with the information transfer rate. A processor with a local memory of M words is considered balanced if it satisfies the requirement.

$$\frac{n_\theta}{\Theta} = \frac{n_\omega}{\Omega} \tag{9}$$

where Θ is the computational bandwidth, that is, the number of operations per second, Ω is the I/O bandwidth, defined as the number of words per second that the processor can communicate with the outside world, and n_θ and n_ω are the total operations and inputs/outputs performed, respectively, Simply stated, Eq. (9) means that a computer architecture is balanced if the time spent in doing computations is the same as for doing the inputs/outputs.

If the computational power is increased by a factor ε without an appropriate increase in the I/O, the architecture can be balanced by using more memory. This problem has been analyzed and it has been shown that in general, different problems have different solutions.[36] Relationships between the new and old memories (M_{new}, M_{old}) for some important applications are as follows:

· *Matrix Computations:* $M_{new} = \varepsilon^2 M_{old}$.
· *Grid Computations (d-Dimensional):* $M_{new} = \varepsilon^d M_{old}$.
· *FFT:* $M_{new} = (M_{old})^\varepsilon$.
· *Sorting:* $M_{new} = (M_{old})^\varepsilon$.
· *I/O Bound Computations Such as Matrix–Vector Multiplication and Solution of Triangular Linear Systems:* Impossible; that is, the processor can-

not be rebalanced by enlarging its local memory, without increasing its I/O bandwidth.

For most of the above-mentioned computation classes, rebalancing the processing element (PE) requires a substantial increase in the memroy size *if the I/O bandwidth is to remain constant.* The local memory size for computation such as fast Fourier transform (FFT) and sorting increases at an exponential rate with increased computational bandwidth. Many PEs are used in parallel architectures to perform a given computation and in these architectures the total amount of the memory and I/O bandwidth should be balanced with total computational bandwidth. Therefore, the VLSI scaling can effectively be used to increase the memory size for balancing the increased computational bandwidth.

7.6 GaAs DIGITAL ICs FOR HIGH-PERFORMANCE COMPUTERS

Superfast computers with subnanosecond cycle time are one of the main applications that force microelectronics into the submicron age. In computer systems, GaAs ICs are generally suited for data paths, sequencers, and fast cache memories. However, recent scaling down of fast silicon processes (ECL, NMOS, and CMOS)[37-39] and the impressive performances achieved have raised the question of the real advantage of GaAs over silicon. For similiar device sizes, GaAs transistors have three to five times greater transconductance those of silicon bipolar and MOS transistors. Similarly, for the same transconductance, GaAs transistors consume much less power than their silicon counterparts. However, GaAs-based ICs are still not used frequently in computer systems, since these ICs have reliability problems. CMOS technology will dominate the low-cost computer area. GaAs MESFET technology[40] will dominate the cost–performance domain because of its low power dissipation, high density, and high performance as compared with that of silicon bipolar technology. It will make initial inroads into computer systems as high-performance cache memory and dominate as the level of integration reaches VLSI level. GaAs technology will also have a future in optical computing systems and in small systems where extremely high speeds are needed. Nonetheless, for a reasonable time during this century we shall continue to have silicon-based systems with some parts done in GaAs technology, rather than full GaAs systems.

7.7 APPLICATIONS OF VLSI TECHNOLOGY

Since the mid-1970s, VLSI technology has been successfully used in many areas, but its effect on computers of all shapes and sizes has been the most dramatic. Some of the application areas got boosts in performance while others became feasible. Some of the important areas of applicaiton influenced by VLSI technology are discussed below.

7.7.1 General-Purpose Computing

Here, we notice the effect of the technology in terms of processing speed, increase in memory size, etc. Mircorpogramming became more prevalent due to the availability of cheap, high-density, and fast memory chips. VLSI circuits have also affected compiler technology through reduced instruction set computers[41] (RISC). Previously, the RISC idea was not frequently used because of increased code size and frequent memory accesses. However, VLSI scaling is rapidly changing the rules of the game. Minisupercomputers have been built using RISC architectures and VLSI technologies.[42,43] The trend to simplify architectures and orchestrate many CPUs to do processing will continue.

7.7.2 Distributed Multiprocessing

VLSI scaling while increasing the processing power also helps to reduce intermodule wiring, power dissipation per operation, etc. Therefore, many functional units can be incorporated into a machine to perform many specialized functions concurrently.[44] Sometimes individual units are identical, but they can be programmed to do different functions.[45–47] The use of either the same or different modules depends on the architecture, but an important consequence is that the technology is rapidly influencing our thinking.

7.7.3 Neural Networks

This is one of the areas in which VLSI technology is quickly making inroads. These networks are highly interconnected, have very simple processing elements, and are ideally suited for VLSI implementations. Progress, however, has been rather slow due to technological limitations and our limited understanding of this multidisciplinary field, but interest in the area is growing rapidly. These networks are treated in detail in Chapter 10.

7.7.4 Artificial Intelligence

Since the early 1950s, the artificial intelligence area has been progressing rapidly, but its progress has been fueled further by the availability of high-speed powerful computing engines. Most of the work has been based on LISP language and almost always interpreted on general-purpose machines. Recently, many compilers and specialized machines have been designed for the language.[48,49] The development of these machines has been influenced by VLSI technology. An artificial intelligence engine based on fuzzy logic has also been designed and shown to achieve speeds in excess of 80,000 fuzzy logical inferences per second.[50] Fifth-generation computer project[51] is another example of the influence of VLSI technology on computer architecture.

7.7.5 Signal Processing

This is one of the earliest areas of applications where an impact was felt and it still continues to be in a state of flux.[52,53] Here, speed and high packaging densities are the main requirements and these are readily provided by VLSI technology. The importance of signal-processor architectures will increase as we progress from information to knowledge to intelligence processing. General-purpose processors are being augmented by these to enhance their processing speeds.

7.7.6 Image Processing and Computer Vision

In this area, the amount of computation if carried out algorithmically is so large that even the fastest computers cannot do all the image processing in real time. It needs massive parallelism and it can be achieved through judicious use of VLSI technology to design special-purpose machines.[54] Through highly concurrent architecture and VLSI scaling, a peak performance of 114 million operations per second has been achieved by this chip.

7.8 SUMMARY

In this chapter, we have examined relationships between computer architectures and VLSI technology. Various computer organizations from single-chip microprocessors to multiprocessors have been discussed to point out the unique requirements they demand from their underlying fabrication technology. It has been shown that VLSI technology has influenced computer architectures the most. However, just like any other technology, VLSI has its own advantages and limitations. Therefore, all the designs must satisfy the constraints imposed by the technology. Important technological constraints due to scaling can be summarized as follows:

1. *Chip Area.* The size of the chip cannot be arbitrarily large, since chip yield decreases exponentially with increasing chip area.

2. *I/O Pads.* The number of available pads for signal transfer between the chip and its environment is limited due to finite and small chip area requirement.

3. *Environmental Mismatch.* By this we mean that loads seen by pad drivers on and off chip are different, generally on-chip load being smaller than off-chip. Smaller signal swings can be tolerated on chip, but from noise considerations we have to interface with relatively large signals outside the chip. This in turn requires different designs to cope with the situation.

4. *Power Dissipation.* Thermal conductivity of silicon is finite and circuits do not work reliably at elevated temperatures. Therefore, we must design to satisfy power-density requirements and this means that one cannot trade power for speed arbitrarily.

5. *Parameter Variations.* Two adjacent devices on the same chip can have different threshold voltages and current driving capabilities due to variations in fabrication processes. This mismatch in device parameters is more pronounced for

smaller-size devices. Therefore, circuit design techniques tolerant of such variations must be used.

6. *Lower Operating Voltage.* Operating voltages are generally lower for high device density technologies. This constrains the type of circuits that can be used to implement needed functions and output drive capabilities of pad drivers.

7. *Higher Leakage Currents.* Leakage current, the current flowing through a device when it is turned off, is higher for submicron VLSI technologies. This factor must be kept in mind while designing a chip.

8. *Faster Device but Slower Interconnects.* VLSI technology scaling yields faster devices relative to interconnections. Now devices are so fast that these can be considered as switching in no time (on the order of tens of picoseconds), but the interconnects used for transporting signal introduce delays that are almost an order of magnitude higher. This is due to the fact that interconnect technology has not progressed at the same rate as device scaling. Fortunately, the technology is improving.

Therefore, computer architectures must be designed such that the above constraints do not degrade performance. The component design is also improving with our understanding of the underlying technology, as reflected in a recent conference.[55]

Since semiconductor memory chips are getting bigger and faster, these will find applications in high-speed cache memories, These are used to enhance machine performance by matching any mismatch between the CPU and slower memories. Thus, a single chip with an increased number of devices on it can be effectively used in balancing the computing and memory bandwidths, as discussed previously.

Looking ahead to where changes in system architecture and component technologies might occur and what new opportunities might present themselves is always speculative. As we get a better understanding of VLSI technology and computer architecture, the following picture will emerge.

There will be a continued effort to push to the device sizes below 0.1 μm, but the interconnection, clock, and signal distribution on chip will be major problems. This is because the relative speed of operation of a part largely will be dominated by the rate of information transfer rather than its processing. Newer technologies such as superconductors and optical interconnections will find more use in high-performance machines. On the architecture side, we should see a heavy emphasis on parallel and highly specialized architectures. It is like going in full circles — before general-purpose processors, most of the systems were hard-wired, not because the people were less smart, but because the technology and economics so dictated. In the near future, we should have a similar situation because technology and economics will favor specialized systems until we invent schemes to control and utilize a large number of processors. VLSI technology is capable of putting more than a million devices on a 1.00 cm^2 chip, but it is crying for good applications aside from memory applications. There will be many such exciting and challenging opportunities in different areas of computer technology.

REFERENCES

1. A. W. Burks, H. H. Goldstein, and J. von Neumann, "Preliminary Discussion of the Logical Design of an Electronic Computing Instrument (May 1946), in A. H. Taub, Ed. *John von Neumann Collected Works*, Vol. 4, Macmillan, New York, 1963.

2. G. Geitz, C. Muller-Schloer, D. Ruth, and H. Schwartzel, "New Structures for Information Processing Machines," *Eur. Conf. Electron., 7th (Eurocon 86)*, p. 10 (1986).

3. J. Backus, "Can Programming be Liberated from the von Neumann Style?" *Commun. ACM*, **8**, 613 (1978).

4. M. J. Flynn, "Some Computer Organizations and Their Effectiveness," *IEEE Trans. Comput.*, **C-21**(9), 948 (1972).

5. P. M. Kogge, *The Architecture of Pipelined Computers*, McGraw-Hill, New York, 1981.

6. T. N. Mudge, J. P. Hayes, and D. C. Winsor, "Multiple Bus Architectures," *IEEE Comput.*, **20**, 42 (1987).

7. K. E. Batcher, "Design of a Massively Parallel Processor," *IEEE Trans. Comput.* **C-29**(9), 836 (1980).

8. H. T. Kung and C. Leiserson, "Systolic Arrays for VLSI," in C. Mead and L. Conway, Eds., *Introduction to VLSI Systems*, Addison-Wesley, Reading, MA, 1980.

9. L. Snyder, "Introduction to the Configurable Highly Parallel Computer," *IEEE Comput.*, **15**(1), 47 (1982).

10. C. Seitz, "The Cosmic Cube," *Commun. ACM*, **28**(1), 22 (1985).

11. H. J. Levenstein, C. J. Bartlett, and W. J. Bertram, Jr, "Multi-Chip Packaging Technology for VLSI-Based Systems," *Int. Solid-State Circuit Conf., Dig. Tech.*, p. 224 (1987).

12. T. Kobayashi et al., "A 6K-Gate CMOS Gate Array," *Int. Solid-State Circuit Conf., Dig. Tech. Pap.*, p. 174 (1982).

13. S.-C. Lee and A. S. Bass, "A 2500-Gate Bipolar Cell Array with 250ps Gate Delay," *Int. Solid-State Circuit Conf. Dig. Tech. Pap.*, p. 178 (1982).

14. Y. Takayam, et al., "A 1ns 20K CMOS Gate Array Series with Configurable 15ns 12K Memory," *Int. Solid-State Circuit Conf., Dig. Tech. Pap.*, p. 196 (1985).

15. S. Kuboki, I. Masuda, and T. Hayashi, "A 4K CMOS Gate Array with Automatically-Generated Test Circuits," *Int. Solid-State Circuit Conf. Dig. Tech. Pap.*, p. 128 (1985).

16. T. Sano et al., "A 20ns CMOS Functional Gate Array with a Configurable Memory," *Int. Solid-State Circuit Conf., Dig. Tech. Pap.*, p. 146 (1983).

17. T. Itoh et al., "A 6,000-Gate CMOS Array," *Int. Solid-State Circuit Conf., Dig. Tech. Pap.*, p. 176 (1982).

18. T. Terada et al., "A 64K GaAs Gate Array" *Int. Solid-State Circuit Conf., Dig. Tech. Pap.*, p. 144 (1987).

19. M. Ueda et al., "A 1.5 μ CMOS Gate Array with Configurable ROM and RAM," *Int. Solid-State Circuit Conf., Dig. Tech. Pap.*, p. 126 (1985).

20. S. Tanaka et al., "A Sub-Nanosecond 8K CMOS/SOS Gate Array," *Int. Solid-State Circuit Conf., Dig. Tech. Pap.*, p. 260 (1984).

21. H. Ulrich et al., "A 100ps 9K Gate ECL Masterslice," *Int. Solid-State Circuit Conf., Dig. Tech. Pap.*, p. 200 (1985).

22. T. Nishimura et al., "A Bipolar 18K Gate Variable-Size Cell Masterslice," *Int. Solid-State Circuit Conf., Dig. Tech. Pap.*, p. 76 (1986).

23. T. Saigo et al., "A 20K-Gate CMOS Gate Array," *Int. Solid-State Circuit Conf., Dig. Tech. Pap.*, p. 156 (1983).

24. Y. Takyama et al., "A 1ns 20K CMOS Gate Array Series with Configurable 15ns 12K Memory" *Int. Solid-State Circuit Conf., Dig. Tech. Pap.*, p. 196 (1985).

25. H. Takahashi, et al., "A 240K Transistor CMOS Array with Flexible Allocation of Memory and Channels," *Int. Solid-State Circuit Conf., Dig. Tech. Pap.*, p. 124 (1985).

26. D. M. DeRuyck, L. Snyder, and J. D. Unruh, "Processor Displacement: Area-Time Tradeoff Method for VLSI Design," *Conf. Adv. Res. VLSI, 1982*, MIT, p. 182 (1982).

27. D. Archer et al., "A 32b CMOS Microprocessor with On-Chip Instruction and Data Caching and Memory Management," *Int. Solid-State Circuit Conf., Dig. Tech. Pap.*, p. 32 (1987).

28. A. Berenbaum, B. W. Colbry, D. R. Ditzel, R. D. Freeman, H. C. McLellan, and M. Shoji, "A Pipelined 32b Microprocessor with 13Kb of Cache Memory," *Int. Solid-State Circuit Conf., Dig. Tech. Pap.*, p. 34 (1987).

29. H. Kadota et al., "A CMOS 32b Microprocessor with On-Chip Cache and Transmission Lookahead Buffer," *Int. Solid-State Circuit Conf., Dig. Tech. Pap.*, p. 36 (1987).

30. J. M. Pedelton, S. I. Kong, E. W. Brown, F. Dunlap, C. Marino, D. M. Ungar, D. A. Patterson, and D. A. Hodges, "A 32b Microprocessor for Smalltalk," *Int. Solid-State Circuit Conf., Dig. Tech. Pap.*, p. 32 (1986).

31. W. R. Heller and W. F. Mikhaikl, "Prediction of Wiring Space Requirements for LSI," *Proc. Des. Autom. Workshop*, p. 32 (1977).

32. A. D. Lopez and H.-C. Law, "A Dense Gate Matrix Layout Method for MOS VLSI," *IEEE Trans. Electron Devices*, **ED-28**(8), 1671 (1980).

33. IBM3081 System Development Technology (Special Issue), *IBM J. Res. Dev.*, **26**(1) (1982).

34. J. E. Price, "On Being the Right Size," *IEEE Des. Automat. Workshop*, Oct. (1981).

35. H. Kanai, "Low Energy LSI and Packaging for System Performance," *IEEE Trans. Components, Hybrids, Manuf. Tech.*, **CHMT-4**(2), 173 (1981).

36. H. T. Kung, "Memory Requirements for Balanced Computer Architectures," *Proc. Int. Symp. Comput. Archit.*, p. 49 (1986).

37. M. Suzuki, M. Hirata, and S. Konaka, "43ps/5.2GHz Bipolar Macrocell Array LSIs," *Int. Solid-State Circuit Conf., Dig. Tech. Pap.*, p. 70 (1988).

38. W. Fichtner, E. A. Hostatter, R. K. Watts, R. J. Bayruns, P. F. Bechtoltd, R. L. Johnston, and D. M. Boulin, "A Submicron NMOS Technology Suitable for Low Power High Speed Circuits," *Tech. Dig. — Int. Electron Devices Meet.*, p. 264 (1985).

39. S. J. Hillenius, R. Liu, G. E. Georgiou, R. L. Field, D. W. Williams, A. Kornblit, D. M. Boulin, R. L. Johnston, and W. T. Lynch, "A Symmetric Submicron CMOS Technology," *Tech. Dig. — Int. Electron Devices Meet.*, p. 252 (1986).

40. Y. Awano, M. Kosugi, T. Mimura, and M. Abe, "Performance of a Quarter-Micrometer-Gate Ballastic Electron HEMT," *IEEE Electron Device Lett.*, **EDL-8**(10), 451 (1987).

41. D. A. Patterson, "Reduced Instruction Set Computers," *Commun. ACM*, **28**(1), 8 (1985).

42. R. Regan-Kelley and R. Clark, "Applying RISC Theory to a Large Computer," *Comput. Des.*, p. 191, Nov. (1983).

43. R. Chang, "Parallel-Processing Computer Overcomes Memory Contention," *Comput. Des.*, p. 113, Sept. 15 (1985).

44. S. Nagashima, Y. Inagami, T. Odaka, and S. Kawabe, "Design Considerations for a High-Speed Vector Processor: The Hitachi S-810," *Proc. Int. Conf. Comput. Des.: VLSI Comput., ICCD84,* p. 238 (1984).

45. *Multimax Technical Summary,* Encore Computer Corporation, Marlboro, MA, 1987.

46. *Butterfly Parallel Processor Overview,* BBN Advanced Computers, Cambridge, MA, 1986.

47. M. Hill et al., "Design Decisions in SPUR," *IEEE Comput.,* **19**(11), 8 (1986).

48. *3600 Technical Summary,* Sybolics Inc., Cambridge, MA, 1983.

49. W. D. Hillis, *The Connection Machine,* MIT Press, Cambridge, MA, 1985.

50. M. Togai and H. Watanabe, "Expert System on a Chip: An Engine for Real-Time Approximate Reasoning," *IEEE Expert,* Fall (1986).

51. T. Moto-oka, "Overview of the Fifth Generation of Computer System Project," *10th Annu. Int. Symp. Comput. Archit.,* p. 413 (1983).

52. N. Ohwada, T. Kimura, and M. Doken, "LSI's for Digital Signal Processing." *IEEE Trans. Electron Devices,* **ED-26**(4), 292 (1979).

53. C. J. Caren et al., "A 60ns CMOS DSP with On-Chip Instruction Cache," *Int. Solid-State Circuit Conf., Dig. Tech. Pap.,* p. 156 (1987).

54. J. P. Norsworthy et al., "A Parallel Image Processor," *Int. Solid-State Circuit Conf., Dig. Tech. Pap.,* p. 158 (1988).

55. International Solid State Circuits Conference, *Int. Solid-State Circuit Conf., Dig. Tech. Pap.,* Feb. (1988).

8 Lateral Surface Superlattices

D. K. FERRY and G. H. BERNSTEIN[†]
Center for Solid State Electronics Research
Arizona State University
Tempe, Arizona

As semiconductor technology continues to pursue the scaling down of integrated circuit dimensions into the submicron and ultrasubmicron regimes, many novel and interesting questions will emerge concerning the physics of charged particles in semiconductors. One of the more important topics to be considered is that of carrier confinement in structures that reduce the dimensionality of the system. Notable among these structures are MOS quantized inversion layers and the heterojunction superlattice discussed in previous chapters. In particular, the fabrication of the quantum well superlattice has been possible because of the advent of molecular-beam epitaxy (MBE) and metal organic CVD (MOCVD) technology, which have been discussed in previous chapters of this book.

The concept of a superlattice, as it is most commonly interpreted, refers to layered structures and was first put forward by Esaki and Tsu.[1] It was theorized that a modified band structure would result in which minibands separated by minigaps would lead to Bragg scattering at the edges of the minizones at relatively low electric fields.[1,2] Indeed, the general potential and energy structure have been verified by careful experimental work so that one can now talk of "bandgap engineering." We want to consider here *lateral surface superlattices*. Lateral surface superlattices, in which the superstructure lies in a surface or heterostructure layer,[3-5] offer considerable advantages for obtaining superlattice effects in a planar technology.

Consider a periodic array of gates. If the periodicity in the gate array can be fabricated with a spacing small compared with the inelastic mean free path of the carriers, then superlattice effects should manifest themselves in the surface conduction channels. While being a distinct limitation to down-scaling of semiconductor device arrays, the quantum collective effects which arise in the surface superlattices are interesting in their own right and offer new device capabilities.

The lateral surface superlattice offers conceptual as well as technological advantages over the layered superlattice in terms of achieving the desired quantum well transport effects. In essence, the reason for this is that the minibands in the

---

[†]Present address: Department of Electrical and Computer Engineering, University of Notre Dame, Notre Dame, IN 46556.

one-dimensional layered superlattices are not separated by real minigaps, but are "connected" by the two-dimensional transverse continuum of states. Electrons can avoid the minigaps by scattering to higher energies in directions parallel to the superlattice planes. To create true minigaps, a multidimensional superlattice or quantization is required. This can be achieved by a lateral superlattice imposed in a quantized inversion layer.

We shall illustrate in this chapter how such lateral surface superlattices can be prepared by a variety of techniques. One formally proposed by Bate[3] is an MOS structure similar to an array of charge coupled devices (CCD) which can be fabricated through the use of fine-line lithography.[6] Effects of naturally occurring lateral superlattices have been observed in MOS structures formed on vicinal planes whose periodicity was on the order of 10 nm.[7] Other approaches use selective area epitaxy, but have not been fabricated to date. These structures have been idealized as either GaAs wells in a GaAlAs matrix,[8] or conversely, GaAlAs barriers in a GaAs matrix, creating a "bump" superlattice.[9] Finally, the complement of the MOS structure is a depletion mode device made in a high electron mobility transistor.[10] This latter device has actually performed as expected, clearly exhibiting the strong negative differential conductivity expected for transport in the surface superlattice devices.

8.1 LATERAL SUPERLATTICES

The concept of a lateral superlattice along a surface has considerable advantages, among which is the ability to control the magnitude of the surface potential and barrier heights seen by an inversion layer. The basic configuration is shown in Fig. 8.1, for an MOS structure.[3] A periodic gate array is buried inside the dielectric and thus differs from a normal CCD array in that the top electrode is a blanket electrode. The top electrode provides energy gap control without requiring critical alignment of successive levels. A one-dimensional (along the surface) implementation of this structure has been achieved,[6] but suffers from the same problems as a normal layered superlattice. If the periodic gates are biased positively, the surface potential for electrons decreases under the gate electrodes and, to a lesser extent, in the gaps. Minority-carrier generation, injection from an FET source, or optical pumping creates the carriers necessary to form the inversion layer under the gates. Thus, in addition to the normal average surface potential, a periodic superlattice potential is seen by the inversion layer electrons. The presence of the top electrode allows for critical control of the relative strengths of both the average potential and the superlattice potential. The one-dimensional MOS structure yielded a modulated density of states indicative of gaps formed in the conduction band, but as expected, showed no negative differential conductivity.

For a two-dimensional structure, we can assume the effective superlattice potential $U(x, y)$ to vary along the structure in a form given by[4]

$$U(x, y) = 4U_0 \cos\left(\frac{2\pi x}{d}\right) \cos\left(\frac{2\pi y}{d}\right), \quad (1)$$

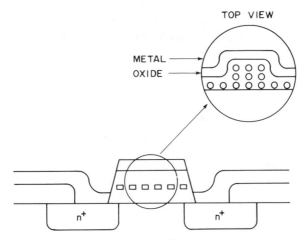

Fig. 8.1 A periodic gate array buried in the MOS oxide is used to induce charge under the elements of the array, thus creating a superlattice in the interfacial plane.

where U_0 is the barrier height formed by the lateral superlattice and d the period of the structure. By introducing a change of variables $u = x + y$, $v = x - y$, with

$$2^{1/2} \frac{\partial}{\partial x} = \frac{\partial}{\partial u} + \frac{\partial}{\partial v}, \qquad 2^{1/2} \frac{\partial}{\partial y} = \frac{\partial}{\partial u} - \frac{\partial}{\partial v}, \tag{2}$$

the effective two-dimensional Schrödinger equation for the interface electrons in the inversion layer is given by

$$\left\{ \frac{d_2}{du^2} + \frac{d^2}{dv^2} + \frac{2m^*}{\hbar^2} \left[E - 2U_0 \cos\left(\frac{2\pi u}{d}\right) - 2U_0 \cos\left(\frac{2\pi v}{d}\right) \right] \right\} \psi(u, v) = 0. \tag{3}$$

This equation is now separable using a product form of the wave function $\psi_1(u)\psi_2(v)$, which yields two effective one-dimensional Schrödinger equations of the form

$$\left\{ \frac{d^2}{du^2} + \frac{2m^*}{\hbar^2} \left[E_1 - 2U_0 \cos\left(\frac{2\pi u}{d}\right) \right] \right\} \psi_1(u) = 0. \tag{4}$$

Introducing the reduced variables,

$$\xi = \frac{gu}{2}, \qquad a = \frac{8m^*E}{\hbar^2 g^2} = 2\frac{m^* d^2 E}{\hbar^2 \pi^2}, \qquad q = \frac{m^* d^2 U}{\hbar^2 \pi^2}, \tag{5}$$

where $g = 2\pi/d$, (4) can be rewritten as

$$\frac{d^2\psi(\xi)}{d\xi^2} + [a - 2q \cos(2\xi)]\psi(\xi) = 0. \tag{6}$$

This latter form is immediately recognized as the Mathieu equation. For $q = 0$, the case for no barriers, all values of a (and hence of energy E) are allowed. However, when $q \neq 0$, gaps open in the spectrum of a. For small q, the lowest gap is centered approximately at the point $a = 1$, and higher gaps occur approximately at $a = 4, 9, \ldots, n^2$. The general energy structure is shown in Fig. 8.2.

It is very important to note that the general solution to the Mathieu equation is of Bloch form

$$\psi(\xi) = \exp(ik\xi)p(\xi), \tag{7}$$

where $p(\xi)$ has the periodicity of the superlattice potential. This is, of course, expected for the periodic potential. For the first minigap to be centered at a particular energy W, we require $a(W) = 1$, and

$$\frac{\hbar^2\pi^2}{2m^*d^2} = W, \tag{8}$$

or

$$d = \left(\frac{\hbar^2\pi^2}{2m^*W}\right)^{1/2}. \tag{9}$$

In Fig. 8.3, we show a plot of the d values required to achieve an energy width of $W = 6k_BT$ in the lowest band. We also show the number of states in the lowest miniband, which is closely related to the surface carrier density required in the inversion layer to fill the first miniband completely. It will be noted that these den-

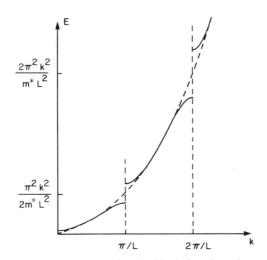

Fig. 8.2 The general energy structure for the Bloch functions that are solutions of the Mathieu equation.

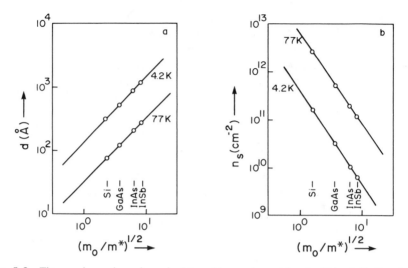

Fig. 8.3 The spacing values d required to achieve a lowest band-width of $6k_BT$ and the density in the inversion layer required to completely fill the lowest sub-band for this case.

sities correspond to a relatively strong inversion layer existing at the surface. For these densities, the Fermi level at low temperature will be well into the conduction band.

The first minigap will have a value given approximately the $\Delta_a = a(\Delta) = 1.9q$ for small values of q. For $U = 0.01$ V, we find that $\Delta_a = 2.9k_BT$ at 77 K and $52.4k_BT$ at 4.2 K, corresponding to rather large gaps when compared with the miniband energy width. It is clear that relatively small induced superlattice potentials are required to produce the miniband/minigap structure in the lowest bands. Indeed, surface band bending corresponding to roughly one trapped electron under each gate could produce sufficient potential to be noticeable at 4.2 K in the case of the very light mass InSb. However, since gaps decrease at higher energies, higher potentials would be required for noticeable effects. In this case, the lowest minibands would be essentially flat, with transport primarily in the upper minibands.

Evidently, from Fig. 8.3, lower effective mass material is favored. Whereas Si requires $d \sim 10$ nm to produce superlattices fully, the effects should be observable in InAs and InSb for $d \sim 50$ nm. One caution, however, must be stated here, and this is that the calculations are done for the size of the bands and gaps and not for whether or not the effects will be washed out by thermal effects. Rather, the important length is not the wavelength, which enters into the equations leading to Fig. 8.3, but the inelastic mean free path which determines the distance over which the electron wave functions remain coherent.

8.2 GaAs STRUCTURES

Other possible structures for the lateral surface superlattice can be developed. One is the complement of the MOS structure discussed above, and is fabricated in a

GaAs MODFET,[10] whereas a second can be fabricated through the use of selective area epitaxy.[8] Although the former is more readily fabricated, the latter was suggested earlier in a historical sense, and we shall discuss it first. A thin GaAs layer grown on top of a semi-insulating or AlGaAs layer can be patterned and have AlGaAs regrown around it (or vice versa through selective epitaxy). Holes can be physically etched in the GaAlAs layer through fine-line lithography and GaAs regrown in these holes. We note that this is just one such paradigm out of a large number of such structures that can be conceived. In such a structure, the electrons are tightly bound into each of the GaAs regions. For an energy barrier of 0.5 V, and a GaAs well diameter of 7.5 nm, only two energy levels are allowed in each well. We can then find these by solving the cylindrical boundary value problem. This structure is probably only of theoretical interest. The important point here is that it is the inelastic mean free path rather than the electron wavelength that is important. This leads us into the more practical approach of the MODFET.

A MODFET is usually a depletion mode device, in which the gate is used to push electrons out of a channel, much like a MESFET, rather than drawing them into the channel as in a MOSFET. Thus, we really need the complement of the dot structure shown in Fig. 8.1. This is shown in Fig. 8.4. In this case, we need a grid, as shown in Fig. 8.5. The grid imposes the periodic potential on the electrons in the channel and, as the device approaches pinch-off, the remaining electrons are left in pockets which are aligned with the holes of the grid. The electrons are thus sitting in small quantum wells induced by the superlattice potential itself. The grid illustrated in Fig. 8.5 was written by electron-beam lithography using single-level PMMA, and the pattern transferred by normal lift-off processing. The grid shown has 40 nm lines on 170 nm spacing. The grid is 28 periods in the source-drain direction and 170 periods in the transverse direction, so that the superlattice gate is approximately $5 \times 30 \ \mu m$. Devices were fabricated from typical modulation-doped heterostructure material commonly employed for MODFETs. The GaAs active layer was an undoped buffer layer that was topped with a 5 nm undoped GaAlAs spacer layer. The Si doping in the GaAlAs layer was $3.5 \times 10^{18} \ cm^{-3}$. This layer was 30 nm thick and was topped with a 7.5 nm GaAs cap layer, comparably doped. All of the fabrication levels were done with electron-beam lithog-

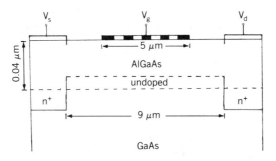

Fig. 8.4 A schematic diagram of the Bloch-HEMT structure.

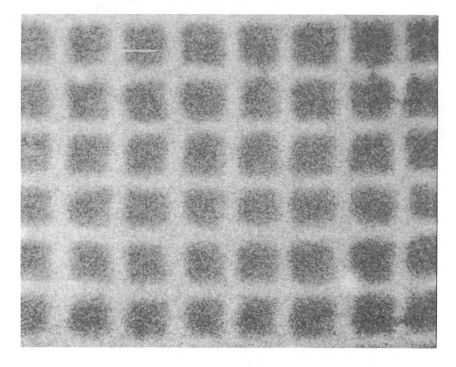

Fig. 8.5 A photomicrograph of the metal grid used for the gate metallization in the Bloch-HEMT.

raphy and the active devices were mesa isolated. Ohmic contacts were Au/Ge/Ni and the Schottky was pure Au. The source-drain spacing was 9 μm.

The Block FETs were tested over a wide temperature range. At 77 K and above, the drain characteristics were typical of standard field-effect transistors of this size. On the other hand, when the temperature was lowered to 4.2 K, pronounced negative differential conductivity (NDC) was observed, as shown in Fig. 8.6. Here, it can be seen that the magnitude of the NDC increased as the devices were pushed toward pinch-off with more negative gate potential. Especially at pinch-off, the curves exhibit exactly the behavior expected for Bloch FETs, as discussed in the next section. As the drain field increases, but remains below that needed for interband tunneling, an increasing number of electrons are expected to undergo Bloch oscillations and be localized in space. These are removed from the conduction process and the current decreases.

The striking features of these curves is this strong NDC evident at high reverse gate voltages. These features can be further explained as follows. When the device is near pinch-off, the electrons in the channel are more fully localized in weakly coupled quantum boxes, which are the sites under the holes in the grid. Thus, there is a strong superlattice potential that is strengthened by the localization of the electrons in the boxes. As the channel density is increased, by making the gate potential more positive, the superlattice potential is weakened by the screening of the

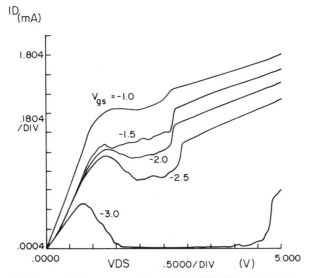

Fig. 8.6 The *I-V* characteristics measured in a Bloch-HEMT at 4.2 K. The grid was composed of 40 nm lines on 170 nm spaces and the gated region measured $5 \times 20 \ \mu$m.

background charge density. To observed the miniband effects, the electrons must have an inelastic mean free path length that is longer than the period of the surface superlattice. The mobility in these layers was such that we estimate the mean free path to be only about 200 nm at 4.2 K, so that the effect is seen only at these low temperatures.

While the experiments are highly suggestive of Bloch oscillations, we must be aware that there are other mechanisms that can provide NDC in devices and which may be the cause of that discussed here. One possible one is real-space transfer, in which electrons are scattered out of the inversion layer and into the GaAlAs, where they may be trapped or contribute to lower-mobility conduction paths. This mechanism is not likely a cause of the effect seen in Fig. 8.6, as this effect should be stronger at more positive values of gate bias, which is opposite to the observations in the figure. Moreover, such an effect should not be sensitive to the grid nature of the gate. In Fig. 8.7, we show the characteristics of a MODFET made in the same manner as the Bloch FET except that the gate metallization is continuous. No NDC is seen in these curves, so that we may conclude that the grid nature of the gate is crucial to the existence of the NDC. On the other hand, the curves in Fig. 8.6 are similar to those of the real-space transfer NERFET device.[11] In this case, the transfer is to a buried-channel layer. Such a layer is not normally included in the MODFET structure. We have profiled our devices via SIMS and find no evidence of a buried layer that could lead to NERFET behavior.

Another possibility for the NDC is sequential "resonant" tunneling, in which the tilting of the minibands in the applied drain potential allows electrons to tunnel from full states in one well to empty states in the next and subsequently to traverse

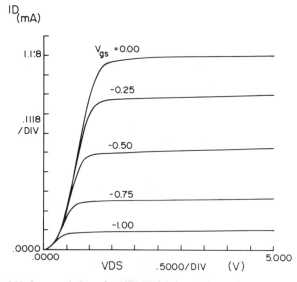

Fig. 8.7 The *I-V* characteristics of a HEMT fabricated in the same fashion as the Bloch-HEMT except that a solid gate metallization is used.

the entire structure. Such effects have been shown to exhibit NDC in layered super-lattices.[12] For sequential tunneling, the curves of Fig. 8.6 would have to fit certain voltage constraints. For the curve with -3.0 V on the gate, we can estimate that the minibands are 50–60 meV wide and have gaps of 60–100 meV. For this band-width and bandgap, the conduction would have to take place at a very high mini-band, based on the size of the minizone in the grid structure used. That is to say, the physical dimensions of the grid impose the positions of the minizone boundaries in momentum space and these set limits on the allowed bandwidths of each mini-band in the reduced zone scheme. The required bandwidth above occurs only for about the 360th miniband. If conduction takes place in this miniband, all of the lower minibands must be full. The number of electrons required to fill a miniband is just $2/d^2$, where d is the periodicity of the superlattice (in each direction). In our case, with $d = 170$ nm typically, there are 6.9×10^9 cm^{-2} electrons in each miniband, so we would have to fill to a very high inversion density, of about 2.5×10^{12} cm^{-2}. However, it is well known that HEMTs can achieve an inversion den-sity of only about 1×10^{12} cm^{-2}. It is not clear, though, that sequential resonant and Bloch oscillations are different quantities. The NDC in the former case arises as the tunneling current decreases when the bands in adjacent wells no longer over-lap. The Bloch localization arises in principle as electrons are accelerated to the top of the band and can no longer tunnel through to vacant states in the adjacent well. In fact, these two effects may supplement each other as possible causes for the observed NDC. Indeed, if a coherent photon is emitted in each well (rather than an inelastic phonon), then

$$eV_\omega = \hbar\omega \tag{10}$$

is required across each well. Since the field is V_ω/d,

$$\omega = \frac{eFd}{\hbar},\tag{11}$$

which is the Bloch frequency. Thus, the Bloch oscillations can be seen to be a coherent sequential tunneling effect.

8.3   TRANSPORT THEORY

As we have seen above, various superlattice structures give rise to energy minibands that vary sinusoidally across the minizone in at least one dimension and that have relatively narrow widths. The shape of such bands results in interesting electrical properties. The one-dimensional superlattice is one such structure, and the lateral surface superlattice is another. In this section, we now want to treat the average velocity and energy of the carriers in the superlattice and the various transport properties that can occur. We will do this in both the steady-state dc conductivity case and in the small signal ac conductivity case.

8.3.1   Steady Transport

The average velocity and energy are found by taking the first and second moments of the Boltzmann transport equation with an assumed form for the distribution function. A constant electric field is assumed to be applied in the plane of the sinusoidal bands, whereas the energy shape in the other two directions is arbitrary. Here, we shall take a simple Wigner function representation as the initial equilibrium distribution.[13] This treatment is valid for a single energy band at low to moderate electric fields for which the sinusoidal band is less than half-filled with carriers. For a band more than half-filled, the Pauli exclusion principle must be taken into account. In addition, we shall use the constant relaxation time approximation, as the effects in which we are interested arise from the properties of the energy bands and not from the energy dependences of the scattering rates. In truth, the relaxation time is not strictly constant, since the density of states for cosinusoidal energy bands in two dimensions show strong Van Hove singularities which lead to strong scattering peaks near midband. Nevertheless, we shall see below that the constant relaxation time approximation is not too bad due to the high scattering rates. Lastly, we assume the distribution function is homogeneous in space.

The form of the time-independent, homogeneous Wigner transport equation in the relaxation time approximation is the same as that of the Boltzmann transport equation, and is

$$\frac{eF}{\hbar}\frac{\partial f(\mathbf{k})}{\partial k_z} = -\frac{f(\mathbf{k}) - f_0(\mathbf{k})}{\tau},\tag{12}$$

where the field F and the direction of the superlattice are taken to be in the z direction. The quantity $f_0(\mathbf{k})$ is found from the equilibrium quantum density distribution by using the Hamiltonian equivalence principle, followed by a Wigner transformation, which leads to the Bloch form

$$f_0(\mathbf{k}) = e^{\beta[\varepsilon - \varepsilon\cos(dk_z)]} f_0(k_x, k_y) \tag{13}$$

in which the energy is

$$E = \varepsilon - \varepsilon\cos(dk_z), \tag{14}$$

where ε is the half-width of the energy band and d the periodic spacing of the superlattice.

Taking the first moment of the velocity in the z direction and the second moment (the total energy) results in the following two equations:

$$\frac{eF}{\hbar} \int\!\!\int_{-L}^{L} v_z \frac{\partial f(k)}{\partial k_z} d^3k = -\Gamma\langle v_z \rangle, \tag{15}$$

$$\frac{eF}{\hbar} \int\!\!\int_{-L}^{L} E \frac{\partial f(k)}{\partial k_z} d^3k = \Gamma n[\langle E \rangle - \langle E_0 \rangle], \tag{16}$$

where $\Gamma = 1/\tau$, $L = \pi/d$, and $\langle S \rangle$ is defined as the average of the quantity S. Here, E_0 is the equilibrium energy without an applied field.

To proceed, we must now introduce the analytical expressions for the velocity and the energy. For these, we use the assumed energy band shape

$$E = \varepsilon - \varepsilon\cos(dk_z) + E(k_x, k_y), \tag{17}$$

$$v_z = \frac{\varepsilon d}{\hbar} \sin(dk_z). \tag{18}$$

The left-hand side of both moment equations may be integrated by parts in the z direction, and the fact that both the energy and the distribution function are periodic in $2L$ and the velocity vanishes at L, $-L$, gives the results that

$$\frac{eFd^2}{\hbar^2} \langle \varepsilon\cos(dk_z) \rangle = \Gamma\langle v_z \rangle, \tag{19}$$

$$-eF\langle v_z \rangle = \Gamma[\langle E \rangle - \langle E_0 \rangle]. \tag{20}$$

The first bracketed average in (19) can be replaced, using (17) by $\langle \varepsilon - E + E(k_x, k_y) \rangle = \varepsilon - \langle E \rangle + \langle E_t \rangle$ [$E_t = E(k_x, k_y)$]. Solving these equations simultaneously gives the expressions for the velocity and energy as functions of the field as

$$\langle v_z \rangle = \frac{(\varepsilon + \langle E_t \rangle - \langle E_0 \rangle) eF\tau d^2}{\hbar^2 (1 + \omega_B^2 \tau^2)}, \tag{21}$$

$$\langle E \rangle = \frac{\langle E_0 \rangle + (\varepsilon + \langle E_t \rangle) \omega_B^2 \tau^2}{1 + \omega_B^2 \tau^2}, \tag{22}$$

where $\omega_B = eFd/\hbar$ is the Bloch frequency. Thus, we see that as the field increases, the average energy rises from its equilibrium value $\langle E_0 \rangle$ ($= 3k_B T/2$) to the half-band energy plus the average transverse energy, and the velocity behaves in a corresponding manner.

The velocity has the same field dependence in (21) as that obtained earlier by Lebwohl and Tsu,[14] except for the energy pre-factor in front of the expression. The difference is caused by the different equilibrium distribution function chosen. In this latter work, the authors assumed the distribution was a zero-temperature Fermi–Dirac. Here, on the other hand, the distribution is a real-temperature one that includes the details of the band shape. Note that the velocity shows an NDC that sets in above $\omega_B \tau = 1$, so that this field corresponds to the current peaks in Fig. 8.6. Monte Carlo calculations have been performed as well, in which the exact details of the energy-dependent scattering processes were included.[15] The general shape of the velocity curves of (21) is found there as well. In fact, for the same set of material (band) parameters, the curves are very close together, which justifies our earlier assumption of a constant relaxation time. The details of the scattering processes just do not make a large difference in the present circumstances and the important aspect for NDC is the band shape itself. We can estimate the mobility in Fig. 8.6, and then the conductivity peak occurs for $\omega_B \tau \sim 1$ (assuming the average mass in the well is about twice the bare GaAs mass). Given the complexity of the device structure, this result is amazingly close to that expected for the condition of Bloch oscillations.

8.3.2 High-Frequency Response

While the above analysis provides limits within which Bloch oscillations may be seen, it does not provide an existence proof on their presence. We now consider several possible routes, besides the normally considered potential NDC, by which Bloch oscillations may be experimentally verified. We pursue this by looking at a number of consequences of superimposing a small ac signal on top of the applied dc bias electric field. Here, we will seek first the ac component of the velocity, and hence of the mobility, that relates to the time-varying frequency ω at which the applied field is oscillating. To achieve this, we must add the time derivative term to (12). We then assume that $\langle E \rangle = \langle E \rangle_0 + E_1$, $\langle v_z \rangle = \langle v_z \rangle_0 + v_1$. Then, (19) and (20), for the ac terms, become

$$(i\omega - \Gamma) v_1 = \frac{eF_1 d^2}{\hbar^2} \langle \varepsilon \cos(dk_z) \rangle_0, \tag{23}$$

$$(i\omega - \Gamma) E_1 = eF_1 \langle v_z \rangle_0 + eF_0 v_1, \tag{24}$$

from which the resulting ac mobility (the real part of the velocity response to F_1) is given by

$$\mu(\omega) = \mu_0 \frac{(1 - \omega_B^2\tau^2)[1 + (\omega_B^2 - \omega^2)\tau^2]}{[1 + (\omega_B^2 - \omega^2)\tau^2] + 4\omega^2\tau^2}, \tag{25}$$

where $\mu_0 = e\tau/m$ is the low-field dc mobility. In Fig. 8.8, we plot this mobility as a function of frequency for a case in which the dc field lies in the NDC regime ($\omega_B\tau = 2$). It is important to note that there is no peak in the conductivity at the Bloch frequency. Rather, there is just a falloff in the negative conductivity to positive values. The lack of a resonance at the Bloch frequency suggests, but does not prove, that the Bloch oscillations are not radiative. In fact, what is probably meant by this result is that the individual Bloch electrons oscillate, but that their phases add incoherently, so that no coherent radiation is produced. On the other hand, the coherent sequential tunneling argument of the previous section suggests a radiative effect. This result is somewhat reinforced by Monte Carlo calculations of the velocity correlation function, which show that the noise spectrum does exhibit a peak at the Bloch frequency.[16] This noise spectrum is shown in Fig. 8.9 for a GaAs-based simulation. In principle, both the ac conductivity and the spectral density of the noise can be measured, although this is difficult for frequencies close to the Bloch frequency, as it lies in the far infrared portion of the spectrum. On the other hand, measurements of the noise emmission with for example, a Fourier transform infrared spectrometer can in principle provide evidence of the Bloch oscillations.

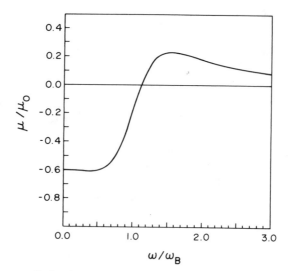

Fig. 8.8 The ac small-signal mobility, according to Eq. (25), is plotted here as a function of frequency. Only the real part is plotted, and the case of $\omega_B\tau = 2$ is taken.

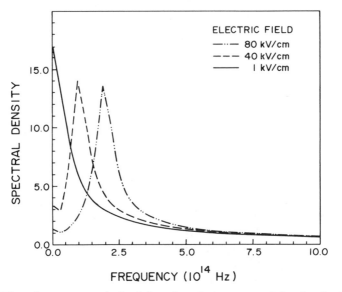

Fig. 8.9 The noise spectrum calculated from the velocity autocorrelation function is shown. The latter is calculated by an ensemble Monte Carlo technique and a peak at the Bloch oscillation frequency is observed.

It may be difficult to couple directly to the amplitude variations of the Bloch oscillations of the electrons themselves. However, it also may be possible to couple to the *phase* of the oscillations. A limitation on amplitude may arise because of the fact that the velocity amplitude is limited by the band structure itself, but the phase is not so constrained and may be the preferrred coupling scheme. In addition, we note that coupling to the phase has a direct analogy with the flux in the Josephson tunnel junctions and the resulting ac Josephson effect, in which steps are produced in the current–voltage characteristics by self-rectification of the ac signal. This is a direct consequence of the cosinusoidal nature of the energy bands. We illustrate this by treating a single electron confined to the cosinusoidal band described by (17), but ignoring the transverse energy for the moment. Under the influence of both a dc electric field F_0 and ac electric field F_1, the time variation of the momentum wave vector, in the absence of scattering, is given by

$$k(t) = k_0 + \frac{eF_0 t}{\hbar} + \frac{eF_1 t}{\hbar} \sin(\omega t),$$ (26)

and the corresponding velocity is just

$$v(t) = v_0 \sin\left[\omega_B t + k_0 d + \frac{\omega_B F_1}{\omega F_0} \sin(\omega t)\right],$$ (27)

where $v_0 = \varepsilon d/\hbar$. This expression can be rewritten as

$$v = v_0 \sum_n J_n(\zeta)\{[\sin(\Theta_B + n\Theta) + (-1)^n \sin(\Theta_B - n\Theta)] \cos(\lambda)$$
$$+ [\cos(\Theta_B + n\Theta) + (-1)^n \cos(\Theta_B - n\Theta)] \cos(\lambda)\},$$

$$(28)$$

where $\zeta = \omega F_1/\omega_B F_0$, $\Theta = \omega t$, $\Theta_B = \omega_B t$, $\lambda = k_0 d$, and J_n is the Bessel function of order n. A dc component of the velocity, and hence of the current, occurs when $\omega_B = \pm n\omega$. For a fixed ω, we change ω_B by changing the dc electric field, and we then expect to see resonance effects at the critical multiples of the ac frequency. The occurrence of such structure would be an unambiguous demonstration of the existence of the Bloch oscillations, but we caution that these will occur only in the NDC region of the dc characteristics.

The resonances above arise because the ac signal is inducing transitions, either emission or absorption, across the individual ladder states of the Stark ladder that coexists with the Bloch oscillations. The presence of a superlattice implies that the Stark ladder is not a real ladder but a virtual one,[17] and the ac effect is coupled to tunneling of the electron from one well to the next. In this case, the electron tunnels through to the adjacent well, emits a photon corresponding to the ac frequency, and drops in energy to a new state described by $E - eF_0 d (\omega - \omega_B)$. The absorption response corresponds to absorbing the photon first and then tunneling to the adjacent well. There is a preferred direction for each of these tunneling processes, as determined by the symmetry breaking of the applied electric field. It is evident from this that the resonances are very closely tied to the concept of sequential tunneling in superlattices themselves.[16,18] The importance of the virtual nature of the Stark ladder was only recently clarified.[17]

8.4 CONCLUSIONS

We have seen that the present theory of Bloch oscillations correlates well with experimental results. These data are, however, only a first step in the study of lateral surface superlattices. In general, as device dimensions shrink with improved lithographic technologies, quantum effects will become increasingly important. Lateral surface superlattices represent something of an upper limit to integrated circuit densities in that device–device interactions and quantum effects will dominate their operation.[19] Overlap of electron wave functions beneath the gates of nearby devices will cause correlation of their states, impeding their individual operation.

As these devices are better understood and their properties controlled, they could prove useful in several respects. The application of a gate potential to control the band structure is especially advantageous. Gaps in the conduction band can be varied for application as a tunable detector of infrared radiation. In addition, if Bloch oscillations are shown to be coherent, emission of electromagnetic radiation would be possible. The oscillation frequency would be tunable through variations in the electric field along the channel. It is conceivable that frequencies as high as tens of terahertz could be generated in this manner. The data already indicate that very large peak-to-valley ratios are attainable with the Bloch FET, making them excellent candidates as high-frequency oscillators and amplifiers.

In addition to these properties, stimulated emission might be achieved by injection of carriers into upper minibands to produce a situation of population inversion between them. When introduced into an appropriate cavity, tunable laser emission would be produced by variation of the minigaps. Higher efficiencies would result by matching the emission peak to the Q of the cavity. Also, the widths of the emission peaks would be decreased through the reduction in dispersion of the minibands. The results could be a reduction in the total number of modes and an increase in the output power in the dominant mode.

Besides the above relatively complicated structures, simpler ones such as double barrier resonant tunneling structures could be improved upon with the introduction of variable barrier heights. It has been shown[20] that variations in barrier height lead to changes in charge storage time of the well, and hence changes in the oscillation frequency. It would be of enormous value to tailor the characteristics of the oscillator to match the circuit parasitics in order to maximize oscillation frequency and power.

We can predict, then, that lateral surface superlattices will eventually result in significant improvements over conventional superlattices. Through increased control of dimensionality and barrier heights, new quantum effects will be manifest, resulting in exciting possibilities for discrete devices and dense circuits.

REFERENCES

1. L. Esaki and R. Tsu, "Superlattice and Negative Differential Conductivity in Semiconductors," *IBM J. Res. Dev.*, **14**, 61 (1970).

2. M. Büttiker and H. Thomas, "Current Instability and Domain Propagation due to Bragg Scattering," *Phys. Rev. Lett.*, **38**, 78 (1977).

3. R. T. Bate, "Electrically Controllable Superlattices," *Bull. Am. Phys. Soc.*, [2], **22**, 407 (1977).

4. D. K. Ferry, "Charge Instabilities in Lateral Surface Superlattices under Conditions of Population Inversion," *Phys. Status Solid B*, **106**, 63 (1981).

5. H. Sakaki, K. Wagatsuma, J. Hamasaki, and S. Saito, "Possible Applications of Surface Corrugated Quantum Thin Films to Negative Resistance Devices," *Thin Solid Films*, **36**, 497 (1976).

6. A. C. Warren, D. A. Antoniadis, H. I. Smith, and J. Melngailis, "Modification of Silicon Electronic Band Structure Using Submicron Period Gate Electrodes," *Tech. Dig. — IEEE Electron Devices Meet.*, p. 866 (1984).

7. P. J. Stiles, T. Cole, and A. A. Lakhani, "Evidence for a Superlattice at Si-SiO$_x$ Interface," *J. Vac. Sci. Technol.*, **14**, 969 (1977).

8. G. J. Iafrate, D. K. Ferry, and R. K. Reich, "Lateral (Two-Dimensional) Superlattices: Quantum-Well Confinement and Charge Instabilities," *Surf. Sci.*, **113**, 485 (1982).

9. G. W. Bryant, D. B. Murray, and A. H. MacDonald, "Electronic Structure of Single Ultrasmall Electron Devices and Device Arrays," *Superlattices Microstruct.* **3,** 211 (1987).

10. G. Bernstein and D. K. Ferry, "Fabrication of Ultra-Short Gate MESFETs and BlochFETs by Electron Beam Lithography," *Superlattices Microstruct.,* **2,** 373 (1986).

11. A. Kastalsky, S. Luryi, A. C. Gossard, and R. Hendel, "A Field-Effect Transistor with a Negative Differential Resistance," *IEEE Electron Device Lett.,* **EDL-5,** 57 (1984).

12. F. Capasso, K. Mohammed, and A. Cho, "Sequential Resonant Tunneling Through a Multi-Quantum Well Superlattice," *Appl. Phys. Lett.,* **48,** 478 (1986).

13. R. K. Reich and D. K. Ferry, "Moment Equations in the Wigner Formulation for Superlattice Band Structures," *Phys. Lett. A.,* **91A,** 31 (1982).

14. P. A. Lebwohl and R. Tsu, "Electrical Transport Properties in a Superlattice," *J. Appl. Phys.,* **41,** 2664 (1970).

15. R. K. Reich, R. O. Grondin, and D. K. Ferry, "Transport in Lateral Surface Superlattices," *Phys. Rev. B: Condens. Matter* [3] **27,** 3483 (1983).

16. R. O. Grondin, W. Porod, J. Ho, D. K. Ferry, and G. J. Iafrate, "On the Existence and Detection of Bloch Oscillations in Superlattices," *Superlattices Microstruct.,* **1,** 183 (1985).

17. J. B. Krieger and G. J. Iafrate, "Time Evolution of Bloch Electrons in a Homogeneous Electric Field," *Phys. Rev. B: Condens. Matter* [3] **33,** 5494 (1986).

18. R. F. Kazarinov and R. A. Suris, "Possibility of the Amplification of Electromagnetic Waves in a Semiconductor with a Superlattice," *Sov. Phys. — Semicond. (Engl. Transl.),* **5,** 707 (1971).

19. J. R. Barker and D. K. Ferry, "On the Physics and Modeling of Small Semiconductor Devides-II. The Very Small Device," *Solid-State Electron.,* **23,** 531 (1981).

20. N. C. Kluksdahl and D. K. Ferry, "Self-Consistent Study of the Resonant Tunneling Diode".

9 Two-Dimensional Automata in VLSI

D. K. FERRY, R. O. GRONDIN, and L. A. AKERS
Center for Solid State Electronics Research
Arizona State University
Tempe, Arizona

9.1 INTRODUCTION

There have been many treatises in the past several years that have addressed the physical requirements of logic elements in very large scale integration (VLSI). For example, it is clear that the requirement of signal restoration at each logic node really means that the devices must be saturating devices, with the saturation levels at "on" and "off" well characterized in terms of the system logic levels. In addition, the need for large noise margins and sharp transitions between logic levels implies that the devices must be highly nonlinear. Certain very high gain semiconductor devices, of course, satisfy both requirements, but this also means that signal levels (the difference between the two logic levels) must be large compared with the thermal voltage $k_B T$, since it is only with these voltage levels that the device is nonlinear. Moreover, the thermal noise in the system is also of this same order of magnitude ($k_B T$), so that the large signal level is necessary to discriminate against noise in the logical computation. Yet, in each case, we are describing the logic requirements as the requirements on a *single electronic device*, rather than on the VLSI chip itself.

What other requirements for logic are there? Certainly, one view is that the logical operations that occur in a computer can be described as the nonlinear interactions between two data streams. This reinforces the above view of the necessity for strongly nonlinear devices, but is not complete since logic also involves the necessity for branching and decision points in the data interaction. Thus, there is a need for uniformity in logic levels throughout the array of devices as well as for ease of measurement of the current state of selected devices. The former is reflected as a limit in logic as nonuniformity in threshold voltage in MOS devices. In fact, a fairly stringent requirement is placed on the control of this quantity in VLSI chips. In some sense, this variation of parameters within the array of devices can be expressed as an additional noise source in the circuit. Again, this result is expressed as a limit on the properties of individual devices, rather than on the chip as a whole.

In conventional descriptions of VLSI circuits, each device is assumed to behave in the same manner within the total system as it does when it is isolated. The full function of the system (or IC) is determined solely by the interconnection metallization specified to join the individual devices together. A different function can only be assigned to the system by redesigning the interconnection metallizations — a practical impossibility for most systems, but in practice accomplished in some programmable systolic arrays, a particular implementation of an iterated tessallation network. The conventional clear separation of device design from system design thus depends on being able to isolate each individual device from the environment of the other devices except for the planned effects occuring through the interconnection network. This simplification is likely to be seriously in error for submicron configured systems, where the devices are packed much closer together. As a result, the isolation of one device from another will be difficult to achieve. Instead, one is driven to consider methods of actually using the interactions between devices as a tool to accomplish *distributed* information processing within the VLSI chip.

Several methods of modeling interdevice coupling at the chip level have been discussed. The most promising involves modeling the chip as a cellular array and employing cellular automata to describe its behavior (see, e.g., Ref. 1). In a cellular array, each element is bidirectionally connected to all of its nearest neighbors. The "state" or status of each element therefore depends on the state of its neighbors. When an input stimulus is applied to the array, the states of all the affected elements fluctuate and feed back to each other until a stable array state is reached. Using cellular automata, the response of an array to a given input can be predicted. Nearest-neighbor coupling in cellular arrays has been shown to behave as an associative memory. Very recently, VLSI design tools have been used to physically implement neural-type circuits for distributed processing of information. The basic structure, the so-called Hopfield model,[2†] is based on neuronlike elements that are heavily interconnected over long distances. Obviously, these long interconnects pose serious problems for the layout on a chip. A much more appropriate basic architecture would be one that is based on interconnections only between nearest or second-nearest neighbors.

In the remaining parts of this introductory section, we will examine the driving forces that provide essential constraints on any architecture that will be utilized in VLSI to provide cooperative processing. We will see that these constraints essentially dictate the ultimate form for such processors. Then, in Section 9.2, we will examine with some care the approaches that have been pursued to achieve such distributed processing, the requirements for general-purpose computation, and where the field must go to progress. In Section 9.3, we will examine a number of the various chips that have actually been fabricated to study these approaches.

---

†While there are certainly earlier uses of this model, the prototypical simplicity and important role of the energy function were popularized by Ref. 2.

Finally, in Section 9.4, we will make a hopefully useful prediction at where the future will lie.

9.1.1 The Driving Forces for VLSI

Over the last two decades, the growth of VLSI has been phenomenal, and the application of special application-specific integrated circuits and microprocessors has blossomed. In fact, the complexity, as defined by the number of devices (or gates) on a single integrated circuit chip, has approximately doubled each year over this time span.[3] There are several factors to this increase in complexity or packing density, including major effects arising from increased die size, reduced device dimensions, and increased circuit cleverness. The latter of these is, of course, the topic of discussion in this chapter. While there have been some indications that this growth is declining, the introduction of the new VHSIC chips in 1988 will clearly indicate that this is not the case and that the rise in complexity is continuing in an unabated fashion. One major reason for this is the impact increased functionality on a single chip has on the cost per function. This increased functionality per chip in fact drives down the cost of a given system and will continue to fuel the drive to increased complexity within a single chip.

Even so, it seems likely that the reduction in growth is due to occur eventually, and perhaps more because we are approaching some limitation in fabrication technology and in our understanding of the physical limits of device performance in integrated systems. With the feature size of VLSI devices, particularly VHSIC devices, dropping below 0.5 μm, many effects become increasingly important. Such behavior as short-channel effects,[4] narrow- and inverse-narrow width effects,[5-12] hot-electron and hole generation with their subsequent substrate and minority-carrier currents,[10,11] and isolation oxide effects must be minimized. Additional properties are also required of devices incorporated into semiconductor chips with VLSI densities. All of these effects suggest that such structures, when coupled to the increased packing densities, can lead to charge injection from one device to another. This device–device interaction has been neglected in most work. As circuit densities are increased, devices will no longer be isolated. Device and circuit behavior resulting from device–device interactions include source–substrate debiasing, dynamic memory node discharging, logic circuit upset, on-chip near-device transconductance modulation, and so on. Any of these effects can seriously degrade device and circuit performance in VLSI systems. It is important to note that these effects have long-range global contributions, but also strong nearest-neighbor correlation effects.[12]

9.1.2 Rent's Rule and Interconnects

It is often suggested that large-scale integration of semiconductor devices will eventually entail a significantly large number of interconnection pin-outs at the periphery (or distributed throughout) the chip. This fact has led to a number of studies that invariably give the results in terms of a relationship that has become known as Rent's rule. Empirically, this relationship gives the number of pins, for a given size module, in terms of a power-law dependence on the number of gates (or the effective number of functional blocks) in the module. The validity of this relationship is based on a number of studies of possible interconnections of present gate arrays or of experiments actually carried out on master-slice-type chips. These latter chips are arrays of devices whose ultimate architectural interconnection is determined by the final mask. Consequently, these chips have become known as gate arrays.

The considerations that lead to ideas such as Rent's rule arise from topographical details of the implementation of a given architecture in an integrated system. In the earliest form, the problem is one of partitioning the system graph into appropriate modules, each of which would probably be a separate chip in early configurations. The partitioning problem is one of assigning the required logic blocks in such a manner that the total number of pins and interconnection wires could be simultaneously minimized. In this context, a logic block is some arbitrary (and usually not specified) primitive function. These blocks are then interconnected by the system graph, called a block graph or net. In early machines, the modules would be composed of a few blocks at most. The choice of assigning blocks and nodes into modules is the partitioning problem.

Since each possible assignment of blocks into modules specifies the necessary module–module interconnections, the total number of pins and interconnections is then dependent on the details of the partition selected. In fact, however, the beauty of Rent's rule is that there is a perhaps general rule that buries the details and concentrates on the more universal aspects. Experimentally, then, Rent's rule has been formulated as

$$P = KG^p, \tag{1}$$

where P is the number of pins, G the number of gates (or blocks in the early work), K the number of pins per block, and p a general exponent. Values of the parameters K and p have been found for early IBM machines, for RCA machines, and for others. One particularly useful study was carried out by Landman and Russo[13] and another by Chiba.[14] They considered graphs of 670–12,700 logic circuits and blocks ranging from a single NOR gate to 30 circuits. Rent's rule was also found to be obeyed, and the parameters varied between 3 and 5 for K and 0.6 and 0.7 for p.

While the requirement on pins and interconnections is significant for predicting growth in future machines, Rent's rule also is significant in that the exponent p is very important in predicting what the average interconnection length *within the chip* will be. Chips with an exponent $p > 0.5$ generally are found to have long interconnection lengths as a rule, whereas chips with an exponent $p < 0.5$ have an interconnection length (in terms of circuit pitches) that is independent of the overall number of gates. We return to this below.

It has generally been the experience in VLSI that the number of pins is much smaller than expected in Rent's rule, at least in terms of the exponents found in

gate arrays. This trend has been studied for the reported number of pins for microprocessor chips and for functional chips (other than gate arrays) of both GaAs and Si.
In addition to microprocessors, these chips include memories, multipliers, and
multiplexers. From this study, the best fit to the pin requirements of highly integrated circuits is found to be[15]

$$P = 7G^{0.21}, \tag{2}$$

which differs appreciably from the earlier forms. These circuits are called *functionally partitioned* circuits. The far fewer number of pins required for such highly
integrated circuitry suggests that future pin intensive circuits, such as supercomputers, will more than likely be found to evolve toward single-chip integration.

Estimating the average interconnection length on an integrated circuit chip has
been a very active area of investigation. It turns out that modern ULSI (ultralarge
scale integration) circuitry is very regular in its layout and usually approximates a
very square array (topographically, the rectangle is essentially similar to the square
for the following arguments). Such structures have been studied by Donath,[16]
whose results give good insight into the roles that the number of gates and Rent's
rule play in this area. Consequently, we shall follow his development quite closely.
These results indicate that, for a square array, the average interconnection length
is given by

$$\langle R \rangle = G^m, \tag{1'}$$

where $m = p - 0.5$, *if* $p > 0.5$. *On the other hand, if* $p < 0.5$, *it is found that*
$\langle R \rangle$ *is independent of the number of gates* in the circuit. This is a significant result,
because we found above that highly integrated functions on a single VLSI chip
generally have $p = 0.2$. It is therefore apparent that ULSI can continue to be developed with ever larger numbers of gates, with the pin-out requirements determined by the function of the chip and not by the increasing number of gates.

In developing a relationship between the average interconnection length $\langle R \rangle$
and the number of gates, we use the empirical form of Rent's rule given above.
The pre-factor K is, in fact, relatively independent of the hierarchical level of the
circuitry. This is evident by the independence of the empirical form on the details
of the module assignments that went into the various data points in the figures.
Consequently, we can adopt a hierarchy that is suitable for the calculations of $\langle R \rangle$
without concern for variations among implementations. The hierarchy we adopt
consists of a number of levels. At the lowest level, we assume that the node contains just four elements, each of which is subsequently connected to its nearest
neighbor. Each adjacent level consists of interconnecting four adjacent nodes to
create the equivalent node for the next level. Thus, there are 4^L gates in the circuit, where L is the number of levels required to accommodate these gates. From
this placement, we can easily find n_k, the average number of connections at the
kth level, and can deduce the average lengths r_k at this level. Then, we may sim-

ply relate these to the total average length. This can then be used to calculate the average interconnect length for $p > 0.5$ as

$$\langle R \rangle = A(p)G^{p-0.5}, \qquad p > 0.5, \tag{3}$$

where

$$A(p) = \left(\frac{14}{9}\right)\left(\frac{1 - 4^{p-1}}{4^{p-0.5} - 1}\right). \tag{4}$$

On the other hand, for small p, we find that $\langle R \rangle$ is independent of C and given by

$$\langle R \rangle = A(p)[B(p) - 1], \tag{5}$$

where

$$B(p) = \frac{4^{p-0.5} - 1}{7(4^{p-1.5} - 1)}. \tag{6}$$

For comparison, let us assume that $G = 2 \times 10^6$ gates and that $p = 0.55$ is a good estimation of the value expected for interconnection-limited circuits such as gate arrays. Then, $A(p) = 10.06$ and $\langle R \rangle = 20.8$ circuit pitches. On the other hand, for $p = 0.21$, as found for functionally integrated circuitry, we find that $\langle R \rangle = 2.95$ circuit pitches. At this level, there is almost an order of magnitude difference in the average interconnection length and consequently in the interconnection capacitance that must be driven by an individual gate. The designer is then free to concentrate on the few long interconnections (such as those running completely across the chip) in optimizing performance.

Finally, when $p = 0.5$, we find that the sums must be reconsidered carefully because of the divergences that appear in the above equations. The result still depends on the number of gates, and the leading terms are

$$\langle R \rangle = \frac{(7/9)\log(G)}{\log(4)}. \tag{7}$$

The pin-to-gate ratio can be interpreted directly in terms of surface-to-volume considerations. Consider a system of functional cells, which are laid on an d-dimensional grid ($d = 1, 2, 3$). Those cells that lie on the interior of the system are gates and those that lie on the periphery are pins (this ignores the scaling difference arising from the fact that pin pads are larger than gate areas). If those cells are all of the same characteristic length, the edge of a cube in $d = 3$, the sides of a square in $d = 2$, and a unit length in $d = 1$, we can easily establish a Rent's rule for each case. For example, in a square array ($d = 2$) of cells, with N^2 blocks, there are $4N - 4$ edge blocks, and $N^2 - 4(N - 1)$ interior blocks, and for large N,

$$P = 4G^{1/2}. \tag{8}$$

Similarly, for $d = 3$,

$$P = 6G^{2/3}, \tag{9}$$

and for $d = 1$,

$$P = 2G^0. \tag{10}$$

These considerations suggest that we interpret Rent's rule as giving us the *information flow dimension* of the integrated circuit chip. This is certainly in keeping with the concept in microprocessors, which are generally von Neumann architectures incorporating a one-dimensional information flow. Another aspect of this is that we are trying to planarize the interconnections. If we have a dimension $d > 2$, we cannot really use a planar graph and the chip will have a much larger number of wire crossings due to the long average interconnect length. As we will see below, this is clearly a detrimental factor in neural networks, where we have total interconnection of each "neuron."

9.1.3 Information and Interconnections

The purpose of the off-chip interconnections, discussed above in regard to pins, is the transmission of information to and from the interior of a chip. These connections can therefore be viewed as a surface across which information must pass in order to reach the enclosing volume of the outside world, in this case the remainder of the system. This analogy is intriguing, since many systems in biology are strongly affected by the surface-to-volume considerations, that is, by the general tendency of the surface area of an object to increase more slowly with increasing radius than does the volume of the object. We now discuss some potential limitations that arise from considering this type of surface-to-volume problem.

As above, we consider a square array of cells, where A_G is the area of an individual cell. Some fraction η of these cells perform a function per clock cycle and each of these dissipates an energy $E_m = P_d t_d$ per operation. Here, P_d is the power dissipation and t_d the delay time, which here will be set to the inverse of the clock frequency f_c as an upper limit. If there are G cells (or gates) in our area, and the heat extraction factor is Q W/unit area, we then demand that the total power dissipated obey

$$P_T = \eta E_m G f_c \leq Q G A_G, \tag{11}$$

or

$$\frac{\eta E_m f_c}{Q} \leq A_G. \tag{12}$$

It is also necessary that we transfer information over some minimum number of cells in each clock cycle. If n denotes that minimum, then

$$n(A_G)^{1/2} \leq \frac{v_s}{f_c}, \tag{13}$$

where v_s is the effective signal propagation velocity. Since A_G appears in both expressions, we can now eliminate it and obtain an upper limit on the clock frequency. This upper bound is

$$f_c \leq \left(\frac{v_s}{n}\right)^{2/3}\left(\frac{Q}{\eta E_m}\right)^{1/3}. \tag{14}$$

The rule (14) merely expresses the idea that we must be able to transmit information a minimum distance simultaneously (this imposes an upper bound on area) and to dissipate the generated heat (which imposes a lower bound on area). These two facts are currently the limiting constraints on the next generation (and even the current generation) of VLSI. To proceed, we now combine this with a simple variant of our idea of surface-to-volume ratio for information. We assume that each time a gate operates, it imposes a need for some information to pass across the surface. This is represented by b, the number of bits of I/O flow required per cell operation. When combined with the previously defined parameters, this yields a total I/O bit rate requirement of

$$B_T = \eta b G f_c. \tag{15}$$

It is interesting to note that Pf_c (P is the number of pins) is an upper limit on B_T. If we demand that B_T be less than this maximum, we can eliminate the clock frequency from (15). Applying Rent's rule (1) to (15) gives the resulting inequality

$$KG^{p-1} > \eta b. \tag{16}$$

For the range of p values of interest here (<0.5), the left-hand is a decreasing function of gate count. This shows that we must lower the net information "metabolism" on the right-hand side with increasing gate count or our system will be limited by this surface-to-volume consideration.

We can also use the above total I/O bit rate to reexpress our upper limit on clock frequency in terms of this total I/O rate. This result is

$$B_T \leq \left(\frac{v_s \eta}{n}\right)^{2/3} bG\left(\frac{Q}{E_m}\right)^{1/3}. \tag{17}$$

If we follow Donath,[16] then for systems with Rent's rule exponents greater than 0.5 we find this upper bound on total I/O rate can actually be a decreasing function of G unless b or η are increasing functions of G. For Rent's exponents less than 0.5, B_T is an increasing function G unless the product $\eta^{2/3}b$ decreases sufficiently rapidly with G.

While this discussion is very general, the important point is that the actual factor that we usually want to maximize as an IC user is not the clock frequency and the gate count (here we exclude memories), but some number such as B_T, which measures how much work the chip does for us on a per unit time basis. In this regard, we can separate our upper bound into two portions in the form

$$B_T \le F_{\text{tech}} F_{\text{arch}} G , \tag{18}$$

where

$$F_{\text{tech}} = \left(\frac{v_s^2 Q}{E_m}\right)^{1/3} \tag{19}$$

is a technology factor that may be largely independent of the actual architecture employed, and

$$F_{\text{arch}} = \eta^{2/3}b \tag{20}$$

is a factor that is purely architectural in nature and may be substantially independent of the underlying technology. Once more, this result points toward adopting an architecture that maintains operands within the confines of the chip for a longer set of processing, or adopts coherency in order to reduce the size of (20) as the number of gates is increased.

9.1.4 Reliability

The introduction of fault tolerance in VLSI and ULSI systems is a means of achieving both a specified system reliability and acceptable production yields. In memory circuits, fault tolerant techniques such as including spare rows and columns have been widely used. At wafer probe, the extra rows or columns are connected if needed as a means of increasing yields. For microprocessors and controllers, traditional approaches such as incorporating error detection and correcting codes, allowing self-test and recovery, have been adopted. Also, the use of redundant sections of the circuit, with removal of the faulty sections at circuit test, offer additional fault tolerant operation. New techniques that use the special properties of a function block, such as an ALU, to provide error detection and correction have been tried, but this approach does not easily extend to other functional blocks.[17] Even if functional level fault tolerance is achieved, the coordination of the different approaches to achieve system-level fault tolerant operation is difficult. These techniques also require additional die area and must be designed into the circuit. What is needed is an architecture incorporating fault tolerant behavior as a *natural* characteristic of the circuit. Neurally inspired architectures offer this property.

Neurally inspired architectures do not store data in fixed, specified locations, but in a distributed fashion as a small variation in the impedances of interconnections spread over a large area. The loss of an interconnection or processing element in most cases will have no effect on system behavioir. This natural robustness to individual device and interconnect failure is coupled with a graceful degradation of system performance as the number of failed devices increases. Instead of the system exhibiting catastrophic failure, one observes reduced associations and accuracy. This natural robustness offers potentially high chip yields with low reliable processing elements without the need for error detection and correcting codes and spare sections of the circuit.

9.1.5   Coherency in VLSI

It is now clear that what we really need, if we are to make a quantum jump forward in the capabilities of microchips, is to develop a coherent, parallel type of processing that provides robustness and is not sensitive to the failure of a few individual gates. The problem of using arrays of devices, highly integrated within a chip and coupled to each other, is not one of making the arrays, but is one of introducing the hierarchical control structure necessary to implement fully the various system or computer algorithms necessary. *In other words, how are the interactions between the devices orchestrated so as to map a desired architecture onto the array itself?* We have suggested in the past that these arrays could be considered as local cellular automata,[18] but this does not alleviate the problem of global control which must change the local computational rules in order to implement a general algorithm. Huberman[19,20] has studied the nature of attractors on finite sets in the context of iterative arrays and has shown in a simple example how several inputs can be mapped into the same output. The ability to change the function during processing has allowed him to demonstrate adaptive behavior in which dynamical associations are made between different inputs, which initially produced sharply distinct outputs. However, these remain only the initial small steps toward the required design theory to map algorithms into network architecture. Hopfield and co-workers,[2,21] in turn, have suggested using a quadratic cost function, which in truth is just the potential energy surface commonly used for Liaponuv stability trials, to formulate a design interconnection for an array of neuronlike switching elements. This approach puts the entire foundation of the processing into the interconnections.

In the following, we discuss the history of these various approaches to developing a design procedure for these networks. It is our general belief that wholly interconnected arrays are not a viable approach for VLSI circuits and that techniques that allow primarily local interconnections are to be sought. This belief arises from previous studies of the interconnection and information flow in VLSI circuits. Finally, we shall describe how a number of these ideas have been implemented in actual VLSI chips in order to test the theories.

9.2 MASSIVELY INTERCONNECTED SYSTEMS

9.2.1 Cellular Arrays

Cellular logic (CL), also called neighborhood logic, is a discipline in the field of computational geometry and is implemented with cellular arrays (CAs). One-dimensional CL is the generation of shift register sequences, two-dimensional CL is the manipulation of two-dimensional topology or images, and three-dimensional CL is concerned with patterns of growth.

CL are digital operations on an array of data such that the array of data is transformed into a new array. Each new state of the element in the transformed array is determined only by the original state of the element and the state of its nearest neighbors. The architecture of a CA is the replication of identical processing elements (PEs), with each PE connected to its nearest neighbors. The nearest-neighbor interconnection pattern is called a cell or tessellation. Although this might seem to be a severe restriction, these arrays are capable of powerful computations. Examples of their applications include parallel image processing and pattern recognition.

The birth of CA actually resulted from von Neumann's interest in neural networks. He was interested in the complexity required for a system to be self-reproducing. Additionally, he was interested in using unreliable components to make highly reliable systems. He intended to use continuous analog elements, but the difficulty of formulating the rules and instructions caused him to redirect his effort toward arrays of digital elements. The array approach interested him since it offered the possibility of parallel operation, such as that in the nervous system. The tessellation formed from connecting the four nearest neighbors (up, down, left, and right) is called the von Neumann interconnection pattern.

CA can be defined by a triple (V, v_0, F), assuming each cell in the array has the same neighborhood. Here, V represents the internal states of the automaton, v_0 the state of a specific element in V, and F a thresholding or transition function.[22] The thresholding function maps V^t into V^{t+1}, where t is the discrete time step. It is not easy to choose a formal description of CA that permits a general description of the automata and still clearly identifies the function that can be implemented with proper programming. Describing the CA as a Turing machine is a popular option. Considering a two-dimensional CA, the internal state of a PE can be represented as a matrix in which each component S^t_{xy} where x and y refer to the Cartesian coordinates of the PE in the array, corresponds to a particular element of the array. Then, S^t_{xy} depends on the values of five independent variables and may be expressed as

$$S^t_{xy} = \{(\alpha, \beta, \gamma)_{xy}, w, z\}^t. \tag{21}$$

The variable α is the binary value of the external input to the PE, β is the next state of the external input and not necessarily present in every computation, γ is the output values of the neighboring PEs and w and z can be considered as instructions. The CA changes the internal state of all the PEs synchronously and in a deterministic manner. The output of each PE is obtained after the new states are reached and is described as

$$0_{xy}^{t+1} = (\alpha, \gamma)_{xy}. \tag{22}$$

Therefore, the operation of a CA is described in terms of a sequence of states,

$$S_{xy} = \{S_{xy}^0, S_{xy}^1, S_{xy}^2, \ldots, S_{xy}^t\}, \tag{23}$$

for any PE having five inputs and a binary output. Each step in the sequence can also be identified with a state transition matrix which maps the "state," defined by the values of the elements, into the next state. In each step, the values of the various variables in (21) define the state transition matrix.

The thresholding function F produces the logical functions by thresholding the signals calculated by the PEs. If $G(x, y)$ is a two-dimensional signal calculated by the PE, $G_L(x, y)$ is a logic function obtained by

$$G_L(x, y) = \begin{cases} 1, & G(x, y) \geq F, \\ 0, & \text{otherwise.} \end{cases} \tag{24}$$

Frequently, multithreshold values are used to generate useful behaviors. Most of the CA architectures use N^4 (von Neumann neighborhood), Fig. 9.1, or N^8 (Moore neighborhood), Fig. 9.2. An extremely important point is that any CA with an arbitrary neighborhood can be shown to be computationally equivalent to an N^4 or N^8 CA network.[23] Using VLSI and ULSI technology, large CA arrays with locally connected PEs offer great potential for computationally powerful parallel processing chips.

The first large-scale CA machine, called the SOLOMON, was constructed in 1960. SOLOMON was a 16×16 computer array built at Westinghouse. Next, a pattern recognition computer, called the ILLIAC, was designed. The ILLIAC was to be a 32×32 array and the first machine capable of processing pictorial information. Its construction was never completed. As the cost per transistor in ICs started to be reduced dramatically in the late 1960s, University College, London, started construction of a family of processors called cellular logic image processors (CLIPs). Since then, a number of CA machines have been constructed, including, for example, the 128×128 massively parallel processors by Goodyear Aerospace. Applications of CA machines are imaging of X-rays, blood-cell recognition, analyzing aerial and bubble chamber photographs, synthetic aperture radar imaging, and for floating point computations.[24] As we shall see below, these machines are effectively isomorphic with the general class of neural networks.

9.2.2 Synthetic Neural Systems

Synthetic neural systems (SNSs), also called artificial neural networks, connectionist networks, or parallel distributed processing networks, constitute the area of

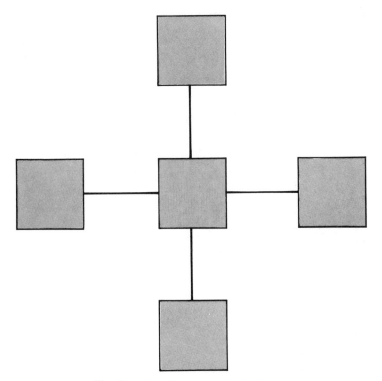

Fig. 9.1   Von Neumann neighborhood.

study that is concerned with the synthesis, design, fabrication, training, and analysis of neuromorphic (i.e., brain-inspired) electronic systems. These systems achieve high performance by having dense interconnections of simple switching elements processing information in parallel. An SNS processes information by its state response to an input (relaxation)[2] or, under certain restricted conditions, by computation.[25] Neuromorphic systems offer the potential of processing information in a manner similar to the methodology developed by nature.

Modern digital computers are excellent at storing and retrieving bits of data in precise locations and at performing complex calculations involving thousands of numbers and millions of operations. These are skills a human brain cannot hope to match. However, as powerful as today's computers are, they are very limited in their abilities. For example, computers are not good at extracting information from the data they store. Hence, they are not good at generalization from information stored in memory, self-organization, autonomous acquisition of knowledge, and learning associations. There are skills a human baby acquires within a few weeks of birth, such as being able to recognize its mother, that require pattern recognition skills far beyond the current capabilities of computers. While no definitive proof exists ruling out the possibility of computers performing such a task or ex-

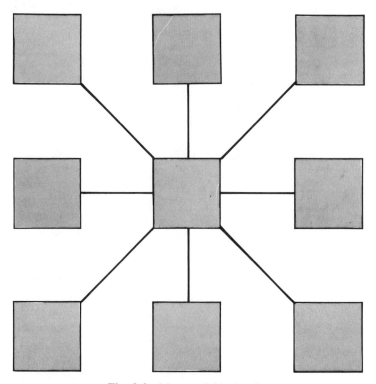

Fig. 9.2   Moore neighborhood.

hibiting other higher cognitive functions such as consciousness,[26] no computer so far has been able to demonstrate such behavior.

SNS is a rapidly growing field primarily because of the promise of solutions to problems that have continued to confound computer science and artificial intelligence. In the next few years, it is hoped that neuromorphic systems can provide solutions to problems requiring parallel searches though spatial and spatial-temporal information, allow implementation of self-organized associative memories, and permit systems to collect knowledge autonomously from observations of its environment.[27] Applications of these systems to problems of sensor processing, knowledge processing, natural language, vision, and real-time control offer new markets for VLSI and ULSI circuits.

The relationship between artificial intelligence (AI), neuroscience, and SNSs needs to be explicitly stated. The goals of both SNSs and AI are the development and applications of systems that can perform the kinds of information processing that humans perform. Although some special-purpose architectures for AI have been developed, AI is mostly the development of heuristic programs to solve problems of a similar nature to the type solved by SNSs. Whereas AI is not overtly concerned with the architecture that runs the software, the architecture is considered an integral part of an SNS. AI uses the didactic method of learning (rule based), whereas an SNS uses the Socratic method (by example). AI has made progress on simple problems in limited environments, but the extension of these programs to more complex problems in more complex environments has in most cases proven to be impractical.[28] Having to define rules explicitly becomes intractable as the problem and the environment grow. The SNS continues to grow naturally with the complexity of the problem by just requiring more processing elements and training sets. Explicit rules do not have to be stated. For example, SNSs have been applied to the question of whether a particular sequence of protein bases within a fragment of DNA is a coding for the production of a protein chain. Traditional approaches require at least 200 base pairs to predict the biological activity of a DNA strand. SNSs, after training, showed remarkable accuracy with only 30 base pairs. The SNS appears to have learned some fundamental rules about genetics that have eluded the biologist.[29]

The neuroscience community is still very unclear on the exact nature used by the brain to process information. Researchers in neural modeling, psychology, cognitive science, and computer science have used models to simulate and study how the brain processes information. Many of these models are closely guided by experimental results and in some cases have offered predictions of human behavior. In all cases, the models are constrained by neural biology. For example, the so-called 100-step rule[30] requires models to have no more than 100 processing levels. This rule exists because experiments on humans indicate that information processing can occur as fast as a few tenths of a second. The delay through a neuron is a few milliseconds, hence no more than a few hundred levels of neurons could have been used. However, many thought processes take considerably longer than a few tenths of seconds. It is thus conceivable that the 100-step rule above should be reinterpreted as the smallest, rather than the largest, number of levels, which points out the "looseness" of such arguments. In any case, SNS is *not constrained* by neural biology, but by VLSI technology. While SNS models can be used to study many of the functions of the brain, the SNS model is self-consistent and does not require the correctness of neural theory to validate its operation. It is only inspired by the tremendous potential, highly parallel operation, and fault tolerant nature of the brain, and is not constrained by the exact details. A brief discussion of biological neural networks will be presented here just to describe the terminology and illustrate the source of the inspiration for the SNS architecture.

The signal-processing element in the brain is the nerve cell or neuron. There are hundreds of different types of neurons, but since most have certain signal-processing characteristics in common, we will represent these properties with one ideal neural model. The internal state of the neuron is characterized by an electrical potential difference across the cell membrane. This voltage is called the membrane potential or activity of the cell. External inputs produce deviations in this voltage from a resting voltage of approximately -70 mV. When the activity level exceeds a certain threshold voltage, a pulse or action potential is transmitted along the output line, called an axon. Each pulse is a large depolarizing signal with a peak-to-peak amplitude of 90 mV and a pulse width of 1–10 ms. The pulse travels along the axon with a constant amplitude and velocity, typically between 10 and

100 m/s, and terminates at the synaptic knob. The neuron emits trains of pulses with a frequency in the range of 2–400 Hz. Single pulses also can be generated, but are random and are not believed to carry information. It is believed that the information resides in the pulse train frequency. Hence, this frequency will be represented by a positive real number. The axons terminate on a very important part of the system, the synapse. The synapse connects to the inputs of the neuron, the dendrites. Figure 9.3 illustrates a neuron with dendrites (inputs), synapses, and an axon (output). The synapse is an extremely complex structure and has critical effects on the behavior of the system. When an incoming signal traveling along the axon reaches the synapse, an electrical to chemical conversion takes place. The input signal causes the release of a substance called a neurotransmitter from small storage vesicles. The released neurotransmitter diffuses across the synaptic cleft to the post-synaptic region, where it alters the potential and causes a signal to be transmitted down the dendrite to the neuron.

The neuron will be assumed to perform a temporal summation of the dendrite signals. Changes in the dendrite voltage are additive and depend on the pulse frequency of the incoming pulse train. Synapses are either excitatory or inhibitory. A synapse is excitatory if it increases the voltage on the dendrite and inhibitory if it decreases this voltage. The neuron then sums all the individual inputs, which in some cases may be over 10^5 but averages 10^4, and causes the neuron to fire if the cell threshold voltage is exceeded. The human nervous system consists of over 10^{11} neurons and more than 10^{15} synapses. This is far more information than can be genetically encoded. Therefore, and more importantly, most synapses are not pre-programmed, but adapt to their environment. Finally, although the neurons are highly interconnected in local regions, full global interconnectivity does not occur. In summary, this has been a simple description of a very complex system, and more extensive details can be found in Ref. 31. In an integrated circuit analogy of the nervous system, the synapses are considered as circuits that are connected in complex ways and whose outputs are integrated by the neuron. Hence, each set of dendrites, synapses, neuron, and axon can be considered a complete processing chip.

To model the nervous system, we let the ith neuron be represented by the node v_i, connected to the jth neuron at node v_j as shown in Fig. 9.4. The path between nodes v_i and v_j is denoted by e_{ij} and has a transmittance value z_{ij}. A signal S_{ij} prop-

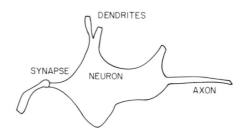

Fig. 9.3   A neuron with dendrites, synapse, and axon.

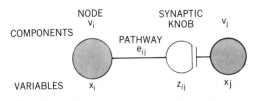

COMPONENTS

NODE v_i SYNAPTIC KNOB v_j

PATHWAY e_{ij}

VARIABLES x_i z_{ij} x_j

Fig. 9.4 Neuron components and variables.

agates along e_{ij}. S_{ij} is a function of the activity level x_i of the node v_i. The magnitude of the signal that reaches the synapse is

$$S_{ij}(t) = f[x_i(t - \tau_{ij})]b_{ij} , \qquad (25)$$

where τ_{ij} is a delay constant and b_{ij} a path strength constant.[32] In general, $f(x_i)$ is any monotonically increasing threshold function, such as

$$f(x_i) = (x_i - \Gamma_i)U_0(x_i - \Gamma_i) , \qquad (26)$$

where Γ_i is an internal noise parameter and U_0 the Heavyside step function defined by $U_0(x) = 1$ for $x \geq 0$ and $U_0(x) = 0$ for $x \leq 0$. Thus, the node v_i emits a signal only when the activity level at the cell exceeds the threshold Γ_i. Notice that we have transformed the burst of pulses into a voltage level which represents the average firing frequency of the neuron. For a network with n nodes, the time evolution of the activity level of node i can be modeled by

$$\frac{\partial x_i}{\partial t} = -A_i x_i + \sum_{k=1}^{n} S_{ki} z_{ki} + I_i(t) . \qquad (27)$$

A_i can be interpreted as a short-term memory decay rate and $I_i(t)$ is an input from either sensory inputs or networks outside the local network. The transmittance term z_{ij}, which can be interpreted as storing long-term memory, changes on a slower time scale and is modeled by

$$\frac{\partial z_{ij}}{\partial t} = -B_{ij} z_{ij} + S'_{ij} f(x_j) , \qquad (28)$$

where $f(x)$ is the function defined in (26), B_{ij} can be interpreted as a forgetting term or long-term memory decay rate, S'_{ij} is the presynaptic learning signal, and x_j is the post-synaptic activity of node v_j. Notice that to change z_{ij}, both S'_{ij} and x_j must be positive simultaneously. Most studies of the use of neural models in systems actually use much simpler forms of these equations, often even ignoring the learning equation (28).

9.2.3   A Computational Model

We approach the general idea that neural networks and cellular automata are iso-morphic by discussing general computation itself. Consider an array of N logic elements, which are supposed to represent all of the central processor and that part of memory so designated (discussed below, as we show that these must be sepa-rated for computation). We can describe the instantaneous state of the machine by one of two possible descriptors: (1) a vector of length N whose elements are zero or one, corresponding to the state of the appropriate logic element; or (2) a vector of length 2^N whose elements are all zero except for a single element whose value is one. This latter description is the one we shall choose, and the single entry iden-tifies which of the possible 2^N combinations is the proper state of the system. This is analogous to the pure state description of quantum mechanics in which the den-sity matrix has a single entry corresponding to the pure state. We call this state vector X, and its value at time step n of the system's evolution is X_n. As the sys-tem is clocked, X evolves according to the iteration

$$X_{n+1} = MX_n, \tag{29}$$

where M is the state transition matrix.

The flow of the computation through the linear vector space defined by the al-lowable states of the system is determined by the set of matrices $\{M_i\}$ that are pos-sible choices for (29). To clarify the nature of these matrices and the requirements on them, let us first consider the case of reversible logic machines. In reversible logic, every state has a single antecedent and a single successor, so that the map-pings represented by M are $1:1$ mappings. Each matrix M_i is a representation of one of the possible invertible operators that define these mappings. Since the oper-ators are invertible and since each state has a unique predecessor and successor, each row and column of M must have a single 1, with all other entries being 0. Thus, there are 2^N 1's and $2^{2N} - 2^N$ 0's for the elements of M, with the 1's distrib-uted so that every row and column has only a single 1. This requirement on the rows arises from each state having a unique successor and similarly for the columns (predecessor). The requirement that *every* row and column have at least one 1 arises from the fact that M is invertible. At this point, we recognize that the set $\{M\}$ is a set of representations for the operators of a subgroup (or the full group) of the permutation group of order 2^N.

Now, one option is to have a stored-program machine in which we consider all of the program store as part of the set of states of the machine. Then, the entire computational system is closed, so that there is no interaction from an external source. Once the program is stored and the system is closed, a single state transi-tion matrix M_{sp} has in effect been selected, and the sequence of progress through the states is set. *However, this system is incapable of "computation."* Indeed, the system is an oscillator and does not fit the basic requirements of a computer. Be-cause of the requirements on M_{sp} discussed above, the set of states that will be vis-ited during execution of the stored program will form a complete ring, which is the logical end of the permutation (also called cyclical) group. Consequently, there is no unique "start state (other than the one in which the system was initially placed) and there is no "stop" state. While this argument is based on intuition, it has its basis in the description of Turing machines. Although no superior definition has been developed for the computer, the concept of the Turing machine is the foundation for the formal theories of computability and computational complexity. Turing's initial discussions determined if a number was computable according to whether or not his machine stopped.[33] Since our construction does not stop, it obviously cannot compute a number. We can draw the important corollary here that a viable Turing machine cannot be a closed system. Even though we based the arguments on invertible logic, the removal of this restriction does not invalidate the argument. We cannot relax the requirement of a single entry per column, since each state must have a unique successor state. On the other hand, several states (or none) may map into one particular state, so that we cannot say anything about the number of entries in any row. Rather, the eigenvalues of M will relate to the convergence properties of the trajectories (in logical phase space) for each M. We can have stable fixed points, loops, or unstable fixed points. In each case of a chosen M, we can determine the nature of these eigenvalues and ascertain whether the process will terminate, thus once again violating the Turing machine concept. Instead, the system must also include an outside measuring body, which determines the current state of the system (Turing's read head) and then modifies the state transition matrix accordingly.

In general, we can then consider the set of states composing X to be just the central processing unit. We can still utilize the reversible logic (as we can also utilize irreversible logic), but there is now an interaction from an outside body—the stored program in memory. The computation is an open system, and we may invoke an entire set $\{M\}$ of allowed state transition matrices. At each stage of the computation, the state of the system is measured and a next M_n is selected according to the results of the measurement (of the current state) and the stored program. This process effectively breaks time-reversal symmetry and introduces a preferred direction (in time) for the system evolution. The only difference now between reversible logic and irreversible logic is that the set $\{M\}$ represents the permutation group or the permutation semigroup, respectively.

The formulation of (29) is not completely abstract, however. This formulation is exactly that described by Peretto and Niez[34] in their investigation of stochastic dynamics of neural networks. In particular, we want to relax our requirement that X have only a single entry, but describe it as a mixed state representation in which each entry is the probability of that state existing. We then immediately recognize that the elements of M are transition probabilities[19,20,34] and we can then use M for discussing the statistical mechanics of the temporal evolution of neural networks. More importantly, each element of M is the transition probability between state i and state j that results from the switching of a single logic element. Thus, the structure and entries of M allow us to formulate (in principle) a network architecture that achieves the desired M. Research is still needed to address this task, that is, it is still necessary to develop a methodology of moving from the specified state transition matrices to a network description that is useful for VLSI.

We can go further in our description, however. It is also important to note that the form of the entries is

$$M_{ji} = W(j|i) = \frac{1}{2N}[1 - f(x)] \tag{30}$$

for the networks of interest. For neural nets, $f(x)$ is usually taken as a simple sigmoidal function, such as in (26) above, which has a smooth, monotonic transition between levels. In general, functions such as arctan(x) and $U_0(x)$ (the Heavyside unit step function) are functions of this class.

9.2.4 Equivalence of the Models

Here, we have found from the above discussions that both cellular array networks and neural systems can be described similarly. The simplified model that is customarily used is

$$x_j = \sum_i C_{ji} y_i - \Theta_j, \tag{31}$$

where y_i is the value of the state i (presumably zero or one, but it could also be a continuous variable in analog systems), C_{ji} the interconnection strength from the output of state i to the input of state j, and Θ_j a threshold value for the state j. The key factor in (31) is that the entry in M can be changed either by changing the interconnection weights C_{ji} or by changing the threshold Θ_j In cellular array networks, $f(x)$ is not a sigmoidal function but a multivalued function that represents a truth-table mapping between input and output, that is, a mod$_2$-type function. *Thus, the primary difference between neural networks and cellular arrays lies in the neighborhood chosen for the interconnection sum in (31) and in the function f(x).* In this regard, the two types of networks are formally equivalent and both are representable by the iteration

$$y_j^{t+1} = f\left[\sum_R C_{ji} y_i - \Theta_j + I_j\right], \tag{32}$$

where R is a neighborhood, and we have added a bias I_j which can represent a learning function. If R is complete and f sigmoidal, we have a neural network. On the other hand, if R is local and f a digital truth table, we have a cellular automata network.

It is important to note that these two paradigms for distributed networks are approaching each other in topology. The Hopfield network is actually a quite primitive form of the neural network. Several more complicated types are described by Hinton and Anderson.[35] Of particular interest are the layered networks, in which information propagates from an input layer, through several buried layers, to an output layer. Learning can be incorporated readily, and one popular approach is the backward propagation algorithm. The layered networks are now formally similar to the iterated array form of cellular automata, in which information propagates from one line of nodes to the next line of nodes. As in the layered neural network, each node is connected to a group of elements of the previous line. In this regard, the two paradigms are rapidly approaching a common architecture. In particular, the iterated arrays of cellular automata have been used for systolic arrays and for signal-processing functions as well.[36] The latter have been designed as programmable, but we are still limited in application of learning techniques in the cellular networks.

9.3   CHIP ARCHITECTURES

Hopfield[2] is generally credited with clearly articulating the fascinating possibilities of neuromorphic architectures to VLSI designers. While many of his ideas and approaches have been previously reported in the literature,[37] Hopfield presented this information in the terminology and in the journals read by designers (see, e.g., Ref. 38. Hopfield also inspired the first chip implementations of these architectures. He promoted the view that highly interconnected networks of nonlinear analog switching elements fabricated on a die could be extremely efficient in generating a good, but not necessary optimal, solution to computationally complex problems.

Equation (27) can be rewritten as

$$\frac{\partial x_i}{\partial t} = -\frac{x_i}{R_i} + \sum_{j=1}^{n} T_{ij} V_j + I_i(t), \tag{33}$$

with $V_j = f[x_i]$. This equation is immediately recognizable as the equation of motion of the analog circuit shown in Fig. 9.5. This circuit sums the inputs V_j, weighted by the conductance of the interconnections T_{ij} and any additional inputs $I_i(t)$, and convolves this input with a nonlinear transfer function to generate an output. The neuromorphic architecture then is just a collection of operational amplifiers (op amps) fully interconnected with resistors. Given an initial condition, the circuit will relax to an output state that satisfies the system constraints. To obtain interesting behaviors from the system, the T_{ij} needs to be able to have both signs. This is easily accomplished by using the noninverting and inverting outputs of the operational amplifiers used to perform the input summations. The inverting output will produce a negative T_{ij}.

The Liapunov or energy function for this system is

$$E = -\frac{1}{2} \sum_{i}^{n} \sum_{j}^{n} T_{ij} V_i V_j + \sum_{i}^{n} \frac{1}{R_i} \int_{0}^{V_i} f^{-1}(V) \, dV + \sum_{i}^{n} I_i V_i. \tag{34}$$

For a symmetric interconnection matrix, that is, $T_{ij} = T_{ji}$, and for a monotonically increasing $f_i^{-1}[V_i]$, it has been shown that the solution for the system is a minimum of E.[39] During relaxation, the system evolves in the interior of a n-dimensional

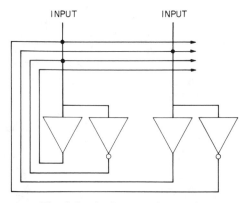

Fig. 9.5 Analog neural network.

hypercube with the corners at $V_i = 0$ or 1. For the high-gain case, the stable states of the system and hence the minimum of the energy function will be at the 2^n corners of the hypercube. While an ideal digital switch could be used in the simulation of this system, relaxing to the minimum in the interior of the hypercube, instead of along the edges, helps the system find a more global, stable solution.[39] This behavior will exist naturally when the circuit is fabricated due to nonideal switching devices. Networks connected in this fashion can be used to relax to the solution of optimization problems.

The classical problem solved by these types of networks is the retrieval of a complete vector given part of the vector, hence an associative or content addressable memory. The initial condition represents part of the complete set of information to be retrieved, and the system will relax to the closest complete set. Figure 9.6 illustrates this behavior. For the associative memory with $T_{ii} = 0$ and $I_i = 0$, the T_{ij}'s can be calculated as

$$T_{ij} = \sum_{s=1}^{m} V_i^s V_j^s, \tag{35}$$

where m is equal to the number of memories to be stored. Other problems that have been implemented with this type of network are the traveling salesman problem[21] and an analog-to-digital converter.[38] For each problem, the formation of the Liapunov function must be determined before the T_{ij} can be calculated.

Inspired by Hopfield's work,[40] a team at Bell Laboratories[41] has designed, fabricated, and tested SNSs that are composed of 22 and 256 neurons. Both designs used resistors for the interconnection elements which could not be changed once they were fabricated. Both inverting and noninverting outputs were available to allow positive and negative weights. The 256 neuron chip was designed with 2.5 μm CMOS device design rules and contained 512 amplifiers placed around the periphery of the chip, as shown in Fig. 9.7. The center of the chip is reserved for the interconnection matrix and it clearly dominates the chip area. Since the 256 I/O

TRANSIENT SOLUTION

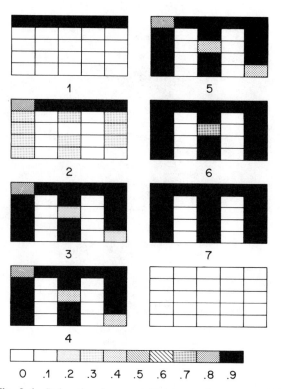

Fig. 9.6 Relaxation from an initial input of — to stored *T*.

lines needed for this design would be impossible with current available packages, the I/O was multiplexed to a 16 bit wide word that was stored in a buffer.

The resistors were not put on during the initial fabrication. Openings in the interconnection matrix were left to provide access to connect a resistor from the silicide line to aluminum line. The resistors were added by covering the die with amorphous silicon and then patterning by electron-beam lithography and reactive ion etching. All the resistors were designed to have the same value. The completed chip had over 25,000 transistors and over 130,000 sites for resistors. The die area was 5.7 mm². This chip is an ROM in the sense that once the resistors are fabricated, they cannot be changed. This means that after the initial programming, no additional information can be stored.

For a system to be able to respond to a changing environment, the interconnection values must change with time. It is believed the human nervous system behaves in this manner. The Bell Laboratories group[42], Mead et al.,[43] and Alspector and Allen[44] have designed chips with programmable interconnection values. As discussed in Section 9.2.2, SNSs are limited by VLSI constraints, not neural constraints. A principal VLSI constraint is cost. The cost of a chip is directly related

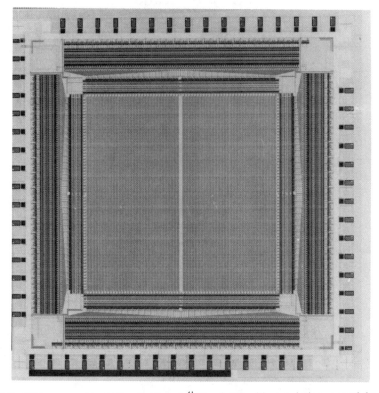

Fig. 9.7   256 neuron chip (After Graf et al.[41] Reprinted with permission, copyright 1986, American Institute of Physics.)

to its die area. By exploiting the natural functions available with analog circuits, such as summation, less die area is consumed than with a digital implementation of the same function. Of course, analog VLSI designs have additional problems not found in digital systems. For example, power dissipation can be a serious problem for large chips. Also, whereas digital circuits are relatively insensitive to varying device parameters, in most cases analog circuits are very sensitive. Analog neural-like circuits must be designed to rely not on the absolute behavior of each individual transistor, but on the cooperative behavior of large numbers of devices. Mead reduced the power consumption problem by operating MOSFETs in the subthreshold mode. To obtain positive and negative transmittance values, as well as symmetrical large signal behavior, NMOS amplifiers with differential input stages and complementary outputs were used. The transmittance elements were designed with four pass transistors operating in the ohmic regime and are capable of three interconnection strengths: -1, 0, and 1. They also designed the transmittance element for dynamic programmability. Figure 9.8 is a diagram of the T_{ij} elements. The chip has 22 active element and 462 elements in the interconnection matrix. The chip was fabricated using 4 μm NMOS technology and measured 6300 $\times$

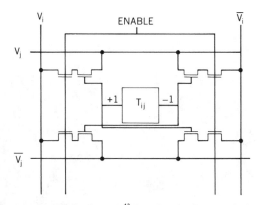

Fig. 9.8 T_{ij} element. (After Sivilotti et al.[43] Reprinted with permission, copyright 1986, American Institute of Physics.)

5700 μm. This chip also explicitly displayed the robustness of these architectures. One die had over 40% of the T_{ij} elements unprogrammable, and the chip could still be trained to recall two memory vectors. Jackel[42] fabricated a 54 neuron chip with approximately 3,000 programmable synapses. This chip is currently being used for handwritten character recognition. Figure 9.9 is a microphotograph of this chip.

In still another investigation of the usefulness of neural networks in performing computation, a digital neuron capable of varying its interconnection neighborhood and interconnection weights has been designed by the authors and fabricated by MOSIS.[45] The custom-designed VLSI chip is among the first implementations of the digital neural networks based on Hopfield's model to have a programmable interconnect pattern. The digital neural network consists of 12 neurons in a systolic array architecture. Each neuron performs the evaluation function as in the Hopfield model, with a programmable threshold. All weights are limited to the range ± 1, with a resolution varying between four and eight bits of representation. Each processing element contains four sections: the router, the memory, the accumulator, and the control unit. Figure 9.10 is a block of one cell; Fig. 9.11 is a microphotograph of the chip. All actions are synchronized by an external clock and control signals. The router is responsible for routing each neuron's output state to its neighbors. A neuron has four connections to each of its four physical neighbors. The router is composed of flip–flops which implement a two-dimensional shift matrix. The input to the flip–flops is controlled by a four-input MUX. This allows each cell to direct its own output to any one of its four nearest neighbors. The selection of the neighbor receiving the signal is controlled by direction control signals. All flip–flops of the shift matrix are controlled by the same signals. On each cycle of the shift clock, information moves in any one of four directions. This signal-routing scheme is similar to that found in Hillis' connection machine. Each cell is designed to be autonomous except for the external clock and control signals. Many cells can be connected together to form a matrix of neurons. The edges of the matrix may be connected to latches that can be read or written by a

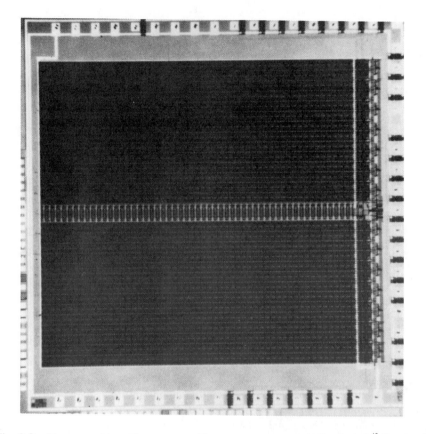

Fig. 9.9   54-neuron chip with programmable synapses memory. (After Jackel.[42] Reprinted with permission.)

microprocessor. This test vehicle is being used to study the effects on hardware of interconnectivity on the transistion from local to distributed data storage, digital learning algorithms, and on system robustness to individual device failure.

SNS chips are direct attacks on the basic problem of ULSI, which, simply put, is what to do with a billion devices. There are three key features that underlie much of the potential success of the SNS. These keys are the use of large numbers of PEs, some form of concurrent processing, and some form of adaptation. It is important to remember, though, that neural networks are not the only systems possible that implement these three key features. For example, reconfigurable systolic arrays generally have these same features. Other possible architectures can be envisioned as well. We next will review one such architecture, a pipelined content addressable memory, and will use a comparison between it and neural networks as a guide in answering the following question: What are the advantages and disadvantages of the SNS when compared with alternative architectures?

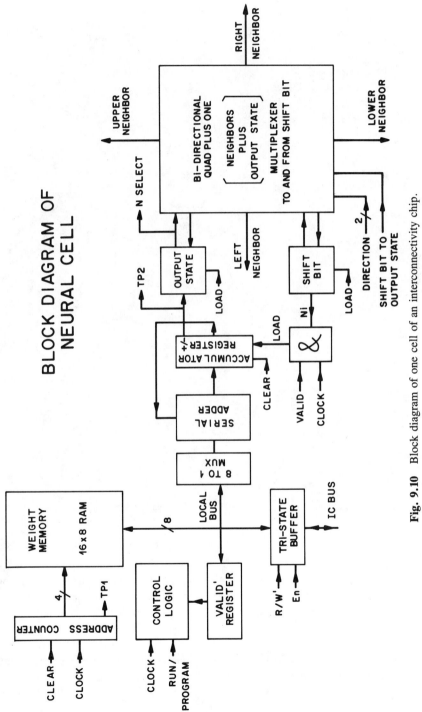

Fig. 9.10 Block diagram of one cell of an interconnectivity chip.

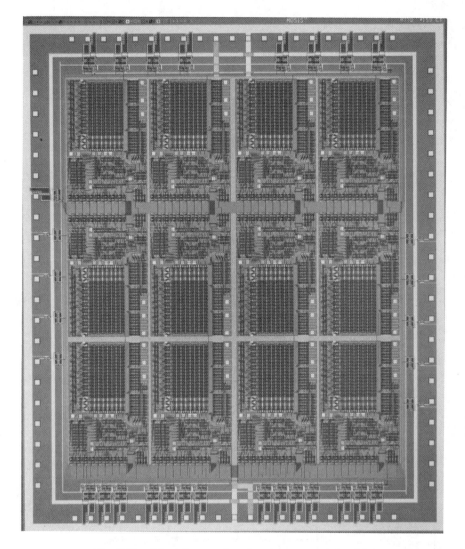

Fig. 9.11   Microphotography of an interconnectivity chip.

The basic structure of this architecture is a pipeline, each stage of which contains a word of memory and some comparison logic. Data flows into the top of the pipeline; at the bottom, we recover this data along with the address of the pipeline stage, which contained in its memory the closest match to the input and a measure of the closeness of this match. Obviously, the comparison logic of each stage compares the input data with the stored data, determines the closeness of this match, and compares this closeness with that of the previous best match. The address and closeness of the best of these two are then fed along with the input data to the next stage of the pipeline. A single pipeline has been fabricated and has passed some crude functional testing.[46]

At this point, we can already begin to compare this scheme with an SNS. First, an SNS usually destroys any input data unless special care is taken, whereas here this data is naturally recovered. Secondly, in an SNS functioning as an associative memory, no measure of the closeness or goodness of the match between the input "key" and the remembered data is provided, whereas here this closeness is supplied. In an SNS, interference can occur between two memories and it is often true that when the system learns something new that previous knowledge is degraded somewhat. No such interference or degradation occurs in the pipelined CAM. This interference between individual memories makes it difficult to predict the actual capacity of an SNS; for the pipelined CAM, however, the capacity is easily determined from knowledge of the number of pipelines, the length of these pipelines, and the word length. While the SNS is resistant to noise degradation of the input data, so is the CAM and largely for the same reason. Since the output is chosen by a "best-match" strategy, noise degradation of the input introduces errors in the CAM only when it has the effect of corrupting the input in a fashion that causes it to appear falsely to be a different portion of the learned corpus. Such effectively false inputs would fool the SNS as well. For that matter, it is difficult to envision any memory that is insensitive to having the input effectively constitute a lie. While an SNS is insensitive to various "soft" errors in the storage process, the error resistance of the CAM in this regard could be improved quite easily by using the error-correcting codes commonly seen in modern memory systems.[47] At this point, it would appear that the SNS is an inferior architecture, but we will now consider an example that shows that although an SNS may in fact be an inferior associative memory there are other applications in which they should outperform systems built around these pipelined CAMs.

One of the more notable successes of neural networks is Sejnowski's NETtalk.[48] This is a software simulation in which a neural net was taught to transcribe English text phonetically. The same task can also be performed by a pipelined CAM.[46] Indeed, a pipelined CAM should be quite effective in this task and can rapidly learn to reproduce a large corpus with 100% accuracy. Here, however, we encounter a feature of an SNS that is very difficult to achieve in another fashion. The trade-off in performance between the two roughly can be described as giving the learning speed advantage to the pipelined CAM while giving an advantage to the neural net in terms of its ability to respond to novel input, that is, words that were not included in the learning corpus. The neural net learns slowly because of the above-mentioned interference between individual pieces of knowledge. However, this interference can be constructive and allow the net to identify common features found in the input without having their existence be explicitly described by an external mentor. Therefore, it tends to form a more generalized response that allows it to respond properly to an input that is novel. For example, after a large number of words have been taught to the neural net, it will tend to transcribe the word "hurling" properly even if it has never seen this word and was only taught very similarly spelled words such as "hurting" as part of the learning corpus. The pipelined CAM, however, under the same circumstances would transcribe the word "hurling" as "hurting" (assuming that this is the closest match from the learning corpus) and therefore make an error. This ability of a SNS to handle novel inputs in an appropriate fashion is not to be underrated and in fact may be its most important attribute, since it is the most difficult one to emulate with other approaches. The central issues are whether or not it is a direct result of the interference between the various items learned and, if so, how to alleviate the difficulties posed by this interference without throwing away this advantage as well.

Another approach that has received some interest also has many of the same virtues as an SNS. This is to utilize fuzzy logic. A structured approach to the design of learning automata that use a fuzzy "neuron" as the PE has recently been proposed. It appears that these can be implemented by VLSI circuits in a regular PLA-like fashion. One goal is to use such elements in the construction of a small chunked neural net in which the interconnection problem is attacked by decomposing a large neural net into several small modules, each of which has a limited internnection space. The problem of constructing an acceptor in particular has been addressed, but the techniques can be applied to a far wider range of automata.[49] These networks are powerful in computing the acceptance of fuzzy automatas and thus well suited to tasks such as pattern or language recognition.

An acceptor is an automaton that decides whether an input string of symbols belongs to a given language.[50] One limitation of the conventional acceptor is that its output is limited to "accept" or "don't accept." By applying fuzzy theory, a fuzzy automaton[51] which determines the degree of acceptance for the symbolic inputs can be constructed. These automata are not based on traditional two-valued logic, but instead on a logic with fuzzy truth, fuzzy connectives, and fuzzy rules of inference.[52] Instead of utilizing a conventional formal language, a fuzzy language will be used.[53] A fuzzy language L is a set of ordered pairs, $L = [(x, u(x))]$, where x is a symbolic string and $u(x)$ is the membership grade of x in L. A fuzzy language can be formed in accordance with a set of production rules, in which each production rule is associated with a weight in the closed interval $(0, 1)$. As is the case for formal language theory, for any regular fuzzy language there is a corresponding fuzzy finite automaton which can be used to determine the fuzzy acceptance $u(x)$. A fuzzy finite automata, FFA, is a six-tuple FFA $= (I, Z, S, M, S_0, T)$ in which I is a finite input alphabet, Z an output alphabet, S a finite set of states, $M: S*I*(0, 1) \rightarrow S$ a fuzzy state transition map, S_0 which is in S a vector of initial states, and T, a subset of S, a vector of final states. The definition is essentially that of nonfuzzy automata with the addition of the fuzzy transistion and distributions for the initial and final state vectors.

The fuzzy acceptance can be defined similarly to that in stochastic automata.[50] Let the input string be $x = a_1, a_2 \ldots, a_n$. We say that the fuzzy automata M accepts x with the fuzzy acceptance $d(x)$, where

$$d(x) = I \circ T(x) \circ F^t = I \circ T(a_1) \circ \cdots \circ T(a_n) \circ F^t. \qquad (36)$$

The symbol $\circ$ denotes the max–min operation; $T(a_i)$ are the fuzzy state transition matrices and I and F are the vectors of initial and final states, respectively. The sequential computation of $d(x)$ is time-consuming. The complexity of computation is further aggravated in the case where a large number of different inputs are involved. However, these difficulties can be eased by exploiting a parallel automaton developed using the following mapping procedure.

First, consider the right-hand side of the production rules. If state X is associated with n input alphabets, then X is represented by a set $\{X_i\}$ of n new states. Second, for each production rule, if state X has been expressed by n states, then the production rule $X \rightarrow pY$ is expanded to $X_1 \rightarrow pY, \ldots, X_n \rightarrow pY$. This procedure produces a matrix $M(x)$ that describes the interconnection of n neurons where each neuron represents a state X_i. A fuzzy neural acceptor can thus be constructed in accordance with such a matrix. Let x be $a_1 \ldots, a_n$. The fuzzy acceptance $d(x)$ can then be expressed as

$$d(x) = I' \circ M(x) * A'(a_1) \circ \cdots \circ M(x) * A'(a_n) \circ F'', \qquad (37)$$

where I' and F' are new sectors of the initial and final states, respectively, $A'(a_i)$ are activation vectors in which the elements are either zero or one, and a_i is an element of the input alphabet. The symbol $*$ denotes an "element-to-element" multiply operation. Note that this multiplication operation is not needed in the hardware implementation. The values of expanded states in I' and F' are simply duplicates of the corresponding original states in I and F, respectively.

The following simple example will help illustrate the basic concepts of this mapping procedure. Consider the following set of production rules. $A \xrightarrow{0.1} bB$, $A \xrightarrow{0.4} aC$, $B \xrightarrow{0.3} aA$, $B \xrightarrow{0.2} bB$, $B \xrightarrow{0.6} aC$, and $C \xrightarrow{0.5} bA$. The rule $A \xrightarrow{0.1} bB$ means that if the system is in state A and is presented with input character b then it with possibility 0.1 will change to state B. We therefore have two transition matrices in this example, one for input a and the other for input b. These two fuzzy state transition matrices $T(a)$ and $T(b)$ are

$$
T(a) = \begin{array}{c} \\ A \\ B \\ C \end{array}
\begin{array}{ccc}
A & B & C \\
0.0 & 0.0 & 0.0 \\
0.3 & 0.0 & 0.6 \\
0.0 & 0.0 & 0.0
\end{array}
\qquad (38)
$$

and

$$
T(a) = \begin{array}{c} \\ A \\ B \\ C \end{array}
\begin{array}{ccc}
A & B & C \\
0.0 & 0.1 & 0.0 \\
0.0 & 0.2 & 0.0 \\
0.5 & 0.0 & 0.0
\end{array}
\qquad (39)
$$

Here, the difficulty is that neither of these two has a simple physical representation. The zero elements for example denote the "nonmembership" of the state transitions and do not correspond to a missing hardware interconnect. The procedure described above, however, produces the following single interconnection matrix $M(x)$:

	A_a	A_b	B_{b1}	B_{b2}	C_{a1}	C_{a2}
A_a	0	0	0.1	0	0.4	0
A_b	0	0	0.1	0	0.4	0
B_{b1}	0.3	0	0	0.2	0	0.6
B_{b2}	0.3	0	0	0.2	0	0.6
C_{a1}	0	0.5	0	0	0	0
C_{a2}	0	0.5	0	0	0	0

The zero elements in the matrix $M(x)$ denote the lack of a hardware interconnection between elements that represent each of the two possible expanded states. Such an element or fuzzy neuron is an element whose output is equal to a number r if it is firing and s if it is not firing, where $0 < r$ and $s < 1$, and is computed using the above max–min operations. This PE will consist of two parts, a performance branch and a learning or adaptation branch. A performance branch contains a minimum unit which performs the fuzzy intersection operation; a maximum unit which performs the fuzzy union operation; registers R1, W, and R2, which hold an internal state value, a weight value (the nonzero value in the associated column of the $M(x)$ represents the weight), and an intermediate result, respectively; and a control block which determines the firing status. The learning branch is capable of updating the values in registers W and R1.

It is the performance element of this fuzzy neuron (FN) that corresponds to the McCullouch–Pitts neuron (MPN). The FN intersection minimum operation corresponds to the MPN weighted multiplication, the FN union-maximum operation to the MPN summation, and the FN trigger-activation to the MPN thresholding. The conventional MPN contains no learning or adaptation behaviors of its own and instead is totally reliant on external processors for the performance of this function.

The system functions in the following fashion. To begin, the initial state and weight value of each neuron are loaded in registers R1 and W, respectively. Then, all the neurons execute concurrently the same sequence of instructions in the following time steps. The neurons will be fired if the corresponding type of input is received. The operation $I' \circ M(x) * A'(a_i)$ is performed by having the neurons (in parallel fashion) first load register R2 with the min of $r1$ (the initial contents of register R1) and w (the weight stored in register W). Then, the maximum of the various inputs I_i is found. If the neuron is firing, that is, if the appropriate input (a or b in our example above) has been presented to the overall network, then this maximum is stored in register R1 and constitutes the new state for the neuron. If the neuron is not firing, then zero is stored in R1.

Simple and regular data flows are the primary concern of VLSI system design. However, the interconnection in itself here will not be regular if a straightforward variation on the processing element is used. One approach to overcome these interconnection difficulties is to decompose the fuzzy neurons into basic functional blocks (e.g., minimum unit). By this functional decomposition approach, all of the minimum and maximum units of a fuzzy acceptor are combined into a MIN plane and a MAX plane, respectively. All the weight and state registers are accordingly combined into two register files. Note that if the weights in the productions are either one or zero, the minimum and maximum operations can simply be realized by AND and OR gates, respectively. Therefore, the structure of a fuzzy acceptor is reduced to a PLA form.

9.4   THE FUTURE

We would like to end this chapter by briefly describing the ultimate integrated circuit. One can envision a combination of wafer scale integration and ultralarge scale integration culminating in a "chip" containing several billion transistors. The central dilemmas are (1) that it is unlikely that all of these devices will work as anticipated by the designer, (2) that we will never be able actually to functionally test this system to a reasonable degree, and (3) to ask what it is that we will do with all of these devices. We would suggest that techniques that answer the first two of these dilemmas will almost certainly lead to an answer to the last.

The chip must be based on a regular array of regularly connected elements or it is unlikely that we will ever be able to complete the initial design. These elements and connections, however, will be plastic. We will be able to reconfigure the system locally to solve any problems posed by a poorly functioning chip. We can even envision systems in which problems such as the tendency of threshold voltages to vary as one moves across a wafer can be handled by allowing various regions of the chip to "learn" that the voltages of the next region are a little low. In short, our chip will contain regional dialects and accents. This type of adaptability and insensitivity to parameter fluctuations will be essential if our yield is to be greater than zero. What is sometimes not mentioned is that they also play a role in long-term reliability as well. There, one is concerned about temporal variations in device parameters through effects such as oxide charging.

While obviously a significant amount of on-chip testing is occurring, it is doubtful that even then the chip will actually be fully testable in a reasonable time frame. We anticipate handling these problems by incorporating a permanent external mentor feature by which the user can inform the chip that the chip is making an error and have the chip correct it. Here, we encounter a new world of yet-undeveloped algorithms for learning new behaviors without destroying the old.

In a certain though far from complete sense, the worry about testability is false. Part of this worry is the observation that we can never actually anticipate the full details of the complicated behaviors and interactions that such a complex system has with its external world. However, our goal is to build systems that, when confronted with novel, unanticipated situations, will usually function in a reasonable if not optimal fashion even if the chip's response to that situation has never been tested. For such a chip, intended to brave new worlds, the time will come when we will have to stop testing it and let it leave the nest. Our challenge, therefore, is to learn how we can build systems capable of being "parented" rather than quality-assured.

The neural system of an insect may contain on the order of a million neurons. Although a neuron is much more complicated than a transistor, our ultimate chip has several orders of magnitude more transistors and its seems to not be totally unrealistic to anticipate achieving similar complexity. Insects such as bees are capable of flight control, fine motor control, genuine navigation, simple pattern recognition, simple communication (the famous bee dance), scheduling the time of day when it is optimal to visit certain locations, developing a map of their surroundings, and a large number of tasks that no robot today can possibly imitate. Attaining similar behaviors as these depends critically on developing adaptability, learning capabilities, and a large measure of self-testing. These, whether obtained through an SNS, a pipelined CAM, AI, or fuzzy logic, are where the frontiers of ULSI will be found.

REFERENCES

1. D. K. Ferry and W. Porod, "Interconnections and Architectures for Ensembles of Microstructures," *Superlattices Microstruct.*, **2**, 41 (1986).

2. J. J. Hopfield, "Neural Networks and Physical Systems with Emergent Collective Computational Abilities," *Proc. Natl. Acad. Sci. U.S.A.*, **79**, 2554 (1982).

3. G. Moore, "Progress in Digital Integrated Circuits," *Proc. Int. Electron Device Meet.*, p. 11 (1975).

4. L. Yau, "A Simple Theory to Predict the Threshold Voltage of Short-channel IGFET's," *Solid-State Electron.*, **17**, 1059 (1974).

5. L. A. Akers, "An Analytical Expression for the Threshold Voltage of a Small Geometry MOSFET," *Solid-State Electron.*, **24**, 621 (1981).

6. L. A. Akers, M. E. Beguwala, and F. Z. Custode, "Model of a Narrow-Width MOSFET Including Tapered Oxide and Doping Encroachment," *IEEE Trans. Electron Devices*, **ED-28**, 1490 (1981).

7. M. Sugino and L. A. Akers, "Subthreshold Current in Oxide Isolated Structures," *IEEE Electron Device Lett.*, **EDL-4**, 114 (1983).

8. M. Shigyo, M. Konaka, and R. L. M. Dand, "Three Dimensional Simulation of Inverse Narrow-Channel Effect," *Electron. Lett.*, **18**, 274 (1982).

9. C. Ji and C. T. Sah, "Two-Dimensional Numerical Analysis of the Narrow Gate Effect in MOSFET" *IEEE Trans. Electron Devices*, **ED-30**, 635 (1983).

10. F. Hsu, P. Ko, S. Tam, C. Hu, and R. Muller, "An Analytical Breakdown Model for Short-Channel MOSFET's," *IEEE Trans. Electron Devices*, **ED-29**, 1735 (1982).

11. J. Mar, S. Li, and S. Yu, "Substrate Current Modeling for Circuit Simulation," *IEEE Trans. Comput.-Aided Des.*, **CAD-1**, 183 (1983).

12. J. R. Barker and D. K. Ferry, "On the Physics and Modeling of Small Semiconductor Devices. II," *Solid-State Electron.*, **23**, 531 (1980).

13. B. S. Landman and R. L. Russo, "On a Pin Versus Block Relationship for Partitions of Logic Graphs," *IEEE Trans. Comput.*, **C-20**, 1469 (1970).

14. T. Chiba, "Impact of the LSI on High-Speed Computer Packaging," *IEEE Trans. Comput.*, **C-27**, 319 (1975).

15. D. K. Ferry, "Interconnection Lengths and VLSI," *IEEE Circuits Devices Mag.*, **1**(4), 39 (1985).

16. W. E. Donath, "Placement and Average Interconnection Lengths of Computer Logic," *IEEE Trans. Circuits Syst.*, **CAS-26**, 272 (1979).

17. M. Sami and R. Stefanelli, "Reconfigurable Architectures for VLSI Processing Arrays," *Proc. IEEE*, **74**, 712 (1986).

18. D. K. Ferry, "Device-Device Interactions," in H. L. Grubin, D. K. Ferry, and C. Jacoboni, Eds., *Physics of Submicron Devices*, Plenum, New York, (in press).

19. T. Hogg and B. A. Huberman, "Attractors on Finite Sets: The Dissipative Dynamics of Computing Structures," *Phys. Rev. A*, **32**, 2338 (1985).

20. B. A. Huberman and T. Hobb, "Collective Computation and Self-Repair in Parallel Computing Structures," *Phys. Rev. Lett.*, **52**, 1048 (1984).

21. J. J. Hopfield and D. W. Tank, "Neural Computation of Decisions in Optimization Problems," *Biol. Cybernet.*, **52**, 141 (1985).

22. K. Preston, M. Duff, S. Levialdi, P. Norgren, and J. Toriwaki, "Basics of Cellular Logic with Some Applications in Medical Image Processing," *Proc. IEEE*, **67**, 826 (1979).

23. E. Codd, *Cellular Automata*, Academic, New York, 1968.

24. K. Preston and M. Duff, *Modern Cellular Automata*, Plenum, New York, 1984.

25. P. Berke, "That Does Not Compute," *Int. Conf. Neural Networks, 1st*, San Diego, CA, Poster (1987).

26. K. Sayre, *Consciousness: A Philosophic Study of Minds and Machines*, Random House, New York, 1969.

27. R. Hecht-Nielsen, "Performance Limits of Optical, Electro-Optical, and Electronic Artificial Neural Processors," Hybrid and Optical Systems, SPIE, 1986.

28. H. Dreyfus, *What Computers Can't Do: The Limits of Artificial Intelligence*, Harper & Row, New York, 1979.

29. S. Weisburd, "Neural Nets Catch the ABCs of DNA," *Sci. News*, **132**, 76 (1987).

30. J. Feldman, "Connectionist Models and Their Applications," *Cognit. Sci.*, **9**, 1 (1985).

31. P. Oosting, "Signal Transmission in the Nervous System," *Rep. Prog. Phys.*, **42**, 99 (1979).

32. S. Grossberg, *Studies of Mind and Brain*, Reidel, Boston, MA, 1982.

33. A. Turing, "On the Computability of Numbers," *Proc. London Math. Soc.*, **42**, 230 (1936).

34. P. Peretto and J. J. Niez, "Long Term Memory Storage Capacity of Multiconnected Neural Networks," *Biol. Cybernet.*, **54**, 53 (1986).

35. G. E. Hinton and J. A. Anderson, Eds., *Parallel Models of Associative Memory*, Erlbaum, Hillsdale, NJ, 1981.

36. C. P. Rialan and L. L. Scharf, "Cellular Architecture for Implementing Projection Operators," unpublished.

37. G. Carpenter and S. Grossberg, "Computing with Neural Networks," *Science*, **235**, 1226 (1987).

38. D. Tank and J. Hopfield, "Simple Neural Optimization Networks: An A/D Converter, Signal Decision Circuit, and a Linear Programming Circuit," *IEEE Trans. Circuits Syst.*, **CAS-33,** 533 (1986).

39. J. Hopfield, *Collective Processing and Neural States,* Contrib. No. 6802, Chemistry Department, California Institute of Technology, Pasadena (1984).

40. J. Hopfield, "Neurons with Graded Response have Collective Computational Properties like those of Two-State Neurons," *Proc. Natl. Acad. Sci., U.S.A.,* **81,** 3088 (1984).

41. G. Graf, L. Jackel, R. Howard, B. Howard, B. Stranghn, J. Denker, W. Hubbard, D. Tennant, and D. Schwartz, "VLSI Implementation of a Neural Network Memory with Several Hundreds of Neurons," *AIP Conf. Proc.,* **151,** 182 (1986).

42. L. Jackel, Personal communication, Oct. 14 (1987).

43. M. Sivilotti, M. Emerling, and C. Mead, "VLSI Architectures for Implementation of Neural Networks," *AIP Conf. Proc.,* **151,** 408 (1986).

44. J. Alspector and R. Allen, "A Neuromorphic VLSI Learning System," *Advanced Research in VLSI,* MIT Press, Cambridge, MA, 1987.

45. L. A. Akers, M. R. Walker, D. K. Ferry, and R. O. Grondin, "Limited Interconnectivity in Synthetic Neural Systems," in R. Eckmiller and C. v.d. Malsburg, Eds., *Neural Computers,* Springer-Verlag, Berlin and New York, 1988.

46. L. T. Clark and R. O. Grondin, "Comparison of a Pipelined 'Best Match' Content Addressable Memory with Neural Networks," *Proc. IEEE Neural Nets Conf., 1st, 1987,* San Diego, CA , pp. III: 411–418 (1987).

47. C. L. Chen and M. Y. Hsiao, "Error-Correcting Codes for Semiconductor Memory Applications: A State-of-the-Art Review," *IBM. J. Res. Dev.,* **28,** 124 (1984).

48. T. J. Sejnowski and C. R. Rosenberg, *NETtalk: A Parallel Network that Learns to Read Aloud,* Tech. Rep. JHU/EECS-86/01, Johns Hopkins University, Electrical Engineering and Computer Science, Baltimore, MD, 1986.

49. L. C. Shiue and R. O. Grondin, "On Designing Fuzzy Learning Neural-Automata," *Proc. IEEE Neural Nets Conf., 1st, 1987,* San Diego, CA , pp. II: 299–307 (1987).

50. D. Hopkin and B. Moss, *Automata,* North-Holland, New York, 1976.

51. W. G. Wee and K. S. Fu, "A Formulation of Fuzzy Automata and its Application as a Model of Learning Systems," *IEEE Trans. Syst. Sci. Cybernet.* **SSC-5**(3), 215 (1969).

52. L. A. Zadeh, "A Fuzzy-Set-Theoretic Interpretation of Linquistic Hedges," *J. Cybernet.,* **2,** 4 (1972).

53. E. T. Lee and L. A. Zadeh, "Note on Fuzzy Languages," *Inf. Sci.,* **1,** 421 (1969).

10 VLSI Electronic Neural Networks

RAMAUTAR SHARMA

AT&T Bell Laboratories
Murray Hill, New Jersey

10.1 INTRODUCTION

Why neural networks? The ever-increasing desire to produce faster and faster computing machines, also known as supercomputers, creates some problems. Although these machines can outperform human beings in doing "exact" calculations or performing a simple algorithmic computation repeatedly, these are no match to the capabilities of human beings in doing some other "intelligent things," such as differentiating a car from a horse, a red-faced monkey from a black-faced one, and so on. Both these tasks can easily be performed by a child in a fraction of a second, but even a supercomputer could take much longer to do these. Computer architects have been trying to design machines that can learn from experience as humans do, but the result is well known: we do not have such computers yet. So, what is wrong with computers? It is not a fault of computers; rather, we do not have a good understanding of the thinking and learning processes to develop algorithms that could be used in designing these machines.

Generally, it is accepted that a human brain has about 10^{10}–10^{11} neurons, each capable of switching at a few thousand times per second. Each of the neurons in the human brain is connected to about 10^4 other neurons. The computational abilities of the brain arise from the large number of neurons and their massive interconnection network. The intelligent behavior results from various highly specialized peripheral parts working in parallel to do some pre-processing of raw information and the collective operation of neurons.

A biological neuron, shown in Fig. 10.1, has an input structure and an output structure, called dendrite and axon, respectively. Inputs from one neuron to another are made at points called synapses. When a neuron is excited by its inputs, it produces a pulse train. The pulsing rate of such a pulse train arriving at a synapse may increase or decrease, depending on whether the connection is excitatory or inhibitory. The pulsing rate eventually saturates to a maximum value with increasing excitation.

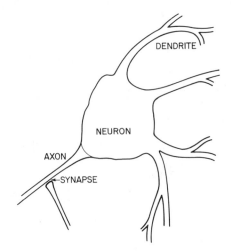

Fig. 10.1 A simple neuron.

Hopfield gives an account of the capabilities of these neurons connected in a network.[1] At present, it is an area of active research for an "intelligent" computer design.[2] However, we will be concerned with the relationship of VLSI technology and neural networks. Basic neural models and their electronic representation will be examined in Section 10.2. Neural networks are described in Section 10.3. Implications of VLSI technology and neural networks will be discussed in Section 10.4. Some of the applications of neural networks are discussed in Section 10.5. Finally, in Section 10.6, we summarize the status of neural network technology and the direction it might take in the near future.

10.2 BASIC NEURAL MODELS

The history of neural networks is rather long and many approaches and representations have been adopted, but the computational model for a neural network in all of these has been similar to the one shown in Fig. 10.2. Recently, Lippmann[3] reviewed the computations with neural networks thoroughly, and our presentation of modeling neurons follows his treatment.

Neural computational elements are nonlinear, and typically these are analog in nature. The simplest node sums N weighted inputs and passes the result through a nonlinearity to produce an output. The node is characterized by an internal threshold t_i and a nonlinearity. Figure 10.2 illustrates three common nonlinearities:

1. *Hard Limiters:* an infinitesimally small deviation of the weighted sum of inputs from the neural threshold changes its state.
2. *Threshold Logic Element:* a neuron's state changes gradually if the weighted sum of inputs exceeds its threshold.
3. *Sigmoid:* a change in neuron state is much smoother and symmetric about threshold value.

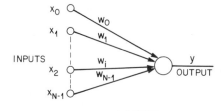

(a) NODE

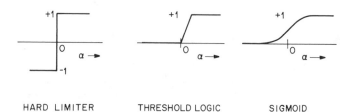

HARD LIMITER   THRESHOLD LOGIC   SIGMOID

(b) NONLINEARITIES

Fig. 10.2 Model of a neuron.

A neuron's behavior does not depend heavily on the details of the nonlinearity. There can be many functional representations of a given type of nonlinearity. For example, a simple sigmoid function can be expressed as follows:

$$y_j = f(S_j) = \frac{1}{1 + e^{-(S_j - t_j)}}. \tag{1}$$

In Eq. (1), S_j and t_j are the weighted sum of inputs and the threshold for j^{th} neuron, respectively. More complex nodes may include time dependencies and other complicated mathematical operations.

An electronic implementation of the neural model of Fig. 10.2 is shown in Fig. 10.3. In Fig. 10.3, w_{ij} are the conductances of the array of input resistors, summing is done on a wire "dendrite," nonlinear function $f()$ is provided by the transfer function of the amplifier, and the output is broadcast to other neurons by a wire

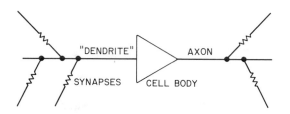

Fig. 10.3 Electronic neuron.

"axon." The neuron pulsing rate is represented by the amplifier output voltage. All neural implementations tend to realize this type of neuron model in one form or another, which will become clear later in this chapter.

10.3 NEURAL NETWORKS

Many neurons can be connected to each other to form a variety of networks. There are two main classes of neural networks, namely, nonrecurrent and recurrent, and these are described below.

10.3.1 Nonrecurrent Neural Networks

In these networks, information flows in the forward direction only, and there are no feedback loops. The network is always stable, and its state depends on the inputs in a simple manner. In the following, we describe the operation of such networks. However, for simplicity and without loss of generality, we assume that M neurons forming a network are connected to N inputs, as shown in Fig. 10.4.

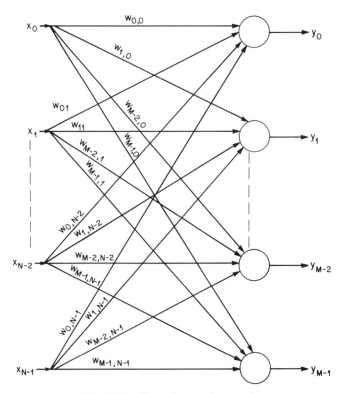

Fig. 10.4 A simple neural network.

For this network, the input–output relationship can be expressed as given by

$$
\begin{bmatrix}
y_0 \\
y_1 \\
\vdots \\
y_{M-2} \\
y_{M-1}
\end{bmatrix}
=
\begin{bmatrix}
w_{0,0} & w_{0,1} & \cdots & w_{0,N-2} & w_{0,N-1} \\
w_{1,0} & w_{1,1} & \cdots & w_{1,N-2} & w_{1,N-1} \\
\vdots & \vdots & & \vdots & \vdots \\
w_{M-2,0} & w_{M-2,1} & \cdots & w_{M-2,N-2} & w_{M-2,N-1} \\
w_{M-1,0} & w_{M-1,1} & \cdots & w_{M-1,N-2} & w_{M-1,N-1}
\end{bmatrix}
\begin{bmatrix}
x_0 \\
x_1 \\
\vdots \\
x_{N-2} \\
x_{N-1}
\end{bmatrix},
$$

or   (2)

$$\mathbf{Y} = \mathbf{W}\mathbf{X},$$

where $\mathbf{X}$ is an N-component column vector of inputs, $\mathbf{W}$ is an $M \times N$ matrix of synaptic weights, and $\mathbf{Y}$ is the resulting M-component column vector of outputs. This type of network has only one layer of neurons and there is no connection between neurons themselves. Naturally, its practical applications are also limited, but it is used here to illustrate the principle and to establish a framework for further discussions. Interesting applications emerge when many neurons interact with each other. This can be achieved through more complex networks formed by cascading two or more layers of such a simple network, as shown in Fig. 10.5. The

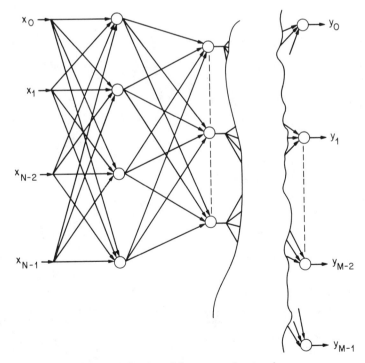

Fig. 10.5   A multilayer neural network.

operation of such networks duplicates that of Fig. 10.4, that is, the outputs of each layer are produced from the weighted sum of the previous layer's outputs. As the number of layers increases, the usefulness of the networks also increases. The outputs from a multilayered neural network can be expressed as

$$\mathbf{Y} = \mathbf{W}^{K-1}\mathbf{W}^{K-2} \cdots \mathbf{W}^1\mathbf{W}^0\mathbf{X}, \tag{3}$$

where $\mathbf{W}^k$ is a matrix of weights for kth layer, the size of this matrix depending on the number of neurons acting as sources and destinations for the layer.

10.3.2 Recurrent Neural Networks

So far we have only discussed nonrecurrent neural networks. The signals flow from the input to output side with no feedback path, and these networks are unconditionally stable. At this point, one can easily see that in nature there are only a few systems that have no feedback involved and, hence, these networks have a limited utilization. However, biological systems have some sort of feedback among various neurons, and therefore a more realistic neural network model should have such feedback paths too. Figure 10.6 shows such a network where the synapse between neurons are bidirectional; however, synaptic strength in forward and reverse directions may be different. The input–output relationship for these networks is given by

$$
\begin{bmatrix} y_0 \\ y_1 \\ \vdots \\ y_{M-2} \\ y_{M-1} \end{bmatrix}^{(p)} =
\begin{bmatrix}
w_{0,0} & w_{0,1} & \cdots & w_{0,M-2} & w_{0,M-1} \\
w_{1,0} & w_{1,1} & \cdots & w_{1,M-2} & w_{1,M-1} \\
\vdots & \vdots & & \vdots & \vdots \\
w_{M-2,0} & w_{M-2,1} & \cdots & w_{M-2,M-2} & w_{M-2,M-1} \\
w_{M-1,0} & w_{M-1,1} & \cdots & w_{M-1,M-2} & w_{M-1,M-1}
\end{bmatrix}
\begin{bmatrix} y_0 \\ y_1 \\ \vdots \\ y_{M-2} \\ y_{M-1} \end{bmatrix}^{(p-1)}
$$

$$
+
\begin{bmatrix}
e_{0,0} & e_{0,1} & \cdots & e_{0,N-2} & e_{0,N-1} \\
e_{1,0} & e_{1,1} & \cdots & e_{1,N-2} & e_{1,N-1} \\
\vdots & \vdots & & \vdots & \vdots \\
e_{M-2,0} & e_{M-2,1} & \cdots & e_{M-2,N-2} & e_{M-2,N-1} \\
e_{M-1,0} & e_{M-1,1} & \cdots & e_{M-1,N-2} & e_{M-1,N-1}
\end{bmatrix}
\begin{bmatrix} x_0 \\ x_1 \\ \vdots \\ x_{N-2} \\ x_{N-1} \end{bmatrix}. \tag{4}
$$

The superscripts (p) and $(p - 1)$ for the column vectors are used to emphasize the recurrent nature. The next state of the network depends not only on inputs to it, but also on its present state, which is brought about by bidirectional paths among neurons. For the network of Fig. 10.6, $M = 6$, $N = 3$, and $w_{ii} = 0$ for $0 \le i \le 5$. On the input side, the matrix elements e_{00}, e_{11}, and e_{22} are w_{i0}, w_{i1}, and w_{i2}, respectively; all other elements are zero.

In general, these recurrent systems may have stability problems. Grossberg[4] and others derived some conditions under which such networks would be stable. Kosko[5] and others have shown that the system is unconditionally stable if the re-

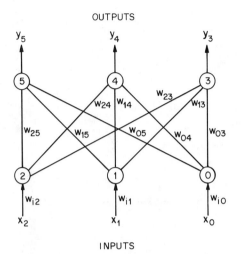

Fig. 10.6 A recurrent neural network.

verse weight matrix is the transpose of the forward matrix, that is,

$$\mathbf{W}_{\text{reverse}} = \mathbf{W}^t_{\text{forward}}. \tag{5}$$

For a recurrent neural network with M maximally connected neurons and N inputs connected to I neurons in the network, the total number of connections is given by

$$C = M(M - 1) + NI. \tag{6}$$

As the number of neurons in the network increases, generally the number of connections increases at less than quadratic rate, shown by Eq. (6). This is because to connect to farther neurons we will need longer dendrites and axons and consequently more energy to keep them active. Therefore, in a real system not all neurons connect to each other.

10.3.3 Network State Evaluation

States of all neurons in a network constitute the network state, and it is evaluated as follows. In a neural network, synaptic connections are undirected and have strength which is represented by fixed, real numbers. The neurons repeatedly examine their inputs and decide to turn ON or OFF by the following scheme. Let w_{ij} be the strength (which can be negative) of the synaptic connection from neuron i to neuron j. Without loss of generality, we assume that $w_{ij} = w_{ji}$ and $w_{ii} = 0$. Let t_i be the threshold value of the ith neuron. Let S_j be the weighted sum of inputs to a neuron, and it is given by

$$S_j = \sum_{i=0}^{N-1} w_{ji} x_i. \tag{7}$$

For simplicity, we also assume that a neuron can be in either of two states:

1. $y_j = -1$ (OFF).
2. $y_j = +1$ (ON).

Under these conditions, the state transitions for the network are governed by the following algorithm:

> **state_transition algorithm** {
>> for each neuron $j \in \{0, 1, \ldots, M - 1\}$
>> {
>>> calculate the weighted sum S_j using Eq. (6);
>>> if $(S_j > t_j)$
>>>> {turn ON neuron; $y_j = +1$;}
>>> else if $(S_j < t_j)$
>>>> {turn OFF neuron; $y_j = -1$;}
>>> else
>>>> {no change in neuron state; y_j remains unchanged;} .
>> }
> }

The network starts in an undefined initial state and runs with each neuron randomly and independently reevaluating itself. The operation continues until a stable state for the network has been reached. The basic operation of the network is to converge to a stable state if we initialize it with a nearby state vector in a Hamming distance sense.[3]

An estimate of convergence time for the network is an important parameter to determine the speed of computations. In general, it is not possible to predict this time accurately; however, an upper bound for the time would also be useful. Abu-Mostafa[6] computed an upper bound on the number of neural state changes $T(m)$ to reach a stable state by minimizing the energy. It is given by

$$T(m) \leq \frac{LM^2}{2}, \tag{8}$$

where L is the number of levels into which w_{ij}'s are quantized and $T(m)$ depends quadratically on the number of the neurons in the network. Now, assuming that an average time for a neuron to change its state is t_0, an upper bound for the convergence time T_{conv} is given by

$$T_{conv} \leq t_0 \left(\frac{LM}{2} \right). \tag{9}$$

Note that the convergence time depends linearly on the number of neurons in the network.

10.4   NEURAL NETWORK IMPLEMENTATIONS

The neural networks including associative memory can be fabricated using electronic components or simulated on a computer. Most of the studies in the area are simulation-based. Kosko[7] has described one such simulator for bidirectional associative memories (BAM) recently and interested readers are referred to his paper. Here, we are concerned more with the hardware aspects of neural networks as related to VLSI technology.

Let us analyze a neural network in order to determine the requirements and the complexity of the problem before considering its implementation details. We will consider the recurrent neural networks, as they are the most useful for practical applications. Generally, for a recurrent neural network the input–output relationship is given by Eq. (4), which can be rewritten as

$$\mathbf{Y}^{(p)} = \mathbf{W}\mathbf{Y}^{(p-1)} + \mathbf{E}\mathbf{X}, \tag{10}$$

where $\mathbf{W}$ is an $M \times M$ matrix with all zeros on its main diagonal, $\mathbf{E}$ band matrix of size $M \times N$ and $\mathbf{X}$ the input vector with N components. The elements e_{ij} of $\mathbf{E}$ are synaptic strengths similar to the elements of $\mathbf{W}$. The superscript (p) and $(p - 1)$ for the column vectors $\mathbf{Y}$ are used to emphasize the interative nature of evaluating components of $\mathbf{Y}$. These indicate the relationship between the present and the next state of the network.

If we assume a maximally connected neural network and each input connected to I neurons, to implement this network we would have to perform $O(M^2)$ multiplications, $O(M \log M)$ additions, and $O(M)$ thresholding operations either in time using single instance of these operators or in space using many instances. These operations are done recurrently until a stable state is reached.

To understand the basic ideas, consider a simple recurrent neural network of Fig. 10.6. The input–output relationship for this network is given by

$$y_0 = w_{00}x_0 + w_{01}y_1 + w_{02}y_2 + w_{03}y_3 + w_{04}y_4 + w_{05}y_5$$

$$y_1 = w_{10}y_0 + w_{11}x_1 + w_{12}y_2 + w_{13}y_3 + w_{14}y_4 + w_{15}y_5$$

$$y_2 = w_{20}y_0 + w_{21}y_1 + w_{22}x_2 + w_{23}y_3 + w_{24}y_4 + w_{25}y_5$$

$$y_3 = w_{30}y_0 + w_{31}y_1 + w_{32}y_2 + 0 + w_{34}y_4 + w_{35}y_5$$

$$y_4 = w_{40}y_0 + w_{41}y_1 + w_{42}y_2 + w_{43}y_3 + 0 + w_{45}y_5$$

$$y_5 = w_{50}y_0 + w_{51}y_1 + w_{52}y_2 + w_{53}y_3 + w_{54}y_4 + 0$$

or

$$y_i = \sum_{j=0, j \neq i}^{5} w_{ij}y_j + w_{ii}x_i, \tag{11}$$

where $i = 0, 1, 2, 3, 4, 5$.

Realization of the network represented by Eq. (11) would require 33 multiplications, 27 additions, and 6 thresholding operations to be performed for each set of output vectors $\{y_0, y_1, y_2, y_3, y_4,$ and $y_5\}$ until a stable state vector according to some criterion has been achieved. Individually, we must evaluate the state of each neuron iteratively, using the signal flow graphs similar to that of Fig. 10.7. This signal flow graph has been particularized for neuron #0. In the figure, n represents the number of bits used for each signal.

Having established the required number of operations for each iteration, we can now examine how these can be implemented. There are essentially the following three approaches for implementing these operations:

1. Serial implementation.
2. Serial-parallel implementation.
3. Parallel implementation.

In serial implementation, we use just one general-purpose processor (micro-/mini-/maxi-/supercomputer) and program these operations on it in a sequential manner. It is the simplest approach, it has been used all along to simulate neural networks. However, it is not much of interest from a VLSI point of view and we will not discuss it further.

In serial-parallel implementation, a set of processors is programmed to perform the computations in parallel for different neurons. Information about neuron states

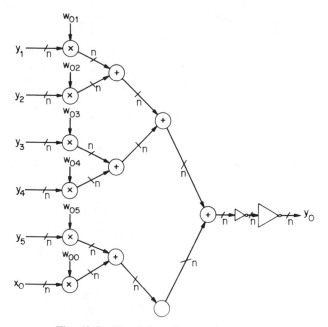

Fig. 10.7   Signal flow diagram for a neuron.

is exchanged through some sort of communication network. Individual processors could be commerically available digital signal processors and their activities coordinated by another single-chip microcomputer. A schematic diagram for this type of implementation is shown in Fig. 10.8.

This approach is an intermediate step toward a fully custom VLSI realization and at present is the most cost-effective way to design neural networks. Penz and Wiggins[8] have built such a system and carried out a comparative study of simulation speeds; their results are shown in Table 10.1. From the table, it is clear that a special-purpose accelerator can provide a substantial increase in performance; it further increases with increasing size of the simulated network.

In the parallel approach, each neuron is implemented in hardware, and the resulting parallelism is enormous. Here, all neurons work in parallel to reach a stable state vector. If we were to implement a neural network such as represented by Eq. (11) in a direct digital manner, we would be implementing the signal flow graph of Fig. 10.7 for each neuron. However, it would lead to a highly inefficient use of silicon area. Therefore, for parallel implementation of neural networks one has to consider not only their models, but also the mechanism by which computation can be carried out efficiently.

As the number of bits to represent signals increases, so does the hardware complexity. But this hardware complexity is needed to keep the asynchronous nature of the operation. However, all these could be replaced by an adder, some memory, and a sequencer to cycle all the inputs sequentially. This analysis is presented to

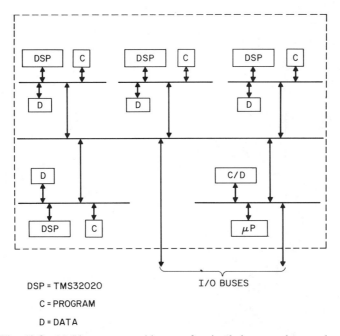

DSP = TMS32020

C = PROGRAM

D = DATA

Fig. 10.8 Multiprocessor architecture for simulating neural networks.

TABLE 10.1   **Performance Comparison of DSP-Based Accelerator and VAX Processors**

Operation	Accelerator Boards	Accelerator Time	VAX		
			8600	11/785	11/750
512 × 512 matrix times 512 vectors	1	39 ms	392 ms	784 ms	1440 ms
Peformance	1	100%	10%	5%	2.7%
1000 × 1000 matrix times 1000 vectors	4	39 ms	1598 ms	3136 ms	5760 ms
Performance	4	100%	2.5%	1.2%	0.7%

indicate the complexities involved with all digital approaches for neural network implementation. Note that the complexity increases almost quadratically with the number of neurons in a network. A plot of number adders to realize a network with different numbers of neurons is shown in Fig. 10.9. Here, we have assumed that an adder tree similar to the one used in Fig. 10.7 would be used. Hardware requirements for a fully parallel all-digital implementation are summarized in Table 10.2,

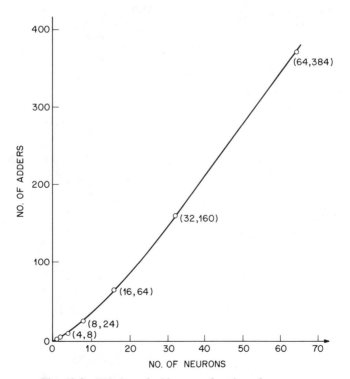

Fig. 10.9   Number of adders as a function of neurons.

TABLE 10.2   Hardware Requirements for Parallel Implementation

No. of Neurons	No. of Multipliers	No. of Adders
1	1	0
2	4	1
4	16	8
8	64	24
16	256	64
32	1,024	160
64	16,384	384
⋮	⋮	⋮
⋮	⋮	⋮
⋮	⋮	⋮
1,024	1,048,576	10,240

which clearly shows that even with submicron VLSI technology we would have difficulty in implementing a reasonably sized neural network. Alternatives using either all analog or memory-based or a combination of several techniques must be examined. It is also important to understand the implications of underlying fabrication technology while designing new architectures for neural networks.

10.4.1   Technological Constraints

Just like any other technology, VLSI technologies have their own advantages and limitations. Therefore, all the designs must satisfy the constraints imposed by the technology. From a single-chip-design viewpoint, restrictions of VLSI technology are many and could form the subject of another book. The important ones are summarized below:

1. *Chip Area.* The size of the chip cannot be arbitrarily large since chip yield decreases exponentially with increasing chip area.

2. *I/O Pads.* The number of available pads for signal transfer between the chip and its environment is limited due to finite and small chip area requirement. This implies that we have to partition and design the network to maximize the throughput.

3. *Environmental Mismatch.* By this we mean that loads seen by pad drivers on and off chips are different, generally, on-chip load being smaller than off-chip. Smaller signal swings can be tolerated on-chip, but, from noise consideration, we must interface with relatively large signals outside the chip. This in turn requires different designs to cope with the situation.

4. *Power Dissipation.* Thermal conductivity of silicon is finite and circuits do not work reliably at elevated temperatures. Therefore, we have to design to satisfy power-density requirements and this means that one cannot trade power for speed arbitrarily.

5. *Parameter Variation.* Two adjacent devices on the same chip can have different threshold voltages and current driving capability due to variations in fabrication processes. This mismatch in device parameters is more pronounced for smaller-size devices. Therefore, circuit design techniques tolerant of such variation, must be used. As seen above, pure digital implementation needs much larger silicon area as compared with analog or some hybrid approach, so there is some trade-off.

6. *Lower Operating Voltages.* Operating voltages are generally lower for high device density technologies. This constrains the type of circuits that can be used to implement needed functions and output drive capabilities of pad drivers.

7. *Higher Leakage Currents.* Leakage current, the current flowing through a device when it is turned off, is higher for submicron VLSI technologies. This factor must be kept in mind while designing a neural network chip.

8. *Faster Device but Slower Interconnects.* Device scaling while satisfying a requirement for higher device densities has created a monster of its own. Now devices are so fast that these can be considered as switching in no time (on the order of a tens of picoseconds), but the interconnects used for transporting signals introduce delays that are almost an order of magnitude higher. This is due to the fact that interconnect technology was not given the same status in the research community as device scaling. Fortunately, things are changing and we should be getting a better deal for interconnections.

10.4.2   VLSI Design Approaches

We have seen that a fully digital design approach for neural networks is not attractive, since the efficiency of silicon area utilization is much less than its analog counterparts that are capable of carrying out much more complex computations using simple circuits. Although analog circuits have much higher computational bandwidths, these are less accurate compared with digital circuits.

Sivilotti et al[9] designed a neural network for vision applications and used mostly analog approach. The basic neural model implemented by them is shown in Fig. 10.10. In this figure, synaptic strength is modeled by the transconductance of MOS transistors on the input side, summing of inputs and timing is provided by the input capacitor C_{in}, and the thresholding function is performed by the amplifier. The control signals $\{CN_1, CN_2, \ldots, CN_n\}$ are used to change the synaptic strengths. Delay associated with state transition τ of such a neuron is bounded by

$$\tau_{best} \leq \tau \leq \tau_{worst}, \tag{12}$$

where τ_{worst} and τ_{best} are given by

$$\tau_{worst} = \left(\frac{1}{g_m}\right)_{worst} C_{in} \quad \text{and} \quad \tau_{best} = \left(\frac{1}{g_m}\right)_{best} C_{in}. \tag{13}$$

The worst case for delay and transconductance g_m is when only one transistor on the input side of the amplifier conducts; g_m is the smallest. On the other hand, the

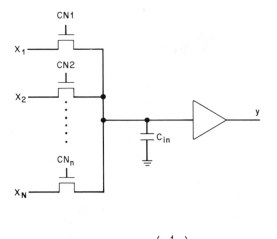

$$\tau_{worst} \ \alpha \ \left(\frac{1}{g_m}\right)_{worst} C_{in}$$

$$\tau_{best} \ \alpha \ \left(\frac{1}{g_m}\right)_{best} C_{in}$$

Fig. 10.10   Analog implementation of a neuron.

best case occurs when all the transistors on the input side are conducting; in this case g_m has maximum value.

In analog implementations, the summation circuit is much smaller compared with digital realization. Besides being a little slower, another major complaint against analog circuits is their accuracy. However, it poses no significant problem for neural networks, as their operation depends only on the collective behavior of all the neurons. The main problem of power dissipation in analog amplifiers is solved by operating individual MOS transistors in the subthreshold region. Analog approach also suffers more with parameter variations, but new circuit configurations are being developed to circumvent this difficulty. Work reported by Sivilotti et al.[9] successfully makes use of this approach and the interested reader is referred to it.

On the other hand, Graf and deVegvar[10] use a hybrid approach based on associative memory model, and the following discussion is based on it. They use a distributed storage with feedback (DSF) type of circuit where an array of amplifiers is fully interconnected through a matix with resistive connections placed at some of the cross points of input and output lines, as shown in Fig. 10.11. In this figure, both inverting and noninverting outputs are fed back into the connection matrix to achieve excitatory and inhibitory connection through simple resistors. A desired set of states can be made stable by proper choice of the connections in the coupling network. However, it has been shown that for a network with M amplifiers the number of stored memories must be limited between 10% and 25% of M vectors of B bits each.[11] It means that one bit of stored memory needs 4 to 10 cross points in the connection matrix.

RESISTIVE INTERCONNECTIONS
MATRIX

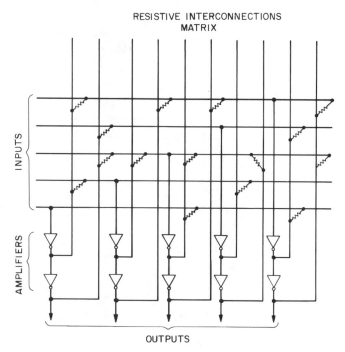

Fig. 10.11   Schematic of an associative memory.

For VLSI implementation, the resistive connections are replaced by MOS transistors, as shown in Fig. 10.12, where the information stored in RAM cells is used to realize proper connectivity. A block diagram to implement an associative memory is shown in Fig. 10.13. In the figure, connection matrix is made of cells similar to those in Fig. 10.12. To store connection information, input data is first shifted and held in the on-chip buffer cells. Appropriate connections are selected through row- and column-address lines. There are two RAM rows and one column associated to each connection, and the information in the buffer cells is written into RAM cells under proper control signals. The buffer cells and shift register are implemented as serial-in parallel-out shift register. Serial writing and reading of information are necessary because of the fixed, small number of input–output pads.

The connection matrix and buffer cells are interfaced to each other through a set of amplifiers, one for every column of connection. Figure 10.14 shows an interconnection of the amplifier and the buffer cells. Note the use of a controlled amplifier; it isolates the interaction between the internal signals and external connection information when memory is being written. Connection between data-input line (DI) and connection matrix is established when the read/write (RW) and input (I) signals are active. A similar connection is established between the data-out (DO) and connection matrix under the control of RW and O signals.

Figure 10.15 shows the details of RAM cells in connection matrix and the amplifiers. These amplifiers are simple inverters that provide an almost hard-limiter

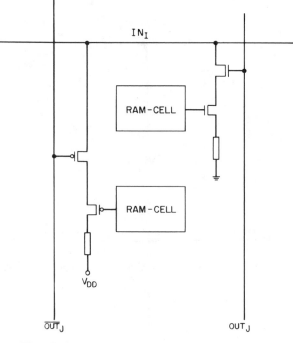

Fig. 10.12   Coupling element for VLSI realization.

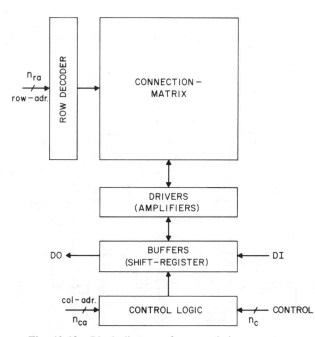

Fig. 10.13   Block diagram of an associative memory.

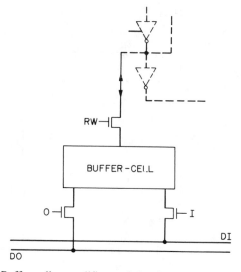

Fig. 10.14 Buffer cells, amplifier, and data input/output interconnections.

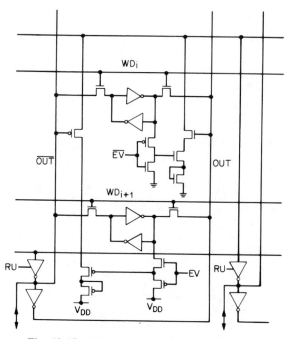

Fig. 10.15 Circuit for RAM cells and amplifiers.

type of nonlinearity, as shown in Fig. 10.2. Signal RU is used to disable one of the inverters of the amplifiers for isolating external inputs and the connection matrix. After presenting a pattern for associating with one of the stored ones, signals RU and EV connect the amplifiers and the matrix. Depending on the contents of two RAM cells, a current flows into or out of the input lines of amplifier I when the output of amplifier J is high, that is, OUT = 1, $\overline{\text{OUT}}$ = 0 (see Fig. 10.12). The voltage of the input line adjusts to a value such that the sum of all currents for the line is zero. When the circuit has stabilized, matched memory pattern is read out through data-out lines. A VLSI chip incorporating these ideas has been described by Graf and deVegvar.[10]

The approach described above to implement is not the only one; many variations with improvements and altogether newer designs are possible. This has been used as an example and sort of existence proof that a fully custom VLSI chip implementing different models of neural networks can be designed and built. Activities will continue to increase in the future as our understanding of models and their implications to VLSI technology improves. We will be entering an exciting era of knowledge processing which requires a close cooperation among various disciplines such as biology, psychology, computer science and engineering, VLSI technology, and, above all, people.

10.5 SOME APPLICATIONS

Neural networks can be used to store information and perform computations. Recently, they have been applied to solve optimization and "traveling salesman" problems and to implement analog-to-digital (A/D) converters.[12,13] Hopfield and Tank,[12,13] for example, have shown that by proper modeling of problems, neural networks can be used very effectively to solve them. The problem of A/D conversion has been cast into an optimization problem, and the final state of the neurons that leads to a minimum energy is the digital representation of the applied analog signal. However, here one can argue that it may not be the best way to do A/D conversion, but Tank and Hopfield[13] just wanted to prove the generality of the neural networks.

Perhaps a better application of neural network is in associative memories, where the network is first "taught" about patterns and later on is asked to recognize correctly under noisy conditions. An associative memory is a mapping from data to data, a mathematical abstraction from the familiar associative structure of human and animal learning. We associate behavioral responses with sensory stimuli. An associative memory is partially distributed when it memorizes by superimposing it on the same memory media — as in neural networks. If used as memories, neural networks differ from conventional computer memories in that the recall is accomplished by associating the data sample with the stored information, not by means of a given address.

In Hopfield model of associative memory,[1] information is stored in the form of binary vectors $V^{(m)}$ whose components $V_i^{(m)}$ i = 0 through B − 1 are either one or zero. The M_v vectors to be stored are used to form a network matrix **W** with elements w_{ij} according to the rule

$$w_{ij} = \sum_{m=0}^{M_v-1} [2V_i^{(m)} - 1][2V_j^{(m)} - 1]. \tag{14}$$

To recall a particular vector, we apply a sample vector $V^{(s)}$ to the inputs of the network. The output vector $V^{(p)}$ is then constructed according to the rule

$$V_i^{(p)} = \begin{cases} 1 & \text{if } \sum_{j=0}^{L-1} w_{ij} V_j^{(s)} > 0, \\ 0 & \text{if } \sum_{j=0}^{L-1} w_{ij} V_j^{(s)} \le 0. \end{cases} \tag{15}$$

Note that this model can reconstruct a given stored vector from a sample of that vector provided that the vectors $V^{(m)}$ of the set of stored information are approximately orthogonal to each other. Associative memories also find applications in many other areas, such as speech and pattern recognition,[3] condensed matter,[11] etc.

10.6 SUMMARY

In this chapter, we have discussed the neural network from the electronic implementation point of view. Important models for neural network and techniques to implement them have been discussed. The modeling area is very active, but hardware realizations have just started surfacing. Array processing techniques are being explored to speed up the neural network simulations. Many array processors using digital signal procesors are being reported and made available commercially.[14] Custom chips using both analog and digital associative memory techniques are being studied. A programmable neural network processor (PNNP) would be a better solution to exploit various algorithms. So far, a hybrid approach using an associative-memory type of implementation seems to be superior in performance. It also has the potential to be used in submicron VLSI technologies. More work is needed to cast applications into some form of neural network problem. Slower applications that is, ≥ 20 μs, will be dominated by mostly analog design approaches, whereas a combination of analog and digital techniques will be used for high-speed applications, that is, ≤ 100 ns.

REFERENCES

1. J. J. Hopfield, "Neural Networks and Physical Systems with Emergent Collective Computational Capabilities," *Proc. Natl. Acad. Sci., U.S.A.*, **79**, 2554 (1982).

2. J. S. Denker, Ed., *Neural Networks for Computing*. Am. Inst. Phys., New York, 1986.

3. R. P. Lippmann, "An Introduction to Computing with Neural Nets," *IEEE ASSP Mag.*, **4**(2), 4 (1987).

4. S. Grossberg, "How Does the Brain Build a Cognitive Code," *Psychol. Rev.,* **87**(1), Jan. (1980).

5. B. Kosko and C. Guest, "Optical Bidirectional Associative Memories," *Proc. SPIE—Int. Soc. Opt. Eng.,* **758,** Jan. (1987).

6. Y. Abu-Mostafa and D. Psaltis, "Optical Neural Computers," *Sci. Am.,* **256**(3), 88 (1987).

7. B. Kosko, "Constructing An Associative Memory," *Byte,* p. 137, Sept. (1987).

8. P. A. Penz and R. Wiggins, "Digital Signal Processor Accelerators for Neural Network Simulations," *AIP Conf. Proc.,* **151,** 345 (1986).

9. M. A. Sivilotti, M. R. Emerling, and C. A. Mead, "VSLI Architectures for Implementation of Neural Networks," *AIP Conf. Proc.,* **151,** 408 (1986).

10. H. P. Graf and P. deVegvar, "A CMOS Associative Memory Chip Based on Neural Networks," *Int. Solid-State Circuit Conf. Dig. Tech. Pap.,* pp. 304 and 437, Feb. (1987).

11. D. J. Amit, H. Gutfreund, and H. Somplinsky, "Storing Infinite Number of Patterns in a Spin-Glass Model of Neural Networks," *Phys. Rev. Lett.,* **55,** 1530 (1985).

12. J. J. Hopfield and D. W Tank, " 'Neural' Computation of Decisions in Optimization Problems," *Biol. Cymbernet.,* **52,** 141 (1985).

13. D. W. Tank and J. J. Hopfield, "Simple 'Neural' Optimization Networks: An A/D Converter, Signal Decision Circuit, and a Linear Programming Circuit," *IEEE Trans. Circuits Syst.,* **CS-33**(5), 533 (1986).

14. *ANZA—Neurocomputing Coprocessor System,* Hecht-Neilsen Neurocomputing Corporation, San Diego, CA, 1987.

11 Rapid Thermal Processing of Silicon

AVID KAMGAR

AT&T Bell Laboratories
Murray Hill, New Jersey

11.1 INTRODUCTION

Fine lithographic techniques along with self-aligned processing provide the resolution needed for the shrinking device dimensions as the trend toward making submicron devices continues. To maintain the lateral geometric integrity and reduce the vertical dimensions, the overall thermal budget has to be kept to a minimum. Any thermal process is best performed at a given temperature for a specific, usually short, length of time. In conventional furnaces, much time is needed for isothermal heating of the massive boats along with the wafers, which is accomplished by convection of heat from the walls of the furnace to the wafer edges and by conduction to the rest of the wafer. It must be done at a slow rate to prevent slippage in the wafer due to temperature nonuniformities. In this type of process, wafers are kept for long times at temperatures below the optimum process temperature, causing unwanted, thermally activated processes such as diffusion to take place. In rapid thermal processing (RTP), on the other hand, the ramp-up times are short (a few seconds compared with ~30 min). The entire surface of the wafer is exposed to the heat source and the wafer heats uniformly by absorbing radiation from an incoherent source in several seconds. Hence, the process can be carried out at its optimum temperature for a well-controlled length of time. For this and several other reasons that help in improved processing, such as the possibility of real-time process monitoring or various in situ processings, the single-wafer rapid thermal processing becomes a necessity in most thermal applications.

RTP began in the late 1960s, when laser heating for making semiconductor diodes[1,2] was used, and in the mid-1970s, with the discovery of implant activation and regrowth of amorphized Si using high-intensity laser pulses.[3-6] Laser annealing turned out to be an interesting and useful research tool, but in wider and more practical applications it soon gave way to incoherent radiation annealing. The use of rapid thermal anneal spread quickly from implant activation and damage removal to virtually all other thermal processes needed for device fabrication.

434

This chapter is an overview of the many applications of RTP to device processing. It also presents the knowledge gained, with the help of short and well-defined RTP cycles, about the kinetics of several processes. We will concentrate on the application of RTP to Si technology, even though it has been extensively studied in the case of III–V compound semiconductors as well.

11.2   ADVANTAGES OF RAPID THERMAL PROCESSING

As mentioned in the introduction, RTP reduces the heating and cooling cycles to several seconds and provides a better control of the peak temperature. A process which may take on the order of 1 h in a conventional furnace can therefore be shortened by two orders of magnitude. This feature is obviously useful in limiting unwanted diffusions, hence controlling feature size. Furthermore, given short, well-controlled anneal times with reduced thermal loading, the actual thermal process can be done at relatively high temperatures. Many of the device properties such as quality of Si–SiO$_2$ interface, dopant activation, defect removal, crystalline quality of deposited layers, silicide resistivity, and glass reflow benefit from high-temperature processing. In addition, because it is a single-wafer processing technique, RTP offers several features, discussed below, that can lead to better device processing.

The RTP chambers are made either with quartz or water-cooled metal. In either case, during the short thermal cycles the furnace walls remain at temperatures below 100°C. This prevents possible contaminants from out-gasing from furnace walls and interfering with the thermal process. This, for example, is particularly important in case of gate oxide growth.

The small volume of RTP chambers allow them to be easily pumped out and new gases admitted in short times. This feature is valuable in carrying out the multistep processes necessary to achieve complex structures. For example, in the case of thin gate structures, it allows the series of processes, including cleaning, oxidation, anneal, and nitridation, as well as polycrystalline silicon deposition, to be performed without the wafer leaving the chamber.

Single-wafer processing allows for real-time process analysis. The temperature of the wafer can be measured by optical or infrared pyrometry. Other, more complex analyses such as ellipsometry and gas analysis can also be performed during the actual process.

RTP systems offer another advantage in response to the demand for the continuously increasing wafer size. Conventional furnaces demand a large expenditure of space and energy both to heat and to dissipate the heat simultaneously. However, in RTP systems, the corresponding increase in size and energy, considering minimal thermal loading and short times, is far less.

11.3   FUNDAMENTALS OF RAPID THERMAL PROCESSING

Incoherent radiation from a tungsten lamp or similar source is absorbed in Si by causing electronic excitations within the conduction band or valence band or by ionization across the gap. The excited electrons ultimately lose their energy by emission of phonons, thereby raising the lattice temperature. The kinetics of heating a Si wafer from room temperature to above 1000°C, using incoherent light, is quite complicated as it involves several temperature and wavelength-dependent absorption mechanisms. It also depends on conductive heating through the wafer thickness when the absorption length of the light is smaller than the sample thickness and on heat losses to the surrounding medium via radiative or convective cooling.

The rate of temperature rise can be written as

$$\frac{dT}{dt} = \frac{P_{abs} - P_{out}}{\rho w c}, \tag{1}$$

where ρ, w, and c are the density, thickness, and specific heat of Si, respectively, $P_{abs} = P_{incident} - P_{reflected} = P_{inc}(1 - R)$ is the power absorbed in Si, and $P_{out} = P_{radiation} + P_{convection}$ is the heat loss due to radiation and convection. Radiation losses are dominant in particular at high temperatures, hence

$$\frac{dT}{dt} = \frac{1}{\rho w c} \{P_{inc}[1 - R(T) - \varepsilon(T)\sigma T^4\} \tag{2}$$

or

$$\frac{dT}{dt} = \frac{1}{\rho w c} \left[\int_0^\lambda P_{abs}(\lambda, T)\, d\lambda - \varepsilon(T)\sigma T^4 \right]. \tag{3}$$

The power absorbed in Si is a strong function of the wavelength of the incident radiation, the sample temperature, and the sample doping. Si is opaque to visible radiation due to bandgap absorption, with an absorption coefficient on the order of 10^5 cm^{-1}. This absorption is cut off, at room temperature, in the infrared region above 1.1 μm wavelength and at longer wavelengths at higher temperatures (2.8 μm at 1400°C) due to the reduction in the bandgap.

The bandgap absorption coefficient for indirect transitions, involving absorption or emission of a phonon with energy E_p, has the following form[7]:

$$\alpha = \frac{A}{E_g^2} \left[\frac{(\hbar\omega - E_g - E_p)^2}{1 - e^{-E_p/kT}} + \frac{(\hbar\omega - E_g + E_p)^2}{e^{E_p/kT} - 1} \right], \tag{4}$$

where A is a constant, $\hbar\omega$ is the incident photon energy, and E_g is the bandgap in Si with the following temperature dependence[8]:

$$E_g(T) = 1.16 - \frac{7.02 \times 10^{-4} T^2}{T + 1108}, \tag{5}$$

where 1.16 eV is the bandgap at 4.2 K and the temperature is in Kelvin.

In the longer wavelength regions free carriers absorb light. This absorption increases linearly with the number of free carriers; hence, it strongly depends on the temperature of the sample and the dopant concentration. The bandgap and free-carrier absorption for Si at $T = 0$ K are shown in Fig. 11.1, as well as radiation spectra for two different sources.

The free-carrier absorption coefficient due to excitations in a single band is

$$\sigma = \frac{e^3 \lambda^2 N}{4\pi^2 \varepsilon c^3 n m^{*2} \mu},$$
(6)

where N is the density of free carriers (either electrons or holes), n the refractive index, m^* the effective mass, and μ the mobility. The density of the free carriers has a strong dependence on temperature. Figure 11.2 shows the calculated free-carrier concentration for n-type Si with an impurity concentration of 1.1×10^{15} cm^{-3} as a function of temperature.[9] There are three distinct regions in the curve: (1) the low-temperature range with partial ionization of impurities, (2) the saturation region with complete ionization the impurities, and (3) the high-

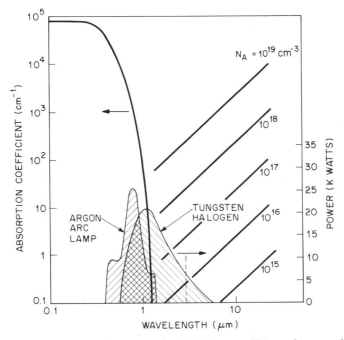

Fig. 11.1 The Si bandgap, and free-carrier absorption coefficients for several different concentrations at $T = 0$ K. For comparison, radiation spectra for two different sources are also shown. The average power for RTP systems using tungsten or arc lamps is indicated on the right axis. The dashed line at 3.5 μm represents the quartz cutoff.

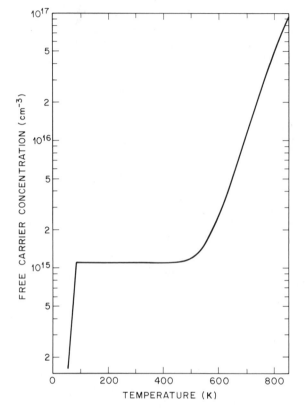

Fig. 11.2   Free-carrier concentration as a function of temperature for n-type, 4 $\Omega \cdot$ cm Si. (After Siregar et al.[9] Reprinted with permission, copyright 1980, *Applied Physics Letters.*)

temperature range (above $\sim$500 K) with increasing free carriers due to the thermal generation of electron-hole pairs.

The rate of temperature rise in the wafer depends on both the light source and the properties of the wafer. The specifics of the wafer, such as the doping concentration, also have an influence on the temperature rise. The dependence of wafer heating rate on doping level has been reported by Seidel et al.[10] However, this effect will not delay achieving high temperatures significantly. As Fig. 11.2 demonstrates, above $\sim$200°C the rate of the thermal generation of carriers increases by about one order of magnitude per 100°C, rapidly dominating the impurity carrier concentration.

The surface condition of the water, including any topography, influences the absorption as well as losses due to emission and reflection. The layered structure on a wafer surface has a large impact as well. For a reproducible temperature cycle, it is better to irradiate the wafer from the back side, making certain that any grown or deposited layer is removed from the back prior to a RTP cycle. For a rapid rise time, it is clearly desirable to utilize the highest lamp power available and then to reduce the power to an appropriate level to produce the desired steady-state temperature.

11.4 RAPID THERMAL PROCESSING EQUIPMENT

RTP equipment for experimental purposes need not be more complicated than a bank of tungsten lamps and a rectangular quartz tubing, with or without a gas-handling system. In industry, on the other hand, high throughput, automation for wafer handling, ease of operator interface with the system, temperature uniformity, etc., become essential. Commercial RTP systems differ in several aspects, some of which are discussed below.

11.4.1 Heating Units

Different manufacturers prefer different heat sources. The most popular seems to be tungsten halogen lamps. Arc lamps, either conventional or water-wall dc arc lamps, are also used. Graphite heaters were used in some older systems. These light sources differ in their spectral output, as shown in Fig. 11.1. Tungsten halogen lamps radiate in the visible and infrared range, arc lamps in the uv, visible, and infrared, and graphite heaters in the infrared. Lamps are turned on and off according to the specifics of the anneal cycle, but the graphite heaters require shutters for short time cycles because of the thermal mass of the heaters themselves.

Another difference in the heating unit between different systems is the arrangement of irradiation from one or both sides. Heating from one side has a major advantage. As we mentioned briefly in the previous section, the reflecting and absorbing properties of a wafer depend on layered structure on its surface. The top surface of the wafer normally has several patterned layers which become more complex at later stages of processing. On the other hand, the back side can always be stripped and cleaned to the bare Si, and if only the back side is exposed to radiation a more predictable and reproducible temperature cycle can be achieved.

11.4.2 Chambers

Since RTP chambers are designed for single-wafer processing, they are generally much smaller than conventional reactors. They are made of quartz or water-cooled metal in rectangular or cylindrical configurations. Rectangular quarts tubes can be used in closed gas systems by purging gases at atmospheric pressures. Cylindrical tubes offer the advantage of pumping capability. With proper gas-handling equipment, these tubes can also be used for chemical vapor deposition of various films such as polycrystalline Si, or tungsten.

11.4.3 Temperature Measurement and Uniformity

Thermocouples and pyrometers are the popular means for measuring temperature. Temperature measurement with thermocouples is usually done on control chips, whereas pyrometry can be done on the actual wafer. Either technique can be used for feedback to a closed-loop operation.

Wafer temperature can be made uniform by several techniques, the most effective of which is uniform illumination. Edge reflectors around the wafer perimeter or light-source reflectors in the chamber are also commonly used, as well as other sophisticated methods. However, some of these techniques may result in a steady-state uniformity by jeopardizing the uniformity during ramp up or ramp down. This is due to overcompensation for the radiation heat loss from the wafer perimeter.

11.4.4 Wafer Holders and Wafer Handling

Wafers are held either with the device side up or down. It is argued that with the device side down the danger of surface contamination by occasional particulates falling on the wafer is reduced. Wafers rest on a low thermal mass holder, such as three or four quartz pins for thermal isolation and stress reduction. In the smaller bench-top units, which are mostly used for experiments, wafer loading is done manually. In the larger, production-type units used in process lines, semiautomatic or cassette-to-cassette handling is now standard.

11.5 TEMPERATURE MEASUREMENTS

Accurate determination of temperature, particularly during transient times, is a difficult task. Both optical and infrared pyrometers have been used for temperature measurements in RTP systems, as well as various kinds of thermocouples. Optical pyrometers measure the visible brightness of the specimen and, hence, are sensitive to temperatures above ~700°C. For determining the actual temperature, the emissivity of the wafer is needed and this may not be accurately known, especially if there are layers of oxide or other material on the wafer surface. Also, particular attention should be taken to shield the pyrometer from sensing stray radiation from the source. Infrared pyrometers give better low-temperature sensitivty, although with the added difficulty that at lower temperatures, where the wafer is transparent to longer wavelengths, the pyrometer may sense the source through the thickness of the wafer.

Thermocouples, when used properly, can be utilized for system calibration or as an approximate temperature monitor when attached to a satellite control chip. The main factor influencing accurate temperature measurements with thermocouples is the degree of thermal contact with the sample. The heat loss via thermocouple wires is another concern. There are a number of ways to attach a thermocouple; various conducting or nonconducting cements and epoxies that do not disintegrate during the high-temperature cycles can be used, as well as welding the thermocouple to the wafer. Sophisticated thin film thermocouples evaporated directly on the sample, such as used in Ref. 11, can be employed to make accurate transient temperature measurements.

Measurements of certain physical parameters, such as oxide thickness or sheet resistance ($R_\square$) of implanted wafers, have been suggested as indirect means of temperature determination or uniformity of temperature. However, the oxide thickness measurements are somewhat long and cumbersome. The $R_\square$ measurements can lead to erroneous results for short anneal times (see Section 11.8.1) and are fairly insensitive for long times.

In short, a standard and accurate technique for measuring temperature has not yet been established. Nevertheless, in using RTP systems in process lines this should not be a deterent so long as the reproducibility of a thermal cycle can be made reliable. With limited effort, an RTP system can be characterized for an optimum thermal cycle for a particular process without an exact knowledge of its detail.

11.6   STRESS DUE TO RAPID THERMAL ANNEAL

With improved temperature uniformity of the RTP systems, the degree of thermal stress and the resulting slippage decrease drastically,[12] However, dynamic temperature gradients through the thickness of a wafer, inherent to RTP, may lead to unexpected problems, in particular where layered structures with different thermal properties are involved.

Several studies of the effects of rapid thermal anneal (RTA) on Si–SiO$_2$ structures have been reported. Weinberg et al.[13] reported a reduction of electron trapping on water-related sites due to a post-oxidation RTA of 50 nm oxides. Lee et al.[14,15] studied the effect of RTA on oxide (80–180 Å) polycrystalline Si gate structures and measured no increase in the interface fixed charges due to RTA processing. They found no significant RTP-induced degradation in the electrical properties of the thermal oxides in capacitor or MOSFET structures except for an unexplained increase in trapping behavior after a 1100°C, 30 s anneal.

In both above studies,[14,15] however, the measurements were carried out after a standard post-metallization anneal in forming gas. A more recent study on Si gate capacitors (4000 Å As-doped polysilicon on 175 Å oxide) showed that due to RTA, interface traps were indeed generated at the Si–SiO$_2$ boundary at temperatures as low as 600°C, but that they were passivated by a standard post-metallization anneal.[16] Figure 11.3a shows the fraction of flat-band voltage shift that is restored by the hydrogen anneal. The generation of interface traps is possibly due to an increase in the disorder of the Si–Si bonds at the Si–SiO$_2$ interface, caused by the difference in the thermal properties of Si and SiO$_2$. During RTA ramp up or down, this difference enhances the dynamic temperature nonuniformity. This study also showed that an increase in either ramp-up or -down rates increased interface trap generation.[16] High-temperature RTA, above 1000°C, induced, in addition, a residual flat-band voltage shift (Fig. 11.3b). This was attributed to either or both changes in the oxide fixed charges, and polysilicon structure, that is, grain size and As redistribution resulting in a change in the polysilicon work function.[16] The RTA effects on polysilicon are discussed in more detail in Section 11.10.

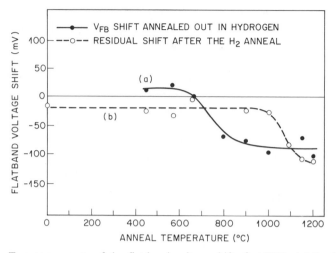

Fig. 11.3 Two components of the flat-band voltage shift after RTA: (*a*) the fraction restored by a H_2 anneal and (*b*) the residual shift. (After Kamgar and Hillenius.[16] Reprinted with permission, copyright 1987, *Applied Physics Letters*.)

11.7 GATE DIELECTRICS

Devices of submicron dimensions require thin gate dielectrics of 40–100 Å, necessitating the development of new dielectrics or new techniques for growing thin oxides. In this section, we will discuss rapid thermal oxidation for growing thin oxides and rapid thermal nitridation of oxides for achieving a promising gate dielectric.

11.7.1 Rapid Thermal Oxidation

The use of RTP for growing thin oxides is attractive in a number of ways. Because of the short, well-controlled process times, thin oxides can be grown at high temperatures. Oxides grown at higher temperatures possess a higher-quality interface with the Si.[17] The feasibility of carrying out *in situ* processes for making complex gate structures is another attractive feature. Such a sequence might include pre-oxidation clean, oxidation, post-oxidation anneal, and nitridation of the oxide, as well as deposition of polysilicon on thin oxides, all performed in a single chamber without the wafer being removed. One example of the thermal cycles used for rapid thermal clean (RTC), oxidation (RTO), and anneal (RTA) is shown in Fig. 11.4*a*[18] The *in situ* processing should improve the yield because of reduced wafer handling. In addition, the chance of contamination from the hot walls of an oxidation furnace[19] is reduced because of the low temperature of the RTP chambers.

Successful growth of thin oxides with RTO has been reported in a number of publications.[20] Figure 11.4*b* shows an example of SiO_2 film thickness as a function of time and temperature for oxidation in dry oxygen. The reported times and temperatures for obtaining a given oxide thickness vary greatly in different publi-

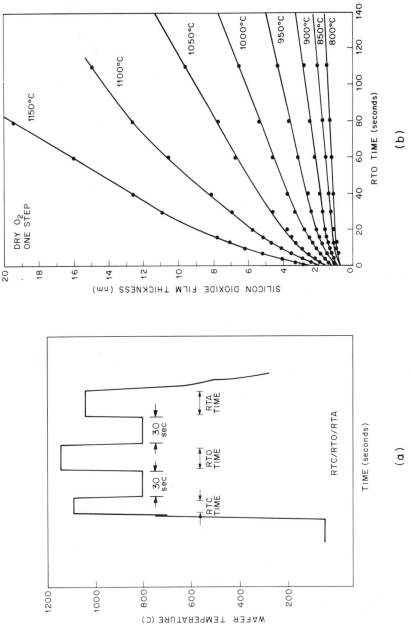

Fig. 11.4 (*a*) An example of a set of sequential thermal cycles used for growing SiO_2. (*b*) Oxide thickness vs. anneal time for various temperatures. (After Nulman.[18] Reprinted with permission, copyright 1987, Electrochemical Society.)

cations. Factors influencing the variations are the thickness of the native oxide, the initial oxide grown during the temperature ramp up, and the lack of a standard method for determination of process time and temperature in RTP systems. Activation energy for oxide growth using RTO is found to be similar to furnace oxidation,[21] indicating the mechanism is the same.

Oxides as thin as 30 Å are reported to have been grown reproducibly on wafers that received a RTC prior to oxidation. Breakdown voltages at 15 MV/cm (measured at 3 nA/cm) and total interface traps of 1×10^{10}–10^{11} cm^{-2} make RTO oxides comparable to low-temperature furnace oxides. The oxide uniformities are reported to be better than ±2% even on 6 in. wafers, with wafer to wafer reproduciblity of better than 1.5%.[18] High-resolution transmission electron microscopy has shown that RTO oxides have a sharper transition with Si compared with oxides grown at 850°C in conventional furnaces.[22]

Another interesting and important application of RTO is growing oxides in trench structures. RTO can be carried out at temperatures above the viscous flow point of SiO_2. Hence, the thinning problem at the corners, observed in conventional low-temperature furnace oxidations, is not observed in the case of RTO.[23]

11.7.2   Rapid Thermal Nitridation of SiO_2

Nitroxides have long been considered promising candidates for gate dielectrics because of their excellent barrier property and high dielectric constant. Furnace operations which yield nitroxides with acceptable properties are carried out in pure ammonia at temperatures of approximately ~1200°C for 1–4 h.[24] To reduce the thermal cycle, other techniques such as plasma, anodic plasma, and fluorine enhancement have been introduced, each with some success along with its own drawbacks. Rapid thermal nitridation offers a more practical solution to the formation of nitroxides by nitriding at temperatures of ≳1150° for times up to a few minutes.[25,26] After a 5 min anneal in ammonia at 1150°C, the nitrogen composition is reported to become quite uniform throughout the oxide.[25] These nitridations were done at relatively low temperatures and long rise times.

RTP offers the possibility of carrying out the nitridation in a different time–temperature regime from those performed in conventional furnaces. Nitridation of oxides involves two steps with different activation energies. The process with the higher activation energy is the barrier formation on the surface of the oxide. The diffusion of nitrogen through the oxide, on the other hand, has a lower activation energy. This nitridation process, which is to a great extent self-limiting, stops further diffusion of N into the SiO_2, hence preventing an excessive buildup of nitrogen at the interface. By taking advantage of the rapid temperature rise in a RTA furnace, formation of a novel dielectric structure, consisting of a thin, ~20 Å barrier layer of nitroxide on the gate SiO_2, has been accomplished.[27]

Figure 11.5 shows the Auger electron spectra for nitroxides formed in two different RTN times at about ~1250°C with less than 10 s rise time to maximum temperature.[28] These spectra illustrate the barrier property of the thin nitroxide film by showing the lack of increase in the amount of nitrogen in the oxide and at

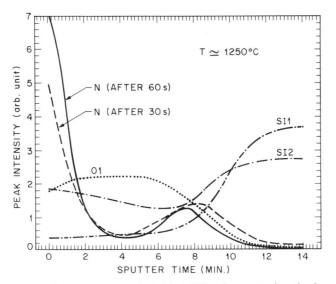

Fig. 11.5 Auger electron spectra of nitrided SiO_2, demonstrating the increase in the amount of nitrogen at the SiO_2 surface for the longer nitridation time, yet the lack of increase in the N content of bulk SiO_2 or Si–SiO_2 interface.

the Si–SiO_2 interface as the nitridation time increases, despite the increase in N concentration at the surface. Such nitroxide films were able to stop Na ion contamination and were resistant to dry oxidation at 900°C for times up to 2 h. Aluminum gate capacitors formed on these nitroxide films (1250°C, 60 s) showed an increase of 7% in the charge storage capacity of 130 Å furnace-grown oxide. Also, the dielectric breakdown (measured at 2 μA/cm) of the furnace oxide was increased from 12.5 MV/cm to 13.5 MV/cm.[28] The flat-band voltage of these capacitors showed a shift to positive voltages due to the inevitable buildup of a small amount of N at the interface. Figure 11.6 shows the corresponding total interface trap density as a function of nitridation time at different temperatures, again illustrating that the pile-up of nitrogen at the interface is limited when nitridation is done at elevated temperatures.

11.8   POST-IMPLANTATION ANNEAL

Annealing of implanted dopants is the most widely studied of all RTP applications. The interest it holds, both scientifically and technologically, lies with the fact that several thermally activated processes with different activation energies are involved. Thermal anneal after implantation crystallizes the implantation-amorphized substrate, activates the dopants, diffuses the dopants, and removes the implantation-induced lattice damage. The crystallization process takes place very rapidly at typical RTA temperatures. Its activation energy is around 2.7 eV, and,

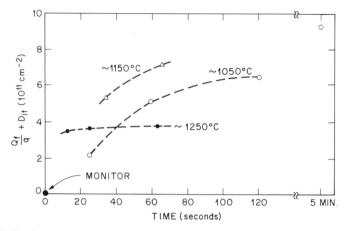

Fig. 11.6 Total interface trap density (oxide fixed charges O_f + interface traps D_{it}) of nitroxide gate capacitors formed at three different temperatures and lengths of time.

for example, at 800°C the amorphized Si crystallizes at a rate of about 1 $\mu m/s$.[29] During this solid-phase epitaxy (SPE), group III and V dopants, because of their similarity with Si, are incorporated on substitutional sites where they are electrically active.

The two competing processes are defect removal and dopant diffusion. The removal of complex defects requires high temperature, and with the long ramp-up/-down times used in conventional furnaces, the dopants can diffuse into the Si beyond an acceptable range. Because these two processes have different activation energies, RTA offers the time and temperature flexibility to remove defects with little or no diffusion of dopants.

11.8.1 Arsenic

Rapid thermal annealing of As-implanted Si is one of the oldest applications of RTP,[30] but it has not yet lost its charm and mystery. Particularly for high doses of As, the anneal kinetics of As-implanted Si is complicated by solid solubility levels, clustering of As atoms, generation of defects, implant damage removal, and transient diffusion effects.

11.8.1.1 Activation and Clustering. The solid solubility and clustering effects are reflected in the sheet-resistance ($R_\square$) data shown in Fig. 11.7. These measurements were carried out on (100) Si wafers implanted with As (40 keV, 5 × 10^{15} cm^{-2}; the corresponding maximum concentration is ~2 × 10^{21} cm^{-3}, centered ~270 Å from the Si surface) and annealed under different conditions. Arsenic amorphizes Si when the implant dose is above ~2 × 10^{14} cm^{-2}. The initial low value for $R_\square$ represents a high degree of electrical activation resulting from the incorporation of As atoms into the substitutional sites during SPE, with a minimum amount of clustering. A certain degree of clustering during the cool-down cycle is inevitable.

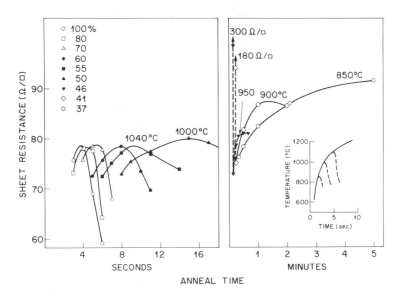

Fig. 11.7 Sheet resistance $R_\square$ as a function of anneal time for several lamp powers (As implanted at 40 keV, 5×10^{15} cm^{-2}). The rise and fall times of the power supply are chosen to be <1 s. The times indicated are the power supply on times. The approximate time–temperature profile for the 100% power is given in the inset. The approximate steady-state temperatures for longer anneal cycles are indicated on the corresponding $R_\square$ curves. (After Kamgar et al.[31] Reprinted with permission, copyright 1986, *Applied Physics Letters*.)

The rise in $R_\square$ for short anneal times is due to the clustering of As atoms. The eventual decrease is due to As diffusion, which lowers the peak concentration, decreases the clustering, and increases the mobility as well. Note that as the anneal temperature is lowered, the maximum in $R_\square$ assumes higher values. This is because of the low activation energy for As clustering ($E_a \sim 1.1$ eV,[31,32] which favors clustering over diffusion, $E_a \sim 4$ eV) at low temperatures.

The dependence of $R_\square$ on anneal time for several As implant doses, shown in Fig. 11.8, illustrates the clustering effects as it becomes more pronounced with increasing As concentration. Clustering of activated As atoms occurs also if Si is given a second anneal at low temperatures for long times. This effect can be reversed by a high-temperature short-time RTA; however, extra diffusion may take place. Figure 11.9 shows the sheet resistance for three sets of short-/long-/short-time anneals. The long-term anneal is done at ~790°C for 5 min. Data in Fig. 11.9 illustrate that the 100%, 3s thermal pulse which is sufficient to activate the dopants fully during the SPE is not able to decluster and electrically activate the clustered As. The 100%, 4 s pulse does decluster the As atoms and restores the $R_\square$ value to its pre-long-term anneal, but the 100%, 5 s pulse lowers $R_\square$ further than its original value due to the diffusion of As atoms into the Si substrate.

11.8.1.2 Defects. An interesting set of cross-sectional transmission electron micrographs illustrating the evolution, coalescence, and eventual disappearance of

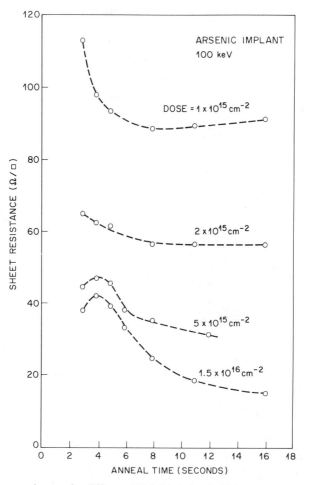

Fig. 11.8 Sheet resistance for different 100 keV As implant doses annealed at 100% lamp power as a function of anneal time. For approximate temperatures, see inset in Fig. 11.7. (After Kamgar and Baiocchi.[32] Reprinted with permission, copyright 1986, Materials Research Society.)

the extended defects as the anneal time and temperature increase is shown in Fig. 11.10. Figure 11.10a shows the 3000 Å thick layer of Si amorphized by the As (40 keV, 5×10^{15} cm^{-2}) implantation. Figures 11.10b–f show the different stages of anneal at 1 s steps as the temperature rises to ~1200°C after 6 s. The corresponding $R_\square$ values are shown in Fig. 11.7 for the 100% power.

As we have already mentioned, the initial stage of anneal is the crystallization of amorphous Si. Subsequently, two layers of extended defects are formed. The dislocations closer to the Si surface are located at the peak of the As concentration and are likely to be related to clustering of As atoms. The As clustering is accompanied by generation of dislocations in the host lattice (as discussed in the next

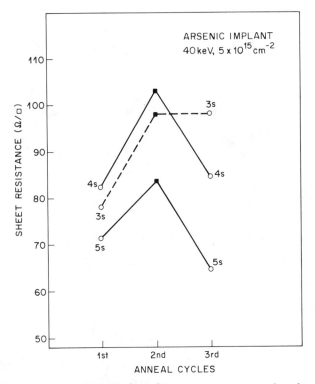

Fig. 11.9 Effect of sequential high-/low-/high-temperature anneal cycles. The second anneal in all cases was done at 790°C for 5 min. The first and third anneals were done at 100% power for the times indicated in the figure. (After Kamgar and Baiocchi.[32] Reprinted with permission, copyright 1986, Materials Research Society.)

paragraph). The second layer is the so-called end-of-range dislocations due to straggling of As atoms into Si. This layer appears just beyond the amorphous/ crystalline (α/c) boundary. Figures 11.10*d–f* show that the two dislocation networks become larger by coalescing and eventually disappear.

The Rutherford backscattering (RBS) studies, however, indicate the presence of residual defects after the disappearance of the extended defects.[32,33] RBS data also indicate, as illustrated in Fig. 11.11, that along with the As clustering a certain disorder is experienced in the position of Si atoms. This disorder can be partially responsible for the decrease in conductivity.[33]

11.8.1.3 Diffusion. RTA, as is the case for conventional furnace anneal, causes thermally assisted dopant diffusion under extreme time and temperature conditions. The controversial issue in the case of short-time RTA is whether or not there is a transient effect associated with As diffusion into the Si substrate. There have been a large number of reports supporting the existence of anomalous diffusion and a large number opposing it. Inaccuracies in temperature measurements are thought

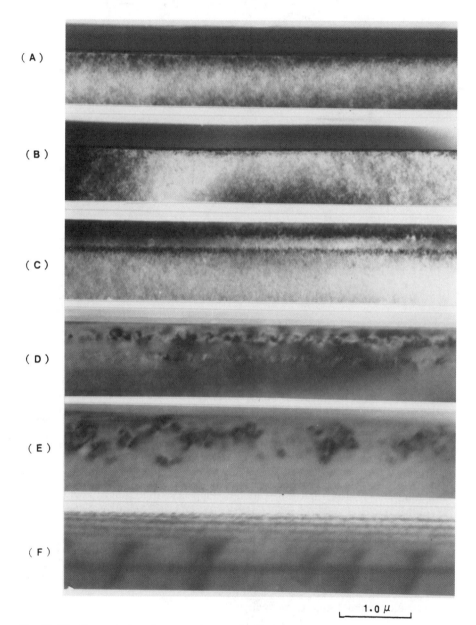

(A)

(B)

(C)

(D)

(E)

(F)

1.0 μ

Fig. 11.10 Transmission electron micrographs illustrating various stages in the anneal cycle of As-implanted Si, that is, crystallization, extended defect formation, their coalescence, and eventual removal. (After Kamgar et al.[31] Reprinted with permission, copyright 1986, *Applied Physics Letters.*)

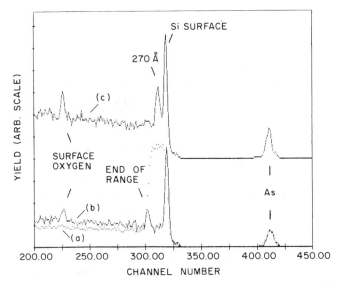

Fig. 11.11 RBS grazing exit angle channeled spectra for a sample As-implanted at 40 keV, 5×10^{15} cm^{-2}. (*a*) Before anneal (*y* axis yield increased by factor 5). (*b*) After 100%, 3 s RTA cycle. (*c*) After 790°C, 10 min anneal. (After Kamgar and Baiocchi.[32] Reprinted with permission, copyright 1986, Materials Research Society.)

to be the main reason for the discrepancies among the experimental results. For a summary of debates over this issue, see, for example, Refs. 34 and 35.

The transient diffusion is believed to take place in the first second or so of rapid anneal at temperatures above 1100°C. It is argued that the anomalous diffusion is facilitated by the implantation amorphized layer and the point defects, possibly vacancies in the Si. The transient diffusion is believed to have a small activation energy, 1.8 eV compared with the normal diffusion with $E_a \simeq 4$ eV. Whether there is a transient dopant diffusion is of importance in developing physical models describing dopant diffusion for computer device modeling. Some of the issues regarding diffusion vs. defect removal in implanted Si are discussed in Section 11.9.1.

11.8.2 Boron

Since boron is a light element, it has a long range when implanted into Si. Moreover, at room temperature it does not deposit the critical damage energy needed for amorphization of the Si substrate; consequently, it gives rise to long channeling tails. The high diffusivity of B in Si and a transient-enhanced diffusion of the channeling tail result in deep junctions when B-implanted Si is thermally annealed. A fraction of B atoms precipitates into the damaged sites at the peak of B concentration and does not contribute to the conductivity. A number of experiments have been performed to eliminate or minimize these problems by combining RTA with all or some of the potentially helpful techniques mentioned below.[36–41]

To minimize the channeling tail and the network of dislocations at the peak of the B profile, the Si substrate can be amorphized prior to the B implantation. As seen in the secondary ion mass spectroscopy (SIMS) data shown in Figs. 11.12a and b, this technique helps to reduce the channeling tail. The additional advantage is that the activation of B occurs during the SPE growth, and, hence, is achieved in shorter times and with higher solubility. Both Si and Ge have been used for pre-amorphization, with the latter being the preferred element for two reasons. Smaller doses of Ge are needed to amorphize the Si because of its heavier atomic weight, therefore reducing the contamination from implantation. It is also reported that compared with the Ge implantation damage the disorder due to Si self-implantation is more difficult to anneal out.[41]

Another technique is using BF_2 ions. In the implantation, BF_2 amorphizes about one-half of its implanted depth distribution. Also, at the same implantation energy, the B atoms have much lower energy and hence a reduced range. However, as illustrated in Fig. 11.13a, the channeling tail remains. In addition, the residual damage remaining just under the α/c boundary getters the fluorine. Very low energy B (1–2 keV) released from 5–10 keV BF_2 implantation at high doses has also been used.[39] Taking advantage of RTA and concentration-enhanced diffusion, junctions as shallow as 1000 Å have been reportedly obtained (Fig. 11.13b).

11.9 JUNCTION FORMATION

Using RTA for shallow junction formation is beneficial for both implanted and diffused junctions. In the case of implanted junctions, the RTA cycle, which is

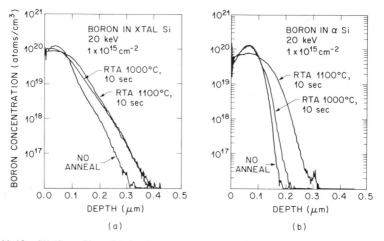

Fig. 11.12 SIMS profiles of B implanted into (a) crystalline Si and (b) into pre-amorphized Si (with Si), before and after several anneals. In the crystalline case, B is implanted through 32 nm of oxide. (After Sedgwick.[37] Reprinted with persmission, copyright 1986, Materials Research Society.)

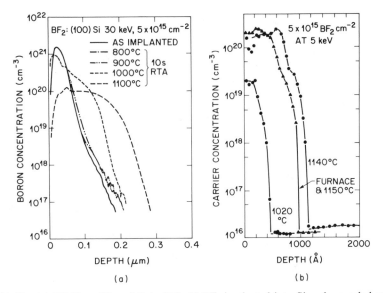

Fig. 11.13 (*a*) SIMS profiles of B in $30\,keV$ BF_2 implanted into Si and annealed at several conditions. (After Mikoshiba et al.[38] Reprinted with permission, copyright 1986, *Japanese Journal of Applied Physics*.) (*b*) Carrier concentration, measured by spreading resistance in 5 keV BF_2-implanted Si after anneal. The effective anneal times are 1 s. In one case, a furnace pre-anneal at 580°C, 20 min is used. (After Davies.[39] Reprinted with permission, copyright 1985, IEEE.)

done at high temperatures for short times, in addition to activating a larger fraction of the activated dopants, can anneal out most of the defects while maintaining the implanted profile. Also, dopants from a spin-on or deposited source can be diffused such that high surface concentrations with shallow diffusion depths are achieved.

11.9.1 Implanted Junctions

In 1982, Sedgwick[42] discussed the importance of short-time anneal in terms of time–temperature space available for shallow junctions. This argument is summarized in Figs. 11.14*a* and *b*. Although these plots show no overlapping region for removal of extended defects and a diffusion limited to 500 Å of either B or As, they indicate a definite trend favoring higher temperatures and shorter anneal times. This advantage was experimentally demonstrated later on in the case of As-doped n^+/p junctions.[43]

In Fig. 11.15, the junction leakage current as a function of As broadening for junctions annealed at a variety of temperatures and for different lengths of time is plotted. These junctions were made by implanting As (60 keV, 4×10^{15} cm^{-2}) through 150 Å sacrificial oxide into 6–8 Ω · cm resistivity, *p*-type (100) Si. It is clear that for a given As broadening, the anneal cycles performed at higher temperatures result in lower junction leakages. This fact is also demonstrated in Fig. 11.16, where the anneal time vs. reciprocal temperature is plotted for As dif-

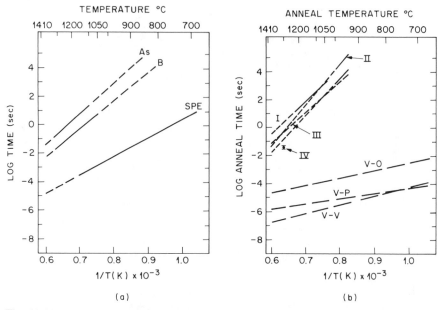

Fig. 11.14 Reciprocal temperature plots showing time required for (*a*) B and As diffusion by 500 Å and SPE of 1000 Å Si layer and (*b*) removing various point defects (lower curves) and extended defects (upper curves). (After Sedgwick.[42] Reprinted with permission, copyright 1983, *Journal of the Electrochemical Society.*)

fusion as well as dislocation removal.[10] In this case, removal of dislocations is limited to the defects that were observed in the TEM.

More recently, a systematic study of shallow p^+/n junctions has been carried out.[44] These junctions were obtained by BF_2^+ (49 keV, 2×10^{15} cm^{-2}) implanted in $15 - 25 \ \Omega \cdot$ cm B-doped (100) Si. A 1050°C, 15 s RTA cycle resulted in low-leakage shallow junctions ($\sim 0.18 \ \mu$m). In addition, it is reported that the junctions made in pre-amorphized Si (by Si self-implantation) resulted in leakage currents three orders of magnitude higher than otherwise.[44] Similar results were found in case of As-doped n^+/p junctions, that is, pre-amorphization by Si increased the junction leakage by one–two orders of magnitude.[45] The extra leakage is presumably due to the residual defects from the Si implant in the active depletion region.

11.9.2 Diffused Junctions

Shallow junctions can also be formed by diffusion from dopants in direct contact with the Si surface. The advantage of this technique is that the ion-implantation damage in the Si is absent, while high surface concentrations and shallow dopant profiles can be achieved by using RTA.

Boron from deposited boron nitride films[46] on the Si surface, or BN disks[47] in contact with the wafer have been diffused into Si. In both cases, shallow junctions of $\sim 0.1 \ \mu$m depth with 10^{19}–10^{20} cm^{-3} surface concentration have been obtained. Reported leakage currents[47] are also reasonable, $\sim 10^{-8}$ A/cm^2.

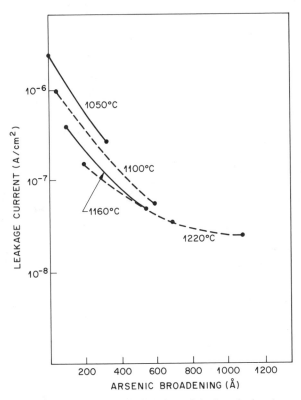

Fig. 11.15 Junction leakage current as a function of As broadening (measured by RBS), caused by RTA at several different temperatures for 60 keV As implanted through 150 Å oxide into Si.

Another technique is to implant the dopant into a silicide formed on the Si surface such that the peak of the implant concentration is either in the silicide or at the Si–silicide interface (see Section 11.11.1). The dopant is then driven into the Si by a suitable RTA cycle. For As implanted at the Si–TiSi$_2$ interface, junctions with $\sim 10^{20}$ cm^{-3} dopant concentration, <0.2 μm junction depth and $\sim 10^{-9}$ A/cm^2 leakage have been reported.[48]

In the case of dopants that are implanted within the silicide, even more encouraging results have been recently reported. Junctions made by implanting either B or A into CoSi$_2$ have resulted in $\sim 10^{20}$ cm^{-3} concentration, 0.05 μm depth, and $\sim 10^{-9}$–10^{-10} A/cm^2 reverse leakage current.[49]

11.10   POLYCRYSTALLINE SILICON ANNEAL

Polycrystalline silicon, or polysilicon, film is used extensively in MOS technology because of its compatibility with other device processing steps. As a gate electrode, it allows for a self-aligned gate and source and drain technology, and as an interconnect it improves layout flexibility. It is also used as high-value resistors

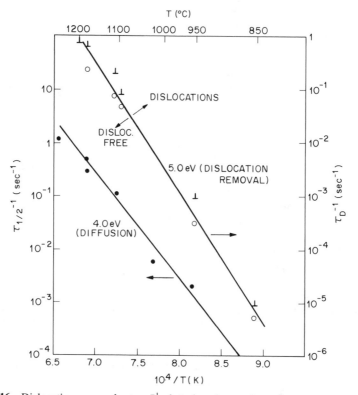

Fig. 11.16   Dislocation removal rate τ_D^{-1} plotted against reciprocal temperature (open circles). The bar symbols are for the shortest times where dislocations are still observed; 5.0 eV corresponds to silicon self-diffusion. The diffusion rates for concentrations to be reduced from a maximum nonannealed value to half that value at $x = R_p$ are also plotted ($\tau_{1/2}^{-1}$). (After Seidel et al.[10] Reprinted with permission, copyright 1985, Material Research Society.)

and as emitters in bipolar transistors. With decreasing device and circuit dimensions, the higher conductivity of doped polysilicon becomes more important, as well as the decrease in the emitter junction depth. In both aspects, RTA is of interest, because whether polysilicon is deposited on SiO$_2$ or on single-crystal Si, RTA enhances the growth rate of polysilicon grains,[50–53] resulting in higher mobility and less grain boundary segregation. RTA also limits the diffusion of dopants from a polysilicon diffusion source.[53,54]

11.10.1   Grain Growth

Polysilicon films are normally deposited by chemical vapor deposition at temperatures ranging from 560°C to 620°C, where at temperatures ≤580°C the film is in an amorphous phase and at ~600–620°C it is in a polycrystalline phase. Compared with the high-temperature polysilicon, the low-temperature polysilicon is known to present superior qualities when annealed in a conventional furnace.[55]

The difference in the final grain size and electrical activation of the two initial phases is, however, negligible when annealed in a RTP furnace.[56] The mechanism responsible for this is the following. In the case of conventional furnace annealing during the long ramp times, small polysilicon grains are formed. To overcome the inter-facial barriers between these small grains and to form large grains, long times or high temperatures are required. In RTA, however, with rapid ramp-up rates the intermediate stage of forming small grains is limited. Figure 11.17 shows an example of retarded grain growth when using conventional furnaces for a 0.2 μm amorphous polysilicon film deposited on 0.2 μm of SiO$_2$. This figure indicates that the times required to obtain a certain grain size are two orders of magnitude larger in conventional furnaces than with RTA.

Another interesting and useful aspect of RTA is that in the case of polysilicon deposited on single-crystal Si, the interfacial native oxide does not impede or retard the epitaxial alignment of polysilicon grain with the underlying Si. The breakup of the interfacial oxide layer into SiO$_2$ islands has been observed in cross-sectional high-resolution electron microscopy.[57] This makes the process less sensitive to variations in the interfacial oxide and reduces the variation in the current gain of bipolar transistors.

11.10.2 Dopant Activation

In most applications, a highly conductive polysilicon is needed. One way to achieve this is by implantation with the desired dopant. The polysilicon is then annealed to activate the dopant. The kinetics of dopant activation in polysilicon is more complicated than in single-crystal Si. There are two fundamental reasons for

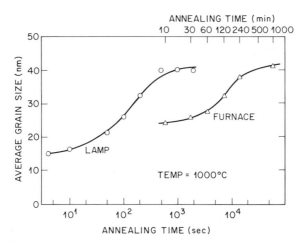

Fig. 11.17 Average polysilicon grain size vs. anneal time for RTA and conventional furnace anneal at 1000°C. (After Arai et al.[52] Reprinted with permission, copyright 1986, *Applied Physics Letters*.)

it: the changes in the polysilicon grain size and the segregation of dopants in and out of the grain boundaries during the heat treatments.

In Fig. 11.18, the sheet resistance of arsenic-implanted (100 keV, 5×10^{15} cm^{-2}) polysilicon (deposited at ~580°C) as a function of anneal time for different lamp powers is shown. The thicknesses of the polysilicon and the underlying oxide are 0.4 and 0.3 μm, respectively.[56] This figure illustrates a behavior very similar to the single-crystal Si shown in Fig. 11.7. However, the changes in $R_\square$ of polysilicon take place in much shorter time spans, which is an indication of faster diffusion times in polysilicon. Diffusion of dopants in polysilicon takes place via grain boundaries and is known to be orders of magnitude faster than the diffusion in single-crystal Si. Another interesting feature evident in Fig. 11.18 is that at higher anneal temperatures $R_\square$ saturates to lower values. This supports the argument made earlier, that RTA, because it is carried out at temperatures higher than the conventional furnace, does result in lower eventual sheet resistance.

Figure 11.19 shows the back-scattering yield in As-implanted polysilicon before anneal and after two different anneal conditions. Even for the 80%, 5 s an-

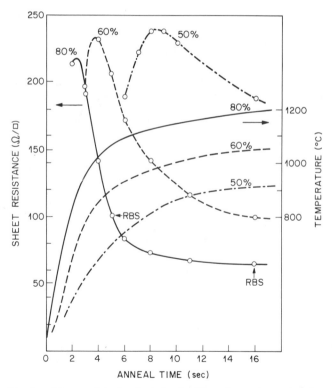

Fig. 11.18 Sheet resistance of As-implanted polysilicon (amorphous phase) as a function of anneal time for several lamp powers. (See Fig. 11.7 for the description of anneal cycles). The approximate peak temperatures are indicated on the right axis. The RBS points referred to are shown in Fig. 11.19.

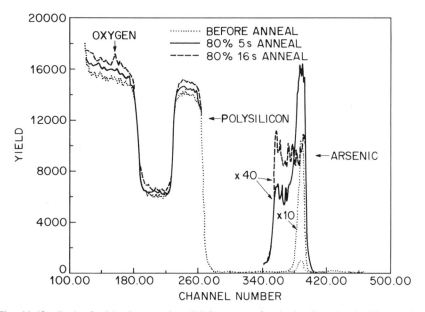

Fig. 11.19 Rutherford back-scattering (RBS) spectra for As-implanted polysilicon before and after two different anneals.

neal condition, there is an appreciable movement in the As atoms, the original peak is broadened by a factor of 2, and the tail of As has diffused 0.4 μm. In the case of single-crystal Si, the same anneal did not cause any measurable movement of As atoms.[31] After 16 s, the As is redistributed uniformly across the polysilicon with a small pile-up at its two interfaces with the oxide. In this experiment, no capping oxide was used; however, the RTA was performed in air.[56] The thermal oxide grown during the RTA of polysilicon is sufficient to prevent the out-diffusion of As from the top surface, even though out-diffusion of As from a polysilicon surface is known to be far more severe than in the case of single-crystal Si. The small oxygen peak denoted in the figure illustrates the gradual growth of the oxide on the surface.

Another complication in the case of polysilicon is that As segregates into the grain boundaries. This is perhaps the major contributing factor to the increase in $R_\square$ seen in Fig. 11.18.

The advantage of RTA in the case of polysilicon is even greater than in the case of single-crystal Si due to the grain growth kinetics of polysilicon. The larger grains result in fewer grain boundaries, hence higher electronic mobility values. There are also fewer sites for As to segregate to. In addition, the initial state of polysilicon, that is, amorphous or polycrystalline, is of little or no consequence. Below are the resulting differences between high- and low-temperature polysilicon, and single-crystal Si that received the same implant (As, 100 keV, 5×10^{15} cm^{-2}) and anneal.

| | | Polysilicon | | Single-Crystal Si |
		Amorphous	Polycrystalline	
Sheet resistance (Ω-cm)	RTA 1150°C, 10 s	70	70	25
	Furnace 950°C, 30 min.	140	290	40

Activation and redistribution of B and P have been studied in detail.[58] Thermal cycles required for satisfactory electrical activation in polysilicon are reported to be similar to those used for single-crystal Si. The existence of a capping layer is essential in the case of P, but B does not seem to out-diffuse from the surface.

11.11 SILICIDE FORMATION

Perhaps the most popular application of RTP in today's technology is in the formation of $TiSi_2$. Self-aligned silicide, or salicide, contacts have become an essential ingredient in VLSI technology. They are used over polysilicon to reduce the gate resistance, on source-drain areas to lessen the contact resistance, and as interconnects for reduced RC delay times. The fabrication of most silicides has turned out to be easier and more reliable when performed in a RTP furnace. We will discuss the RTP of $TiSi_2$ and $CoSi_2$ in some detail; however, RTP has successfully been used in the formation of many silicides, such as W[59,60] Mo,[60–63] Ta,[61] and Ni[64] silicides.

11.11.1 Titanium Silicide

$TiSi_2$ has the lowest resistivity of all silicides[65] and is therefore most suitable for VLSI technology. However, its formation presents more problems than most other silicides. Ti is very reactive with oxygen. Also, in $TiSi_2$ formation Si is the dominant diffuser, hence linewidth control and bridging of silicide onto SiO_2 become problems. In conventional furnaces, an ultrapure N_2 atmosphere (i.e., no oxygen) is needed to prevent the reaction of Ti with oxygen and to stop Si migration. In RTA, the small volume of the chamber helps to purge out the oxygen with nitrogen more efficiently, and the rapid ramp-up times and better control of the process temperature help in preventing the oxidation of Ti. RTP, in addiiton, allows loading and unloading at sufficiently low temperatures, and this reduces the change of Ti oxidation.

Figure 11.20 shows a series of RBS spectra illustrating the dependence of the silicidation process on the anneal temperature. 100 nm Ti film has been deposited on Si and, using halogen lamps, reacted for 90 s at different temperatures.[66] Formerly, it was believed that in the first stage of silicidation a metal rich compound is formed, that is, TiSi, and in the later stage $TiSi_2$ formation takes place. It has

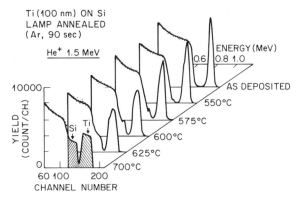

Fig. 11.20 RBS spectra for as-deposited Ti on Si and after RTA at different temperatures for 90 s. (After Okamoto et al.[66] Reprinted with permission, copyright 1985, *Journal of Applied Physics*.)

recently been discovered that the initial phase is also $TiSi_2$; however, it is a metastable phase with a C49 structure.[67] Upon further processing, this phase transforms to the equilibrium phase of $TiSi_2$, which has a C54 structure. The C49 crystals are heavily faulted and have a high resistivity. The C54 crystals are fault-free, have a lower resistivity, and are 10–100 times larger.

A procedure for silicided shallow junction formation using $TiSi_2$ is depicted in Figs. 11.21*a–d*. Ti film is deposited on patterned wafers. Gate areas are likely to have a polysilicon structure, whereas the source-drain contacts are single-crystal Si with an oxide sidewall separating them. Ion mixing to break up the interfacial oxide between the Ti and Si may or may not be used at this stage. Ti is then reacted with Si at temperatures of ~600°C to form C49 $TiSi_2$. The unreacted Ti and the residual Ti nitride and Ti oxide are removed by a selective wet etch. The appropriate dopant, As or P for n^+ junction, or B for p^+ junction, is implanted into $TiSi_2$ such that the implant damage remains within the silicide. An oxide cap is deposited on the surface to prevent the dopant out-diffusion. Finally, a high-temperature RTA ($>800°C$) is performed to out-diffuse the dopant and form C54 $TiSi_2$.[68]

RTP is shown to be even more advantageous in the case of silicide over polysilicon anneal. It is reported that the $TiSi_2$ formed at 900°C, 10 s on polysilicon remains stable at higher temperatures (1100°C, 10 s) with negligible or no diffusion of Ti into the polysilicon grain boundaries.[69]

11.11.2 Cobalt Silicide

Next to $TiSi_2$, $CoSi_2$ has the lowest resistivity, and it has recently been attracting more attention for use in self-aligned contact and interconnect technologies.[70–72] Co is not as reactive with oxygen as Ti, and in the silicide formation Co is the dominant diffusing species. Therefore, the need for using RTA in case of $CoSi_2$ is

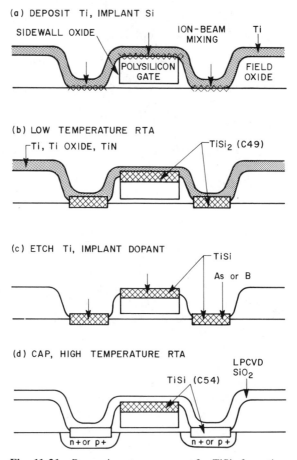

Fig. 11.21 Processing step sequence for TiSi$_2$ formation.

more difficult to justify; however, very successful results have been reported in CoSi$_2$ formation using RTA. These include silicide and interconnect,[70] as well as shallow junction formation by rapid thermal out-diffusion of implanted dopants from CoSi$_2$ into the junction.[49,70]

The rapid ramp up and short, well-controlled anneal cycles have made it possible to study the kinetics of CoSi$_2$ formation. Figure 11.22 shows the sheet resistance of a 80 nm Co film deposited on Si and annealed for 30 s in N$_2$ at different temperatures. X-ray diffraction studies on these samples indicated that the silicidation process occurs with a Co $\rightarrow$ Co$_2$Si $\rightarrow$ CoSi $\rightarrow$ CoSi$_2$ phase sequence.[71] The peak of the sheet resistance around 500°C corresponds to the CoSi phase. This phase formation is obviously a function of anneal time as well as temperature. In Ref. 72, similar results are reported for 10 s anneal times, where the peak in $R_\square$ occurs at ~700°C.

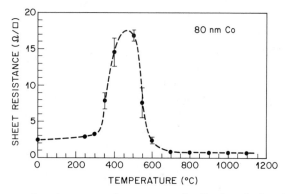

Fig. 11.22   Sheet resistance vs. anneal temperature for Co deposited on Si and annealed for 30 s. (After Van den Hove et al.[71] Reprinted with permission, copyright 1986, *Journal of Vacuum Science Technology.*)

11.12   GLASS REFLOW

Chemical vapor deposited dielectrics such as phosphosilicate glass (PSG), or borophosphosilicate glass (BPSG) are used to provide electrical isolation between metal interconnects and the underlying structures. The deposited glass needs to be thermally annealed to smooth out the sharp step profiles which are difficult to cover uniformly by overlying metal. This anneal cycle, glass reflow, is the last high-temperature heat-treatment cycle in VLSI technology and is often used to activate the source-drain implants as well. Hence, in a fabrication line designed to operate with RTP, one has to verify that the RTA cycle necessary for glass reflow is suitable for shallow junction formation and is compatible with the earlier thermal processes.

A number of experiments to verify the compatiblity have been done.[73–75] It has been shown that anneal cycles anywhere from 950°C, 80 s to 1200°C, 5 s can reflow the glass and preserve the junction integrity and depth. It is also shown that addition of H_2 to the ambient,[76] or dopants such as As,[77] or increasing P[76] content induce greater flow, hence require lower temperatures for the reflow.

It is also essential to note that the increase in the sheet resistance of implanted single-crystal Si or polysilicon, due to the clustering or grain boundary segregation caused by a low-temperature anneal, makes the use of high temperatures for glass reflow a necessity.

11.13   *IN SITU* PROCESSING

In situ fabrication of multilayer structures is extremely useful in ULSI technology for avoiding particulate or chemical contamination which may occur when moving wafers from one furnace to another. The RTP makes various *in situ* processes possible by the virtue of its small chamber size and rapid thermal cycles. The chamber allows rapid and effective change of ambient. It also stays relatively cool during the high-temperature cycles and hence has a reduced memory of the gases previously used. The *in situ* cleaning of the wafers and growth and nitridation of oxides were discussed earlier. Rapid thermal furnaces can, in addition, be used for chemical vapor deposition of various materials. Several successful *in situ* processes and depositions have been reported in the literature.[78–82] These processes are accomplished by changing the gases in the chamber while the wafer is cold, and heating the wafer only when the gas flow is stabilized.

11.13.1   *In situ* MOS Gate Structures

One particularly useful process is the *in situ* CVD of polysilicon on thin gate dielectrics. In conventional furnace processing, in order to reduce the chance of contamination of the thin oxides the time between the polysilicon deposition and gate oxide growth is kept to a minimum. Nevertheless, the wafers must be moved from the oxidation furnace to the CVD station. Using RTP, the oxidation and the polysilicon deposition can be accomplished in the same furnace, thus potentially improving the device yield. Sturm et al.[79] have reported *in situ* growth of 30 nm oxide and LPCVD of 0.4 μm of boron-doped polysilicon with acceptable CV characteristics.

Tungsten gates have also been deposited *in situ* on thin oxides or nitroxides. This process has been carried out in conjunction with plasma processing, with the potential advantage of low-temperature RTP.[80]

11.13.2   Epitaxial Multilayer Structures

The combination of RTP and CVD processing which has been developed by Gibbons et al.[78] and named LRP (limited-reaction processing) has also been used for epitaxial growth of multilayer structures. Doped and undoped $p/i/p/i$ epitaxial Si have been grown on Si substrates.[81] Also, $Si/Si_{0.9}Ge_{0.1}/Si/Si_{0.9}Ge_{0.1}$ superlattice structures are made. The $Si_{0.9}Ge_{0.1}$ layers have a thickness of 15 nm, and the abruptness and the quality of these structures are reported to be comparable to molecular-beam epitaxy (MBE) material.[82]

11.14   CONCLUSIONS

0.5 μm or smaller device technology requires precise control of feature size. This is provided with more ease and greater chance of success in rapid thermal processing. In RTP, the entire surface of the wafer heats uniformly and the need for long stabilizing times is thus eliminated. Since it is a single-wafer processing, RTP allows high-temperature processing along with short, well-defined times performed in small, cold-wall furnaces.

The resulting advantages are many fold. Thinner and more reproducible oxides can be grown with higher-quality $Si–SiO_2$ interfaces. Nitroxide gate dielectrics can be grown in a time–temperature regime not available in conventional furnaces. Higher solubility of dopants with less diffusion and more complete removal of the implant damage can be achieved. High-quality silicides can be formed with more ease in processing. Larger-grain, lower-resistivity polysilicon and more reliable glass reflow are also made possible.

The short well-defined temperature cycles of RTP have also made possible the study of the kinetics of various processes. This helps select the process window most suitable to particular applications. In addition, in RTP systems, various *in situ* processings can be performed.

Numerous reports and publications have been made (see, e.g., several proceedings of the Materials Research Society Symposia) on the many applications of RTP. However, in process-line applications, RTP is still in its infancy. More experiments on sequential temperature treatments need to be carried out. Thermal process sequences tailored to the advantages of RTP must be laid out. Furthermore, RTP needs to be incorporated into process modeling schemes.

REFERENCES

1. J. M. Fairfield and G. H. Schwuttke, "Silicon Diodes Made by Laser Irradiation," *Solid-State Electron.*, **11,** 1175 (1968).

2. F. E. Harper and M. I. Cohen, "Properties of Si Diodes Prepared by Alloying Al into n-Type Si with Heat Pulses from a Nd:Yag Laser," *Solid-State Electron.*, **13,** 113 (1970).

3. G. A. Kachurin, N. B. Pridachin, and L. S. Smirnov, "Annealing of Radiation Defects by Laser Radiation Pulses," *Sov. Phys. — Semicond.,(Engl. Transl.),* **9,** 946 (1976).

4. E. I. Shtyrkov, L. B. Khaibullin, M. M. Zaripov, M. F. Galyatudinov, and R. M. Bayazitov, "Local Laser Annealing of Implantation Doped Semiconductor Layers," *Sov. Phys. — Semicond., (Engl. Transl.),* **9,** 1309 (1976).

5. A. G. Klimenko, E. A. Klimenko, and V. I. Donin, "Use of Argon Laser Radiation in Restoration of Ion-Implantation-Amorphised Silicon Surface," *Sov. J. Quantum Electron. (Engl. Transl.),* **5,** 1289 (1976).

6. J. C. C. Fan and H. J. Zeiger, "Crystallization of Amorphous Silicon Films by Nd:Yag Laser Heating," *Appl. Phys. Lett.,* **27,** 224 (1975).

7. R. A. Smith, *Semiconductors,* 2nd ed., p. 321, Cambridge University, London and New York, 1978.

8. Y. P. Varshni, "Temperature Dependence of the Energy Gap in Semiconductors," *Physica (Amsterdam),* **34,** 149 (1967).

9. M. R. T. Siregar, W. Luthy, and K. Affolter, "Dynamics of CO_2 Laser Heating in the Processing of Si," *Appl. Phys. Lett.,* **36,** 787 (1980).

10. T. E. Seidel, D. J. Lischner, C. S. Pai, R. V. Knoell, D. M. Maher and D. C. Jacobson, "A Review of RTA of B, BF_2, and As Ions Implanted into Si," *Nucl. Instrum. Methods Sect. Phys. Res., B,* **7/8,** 251 (1985).

11. P. Baeri, S. W. Campisano, E. Rimini, and J. P. Zhang, "Time Resolved Temperature Measurement of Pulsed Laser Irradiated Ge by Thin Film Thermocouple," *Appl. Phys. Lett.*, **45**, 398 (1984).

12. J. Blake, J. C. Gelpey, J. F. Moquin, J. Schlueter, and R. Capodilupo, "Slip Free Rapid Thermal Processing," *Mater. Res. Soc. Symp. Proc.*, **92**, 265 (1987).

13. Z. A. Weinberg, D. R. Young, J. A. Calise, S. A. Cohen, J. C. Deluca, and V. R. Deline, "Reduction of Electron and Hole Trapping in SiO_2 by RTA," *Appl. Phys. Lett.*, **45**, 1204 (1984).

14. S. K. Lee, D. K. Shih, D. L. Kwong, N. S. Alvi, N. R. Wu, and H. S. Lee, "Effects of Rapid Thermal Processing on Thermal Oxides of Si," *Mater. Res. Soc. Symp. Proc.*, **71**, 449 (1986).

15. S. K. Lee, D. L. Kwong, and N. S. Alvi, "Effects of Rapid Thermal Processing on Thermal Oxides of Si," *J. Appl. Phys.*, **60**, 3360 (1986).

16. A. Kamgar and S. J. Hillenius, "Rapid Thermal Anneal Induced Effects in Polycrystalline Si Gate Structures," *Appl. Phys. Lett.*, **51**, 1251 (1987).

17. R. R. Razouk and B. E. Deal, "Dependence of Interface State Density on Si Thermal Oxidation Process Variables," *J. Electrochem. Soc.*, **126**, 1573 (1979).

18. J. Nulman, "Rapid Thermal Growth of Thin Si Dielectric for ULSI Applications," in S. Broydo and C. M. Osburn, Eds., *Proceedings of the First International Symposium on ULSI Science and Technology*, p. 141, Electrochem. Soc., Pennington, NJ, 1987.

19. P. F. Schmidt, "Furnace Contamination and its Remedies," *Solid State Technol.*, June, p. 147 (1983).

20. See, for example, A. M. Hodge, C. Pickering, A. J. Pidduck, and R. W. Hardeman, "Silicon Oxidation by RTP," p. 313; J. G. Gelpey, P. O. Stump, and R. A. Capodilupo, "Oxide Growth Using the Water-Well Arc Lamp," p. 321; and Z. A. Weinberg, T. N. Nguyen, S. A. Cohen, and R. Kalish, "SiO_2 Growth and Annealing by Lamp Heating," p. 327, *Mater. Res. Soc. Symp. Proc.*, **52** (1986).

21. Y. Sato and K. Kiuchi, "Oxidation of Si Using Lamp Light Radiation," *J. Electrochem. Soc.*, **133**, 652 (1986).

22. J. Nulman, J. P. Krusius, and P. Renteln, "Material and Electrical Properties of Gate Dielectrics Grown by RTP," *Mater. Res. Soc. Symp. Proc.*, **52**, 341 (1986).

23. H. Wendt and E. Bußmann, "Rapid Thermal Processes for VLSI," *Tech. Proc. Semicon/Europa*, Zurich, p. 120 (1987).

24. T. Ito, T. Nozaki, and H. Ishikawa, "Direct Thermal Nitridation of SiO_2 Films in Anhydrous Ammonia Gas," *J. Electrochem. Soc.*, **127**, 2053 (1980).

25. J. Nulman, J. P. Krusius, and L. Rathbun, "Electrical and Structural Characteristics of Thin Nitrided Gate Oxides Prepared by Rapid Thermal Nitridation," *Int. Electron Devices Meet. Conf. Abstr., 1984*, San Francisco, CA, p. 169 (1985).

26. M. M. Moslehi and K. C. Saraswat, "Thermal Nitridation of Si and SiO_2 for VLSI," *IEEE Trans. Electron. Devices*, **ED-32**, 106 (1985).

27. C. C. Chang, A. Kamgar, and D. Kahng, "High Temperature RTN of SiO_2 for Future VLSI Applications," *IEEE Electron. Device Lett.*, **EDL-6**, 476 (1985).

28. A. Kamgar, D. Kahng, H. S. Luftman, and W. S. Lindenberger, unpublished.

29. G. L. Olson, "Kinetics and Mechanisms of Solid Phase Epitaxy and Competitive Processes in Si," *Mater. Res. Soc. Symp. Proc.*, **35**, 25 (1985).

30. R. L. Cohen, J. S. Williams, L. C. Feldman, and K. W. West, "Thermally Assisted Flash Annealing of Si and Ge," *Appl. Phys. Lett.*, **33**, 751 (1978).

31. A. Kamgar, F. A. Baiocchi, and T. T. Sheng, "Kinetics of As Activation and Clustering in High Dose Implanted Si," *Appl. Phys. Lett.*, **48**, 1090 (1986).

32. A. Kamgar and F. A. Baiocchi, "Metastable Activation in Rapid Thermal Annealed As Implanted Si," *Mater. Res. Soc. Symp. Proc.*, **52**, 23 (1986).

33. F. A. Baiocchi and A. Kamgar, "Rutherford Backscattering and Channeling Studies of Defects in As Implanted Si Induced by As Clustering," *Mater. Res. Soc. Symp. Proc.*, **82**, 115 (1987).

34. R. B. Fair, "Impurity Diffusion During RTA," *Mater. Res. Soc. Symp. Proc.*, **35**, 381 (1985).

35. T. E. Seidel, C. S. Pai, D. J. Lischner, D. M. Maher, R. V. Knoell, J. S. Williams, B. R. Pennmalli, and D. C. Jacobsen, "Rapid Thermal Annealing in Si," *Mater. Res. Soc. Symp. Proc.*, **35**, 329 (1985).

36. T. E. Seidel, "RTA of BF_2^+ Implanted Preamorphised Si," *IEEE Electron Device Lett.*, **EDL-4**, 353 (1983).

37. T. O. Sedgwick, "RTP and the Quest for Ultra Shallow B Junctions," *Mater. Res. Soc. Symp. Proc.*, **71**, 403 (1986).

38. H. Mikoshiba, H. Abiko, and M. Kanamori, "Junction Leakage Current in BF_2^+-Implanted Rapid Thermal Annealed Diodes," *Jpn. Appl. Phys.*, **25**, L631 (1986).

39. D. E. Davies, "1-2 keV B Implants into Si," *IEEE Electron Device Lett.*, **EDL-6**, 397 (1985).

40. A. C. Ajmera and G. A. Rozgonyi, "Elimination of End-of-Range and Mask Edge Lateral Damage in Ge^+ Preamorphised, B^+ Implanted Si," *Appl. Phys. Lett.*, **49**, 1269 (1986).

41. T. E. Seidel, R. V. Knoell, G. Poli, B. Schwartz, F. A. Stevie, and P. Chu, "RTA of Dopants Implanted into Preamorphised Si," *J. Appl. Phys.*, **58**, 683 (1985).

42. T. O. Sedgwick, "Short Time Annealing," *J. Electrochem. Soc.*, **130**, 484 (1983).

43. A. Kamgar, W. Fichtner, T. T. Sheng, and D. C. Jacobsen, "Junction Leakage Studies in Rapid Thermal Annealed Diodes," *Appl. Phys. Lett.*, **45**, 754 (1984).

44. I.-W. Wu, R. T. Fulks, and J. C. Mikkelsen, Jr., "Optimization of BF_2^+ Implanted and Rapidly Annealed Junctions in Si," *J. Appl. Phys.*, **60**, 2422 (1986).

45. A. Kamgar and B. Schwartz, unpublished.

46. W. S. Lindenberger, A. Kamgar, and S. Muller, "Using Boron Nitride Thin Films to Dope Si Surfaces," unpublished.

47. K.-T. Kim and C.-K. Kim, "Formation of Shallow p^+-n Junctions using BN Solid Diffusion Source," *IEEE Electron Device Lett.*, **EDL-8**, 569 (1987).

48. D. L. Kwong and N. S. Alvi, "Electrical Characterization of Ti-Silicided Shallow Junctions Formed by Ion-Beam Mixing and RTA," *J. Appl. Phys.*, **60**, 688 (1986).

49. B. M. Ditchek, M. Tabasky, and E. S. Bulat, "Shallow Junction Formation by the Redistribution of Species Implanted into $CoSi_2$," *Mater. Res. Soc. Symp. Proc.*, **92**, 199 (1987).

50. S. J. Krause, S. R. Wilson, W. M. Paulson, and R. B. Gregory, "Grain Growth During Transient Annealing of As Implanted Polycrystalline Si Films," *Appl. Phys. Lett.*, **45**, 778 (1984).

51. H. B. Harrison, S. T. Johnson, Y. Komem, C. Wong, and S. Cohen, "Using Rapid Thermal Processing to Induce Epitaxial Alignment of Polycrystalline Si Films on (100) Si," *Mater. Res. Soc. Symp. Proc.*, **71**, 455 (1986).

52. H. Arai, K. Nakazawa, and S. Kohda, "Recrystallization of Amorphous Si Film By Tungsten Halogen Lamp Annealing," *Appl. Phys. Lett.*, **48**, 838 (1986).

53. J. L. Hoyt, E. Grabbe, J. F. Gibbons, and R. F. W. Pease, "Epitaxial Alignment of As Implanted Polycrystalline Si Films on ⟨100⟩ Si Obtained by Rapid Thermal Annealing," *Appl Phys. Lett.*, **50**, 751 (1987).

54. C. Y. Wong, Y. Komem, and H. B. Harrison, "Electrical Activation of As Ion-implanted Polycrystalline Si by Rapid Thermal Annealing," *Appl. Phys. Lett.*, **50**, 146 (1987).

55. G. Harbeke, L. Krausbauer, E. F. Steigmeier, A. E. Widmer, H. F. Kappert, and G. Neugebauer, "Growth and Physical Properties of LPCVD Polycrystalline Si Films," *J. Electrochem. Soc.*, **131**, 675 (1984).

56. A. Kamgar, F. A. Baiocchi, T. T. Sheng, and J. M. Brown, unpublished.

57. H. J. Böhm, H. Kabza, T. F. Meister, and H. Wendt, "Formation of Very Shallow Junction Using Self-Aligned Bipolar Transistors Using Rapid Optical Annealing," in S. Broydo and C. M. Osburn, Eds., *Proceedings of the First International Symposium on ULSI Science and Technology*, p. 347, Electrochem. Soc., Pennington, NJ, 1987.

58. R. Chow and R. A. Powell, "Activation and Redistribution of Implants in Polysilicon by RTP," *Semicond. Int., 1985,* May, p. 108 (1985).

59. M. Y. Tsai, F. M. d'Heulre, C. S. Petersson, and R. W. Johnson, "Properties of W Silicide Film on Polycrystalline Si," *J. Appl. Phys.*, **52**, 5350 (1981).

60. M. I. J. Beale, V. G. I. Deshmukh, N. G. Chew, and A. G. Cullis, "The Use of Ion Beam Mixing and Rapid Thermal Annealing in the Formation of W and Mo Silicides," *Physica (Amsterdam)*, **129B**, 210 (1985).

61. C. B. Cooper, III, and R. A. Powel, "The Use of RTP to Control Dopant Redistribution during Formation of Ta and Mo Silicide/n$^+$ Polysilicon Bilayers," *IEEE Electron Device Lett.*, **EDL-6**, 234 (1985).

62. D. L. Kwong, D. C. Meyers, and N. S. Alvi, "Simultaneous Formation of Silicide Ohmic Contacts and Shallow p$^+$-n Junctions by Ion-Beam Mixing and Rapid Thermal Annealing," *IEEE Electron Device Lett.*, **EDL-6**, 244 (1985).

63. S. A. Agamy, V. Q. Ho, and H. M. Naguib, "As$^+$ Implantation and Transient Annealing of MoSi$_2$ Thin Films," *J. Vac. Sci. Technol.*, A [2], **3**, 718 (1985).

64. P. K. John, S. Gecim, Y. Suda, B. Y. Tong, and S. K. Wong, "Silicide Formation by Pulsed-Incoherent-Light Annealing," *Can. J. Phys.*, **63**, 876 (1985).

65. S. P. Murarka, *Silicide for VLSI Applications*, Academic, New York, 1983.

66. T. Okamoto, K. Tsukamoto, M. Shimizu, and T. Matsukawa, "Ti Silicidation by Halogen Lamp Annealing," *J. Appl. Phys.*, **57**, 5251 (1985).

67. R. Beyers and R. Sinclair, "Metastable Phase Formation in Titanium-Silicon Thin Films," *J. Appl. Phys.*, **57**, 5240 (1985).

68. D. L. Kwong, Y. H. Ku, S. K. Lee, E. Louis, N. J. Alvi, and P. Chu, "Silicided Shallow Junction Formation by Ion Implantation of Impurity Ions into Silicide Layers and Subsequent Drive-In," *J. Appl. Phys.*, **61**, 5084 (1987).

69. J. Narayan, T. A. Stephenso, T. Brat, D. Fathy, and S. J. Pennycook, "Formation of Silicides by Rapid Thermal Annealing Over Polycrystalline Si," *J. Appl. Phys.*, **60**, 631 (1986).

70. L. Van den Hove, R. Wolters, K. Maer, R. F. De Keersmaecker, and G. J. Declerck, "A Self-Aligned CoSi$_2$ Interconnection and Contact Technology for VLSI Application," *IEEE Trans. Electron Devices,* **ED-34,** 554 (1987).

71. L. Van den Hove, R. Wolters, K. Maex, R. F. De Keersmaecker, and G. J. Declerck, "A Self-Aligned CoSi$_2$ Technology Using RTP," *J. Vac. Sci. Technol., B,* [2] **4,** 1358 (1986).

72. M. Tabasky, E. S. Bulat, B. M. Ditcheck, M. A. Sullivan, and S. C. Shatas, "Direct Silicidation of Co on Si by RTA," *IEEE Trans. Electron Devices,* **ED-34,** 548 (1987), or in *Mater. Res. Soc. Symp. Proc.,* **52,** 271 (1986).

73. T. Hara, H. Suzuki, and M. Furukawa, "Reflow of PSG Layers by Halogen Short Duration Heating Technique," *Jpn. J. Appl. Phys.,* **23,** L452 (1984).

74. N. Shah, J. McVittie, N. Sharif, J. Nulman, and A. Gat, "Characterization of PSG Films Reflowed in Steam Using RTP," *Mater. Res. Soc. Symp. Proc.,* **52,** 231 (1986).

75. J. S. Mercier, L. D. Madsen, and I. D. Calder, "Rapid Isothermal Fusion of BPSG Films," *Mater. Res. Soc. Symp. Proc.,* **52,** 251 (1986).

76. N. S. Alvi and D. L. Kwong, "Reflow of Phosphosilicate Glass by Rapid Thermal Annealing," *J. Electrochem. Soc.,* **133,** 2627 (1986).

77. M. Furukawa and T. Hara, "Rapid Heating Reflow of Phosphosilicate Glass Enhanced by As Ion Implantation," *Jpn. J. Appl. Phys.,* **25,** L795 (1986).

78. J. F. Gibbons, C. M. Gronet, and K. E. Williams, "Limited Reaction Processing: Si Epitaxy" *Appl. Phys. Lett.,* **47,** 721 (1985).

79. J. C. Sturm, C. M. Gronet, and J. F. Gibbons, "In-Situ Epitaxial Si-Oxide-Doped Polysilicon Structures for MOS Field-Effect Transistors," *IEEE Electron Device Lett.,* **EDL-7,** 577 (1986).

80. M. M. Moslehi, M. Wong, K. C. Saraswat, and S. C. Shatas, "In-Situ MOS Gate Engineering in a Novel Rapid Thermal/Plasma Multiprocessing Reactor," *Symp. VLSI Technol., 1987,* Japan (1987).

81. C. M. Gronet, J. C. Sturm, K. E. Williams, J. F. Gibbons, and S. D. Wilson, "Thin Highly Doped Layers of Epitaxial Si Deposited by LRP," *Appl. Phys. Lett.,* **48,** 1012 (1986).

82. C. M. Gronet, C. A. King, and J. F. Gibbons, "Growth of GeSi/Si Strained-Layer Superlattices Using LRP,"*Mater. Res. Soc. Symp. Proc.,* **71,** 107 (1986).

12 Lithography

R. K. WATTS

AT&T Bell Laboratories
Murray Hill, New Jersey

12.1 INTRODUCTION

A larger fraction of the cost of a modern production facility for integrated circuits is spent for lithographic systems than for any other type of process equipment. The lithographic equipment is the most complex — and therefore should be considered the most likely to fail — of the many components of the production line. In these aspects, it is the most important part of the line.

This chapter examines the status of each of a variety of high-resolution lithographic techniques in the late 1980s. Representative commercially available systems are mentioned along with systems under development. Projections are made for each technique to show what, if anything, can be done to allow it to meet the future needs of the industry. Resolution is emphasized because it lends itself to analysis more readily than any other facet of the subject. For example, advances in the precision of level-to-level registration depend on progress in optical and mechanical engineering practice; there is no theoretical underpinning.

12.2 OPTICAL LITHOGRAPHY

Projection printing is the imaging of a mask pattern onto a wafer by means of an optical projection system. Because it combines high resolution with low defect density, it is the favored optical method. For a given number of resolution points or smallest features resolvable by the system (typically 10^8–10^9), smaller features are obtained in the image if the image field is kept small.

Available systems are of three types. In the reduction wafer stepper[1] an image of a mask reticle, reduced typically five or ten times, is exposed on the wafer, then stepped to a neighboring position and exposed again, and so on until the whole wafer is patterned. The large reticle pattern is easy to produce and to inspect for defects. Refractive optics are employed. In the 1:1 Ultratech stepper, the reticle pattern is the same size as the image pattern, and the optics are refractive and reflective.[2] In the Perkin Elmer printer, mask and image patterns are also the same size, but mask and wafer are scanned continuously and synchronously through highly corrected portions of the object and image fields.[3] The smaller numerical aperture of this last system makes it unsuitable for submicron imaging.

12.2.1 A Benchmark: 1 μm Photolithography

Before considering submicron lithography, we shall first examine, as a point of reference, the requirements and the practice of photolithography for MOS circuits with design rules of 1 μm. Such a design rule generally means that the minimum linewidth and the minimum spacing between lines is 1 μm; the minimum contact window is 1 × 1 μm, and the maximum allowed misregistration between levels is ±0.25 μm. Let us see how several groups of workers have put 1 μm lithography into practice.

In 1980, Sigusch et al.[4] of Siemens used a wafer stepper with a Zeiss 10× reduction lens of numerical aperture (NA) 0.28 to expose AZ1450J positive photoresist 1.5 μm thick. Test circuits, including a 16 kbit SRAM, with minimum linewidths of 1 μm were fabricated. Linewidth errors were within ±0.1 μm for metal and field oxide levels and within ±0.2 μm for polysilicon and contact window levels. Registration errors were less than ±0.5 μm. In 1982, workers at Siemens[5] reported extending this process to 0.75 μm design rules using a higher resolution Zeiss lens of NA 0.42, again exposing 1.5 μm of resist with 436 nm radiation. In this case, linewidth variations were kept within ±0.15 μm for the metal level and within ±0.1 μm for the other levels. Registration errors are not quoted, but the maximum error allowed by the design rule is ±0.3 μm.

Hillis et al.[6] of Hewlett-Packard have reported implementation of a 1 μm process for the production of chips for the HP9000 computer. Some chips of the set contain as many as 6.6×10^5 transistors. With a wafer stepper and standard photoresist, it was not possible to keep linewidth variations within the required ±0.1 μm and the authors were forced to employ a bilayer resist to meet the specifications. This bilayer resist system, shown in Fig. 12.1b, incorporates a thick lower layer of poly methyl methacrylate (PMMA), which provides a more nearly planar surface than the underlying circuit topography on which to apply the upper thinner layer of Kodak 809 positive photoresist. The top layer is exposed by the 436 nm radiation of the stepper, to which the PMMA is insensitive. A dye was added to the PMMA to minimize reflections of the 436 nm radiation. After development, the top layer serves as a conformal mask for flood exposure of the PMMA with shorter wavelength radiation.

In 1983, Orlowsky et al.[7] of AT&T Bell Laboratories reported a 1 μm NMOS process. A wafer stepper with a Tropel 5× reduction lens of NA 0.35 was used to expose a trilayer resist structure, shown in Fig. 12.1c, with 405 nm radiation. After resist exposure, a thin layer of SiO_2 (plasma-deposited) beneath the resist is patterned by reactive ion etching (RIE). The SiO_2 then serves as a stencil mask for reactive ion etching of the underlying thick polymer layer, usually hard-baked HPR206 photoresist. Linewidth errors for the important polysilicon level were less than ±0.1 μm. Level-to-level registration errors were typically ±0.3 μm or less.

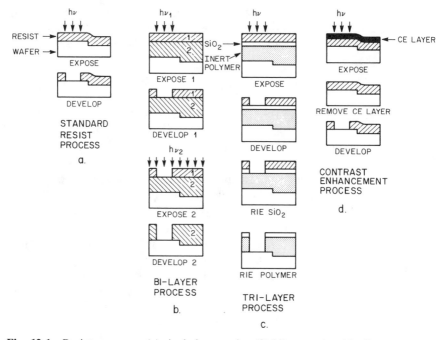

Fig. 12.1 Resist processes: (*a*) single-layer resist, (*b*) bilayer resist, (*c*) trilayer structure, (*d*) contrast enhancement process.

The trilayer process is not well suited to high-volume production because of the slow and rather involved processing.

A fourth popular resist process is shown in Fig. 12.1*d*. In this scheme, called contrast enhancement (CE), a thin layer of bleachable material is spun over the positive photoresist.[8] Upon exposure, the illuminated areas of the bleachable CE layer become more transparent as exposure progresses. Ideally, a part of the pattern that should receive no exposure, corresponding to an opaque feature on the mask, would receive none. But, of course, some light from nearby features corresponding to transparent parts of the mask spills over to these regions. The transmission of the CE layer is greater the higher the intensity of the light incident on it. Thus, the contrast, or ratio of maximum light intensity to minimum intensity in the pattern, of the radiation exposing the underlying resist is increased or enhanced. This process increases resolution and linewidth control.

The following conclusions can be drawn from these reports. Wafer steppers available in 1980–1983 provided adequate resolution for 1 μm lithography. Some workers found it necessary to employ a more complex multilayer resist scheme to obtain adequate linewidth control.

The level-to-level registration provided by the full wafer alignment scheme, wherein the reticle is first aligned to the wafer and then each field is exposed in turn, was barely adequate. A more accurate method in which each exposure field is separately aligned is now common practice. The registration errors are not due to pattern placement errors on the reticle. Figure 12.2 shows a typical error map of a reticle, formed by plotting the placement errors measured at 100 locations in a 100×100 mm array. The maximum error is only 0.099 μm. The corresponding error in the 5× or 10× reduced image of the reticle is negligible compared with the requirements of 1 μm lithography.

12.2.2 Limits of Photolithography

Figure 12.3 shows a general schematic representation of an imaging system. The object plane is the mask or reticle. It is illuminated by an illumination system (not shown) and imaged by the projection optics onto the image plane, which will be the wafer. We consider in turn two extreme cases: spatially coherent illumination of the mask and spatially incoherent illumination of the mask in the diffraction limit. This situation has been treated by Born and Wolf,[9] among others; we shall quote some of their results.

If the mask is illuminated by a point source, then there is a definite phase relation between the field amplitudes $U_0(x_0, y_0)$ at different points on the object plane. (The phase at each point x_0, y_0 on the object plane is simply given by $\mathbf{k} \cdot \mathbf{r}$, where $\mathbf{k}$ is the propagation constant of the wave and $\mathbf{r}$ is the path from the source to x_0, y_0.)

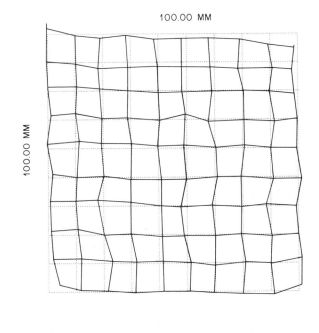

HSCALE = 0.191 UM
VSCALE = 0.191 UM
ERR   = 0.099 UM MAX
ERR   = 0.034 UM RMS

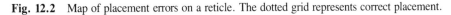

Fig. 12.2 Map of placement errors on a reticle. The dotted grid represents correct placement.

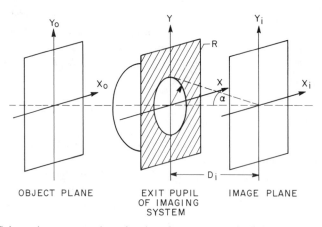

OBJECT PLANE   EXIT PUPIL   IMAGE PLANE
                     OF IMAGING
                       SYSTEM

Fig. 12.3   Schematic representation of an imaging system. R is the radius of the exit pupil.

The illumination is said to be spatially coherent. Then the Fourier transform $G_0(u, v)$ of $U_0(x, y)$ is related to $G_i(u, v)$, the Fourier transform of the amplitude $U_i(x_i, y_i)$ in the image plane, by the relation

$$G_i(u, v) = H(u, v)G_0(u, v),$$ (1)

where $H(u, v)$ is the coherent transfer function. $H(u, v)$ is the (spatial) frequency response of the optical system. For a round pupil of radius R (see Fig. 12.3), $H(u, v)$ is given by

$$H(u, v) = 1 \quad \text{if } \sqrt{u^2 + v^2} \le \frac{R}{\lambda D_i}$$ (2)

$$= 0 \quad \text{if } \sqrt{u^2 + v^2} > \frac{R}{\lambda D_i}.$$

λ is the wavelength of the illuminating light. H is plotted in Fig. 12.4, where the spatial frequency u is normalized by dividing it by the spatial frequency u_M, where

$$u_M = \frac{2R}{\lambda D_i} = \frac{1}{\lambda F}$$ (3)

$$= 2 \left[\frac{1}{\lambda \sqrt{(NA)^{-2} - 1}} \right].$$

where F is the f number of the projection system, $F \equiv D_i/2R$, and NA is the numerical aperture, $NA \equiv \sin \alpha = R/\sqrt{D_i^2 + R^2}$. The spatial frequency u is usually measured in line pairs, or cycles, per mm. The inverse of u is a feature size. In this case, the feature is a grating of equal lines and spaces; u^{-1} is the period of the grating.

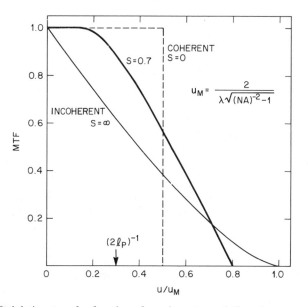

Fig. 12.4 Modulation transfer functions for coherent, partially coherent, and incoherent imaging plotted vs. normalized spatial frequency.

In the opposite extreme, where the light source is very large and there is therefore no definite phase relation between two points on the object plane (mask), the illumination is said to be incoherent. In this case, there is a relation such as that of Eq. (1) between the Fourier transforms G_0 and G_i of the object and image plane *intensities*, the squares of the wave amplitudes, given by

$$G_i(u, v) = H(u, v)G_0(u, v) ,$$ (4)

where H is called the optical transfer function. The absolute value $|H|$ is the modulation transfer function (MTF). It is also plotted in Fig. 12.4. The analytical expression analogous to equation (2) can be written for H. It is

$$H(u) = \frac{2}{\pi} \left[\cos^{-1}\left(\frac{u}{u_M}\right) - \frac{u}{u_M} \sqrt{1 - \left(\frac{u}{u_M}\right)^2} \right] ,$$ (5)

for $u \leq u_M$. The limiting frequency is twice as great ($u/u_M = 1$) as in the case of coherent illumination ($u/u_M = 0.5$).

The transfer function describes how the optical projection system degrades the image of a sinusoidal grating object pattern of equal lines and spaces. The modulation in a grating of spatial frequency u is $M(u)$, where

$$M(u) = \frac{I_M(u) - I_m(u)}{I_M(u) + I_m(u)} .$$ (6)

I_M and I_m are the maximum and minimum intensities, respectively, in the grating pattern. The MTF at frequency u is the ratio of image modulation to object modulation,

$$MTF = \frac{M_i(u)}{M_0(u)}. \tag{7}$$

Although the sine wave formulation is more convenient mathematically, the MTF is always measured experimentally for an equal line/space or clear/opaque grating pattern (of periodicity u^{-1}) on a mask. That is, the object is a square wave, not a sine wave. For the incoherent case, there is a simple relation between the sine wave and the square wave MTFs.[10]

In modern printers, the illumination is intermediate between the coherent and incoherent limits. The separation between points in the object that have correlated phases is neither infinite nor near zero. It can sometimes be varied by means of a diaphragm in the condenser illuminating the mask. This is a familiar phenomenon to anyone who has used a microscope. As the aperture stop diaphragm is closed, the image looks "sharper," but interference rings finally appear around edges, and the light level drops. This is the near-coherent case. Opening the diaphragm causes the rings to disappear and the image becomes brighter. The ratio S of the numerical aperture of the condenser to that of the projection optics is often used as a measure of coherence. $S = 0$ implies near-coherence; if $S = 1$, the apertures are matched and the entrance pupil of the projection optics is filled to give near incoherent illumination. Further increase of S just causes more light scattering. $S = \infty$ corresponds to the incoherent limit. Partial coherence ($0 < S < 1$) has some advantages over incoherent illumination. The useful range,[11] MTF > 0.6, is extended to higher spatial frequencies. Edge gradients in the image become steeper, and the image is a little less sensitive to defocus. Also plotted in Fig. 12.4, is the MTF corresponding to the partial coherence $S = 0.7$. For most projection printers, S has a value between 0.4 and 0.8.

Although most printers have nearly diffraction-limited optics, the aberrations are never zero.[12] Aberrations reduce the MTF. A very important aberration is the focus error. This error might be caused by improper operation of the automatic focus tracking or by departure of the wafer surface from the proper plane. The Rayleigh unit of defocus,

$$w_R = \frac{\lambda}{2(NA)^2}, \tag{8}$$

corresponds to a phase error of $\pi/2$ at the edge of the pupil and is supposed to be the largest focus error which is "imperceptible" to the eye.

Since the transfer functions with round pupil are symmetrical under rotation about the vertical axis, they can be written and displayed as functions of a single variable u, as in Eqs. (5)–(7). However, it is understood that u stands for $\sqrt{u^2 + v^2}$. This two-dimensional nature explains why a small contact window with predominant spatial frequency components $u = v = u_0$ requires a different exposure from a long line of the same width with $u = u_0$, $v \approx 0$, for $|H(\sqrt{2}\,u_0)|$ is less than $|H(u_0)|$. If both types of features occur on the same mask, both types of resist image will not have correct dimensions. In general, for very small features representing high spatial frequencies, the required exposure depends on the shape of the feature and proximity to other features, setting a practical resolution limit much less than u_M. This is called a "proximity effect."

Wafer steppers are used under many different conditions, ranging from tests by the vendors to operation by scientists and engineers intent on obtaining the highest possible resolution for a critical experiment in device fabrication to operation by less highly skilled personnel in high-volume chip production. As a rough measure of the best resolution obtainable, we take the linewidth $\ell_{0.6} \equiv 0.5\, u_{0.6}^{-1}$, where $u_{0.6}$ is the value of u at which the MTF curve of the stepper lens is 0.6. For the curve $S = 0.7$ of Fig. 12.4, this is $u_{0.6}/u_M = 0.48$ giving

$$\ell_{0.6} = 0.52\lambda\sqrt{(NA)^{-2} - 1} \approx \frac{\lambda}{2NA}. \tag{9}$$

In the literature, one often finds the expression for best resolution written as

$$\ell = \frac{K\lambda}{NA}, \tag{10}$$

where $K = 0.8$ for production conditions, and K is some smaller value near 0.5 for better conditions.[13]

Table 12.1 gives specifications of some lenses and the resolutions and depths of focus obtained with them and reported in 1986.[14-18] The field diameter ranges from 7 mm for the lens with largest aperture ($NA = 0.6$) to 20 mm. A larger diameter of the exposure field implies higher throughput, since fewer exposures are necessary. (Throughput, or number of wafers patterned per hour, depends also on wafer size, exposure time, and alignment or registration time. It typically ranges from 10 to 50 wafers per hour for wafer diameter 125 mm or 150 mm.) ℓ_{1L} denotes the resolution obtained over the entire field reported in single-layer positive photoresist of thickness slightly greater than 1 μm, and w_{1L} is the depth of focus measured under these conditions. ℓ_{3L} and w_{3L} are the corresponding quantities measured with trilayer resist. For the second entry in the table, the authors observed the same values for tri-layer resist and for the contrast-enhancement method of Fig. 12.1d. As a measure of the quality of the optics, lines 9 and 12 of Table 12.1 give the number of resolution points in the exposure field. Thus, Equation (9) gives a fairly good description of ultimate resolution obtainable in trilayer resist. Also tabulated is the value of $\ell_P = 0.8\lambda/NA$, the best resolution to be expected under production conditions. It is nearly equal to the experimental resolution obtained in thicker single-layer resist, ℓ_{1L}. (The imaging layer in the trilayer structures in Table 12.1 was 0.5 to 0.7 μm in thickness.)

There is another reason for being conservative in estimating resolution. The MTF curves shown are for grating features consisting of equal lines and spaces;

TABLE 12.1   Resolution and Depth of Focus Obtained With Wafer Steppers

Lens	5A3 Nikon	Nikon	10.78.48 Zeiss	Silica Tropel	Silica Matsushita
Reduction	5×	10×	10×	5×	5×
λ(nm)	436	436	365	248	248
NA	0.42	0.6	0.42	0.38	0.35
Field dia. (mm)	20	7	14.1	14.5	14.1
$\ell_{0.6}$ (μm)	0.5	0.4	0.4	0.3	0.4
w_R (μm)	1.2	0.6	1.0	0.9	1.0
ℓ_p (μm)	0.8	0.6	0.7	0.5	0.6
ℓ_{1L} (μm)	0.8	0.6	0.7		0.5
w_{1L} (μm)	2	1			
Resolution points	5×10^8	1×10^8	3×10^8		6×10^8
ℓ_{3L} (μm)	0.6	0.4	0.5	0.5	
w_{3L}	1.5		1	0.5	
Resolution points	9×10^8	2×10^8	6×10^8	7×10^8	
Reference	14	15	16	17	18

they do not apply to other features. A small square feature of size $\ell \times \ell$ is more difficult to resolve than a grating of period 2ℓ, and an isolated line of width ℓ is easier to resolve. For example, with the process reported in Ref. 7, 0.75 μm gates are easily printed and 0.5 μm gates can be printed with somewhat lower yield. However, 0.75 × 0.75 μm contact windows are difficult to print, and 0.5 × 0.5 μm windows are impossible.

The last two columns in the table describe systems that use a KrF excimer laser as light source. Because chromatic aberration of the silica lens is large, the 248 nm radiation must be spectrally narrowed. Figure 12.5 shows 0.5 μm lines produced in trilayer resist with the system of column 4 of Table 12.1. Table 12.2 shows projected specifications of future such systems, kindly provided by V. Pol. Clearly, the last system would need better registration capability if the potentially high resolution is to be useful.

Since resolution of an optical system $\sim \lambda/NA$ and depth of focus $\sim \lambda/(NA)^2$, it is better to reduce λ than to increase NA to improve resolution. A disadvantage

TABLE 12.2   Projected Specifications of Future Deep UV Wafer Steppers[a]

	248/KrF	193/ArF	193/ArF
λ (nm)/laser	248/KrF	193/ArF	193/ArF
NA	0.5	0.38	0.5
Field dia. (mm)	15	20	15
Resolution with trilayer resist (μm)	0.3	0.3	0.2
Focus depth (μm)	±0.5	±0.6	±0.4
Registration, 2σ (μm)	±0.1	±0.1	±0.1

[a]*Source:* Provided by V. Pol, AT&T Bell Laboratories.

Fig. 12.5   A 0.5 μm pattern imaged in trilayer resist with the 248 nm stepper of Table 1, column 4.

of changing the wavelength is that entirely different resists may be required. Let us see how far this reduction of λ can be carried. For $\lambda < 193$ nm, silica lenses absorb too strongly; reflective optics must be used.[19] The optical path must be in vacuum for $\lambda < 185$ nm. For $\lambda < 170$ nm, the silica reticle absorbs too stongly. Other possible materials for reticle substrates and the spectral regions in which they are transparent are CaF_2 or LaF_3, $\lambda > 130$ nm, and MgF_2, $\lambda > 120$ nm.[20] These fluoride materials are not now available in the form and quality needed for reticle substrates. Resolution of a system exposing with 130 nm radiation would depend on the value of NA that could be achieved and the degree to which aberrations could be corrected — a more difficult task than at longer wavelengths. In addition, the increased Rayleigh scattering ($\sim\lambda^{-4}$) would degrade contrast. It is not likely that systems requiring a stencil reticle for use at even shorter wavelengths will be developed. However, if a mask is used in reflection rather than transmission, then even shorter wavelengths could be used.

At 248 nm, an MP2400 photoresist can be used.[17,21] In the deep ultraviolet spectral region, MP2400 is not as good an image replication medium as it is at slightly longer wavelength. Figure 12.6 shows 0.35 μm lines, which can only be produced in the center of the exposure field, formed in MP2400 by the system of

Fig. 12.6   SEM photo of 0.35 μm line/space patterns in an MP2400 resist recorded by the 248 nm stepper of Table 12.1, column 4. The resist is the uppermost layer in a trilayer resist structure.

Table 12.1, column 4. Figs. 12.7a and b show simulated profiles[21] for this system with an MP2400 resist. The sloped profile of Fig. 12.7b matches the experimental profile of Fig. 12.6. The lower part of Fig. 12.7 (c, d) shows simulated profiles for a fictitious resist with a factor of 3 less absorption at 248 nm than an MP2400. The reduced absorption leads to better profiles. Depth of focus, at least as measured experimentally (the most important measure), depends strongly on resist properties, as can be seen by comparing w_R, w_{1L}, and w_{3L} of Table 12.1.

MP2400 is a two-component positive photoresist. One component is a novolac resin, and the other component is a naphthoquinone diazide dissolution inhibitor. The exposing radiation destroys the inhibition, in this case by an excitation transfer mechanism, and allows the resin to dissolve in the aqueous developer. Because the deep ultraviolet photon has sufficient energy to break bonds, it is also possible to find one-component positive resists that function in the following way. The chain scission produced by the radiation leads to a reduction in molecular weight for some of the molecules. The developer selectively dissolves the part with lower molecular weight. Wolf et al.[22] have studied several resists of this kind. They found that they are less sensitive at 248 nm than an MP2400. Another disadvantage of these resists is their generally poorer resistance to dry etches. An example is poly butene sulfone (PBS). It can be used only with wet etching, as in photomask making.

Most polymeric materials absorb too strongly for $\lambda < 190$ nm.[23] In principle, the new surface sensitization techniques, which do not require deep penetration of

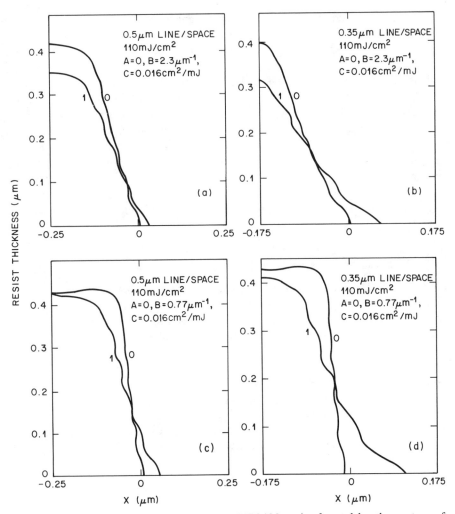

Fig. 12.7   Simulated profiles in (a, b) an MP2400 resist formed by the system of Table 12.1, column 4, with $\lambda = 248$ nm, $NA = 0.38$, $S = 0.7$. In (c) and (d), the absorption coefficient of the resist is reduced by a factor of 3. Ref. 21 gives details of the calculation. A, B, and C are model parameters explained in Ref. 20. The photomask covers the portion $x \leq 0$. The 0 and 1 beside the curves indicate perfect focus and defocus by 1 μm, respectively.

the exposing radiation, could be employed.[24,25] A thinner resist, or thinner sensitive top layer, leads to higher resolution (see Table 12.1) and increased depth of focus for lines of given width. Thinner resist also leads to higher defect densities. No concentrated sophisticated effort has been made to reduce defect density in very thin films.

Because of the almost vanishingly small depth of focus accompanying very high resolution, the distance between the imaging system and the wafer must be very well controlled. This will require heroic engineering efforts and may impose a limit on the useful resolution.

12.3 ELECTRON LITHOGRAPHY

Scanning electron beam systems have been under development since the mid-1960s. They have had a considerable impact on mask making and are to be found in every large mask shop. The mask of Fig. 12.2 was made with a MEBES electron-beam pattern generator, and the placement errors were measured with a similar machine. The other main uses of these systems are exploratory device research and low-volume manufacture of high-performance chips. Excellent commercial systems are available for both applications. Electron proximity printing is still being developed.

12.3.1 Scanning Systems

Scanning beam systems are similar to scanning electron microscopes, from which they are derived. Figure 12.8 shows a simplified drawing of a typical electron optical column. There are several magnetic lenses which project an image of the source onto the image plane (a mask blank or a wafer on an x–y interferometrically controlled stage.) The beam can be blanked—deflected rapidly past the edge of an aperture—by electrostatic blanking plates. It is scanned over the image plane magnetically by orthogonal coils, as shown, or electrostatically. The stigmator coils compensate for any astigmatism introduced by departures from cylindrical symmetry in the column.

The source of electrons, called the electron gun, is of one of two types: a thermionic cathode, usually LaB_6, or a thermal field emitter (TFE) of impregnated

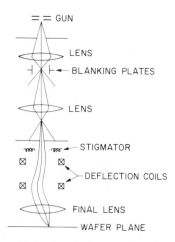

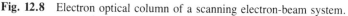

Fig. 12.8   Electron optical column of a scanning electron-beam system.

tungsten. For the thermionic cathode, the current density at the cathode J_c is given by

$$J_c = AT^2 \exp\left(\frac{-E_w}{kT}\right), \tag{11}$$

where E_w is the work function of the cathode, ~ 2.5 eV for LaB_6, and A is Richardson's constant. Typically, $T \approx 2000$ K. In the gun, the electrons are accelerated by a high voltage V (usually 10–50 kV), and after leaving the gun they traverse the column with constant velocity. The maximum current density that can be focused with convergence half-angle α is

$$J_m = J_c\left(\frac{qV}{kT}\right) \sin^2 \alpha. \tag{12}$$

The beam brightness β, the current density per unit solid angle, is the same throughout the column as the beam diverges and converges through the various solid angles indicated in Fig. 12.8. The maximum value of the brightness is

$$\beta = \frac{J_c qV}{kT}. \tag{13}$$

For LaB_6, β is $\sim 10^6$ A/cm^2/str. The current density J_i and current I_i in the image plane are

$$J_i \le \pi\beta\alpha^2, \qquad I_i = J_i\left(\frac{\pi d_i^2}{4}\right), \tag{14}$$

In Section 12.2, α was called the numerical aperture or NA; it is much smaller here than in the light optical case.

The resolution, or diameter d of the writing beam is determined by aberrations. The beam diameter one would measure experimentally is found by taking the square root of the sum of squares of d_i and the various aberration blurs. The main aberrations of an undeflected beam are spherical and chromatic aberration, which are considered to contribute blurs d_s and d_c, respectively, to the spot size:

$$d_s = \frac{1}{2}C_s\alpha^3, \qquad d_c = C_c\left(\frac{\Delta V}{V}\right)\alpha, \tag{15}$$

where C_s and C_c are constants characterizing the aberrations and $q\,\Delta V$ is the energy spread resulting from the high temperature of the cathode. Another aberration, important only for high beam current, is due to the mutual Coulomb repulsion of the electrons. The resulting blur d_{ee} is approximately given by

$$d_{ee} \approx \frac{LI}{\alpha V^{3/2}} \times 10^8 \ \mu m, \tag{16}$$

where L is the column length. Deflection aberrations, which we lump together to contribute a blur d_{df}, are proportional $r\alpha^2$ or to $r^2\alpha$, where r is the distance of deflection in the image plane. Field distortion and some of the deflection aberrations can be reduced by the addition of correction signals to the scan coils. Table 12.3 shows aberrations for the IBM VS1 machine.[26] More information of a general nature on the electron optics of scanning electron-beam machines and a little discussion of software and control systems can be found in Refs. 27 and 28. We will now briefly discuss a few particular machines.

The EBES (for electron-beam exposure system) machine,[29] developed by Bell Laboratories, is of the raster scan type, wherein the beam is deflected repetitively over the exposure field, as in a televison raster. It is available commercially from Perkin Elmer as the MEBES system. This machine is the premier mask making system available, being designed for highly accurate pattern placement. Beam deflection is (mainly) in one dimension, as shown in Fig 12.9. The stage moves continuously in a direction perpendicular to the writing direction. The pattern data are decomposed into a number of stripes, and one stripe is written over all chips of the same type before the next stripe is begun.

The stripe is 2048 addresses wide, an address corresponding to 0.1–1.0 μm. Beam diameter can be varied from 0.1 to 1.0 μm. The pattern information comes to the blanking plates from a shift register at a 40 or 80 MHz rate. Since total time to write the 2048 address scan is 31.6 μs–12.5 ns per spot + 6 μs for flyback — the writing time is approximately 8 cm^2/min with the 0.5 μm adddress and 2.4 cm^2/min with 0.25 μm address.

Most other machines employ the more efficient vector scan technique, where the beam scans only where a pattern element is to be exposed. Examples of machines with such scanning are EBES4,[30] JBX6A3, EB60,[31] EL3[32] and VS1.[26] It is more difficult to calculate throughput (wafers or masks patterned per hour) for these systems. Introductions to these calculations may be found in Refs. 27 and 28.

The EBES4 machine has a TFE cathode and a continuously moving stage. The coordinate system can be distorted digitally to correct errors or improve registration to existing levels. Column length was kept to 0.6 m to reduce Coulomb interaction effects. [See Eq. (16)] Three hierarchical deflection systems are used: magnetic deflection over a 0.28 mm square field, electrostatic deflection over a

TABLE 12.3   Aberrations at the Corners of 2 × 2 mm and 4 × 4 mm Exposure Field for VS1[a]

	2 mm field (μm)	4 mm field (μm)
Spherical, d_s	0.01	0.01
Chromatic, d_c	0.04	0.04
Deflection, d_{df}	0.087	0.35
Total, no dynamic correction	*0.095*	*0.35*
Total, dynamic correction	*0.043*	*0.050*

[a]From Ref. 26.

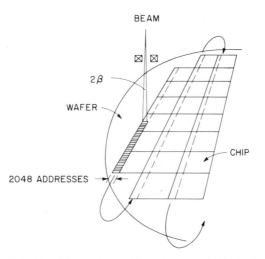

Fig. 12.9 EBES/MEBES writing scheme. Curved arrows indicate the serpentine stage motion.

32 μm square field, and electrostatic deflection for drawing microfigures over a 2 μm square field.

For machines with variable-shaped beams, the calculation of throughput is straightforward. We take the JEOL JBX6A3 as an example. In the JBX6A3 machine, shown schematically in Fig. 12.10, the spot is rectangular of variable size and aspect. The image of the first square aperture may be shifted in two dimensions to cover various portions of the second aperture, which is the object imaged on the mask or wafer. Minimum rectangle width is 0.3 μm; maximum dimension is 12.5 μm. Maximum current density is 2 A/cm^2 with the LaB_6 cathode. A tungsten cathode is also available. Electrostatic beam deflection is employed. Proximity effect corrections are easily made by variation of the exposure time of individual rectangles. Exposure field is at most 2.5 × 2.5 mm. Thus, many fields must be stepped to expose a mask or wafer. For a chip containing 9×10^4 rectangles, the patterning time, assuming a shot time of 1.6 μs per shot, is 0.14 s. The time to step and align each exposure field, including a 0.25 s alignment time, is 0.39 s. Thus, for each chip containing four fields we have

$$T_e = 0.14 \text{ s}$$
$$\underline{T_{sr} = 1.56 \text{ s}}$$
$$1.7 \text{ s}$$

Total time to pattern a 100 mm wafer containing 282 chips would be 8.0 min. Thus, 7.5 wafers/h could be patterned. Many wafers can be held at one time in the vacuum system so that the load/unload time per wafer is small.

Two machines specially configured for high-resolution work are the JEOL JBX5D2[33] and the Cambridge EBMF2.5UHR.[34] Table 12.4 gives some specifica-

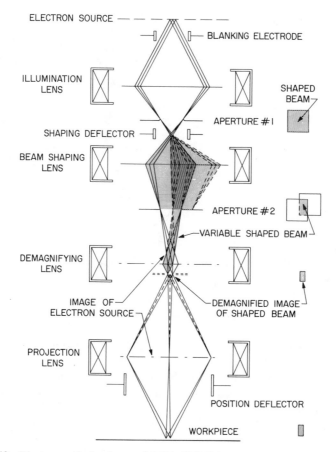

Fig. 12.10   Electron optical column of JEOL JBX6A3 scanning electron-beam system.

TABLE 12.4   Some Specifications of Cambridge EBMF2.5UHR and JEOL JBX5D2 High-Resolution Electron-Beam Machines.[a]

	JBX5D2	EBMF2.5UHR
Electron energy (keV)	25 or 50	1–50
Scan type	Vector	Vector
Field size (mm²)	0.08×0.08–1.5×1.5	0.05×0.05–2×2
Minimum spot dia. (nm)	8	8
Registration error (nm)	<40	± 30 (1σ)
Stage motion	Step–repeat	Step–repeat
Cathode	LaB_6	LaB_6

[a]From Refs. 33 and 34.

tions. These machines are intended for device development work in which short patterning time is not a primary consideration. Both can operate at 50 kV. With increasing electron energy, resists become less sensitive. However, this effect is compensated by the larger beam currents available, as shown in Eq. (12)–(14).

12.3.2 Proximity and Projection Systems

An electron proximity printing system has been under development for a number of years by IBM Deutschland.[35] This is a step–repeat system in which a silicon membrane stencil mask containing one chip pattern is shadow-printed onto the wafer. Since the mask cannot accommodate reentrant geometries (e.g., doughnut), these are printed with two masks. Registration is accomplished by reference to alignment marks on each chip. Overlay error is less than 0.05 μm (3σ). An advantage of this system is its ability to measure and compensate mask distortions. Proximity effects must be treated by size changes of pattern elements. The major disadvantage of the system is the need for two masks for some patterns.

In a 1:1 projection system, parallel electric and magnetic fields image electrons onto the wafer.[36] The mask is of quartz patterned with chrome and covered with CsI on the side facing the wafer. Photoelectrons are generated on the mask/cathode by back-side UV illumination. Westinghouse, Thomson CSF, Philips, Toshiba, and Radiant Energy Systems have developed systems. The advantages of the system include stable mask, good resolution, fast step–repeat exposure with low-sensitivity electron resists, large field, and fast alignment. Proximity effects can be compensated for by undersizing or oversizing the features on the mask, or the electron energy can be increased to 50 keV or more. Apparently, neither method is entirely satisfactory. The bias-exposure method is another possiblity.[37] An electrostatic chuck holds the wafer.

Another problem remains to be solved before this system can become a strong contender for the high-volume production of advanced chips. The cathode has an unacceptably short life before the CsI must be replenished by evaporation of fresh material. It lasts for only 50 exposures.

In the projection reduction systems developed at Tübingen University and at IBM, a reduced image of a foil stencil mask is stepped to fill the wafer area. The exposure field of the IBM machine is 3 × 3 mm.[38] Special problems include long alignment times and a difficult mask technology, the latter being the more serious.

12.3.3 Resolution Limits

In Section 12.2.2, resolution was examined with the aid of the modulation transfer function, or MTF. The MTF can be written as

$$MTF(u) = MTF_s(u) \times MTF_r(u),\qquad(17)$$

where MTF_s represents the imaging system and MTF_r the exposure (but not the development) of the resist. In the optics case, $MTF_r \approx 1$[39] and MTF_s is determined by diffraction with a perfect lens. In the electron case, MTF_s is determined by aberrations. (For a 50 keV electron and $NA = \alpha = 10^{-2}$ rad, $\lambda/NA = 3$ Å. Thus, diffraction is negligible.) Table 12.4 shows that commercial machines are available with resolution 80 Å, and TEMs have even smaller beams. In the electron case, scattering in the resist is of chief importance.

An electron beam of zero width incident at the origin on a resist film spreads out because of scattering in the film and from the substrate below. The result is that the energy per unit volume $U(x, z)$ deposited by the beam in the resist at position x, z is given approximately by

$$U(x, z) = a_1 \exp\left(\frac{-x^2}{\beta_f^2}\right) + a_2 \exp\left(\frac{-x^2}{\beta_b^2}\right). \tag{18}$$

Equation (18) is the analog of the point-spread function or Airy pattern of light optics. β_f is a width due to forward scattering in the resist. It is given approximately by[40]

$$\beta_f(z) = \left[\frac{9.64z(\mu m)}{V(kV)}\right]^{1.75}. \tag{19}$$

For a 20 keV electron at the bottom of a 0.5 μm resist film, $\beta_f = 0.08$ μm. At 50 keV, $\beta_f = 0.02$ μm. The width β_b due to backscatter in the resist and from the substrate is roughly proportional to the electron range in the substrate material. For 20 keV and 50 keV electrons and a silicon substrate, β_b is 2.3 μm and 6 μm, respectively.[40] With a thick polymer substrate under the top layer of a multilayer resist structure, β_b and a_2 are smaller.

From measurements of exposures of thin PMMA on a membrane substrate ($a_2 = 0$) with a 50 kV STEM, Broers[41] has established that the point $MTF = 0.6$ occurs at a linewidth of 22 nm. This represents a resolution limit, which is probably due to the low-energy secondary electrons produced in the resist. On thick substrates, best resolution is also obtained at high voltage. The large backscatter range just contributes a uniform background fog. Proximity effects are reduced. For a thick substrate, $MTF = 0.5$ for lines of width 50 nm in thin resist (0.1 μm). Howard et al.[42] have patterned 40 nm metal lines on a silicon substrate using 0.12 μm thick PMMA in a multilayer structure and 30 keV electrons. The smallest features patterned by electron beam are the 1.5 nm lines written in NaCl by Issacson and Murray[43] using a 100 kV STEM. In what may be the last word on the subject of high voltage, Jones et al.[44] have done lithography at 500 keV. They obtain 0.1 μm line/space grating patterns in 1 μm thick PMMA using a 0.1 μm beam. Walls of the pattern are vertical and straight. They expect to obtain this 10:1 aspect ratio also in even finer patterns written with a smaller beam.

12.4 X-RAY LITHOGRAPHY

X-ray lithography was first proposed by Spears and Smith[45] in 1972. Since that time, it has been developed to the point where commercial systems are now available. In Section 12.2.2, the advantage of reducing the wavelength in photolithography was pointed out. However, a limit is soon reached where all materials become opaque because of the fundamental absorption. Transmission increases again in the soft X-ray region. Because of the absorption spectra of materials used for masks and resists, the useful spectral region is 4–50 Å, with most systems using radiation nearer the short wavelength limit of the range.

12.4.1 Proximity Printing

There are no efficient projection optics for the soft X-ray spectral region. Proximity printing with uncollimated illumination must be used. The arrangement is shown in Fig. 12.11, where the symbols S, D, g, and R are defined as the diameter of the X-ray source, the spacing between source and mask, the gap between mask and wafer, and the radius of the exposure field, respectively. The magnification in the image is $1 + g/D$. The run-out at the edge of the field is called Δ, where

$$\Delta = \frac{gR}{D}. \tag{20}$$

Equation (20) shows that the gap g must be closely controlled so that Δ will be nearly the same for all wafer levels; otherwise, level-to-level registration would suffer. The penumbral blur δ is caused by the nonzero extent of the source:

$$\delta = \frac{gS}{D}. \tag{21}$$

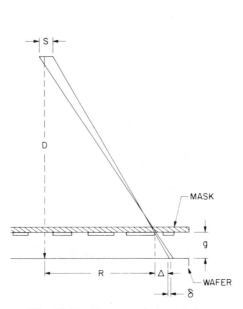

Fig. 12.11 X-ray proximity printing.

Figure 12.12 illustrates the role played by diffraction in proximity printing. The image of an edge on the mask is blurred by diffraction over a distance dx, where

$$dx = 1.3 \sqrt{\frac{g\lambda}{2}}, \qquad D \gg g, \qquad (22)$$

where dx is the distance between the 90% and 10% intensity points of Fig. 12.12. The edge width of a pattern in resist, if it is determined largely by this blur, will be considerably smaller than dx because of the nonlinear nature of resist development. For an equal line/space grating pattern, the distortion produced by diffraction can be much worse than that shown in Fig. 12.12 or the image of the edge of a wide line. As the spacing between lines decreases, the minimum intensity I_m (between the lines) increases, reducing contrast I_M/I_m. The first two columns of Table 12.5 give values of these parameters for two commercial X-ray steppers.[46,47] For such systems, $\delta \gg dx$.

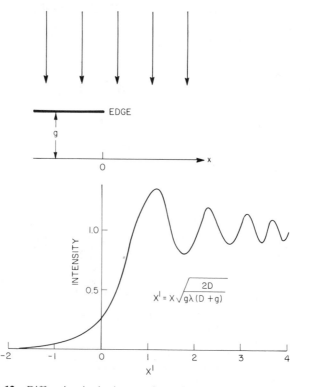

Fig. 12.12 Diffraction in the image of an edge with proximity printing.

TABLE 12.5   Some Specifications of Micronix MX1600, Hampshire Instruments XRL5000, and Perkin Elmer XLS1000 X-Ray Systems and Projected Specifications of COSY Exposure System.

	XLS1000	MX1600	XRL5000	COSY
Wavelength, λ (Å)	7.0	4.4	12–14	4–20
Source dia., S (mm)	1.5	3.5	0.2	0.5
Source–mask dist., D (cm)	18	25	7–10	400
Mask–wafer gap, g (μm)	20	25	20	~50
Field size (mm^2)	≤30 × 30	≤50 × 50	14 × 14	~40 × 40
Penumbral blur, δ (μm)	0.17	0.35	0.05	0.006
Diffraction blur, dx (μm)	0.1	0.09	0.15	0.2
Runout, Δ (μm)	2.4	3.5	2.5	0.01
Registration error (μm)	±0.15	±0.1 (2σ)	±0.1 (3σ)	~±0.02
Throughput (150 mm wafers/h)	40[a]	10[b]	15[c]	100[d]

[a]10 mJ/cm^2 resist assumed.
[b]20 mJ/cm^2 resist (DCOPA) assumed.
[c]MP2400 resist assumed.
[d]Per exposure station, 100 mJ/cm^2 resist assumed.

12.4.2   Masks

An X-ray mask consists of a heavy metal absorber on a transmissive membrane substrate. In contrast with a photomask, the ratio of metal thickness to substrate thickness is not so small, making fabrication more difficult. These thicknesses are determined by the transmission of the materials for the X-ray wavelength of interest.

Of the heavy metals, gold has been widely used because it is easily patterned. The thickness of gold necessary for absorption of 90% of the incident X-ray flux is 0.7 μm, 0.5 μm, 0.2 μm, or 0.08 μm for the X-ray wavelengths 4.4 Å (PdL), 8.3 Å (AlK), 13.3 Å (CuL), or 44.8 Å (CK), respectively. Methods for patterning the metal with high resolution include ion milling, sputter etching, and electroplating. Electroplating produces excellent definition with vertical walls, but requires a vertical wall primary pattern in resist of thickness at least equal to that of the metal to be plated unless a complex process of mask replication is employed. More often, a subtractive process has been selected in which a thinner resist layer is used to pattern a thin layer of a refractive metal; the refractive metal serves as a mask for ion-milling the underlying gold.

The membrane forming the mask substrate should be as transparent to the X-rays as possible, smooth, flat, dimensionally stable, reasonably rugged, and transparent to visible light if an optical registration scheme is used. Materials that have been used include polymers such as polyimide and polyethylene terephthalate, silicon, BN, SiC, Si_3N_4, Al_2O_3, and a $Si_3N_4/SiO_2/Si_3N_4$ sandwich structure. Although different mask substrates are appropriate for different portions of the soft

X-ray spectrum, there is not yet general agreement on the best material for any particular wavelength.

The major questions remaining about X-ray masks concern their dimensional stability, minimum defect densities attainable, and ease of handling. Dimensional stability can be degraded by radiation damage produced by the X-ray flux. This also makes the mask substrate optically opaque. Polymer membrane substrates can be distorted locally by the absorber metallization. For greater dimensional integrity, a stiffer substrate material is favored. In Fig. 12.13, the dimensional stability of a mask with substrate consisting of a BN membrane covered with a polyimide layer is shown. The figure shows the displacement due to mask processing of a set of fiducial marks written originally on the mask with an EBES pattern generator and read after processing. In this typical result for a 27 mm array, maximum pattern displacement is 0.18 μm and average placement error is 0.07 μm, about twice as bad as for the photomask of Fig. 12.2. Moreover, the photomask array is four times as large. For these masks, defect density is typically $\sim 2/\text{cm}^2$. Approximately half of these mask defects do not lead to defects in the resist image on the wafer.

27.0000 MM

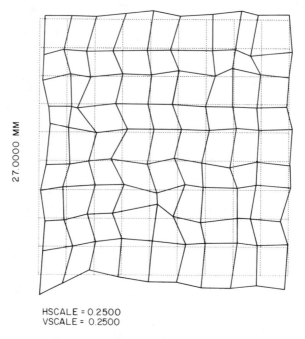

HSCALE = 0.2500
VSCALE = 0.2500

ERR   = 0.1801   UM MAX
ERR   = 0.0663   UM RMS

Fig. 12.13   Placement error map for a gold/BN X-ray mask. This map should be compared with that of Fig. 12.2 for a Cr/silica photomask.

12.4.3   The Source Brightness Problem

In the conventional X-ray tube, X-rays are produced when electrons strike a metal target. The X-ray spectrum depends on the target and consists of one or more sharp characteristic lines and a broad continuum. The X-ray flux emitted is limited by the ability of the target to dissipate the heat produced. Higher flux is available from tubes with a rotating anode in which a fresh surface is continually presented to the impinging electrons.[48] Even these sources are insufficiently bright. In systems such as the first two of Table 12.5, flux available at the wafer is typically two to three orders of magnitude less than that supplied by an optical wafer stepper. Because of this, highly sensitive resists, which have low resolution, must be used.[49]

Another type of source, which is capable of an order of magnitude greater flux, is the plasma-discharge source. There are several versions, but all function by heating a plasma to a high enough temperature that X-radiation is given off. The radiation consists of strong lines superimposed on a weak continuum. The source is pulsed with a low repetition rate. In one embodiment, the source size is 2 mm.[50] Repetition rate is 3 Hz. For an FBM-G resist, the total time to expose nine 2 × 2 cm fields is 120 s. If overhead is 30 s, throughput is projected to be 24 wafers/h. Special problems with such a source are reliability and contamination produced in the plasma chamber.

Column 3 of Table 12.5 shows specifications of an X-ray stepper with a laser-driven plasma source.[51] The small size of the source leads to a small penumbral blur. In this system, resolution is limited to ~0.4 μm by the alignment precision, which is in turn largely determined by the control (± 0.5 μm) of the mask–wafer gap. (Typical design rules require that the registration error be no larger than one-fourth the minimum feature size.) The contamination problem has been solved by the designers of this machine. The relatively longer wavelength makes possible the use of high-resolution, less sensitive positive resists because absorption in the resist is proportional to λ^3.

12.4.4   Storage Ring Source

An electron storage ring is a bright source of X-rays, providing more than two orders of magnitude greater flux than an X-ray tube. A single storage ring can provide radiation to a large number of exposure stations. The vertical angular divergence of the radiation is only ~$(1957E)^{-1}$ radians, where E is the electron energy in GeV. Although an exposure station is typically located 4–10 m from the ring, the beam is still of such small vertical extent that it must be scanned across the wafer exposure field.

High-energy electrons are provided to the storage ring by a microtron, a small synchrotron, or a small linear accelerator. The ring may be briefly operated as a synchrotron to boost the electron energy to the final value. Then the electrons may circulate for several hours in a stable orbit, the loss due to the power radiated as synchrotron radiation being compensated by one or more acceleration cavities around the ring. The peak of the power spectrum of the synchrotron radiation occurs at wavelength λ_p. This is related to the electron energy E (GeV) and magnet bending radius R (m) by

$$\lambda_p = 2.35 \frac{R}{E^3}. \tag{23}$$

During the last few years, a Fraunhofer Institute in Berlin has been developing a small storage ring called COSY for X-ray lithography.[52] Total floor space required is 30 m². Other components are under development elsewhere. For example, the stepper has been constructed by Suss. Alignment accuracy is ±0.02 μm. COSY will provide a flux density of 250 mW/cm². There will be ten beam lines, each with its X-ray stepper. Some parameters of this system are shown in Table 12.5, column 4. Because of the high flux available, high throughput is possible with high-resolution, less sensitive positive resists (last line, Table 12.5). A similar effort is under way in Japan. The technology of electron storage rings is old and well known. The major problem remaining is the mask.

12.4.5 Limits

Describing an X-ray system with the formalism of the modulation transfer function is difficult because so many effects must be taken into account: the penumbral blur, the diffraction, the transmission of "clear" and "opaque" parts of the mask, and the spectral distribution of the radiation. A definitive study of the resolution limit has been made by Betz et al.[53] Using a three-dimensional simulator, they have included all relevant effects. They find a resolution limit of ~0.2 μm. Linewidth control deteriorates as the limit is approached.[54]

The dominant effect is diffraction, as can be seen from Table 12.5, line 7. The minimum diffraction blurring is set by the minimum gap g, which can be used reliably[53] (~30 μm). The range of photoelectrons produced in the resist is only about 0.01 μm. Nothing would be gained by going to longer wavelengths; this would only increase the diffraction blur. Their simulation beautifully demonstrates that the most difficult pattern to resolve is an array of small squares with separation equal to edge length. Squares with edges less than 0.2 μm cannot be resolved.

12.5 ION LITHOGRAPHY

As with electron lithography, ion lithography systems are of two types: scanning beam systems and mask-based systems. Mask-based systems promise higher throughput because of the parallel nature of the exposure, in contrast with the serial exposure of pattern elements by a writing beam.

Ions are scattered less in material than electrons of the same energy. Ions deposit energy more efficiently than electrons. For this reason, resists are about two orders of magnitude more sensitive for ion exposure than for electron exposure. However, the sensitivies are about the same if measured in terms of energy deposited in the resist rather than incident dose (number of particles incident per unit area).

12.5.1 Scanning Systems

The problems of ion optics for scanning ion systems are more severe than for electron optics. The brightest sources are the two types of field ionization source in which ions are produced in the strong field near a pointed tungsten tip. The source of ionized material is a gas surrounding the tip or a liquid metal that flows to the tip from a reservoir. The largest current densities obtained in the focused image of such a source are 1.5 A/cm^2 for Ga$^+$ in a 0.1 μm spot and 15 mA/cm^2 for H$^+$ ion in a 0.65 μm spot.[55,56] Total beam current is severely limited. Electrostatic lenses rather than magnetic must be used for focusing ion beams. Similarly, magnetic deflection is much less practical than electrostatic. Electrostatic optical systems generally have higher aberrations, necessitating small aperture α and small scan field. As in electron optics, resolution is limited by the aberrations.

Seliger et al.[55] have reported a scanning system in which a beam of 57 keV Ga$^+$ ions is focused to a 0.1 μm diameter spot with current density 1.5 A/cm^2. Spot size is apparently limited by chromatic aberration of the electrostatic lens and the large 14 eV energy spread of the source. Figure 12.14 is a sketch of the optical

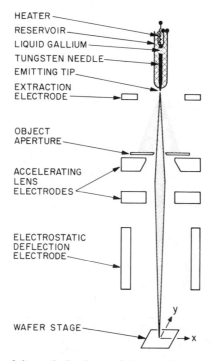

HEATER
RESERVOIR
LIQUID GALLIUM
TUNGSTEN NEEDLE
EMITTING TIP
EXTRACTION ELECTRODE

OBJECT APERTURE

ACCELERATING LENS ELECTRODES

ELECTROSTATIC DEFLECTION ELECTRODE

WAFER STAGE

Fig. 12.14 Schematic of the optical column of the scanning ion-beam system of Seliger et al.[55]

column. Superficially, it resembles the electron optical column of Fig. 12.6. A system with a H_2^+ source being constructed at Cornell University will have a beam diameter of 10 nm and a 0.2×0.2 mm scan field.[57] Current density will be about 200 A/cm^2. Commercial scanning systems available from JEOL and Seiko have beam diameters of ~0.1 μm. They are used for repair of photomasks.

12.5.2 Proximity Printing

In an ion proximity printer, a collimated ion beam is incident on a mask that is in proximity to the wafer to be exposed. Masks are of two types: a stencil mask and a mask with a thin membrane covering the "clear" or more transmissive areas. Of course, the stencil mask provides higher resolution. The stencil mask developed at Lincoln Laboratory[58] is a SiN$_x$ membrane with holes reactively ion-etched. 100 keV H_2^+ ions passing through the mask diverge by less than 0.03°. 80 nm lines and spaces have been defined in 0.3 μm thick PMMA. Since a single stencil mask cannot accommodate reentrant geometries, a supporting grid has been employed in a later version of this mask.[59] The beam is caused to wobble slightly to blur the image of the grid. This degrades resolution to 0.3–0.4 μm. The silicon stencil mask developed for electron proximity printing has also been used for ion printing.[60]

In the second type of mask, 300 keV protons are projected through the 0.5 μm thick portions of an all-silicon mask.[61] The mask is aligned so that the ions travel along the channels in the (100) direction. Throughput is projected to be 60 100 mm wafers/h and overlay error is ~0.1 μm—barely adequate for 0.5 μm lithography. Resolution is set by scattering of the ions as they emerge from the channels. Edge resolution is 0.1 μm for the 0.3° beam scattering angle.

12.5.3 Projection Printing

A 10× reduction projection stepper has been developed by Sacher Technik.[62] Light ions of energy 60–100 keV are projected through a thin foil stencil mask. Reentrant geometries force the use of two masks for each level imaged. Registration errors are ~0.05 μm and field size is 2.5×2.5 mm. Exposure time for PMMA resist is 0.2 s per field.

12.5.4 Limits

The average path of an energetic ion incident parallel to the z axis and moving through a solid is characterized by three parameters[63]: R_p, the distance the ion moves along the z axis, ΔR, the straggle in the z direction, and ΔY, the straggle perpendicular to the z axis. The last of these, ΔY, is a measure of the resolution to be expected in ion-beam lithography. The incident ion will give up energy to the electrons of the solid by inelastic scattering and to the nuclei. Since the ion is much more massive than the electrons, the energy loss to the electrons does not cause much deviation of the ion from its initial trajectory. Which mechanism predominates depends on the mass and energy of the ion and on the masses of the constituent atoms of the solid. For energies of interest, the electronic stopping is by far the more important for light ions in resist.

All the parameters, R_p, ΔR, and ΔY, increase with increasing ion energy. R_p should be greater than the resist thickness and ΔY should be as small as possible to give best resolution. Thus, the ratio $\Delta Y/R_p$ is of interest. The H^+ ion has the lowest value of this ratio,[63–65] as shown in Fig. 12.15. And for a given energy, H^+ has the longest range in PMMA. Therefore, highest resolution for resist exposure should be obtained with protons.

Figure 12.16 shows contours of equal energy dissipation for protons of energy 70 keV incident on PMMA. The half-plane $x < 0$ is masked off. A small amount of beam divergence is included in order to simulate a real system. The calculations were carried out by M. D. Giles by the method of Ref. 66. The dose necessary to expose PMMA is 2.9 times larger than a dose that causes essentially no dissolution of the resist in the developer.[67] Thus, the distance between the two contours corresponding to maximum energy dissipation and 0.35 × maximum can be called the "edge width" W of the pattern. This is not necessarily the width that would be measured for a profile in resist. The profile should be calculated by including resist development in the simulation, as in Fig. 12.7. However, W will be a convenient measure for comparing contours such as those of Fig. 12.16 for different ion

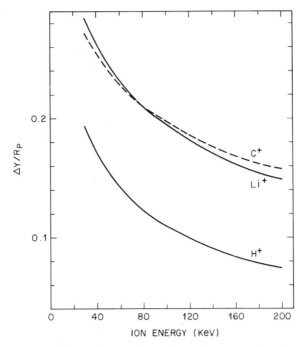

Fig. 12.15 Ratio of lateral straggle to projected range for protons in resist as a function of proton energy. The data were taken from Ref. 62–64.

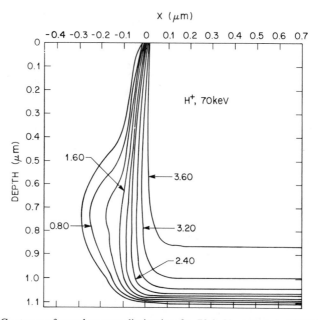

Fig. 12.16   Contours of equal energy dissipation for 70 keV protons in PMMA, as calculated by M. D. Giles. The left half-plane is masked off; the mask edge is at $x = 0$. The numbers on the contours are $\log U$, where U is the energy loss per unit volume.

energies. This factor of 2.9 is less than the factor $I_M/I_m = 4$ [or $MTF = 0.6$, from Eq. (6)] necessary for positive photoresists. (See Section 12.2.2) This width is shown in Table 12.6 for several ion energies for three different depths that would be of interest for practical resist exposure. From the table, 70 keV seems to be the best choice for resist thicknesses near 0.4 μm. To prevent radiation damage to devices, a multilayer resist structure can be employed.

If the beam is not perfectly collimated, but diverges with angle θ, then there will be a penumbral blur given by $g \tan \theta$. If $g = 50$ μm and $\theta = 0.03°$, then this blur is 26 nm. The effect of the blur in the resist image will be smaller than this, as in the X-ray case (Section 12.4.1). We can neglect the effect of the X-rays and energetic electrons produced along the ion path. The exposure of resist by the ions is much more important than the negligible exposure produced by them. Lines as small as 40 nm have been replicated.[68]

It has been demonstrated experimentally that there is no effect on the ion exposure caused by backscatter from the substrate, in contrast with electron-beam lithography.[67,69] The analog of the point-spread function of Eq. (18) represents the energy deposition by a point ion beam. It can usually be fit rather well by an exponential or a Gaussian function. For example, for 70 keV protons it is given by

$$U(x, z) = A \, \exp\!\left(\frac{-x}{\beta}\right). \tag{24}$$

For PMMA resist of 0.4 μm thickness, $\beta(z = 0.4\ \mu\text{m}) = 3.5$ nm. The MTF can be calculated from $U(x, z)$. This is shown in Fig. 12.17, where the MTF for the deep UV Tropel stepper lens of Table 12.1 is also plotted for comparison.

Ion doses required for resist exposure are 10–100 times less than for electrons of practical energies.[70] If the resist is too sensitive, there may be so few ions present in a resolution element that statistical fluctuations in this small number of ions could lead to sizable exposure variations. For PMMA with sensitivity 2×10^{13} ions/cm^2, there are 500 ions in a $0.05 \times 0.05\ \mu$m^2 cell. Clearly, a resist ten times as sensitive (2×10^{12} ions/cm^2) would have too few ions in this cell for comfort.

A machine such as that described by Bartelt[61] can supply 6.3×10^{12} ions/s · cm^2 onto a 2 cm^2 surface. This flux would make possible exposure of 2 cm^2 of PMMA resist in 3 s. The Li$^+$ beam used in the machine of Ref. 60 exposes a 5×5 mm field in 22 ms.[71] Thus, high throughput and high resolution are available simultaneously.

12.6 METROLOGY

Metrology on masks and wafers characterizes the lithographic process by measurement of linewidths, pattern placement, level-to-level registration, and resolution. Pattern placement accuracy on masks is measured by an optical measuring ma-

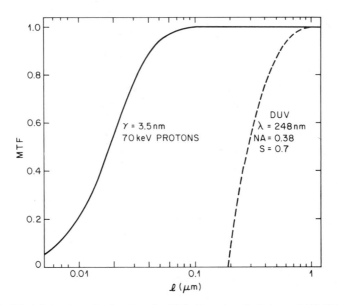

Fig. 12.17 Modulation transfer function for 70 keV protons in 0.4 μm thick PMMA plotted against linewidth of an equal line/space grating pattern. The MTF of a deep UV wafer stepper is also shown for comparison.

TABLE 12.6 "Edge Widths" for Lines in PMMA Defined by a Beam of Protons of Various Energies.[a]

Energy (keV)	$W(0.2 \ \mu m)$ (nm)	$W(0.3 \ \mu m)$ (nm)	$W(0.4 \ \mu m)$ (nm)	R_p (μm)
30	13	19	35	0.58
70	16	18	22	1.0
120	18	22	30	1.6
200	30	35	40	2.5

[a]$W(z)$ is the lateral separation at depth z between the maximum energy dissipation contour and the contour representing $0.35\times$ maximum. The widths W were taken from plots such as Figs. 12.16 and 12.17. The calculated range R_p is in excellent agreement with the measurements of Ref. 64.

chine or by an electron-beam pattern generator. In both cases, an interferometrically controlled stage moves the mask in two dimensions to bring each mark to be measured under the measuring tool. The tool reads the mark line by generating a profile of light intensity or intensity of backscattered electrons $I(x)$ as a function of x, the distance perpendicular to the line edge. $I(x)$ represents a somewhat distorted profile of the cross section of the line. The two edges are detected when $I(x)$ exceeds a threshold value, and the line center midway between the edges is calculated and registered as the position of the mark in the x direction. If the edges are symmetrical, the center position does not depend on the threshold value. Figure 12.2 is a plot of such measurements, in this case made with an electron-beam pattern generator.

Since two edges are detected for each line arm of each mark, the "width," or difference of the two edge positions, is also available. Because this number depends on the threshold value, a careful calibration is necessary. The calibration is different for each type of line measured (chromium on glass, $TaSi_2$ on silicon, etc.). Figure 12.18 shows such output, again measured with an electron-beam pattern generator, for the gold lines on a BN X-ray lithography mask. Deviations from the average linewidth are plotted. Reference 72 discusses linewidth measurements on wafers using a low-voltage SEM and digital image processing. Accuracy and precision are within 0.05 μm. This method is slower than that employing an e-beam pattern generator.

Lithography on wafers is most easily characterized by measuring electrically with a high-speed prober. The chip array on wafers processed on a small line dedicated to experimentation and process development will contain many test chips in addition to the primary circuit chip. The test chips incorporate special testers for evaluation of lithography in addition to the usual process testers, such as capacitors for measurement of oxide thickness, MOSFETs of various dimensions, and the like. In production, these testers can be placed in the saw lanes between chips. Linewidth W is measured for a long line of length L by first measuring the sheet resistance $R_\square$ of the material of which the line is composed by means of a van der

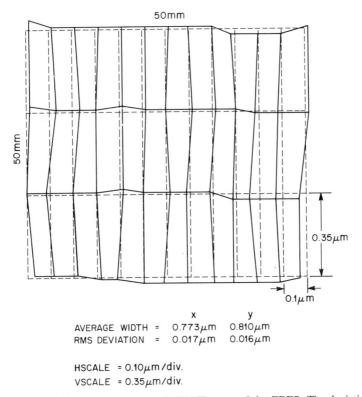

AVERAGE WIDTH = 0.773 μm   0.810 μm
RMS DEVIATION = 0.017 μm   0.016 μm

HSCALE = 0.10 μm/div.
VSCALE = 0.35 μm/div.

Fig. 12.18 Linewidths measured on a gold/BN X-ray mask by EBES. The deviation from the average width is plotted.

Pauw cross. Then, the width is given by

$$W = \frac{R_\square L}{R},$$ (25)

where R is the measured resistance of the line. W and $R_\square$ can be displayed as maps to show their variation over the wafer.[73]

Level-to-level registration measurement can also be reduced to a resistance measurement. Wafer maps of registration errors measured in this way are shown in Ref. 73. Examination of a long series of such maps shows a steady improvement in registration over the years.

Figure 12.19 illustrates a way to measure resolution electrically. The measurement is made with the aid of a special test chip containing nine meander patterns.[74] Each meander pattern consists of a 3 cm long meander line interleaved with two comb structures. In the example shown, the width and meander-comb spacing have nine values ranging from 0.5 μm to 2.5 μm. The wafer array consists of 300

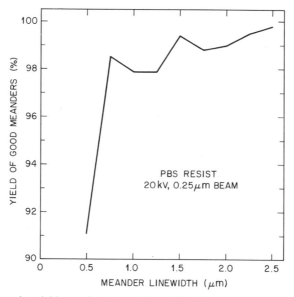

Fig. 12.19 Meander yield as a function of linewidth. The patterns were defined in PBS resist by electron beam and wet-etched into a chromium film.

such meander test chips with a total active area of 8.0 cm². The total area of each of the nine meander patterns is 0.88 cm². The pattern is etched into a conductive material on an insulating layer (doped polysilicon on oxide, aluminum on oxide, etc.) and the resistance of each meander is measured. A meander is considered good if it is not shorted to one of the combs and is not broken. The type of defect gives information about the resist process and the etching. For example, resist bridging will lead to meander-comb shorts. Figure 12.19 indicates a resolution of barely 0.5 μm. Since the area of the meander pattern is known, defect densities can be calculated. However, because the meander pattern is extremely dense — much denser than any level of a real circuit — the resulting defect densitities are quite high. It is a very stringent test. The variety of these electrial testers is limited only by the imagination of the designer, who always finds them great fun to concoct.

In addition to the measurements mentioned, there are specialized tests for characterizing particular lithographic tools.[75] For example, butting errors between stripes are measured for EBES/MEBES electron-beam pattern generators (see Fig. 12.9); field distortion is measured for stepper lenses; pattern magnification is measured for X-ray proximity printing, etc. The three trends that will probably strengthen as lithography advances to smaller dimensions are : (1) the supplanting of optical measurements by electron-beam measurements; (2) increasing use of electrical measurements on wafers; and (3) automatic acquisition of large numbers of measurement data for reliable statistical analysis, especially for the tracking of the performance of lithographic tools over time. Reference 76 is a guide to metrology for high-resolution lithography.[†]

12.7   CONCLUSION

Sections 12.2.2 and 12.4.5 have shown that the ultimate resolution limits of light optical lithography and X-ray lithography are rather similar—0.2–0.3 μm. This is a consequence of the superiority of the diffraction-limited optical projection system over X-ray proximity printing. The optical limit is roughly λ. The X-ray limit is roughly $\sqrt{g\lambda}$. Taking $\lambda = 0.2$ μm for the optical case, and $g = 30$ μm, $\lambda = 0.001$ μm for the X-ray case, we see that $\lambda^2_{\text{optical}} \approx g\lambda_{\text{x-ray}}$. (We are considering the storage ring source or the plasma source, not the X-ray tube. For the X-ray tube, resolution is limited by penumbral blur and the resist.) What can change this condition? The storage ring could operate at a shorter wavelength, say 5 Å. However, this would reduce resist sensitivity, which is proportional to λ^{-3}, in the absence of special dopants which have absorption spectra tuned to the X-radiation. If in addition a smaller gap could be used reliably—say 15 μm instead of 30 μm—then the X-ray resolution could be improved by about a factor of 2.

In the case of high resolution optical systems, control of the focus is a formidable problem. For X-ray printers with tube sources, the proximity gap g must be controlled just as closely to maintain good level-to-level registration. This is not the case with the storage ring source because of the smaller angular divergence of the radiation. In this case, g must be controlled in order to control resolution ($\sim\sqrt{g}$), and this is a much less stringent requirement.

Optical lithography would seem to be favored over X-ray lithography, but it does have three problems: (1) the depth of focus is small at high resolution; (2) linewidth control should be worse at high resolution than for X-ray lithography, as shown by Smith[77]; and (3) new resists will be required.

The highest resolution has been demonstrated by scanning electron lithography. The use of higher voltages leads to reduced proximity effects and better definition in thicker resist. Higher voltage does not necessarily imply more radiation damage in direct patterning of wafers. There are two compensating effects. At higher voltage, more beam current is required for the same resist. However, the damage occurs deeper in the substrate, where it does less harm. Scanning systems provide throughput that is very roughly proportional to the square of feature size. Thus, they are useful only for mask making, low-volume production, and device research. Electron projection and promixity systems suffer from several drawbacks, as detailed in Section 12.3.2, but this situation could change.

Ion proximity printing with a stencil mask combines high resolution without proximity effects with the ability to pattern usefully thick resists. Throughput can be high even if two masks must be used to pattern some levels. The only problem

[†]A conference on metrology for integrated circuits is held annually the the Society of Photo-Optical Instrumentation Enegineers, which publishes the proceedings.

seems to be the mask. Ion reduction projection is also attractive. The stencil mask is fragile, but larger than masks for proximity printing. The same issues discussed for X-ray masks in Section 12.4.2 are also of concern of ion-beam masks. Some studies of the effect of the elevated operating temperature of these masks have been reported.[71,78] Dimensional stability may be affected by radiation damage; little is yet known about the stability of these masks.

REFERENCES

1. H. E. Mayer and E. W. Loebach, "A New Step-by-Step Aligner For Very Large Scale Integration (VLSI) Production," *Proc. Soc. Photo-Opt. Instrum. Eng.*, **221,** 9 (1980).

2. R. Hershel, "Optics in the Model 900 Projection Stepper,"*Proc. Soc. Photo-Opt. Instrum. Eng.*, **221,** 39 (1980).

3. D. A. Markle, "A New Projection Printer," *Solid State Technol.*, **17,** 50 (1974).

4. R. Sigusch, K. H. Horninger, W. A. Müller, and D. Widman, "A 1 μm Process: Linewidth Control Using 10:1 Projection Lithography," *Tech. Dig. — Int. Electron Devices Meet.*, p. 429 (1980).

5. W. Arden, W. Beinvogl, and W. Müller, "Submicron MOS Process With 10:1 Optical Projection Printing and Anisotropic Dry Etching," *Tech. Dig. — Int. Electron Devices Meet.*, p. 403 (1982).

6. G. Hillis, K. Bartlett, M. Chen, R. Trutna, and M. Watts, "VLSI Production With a Multi-Layer Photolithography Process," *Tech. Dig. — Int. Electron Devices Meet.*, p. 554 (1983).

7. K. J. Orlowsky, D. V. Speeney, E. L. Hu, J. V. Dalton, and A. K. Sinha, "Fabrication Demonstration of 1–1.5 μm NMOS Circuits Using Optical Tri-Level Processing Technology," *Tech. Dig — Int. Electron Devices Meet.*, p. 538 (1983).

8. P. R. West and B. F. Griffing, "Contrast Enhancement — A Route to Submicron Optical Lithography," *Proc. Soc. Photo-Opt. Instrum, Eng.*, **394,** 33 (1983).

9. M. Born and E. Wolf, *Principles of Optics,* Pergamon, Oxford, UK, 1980.

10. M. C. King and M. R. Goldrick, "Optical MTF Evaluation Techniques for Microelectronic Printers," *Solid State Technol.*, **20,** 37 (1977).

11. W. G. Oldham, S. Subramanian, and A. R. Neureuther, "Optical Requirements for Projection Lithography," *Solid-State Electron.*, **24,** 975 (1981).

12. W. G. Oldham, "Contrast in High Performance Projection Optics," *Kodak Microelectron. Semin. Interface,* **81,** 75 (1982).

13. H. L. Stover, "A Glimpse into the Future of Optical Lithography," *Proc. Soc. Photo-Opt. Instrum. Eng.*, **633,** 2 (1986).

14. M. J. Grieco, D. W. Peters, and A. Kornblit, "A Comparison of Multi-Level and Single-Level Resist Processes and Their Impact on Stepper Lens Resolution," *Kodak Microelectron. Semin. Interface,* **86,** 193 (1987).

15. K. Ushida, M. Kameyama, and S. Anzai, "New Projection Lenses for Optical Stepper," *Proc. Soc. Photo-Opt. Instrum. Eng.*, **633,** 17 (1986).

16. J. Biesterbos, A. Bouwer, G. van Engelen, G. van der Looij, and J. van der Werf, "A New Lens for Submicron Lithography and its Consequences for Wafer Stepper Design," *Proc. Soc. Photo-Opt. Instrum. Eng.*, **663,** 34 (1986).

17. V. Pol, J. H. Bennewitz, G. C. Escher, M. Feldman, V. A. Firtion, T. E. Jewell, B. E. Wilcomb, and J. T. Clemens, "Excimer Laser-Based Lithography: A Deep Ultraviolet Wafer Stepper," *Proc. Soc. Photo-Opt. Instrum. Eng.*, **663**, 6 (1986).

18. M. Sasago, M. Endo, Y. Hirai, K. Ogawa, and T. Ishihara, "Half-Micron Photolithography Using A KrF Excimer Laser Stepper," *Tech. Dig. — Int. Electron Devices Meet.*, p. 316 (1986).

19. D. A. Markle, "The Future and Potential of Optical Scanning Systems," *Solid State Technol.*, **27**, 159 (1984).

20. W. R. Hunter and S. A. Malo, "The Temperature Dependence of the Short Wavelength Transmittance Limit of Vacuum Ultraviolet Window Materials I. Experiment," *J. Phys. Chem. Solids*, **30**, 2739 (1969).

21. R. K. Watts, T. M. Wolf, M. Y. Hellman, and L. E. Stillwagon, "Simulation of Microposit 2400-17 Resist Profiles for 0.5 μm Photolithography at 248 nm," in R. E. Howard, E. L. Hu, S. Pang, and S. Mamba, Eds., *Microfabrication*, Materials Research Society, Pittsburgh, PA, 1987.

22. T. M. Wolf, R. L. Hartless, A. Shugard, and G. N. Taylor, "The Evaluation of Positive Acting Resists for Lithography at 248 nm," *J. Vac. Sci. Technol.*, B [2], **5**, 396 (1987).

23. H. R. Philipp, H. S. Cole, Y. S. Liu, and T. A. Sitnik, "Optical Absorption of Some Polymers in the Region 240-170 nm," *Appl. Phys. Lett.*, **48**, 192 (1986).

24. G. N. Taylor, L. E. Stillwagon, and T. Venkatesan, "Gas-Phase Functionalized Plasma-Developed Resists: Initial Concepts and Results for Electron Beam Exposure," *J. Electrochem. Soc.*, **131**, 1658 (1984).

25. F. Coopmans and B. Roland, "DESIRE: A Novel Dry-Developed Resist System," *Proc. Soc. Photo-Opt. Instrum. Eng.*, **633**, 34 (1986).

26. T. H. P. Chang, A. D. Wilson, C. H. Ting, R. Viswanathan, M. Parikh, and E. Munro, "The Probe-Forming and Deflection System for Vector Scan 1 E-B Lithographic System," *Proc. Symp. Electron Ion Beam Sci. Technol.*, p. 377 (1976).

27. G. R. Brewer, Ed., *Electron Beam Technology in Microelectronic Fabrication*, Academic, New York, 1980.

28. R. K. Watts, "Advanced Lithography," in D. F. Barbe, Ed., *Very Large Scale Integration, VLSI, Fundamentals and Applications*, Chapter 3, Springer-Verlag, Berlin and New York, 1982.

29. D. R. Herriott, R. J. Collier, D. S. Alles, and J. W. Stafford, "EBES: A Practical Electron Lithographic System," *IEEE Trans. Electron Devices*, **ED22**, 385 (1975).

30. D. S. Alles, C. J. Biddick, J. H. Bruning, J. T. Clemms, R. J. Collier, E. A. Gere, L. R. Harriott, F. Leone, R. Liu, T. J. Mulrooney, R. J. Nielsen, N. Paras, R. M. Richman, C. M. Rose, D. P. Rosenfeld, D. E. A. Smith, and M. G. R. Thomson, "EBES4 — A New Electron Beam Exposure System," *J. Vac. Sci. Technol.*, B [2], **5**, 47 (1987).

31. M. Fujinami, N. Shimazu, T. Hosukawa, and A. Shibayama, "EB60: An Advanced Direct Wafer Exposure Electron Beam Lithography System For High-Throughput, High Precision Submicron Pattern Writing," *J. Vac. Sci. Technol.*, B [2], **5**, 61 (1987).

32. D. E. Davis, S. J. Gillespie, S. L. Silverman, W. Stickel, and A. D. Wilson, "EL3 Application to 0.5 μm Semiconductor Lithography," *J. Vac. Sci. Technol.*, B [2], 1003 (1983).

33. N. Goto, K. Tanaka, and S. Miyauchi, "Development of A Nanometric E-Beam Lithography System (JBX5D2)," *Proc. Soc. Photo-Opt. Instrum. Eng.*, **537**, 7 (1985).

34. I. A. Cruttwell, W. V. Colbran, and B. A. Wallman, "E-Beam Tool Requirements for Nanolithography," *Proc. Soc. Photo-Opt. Instrum. Eng.*, **537**, 2 (1985).

35. P. Nehmiz, W. Zapka, V. Behringer, M. Kallmeyer, and H. Bohlen, "Electron Beam Proximity Printing: Complementary Mask and Level-to-Level Overlay with High Accuracy," *J. Vac. Sci. Technol., B* [2], **3**, 136 (1985).

36. R. Ward, A. R. Franklin, I. H. Lewin, P. A. Gould, and M. J. Plummer, "A 1:1 Electron Stepper," *J. Vac. Sci. Technol., B* [2], **4**, 89 (1986).

37. E. Nishimura, T. Takigawa, T. Abe, and Y. Katoh, "Proximity Effect Reduction in High Voltage Electron Beam Lithography by a Bias Exposure Method," *J. Vac. Sci. Technol., B* [2], **4**, 164 (1986).

38. M. B. Heritage, "Electron-Projection Microfabrication System," *J. Vac. Sci. Technol.*, **12**, 1135 (1975).

39. M. Nakase, "Determination of the MTF of Positive Photoresists Using the Monte Carlo Method," *Photogr. Sci. Eng.*, **27**, 254 (1983).

40. A. N. Broers, "Resolution, Overlay, and Field Size for Lithography Systems," *IEEE Trans. Electron Devices*, **ED-28**, 1268 (1981).

41. A. N. Broers, "Resolution Limits of PMMA Resist for Exposure with 50 kV Electrons," *J. Electrochem. Soc.*, **128**, 166 (1981).

42. R. E. Howard, E. L. Hu, L. D. Jackel, P. Grabbe, and D. M. Tennant, "400Å Linewidth E-Beam Lithography on Thick Silicon Substrates," *Appl. Phys. Lett.*, **36**, 592 (1980).

43. M. Isaacson and A. Murray, "In Situ Vaporization of Very Low Molecular Weight Resists using 0.5 nm Diameter Electron Beams," *J. Vac. Sci. Technol.*, **19**, 1117 (1981).

44. G. A. C. Jones, S. Blythe, and H. Ahmed, "Very High Voltage (500 kV) Electron Beam Lithography for Thick Resists and High Resolution," *J. Vac. Sci. Technol., B* [2], **5**, 120 (1987).

45. D. L. Spears and H. I. Smith, "High Resolution Pattern Replication Using Soft X-rays," *Electon. Lett.*, **8**, 102 (1972).

46. B. S. Fay and W. J. Novak, "Advanced X-ray Alignment System," *Proc. Soc. Photo-Opt. Instrum. Eng.*, **632**, 146 (1986).

47. R. B. McIntosh, Jr., G. P. Hughes, J. L. Kreuzer, and G. R. Conti, "X-Ray Step-and-Repeat Lithography System for Submicron VLSI," *Proc. Soc. Photo-Opt. Instrum. Eng.*, **632**, 156 (1986).

48. M. Yoshimatsu and S. Kozaki, "High Billiance X-Ray Sources," in H. J. Queisser, Ed., *X-Ray Optics,* Chapter 2, Springer-Verlag, Berlin and New York, 1977.

49. F. L. Hause, P. L. Middleton, and H. L. Stover, "COP X-Ray Resist Technology for 1 μm LSI," *Proc. Kodak Microelectron. Semin. Interface 79*, p. 166 (1980).

50. I. Okada, Y. Saitoh, S. Itabashi, and H. Yoshihara, "A Plasma X-ray Source for X-Ray Lithography," *J. Vac. Sci. Technol. B* [2], **4**, 243 (1986).

51. R. D. Frankel, J. P. Drumheller A. S. Kaplan, and M. J. Lubin, "X-Ray Lithography Process Optimization Using a Laser-Based X-Ray Source," *Proc. Kodak Microelectron. Semin. Interface 86*, p. 82 (1987).

52. A. Heuberger, "X-Ray Lithography," in K. B. van der Mast and S. Radelaar, Eds., *Microcircuit Engineering 85*, Elsevier, Amsterdam, 1985.

53. H. Betz, K. Heinrich, A. Heuberger, H. Huber, and H. Oertel, "Resolution Limits in X-Ray Lithography Calculated by Means of X-Ray Lithography Simulator XMAS," *J. Vac. Sci. Technol., B* [2], **4**, 248 (1986).

54. D. L. White, AT&T Bell Laboratories, private communication.

55. R. L. Seliger, J. W. Ward, V. Wang, and R. L. Kubena, "A High Intensity Scanning Ion Probe with Submicrometer Spot Size," *Appl. Phys. Lett.,* **34**, 310 (1979).

56. J. Orloff and L. W. Swanson, "A Scanning Ion Microscope with a Field Ionization Source," *Scanning Electron Mircosc.,* p. 57 (1977).

57. G. N. Lewis, H. Paik, J. Mioduszewski, and B. M. Siegel, "A Hydrogen Field Ion Source with Focusing Optics," *J. Vac. Sci. Technol., B* [2], **4**, 116 (1986).

58. J. N. Randall, D. C. Flanders, and N. P. Economou, "Masked Ion Beam Lithography Using Stencil Masks," *Proc. Soc. Photo-Opt. Instrum. Eng.* **471**, 47 (1984).

59. J. N. Randall, L. A. Stern, and J. P. Donnelly, "The Contrast of Ion Beam Stencil Masks," *J. Vac. Sci. Technol., B* [2], **4**, 201 (1986).

60. R. Speidel and V. Behringer, "Lithium Ion Beam Exposure of PMMA Layers Without Proximity Effect," *Optik,* **54**, 439 (1979/80).

61. J. L. Bartelt, "Masked Ion Beam Lithography: An Emerging Technology," *Solid State Technol.,* **28**, 215 (1986).

62. G. Stengl, H. Löschner, W. Maurer, and P. Wolf, "Ion Projection Lithography Machine IPLM-01: A New Tool for Sub 0.5 μm Modification of Materials," *J. Vac. Sci. Technol., B* [2], **4**, 194 (1986).

63. A. F. Burenkov, F. F. Komarov, M. A. Kumakhov, and M. M. Temkin, *Tables of Ion Implantation Spatial Distribution,* Gordon & Breach, New York, 1986.

64. I. Adesida, L. Karapiperis, C. A. Lee, and E. D. Wolf, "A Study of Ion Beam Exposure Profiles in Polymethyl Methacrylate (PMMA)," in A. Oosenbrug, Ed., *Microcircuit Engineering 81,* p. 322, Swiss Fed. Inst. Technol., Lausanne, 1981.

65. I. Adesida and L. Karapiperis, "The Range of Light Ions in Polymeric Resists," *J. Appl. Phys.,* **56**, 1801 (1984).

66. M. D. Giles, "Ion Implantation Calculations in Two Dimensions Using the Boltzmann Transport Equation," *IEEE Trans. Comput.-Aided Des.,* **CAD-5**, 679 (1986).

67. K. Moriwaki, H. Aritome, and S. Namba, "Ion Beam Exposure of FPM and AZ1350 Resists for An Application to Nanometer Lithography," in A. Oosenbrug, Ed., *Microcircuit Engineering 81,* p. 329, Swiss Fed. Inst. Technol., Lausanne, 1981.

68. J. N. Randall, D. C. Flanders, N. P. Economou, J. P. Donnelly, and E. I. Bromley, "Masked Ion Beam Resist Exposure Using Grid Support Stencil Masks," *J. Vac. Sci. Technol., B* [2], **3**, 58 (1985).

69. K. Moriwaki, H. Aritome, and S. Namba, "High Resolution Fabrication of Submicron Structures by Ion Beam Lithography" *Jpn. J. Appl. Phys.,* **20**, Suppl. 20-1, 69 (1981).

70. R. G. Brault and L. J. Miller "Sensitivity and Contrast of Some Proton-Beam Resists," *Polym. Eng. Sci.,* **20**, 1064 (1980).

71. U. Behringer and R. Speidel, "Investigation of the Radiation Loads of a Self-Supporting Silicon Mask in an Ion Beam Proximity Printer," in A. Oosenbrug, Ed., *Microcircuit Engineering 81,* p. 369, Swiss Fed. Inst. Technol., Lausanne, 1981.

72. J. Frosien, "Digital Image Processing for Micrometrology," *J. Vac. Sci. Technol., B* [2], **4**, 261 (1986).

73. C. P. Ausschnitt, T. A. Bruner, and S. C. Yang, "Application of Wafer probe Techniques to the Evaluation of Projection Printers," *Proc. Soc. Photo-Opt. Instrum. Eng.*, **334,** 17 (1982).

74. J. H. Bruning, "Performance Limits in 1:1 UV Projection Lithography," *J. Vac. Sci. Technol.*, **16,** 1925 (1979).

75. R. M. Sills and K. P. Standford, "E-Beam System Metrology," *Solid State Technol.*, **20,** 191 (1983).

76. *Proc. Soc. Photo-Opt. Instrum. Eng.*, **565** (1985).

77. H. I. Smith, "A Model for Comparing Process Latitude in UV, Deep UV, and X-Ray Lithography," *J. Vac. Sci. Technol.*, B [2], **6,** 346 (1988).

78. J. L. Bartelt, C. W. Slayman, J. E. Wood, J. Y. Chen, and C. M. McKenna, "Masked Ion-Beam Lithography: A Feasibility Demonstration for Submicrometer Device Fabrication," *J. Vac. Sci. Technol.*, **19,** 1166 (1981).

Index

Aberration, 483
Abrupt heterostructure, 162
Absorption, 436
ac bias, 62
Accumulation layer resistance, 37
Activation energy, 444, 445–447, 451
Addition, 421
Addresses, 484
Airy pattern, 488
AlAs, 122, 123
   high electron mobility transistor, 179, 188
   resonant tunneling bipolar transistor, 239, 240, 241
   resonant tunneling diode, 213, 219
AlAs/GaAs:
   incoherent resonant tunneling, 210
   resonant tunneling, 204, 214
AlAs/GaAs/AlAs, 236
AlGaAs:
   high electron mobility transistor, 177, 179, 181, 182, 185–186, 188, 189
   lateral surface superlattice, 365
   resonant tunneling, 204, 219, 220
   resonant tunneling unipolar transistor, 256, 260
   short-channel effects, 146–152
$Al_{0.07}Ga_{0.93}As$, 240
$Al_{0.20}Ga_{0.80}As$, 245–246
$Al_{0.3}Ga_{0.7}As$, 208, 209, 213, 236
$Al_{0.33}Ga_{0.67}As$, 245
$Al_{0.45}Ga_{0.55}As$, 229
$Al_{x}Ga_{1-x}As$, 122, 210, 213
AlGaAs/GaAs, 125
   bipolar junction transistor, 158–164
   field-effect transistor, 128, 130, 131–146
   resonant tunneling bipolar transistor, 228, 238
   resonant tunneling diode, 210
   vertical FET, 163
$Al_{0.325}Ga_{0.675}As/GaAs$, 214
$Al_{0.35}Ga_{0.65}As/GaAs$, 219, 220
$Al_{0.50}Ga_{0.50}As/GaAs$, 215
Algorithmic computation, 413

$Al_{x}In_{1-x}As$, 210
AlInAs/GaInAs, 243
$Al_{0.48}In_{0.52}As/Ga_{0.47}In_{0.53}As$, 247
AlN, 123, 125
ALU, 385
Aluminum, 445
Amorphization, 451
Amplifier, 415–416
Analog amplifier, 427
Analog-to-digital converter:
   resonant tunneling bipolar transistor, 231–232
   synthetic neural system, 431
Application specific chip, 4
Application-specific integrated circuit, 379
Architecture, see Chip architecture; Computer architecture
Arc lamp, 439
Array processor, 341
Arsenic, 446–451, 458
Artificial intelligence, 338
   synthetic neural system, 390–391
   VLSI technology, 354
As, 459, 463
As-doped n^{+}/p junction, 453
Associative memory, 421, 427, 431–432
Attenuation constant, 284
Auger electron spectra, 444
Auger recombination, 59
Automaton, 387
Average interconnect length:
   interconnections, 270
   Rent's rule, 382
Average wire length, 275
Axial field component, see Longitudinal field component
Axon, 391–392, 413

Backplane wiring, 345
Backward crosstalk, 314, 318
Ballistic effect:
   current-voltage characteristics, 134
   field-effect transistor, 127, 130
   unipolar vertical device, 162

Ballistic electron, 165
Ballistic transistor, 126
Ballistic transport:
  resonant tunneling bipolar transistor, 238
  submicron transistor, 124
  unipolar vertical device, 165
Band diagram, 156
Band discontinuity, 145
Bandgap:
  rapid thermal processing, 436
  submicron transistor, 122
Bandgap narrowing, 99–100
Band structure, 122
Barrier height, 165
Barrier lowering:
  CMOS design, 23, 33, 35, 36–37, 63
  hot-carrier degradation, 54–55, 58
  response time improvement design, 72
Base-conductivity modulation, 92–93
Base punchthrough, 90. *See also* Punch-
  through
Base spreading resistance, 158
Beam brightness, 483
BF_2 ion, 452
Bidirectional associative memory, 421
Bidirectional synapse, 418
Bilayer resist, 471
Bimodal distribution, 275
Biological system, 390–393, 418
Bipolar transistor, 87–116
  circuits, 107–113
  design, 89–96
  future possibilities of, 113–116
  nonscalable components, 98–99
  rapid thermal processing, 457
  scaling, 96–98
  scaling properties/limits, 99–104
  structures, 104–107
Bit rate, 292
Bloch field effect transistor, 366, 367
Bloch frequency, 372
Bloch localization, 368
Bloch oscillation, 368, 369, 371, 372, 373,
  374
BN, *see* Boron nitride
Boltzmann equation, 46, 47–48, 49, 50
BOMOS, *see* Buried oxide MOSFET
  (BOMOS)
Book-to-bill ratio, 3–4
Boron, 451–452
Boron nitride, 454
Borophosphosilicate glass (BPSG), 463
Brain, *see* Human brain

Breakdown voltage:
  MOSFET, 39, 63
  rapid thermal oxidation, 444
Buffer cell, 428
Built-in voltage, 135
Bulk punchthrough, *see also* Punchthrough
  CMOS design, 31, 33
  MOSFET, 63, 75
Buried-channel devices
  CMOS design, 23, 33
  hot-carrier degradation, 62
  response time improvement design, 73
Buried drain, 68
Buried oxide MOSFET (BOMOS), 70

CA, *see* Cellular array
Capacitance, *see also* Line capacitance
  interconnection, 277. *See also* RC line
    model
  interconnection parameters calculation,
    303, 305
Capacitance matrix, 301, 302
Capacitive noise, 317, 318
Capacitor, 126
Carrier density fluctuation, 44
Carry look-ahead circuit, 191
CBL, *see* Charge buffer logic
CCD array, 361
Cellular array, 378, 387–388, 396
Cellular automata:
  systolic array, 397
  two-dimensional automata, 378, 386
Cellular logic, 387–388
Cellular network, 397
Centroid, 25, 29, 30
Chambers, 439
Channel depth, 144
Channel doping:
  current-voltage characteristics, 134
  MOSFET, 75
  NMOS design, 20
Channeling tail, 451, 452
Channel length:
  current/voltage characteristics, 142
  MOSFET, 72–75
Channel length modulation, 147
Channel profile:
  CMOS, 22–23, 25, 28, 31, 37
  NMOS, 21
Characteristic impedance:
  crosstalk, 314, 318, 320
  interconnection parameter calculation, 298,
    299, 300
  pulse propagation, 311–312
  RLC line models, 288, 293

Charge buffer logic, 114
Charge control concept, 94
Charge pumping, 62
Chemical vapor deposition, 456, 463
Chip architecture, 397–409. *See also* Computer architecture
Chromatic aberration, 483
Circuit speed, 17
Clock frequency, 384, 385
Clustering, 446–447, 448–449, 463
CMOS, *see* Complementary MOS design
Cobalt silicide, 461–462
Coherency, 386
Coherent (Fabry–Perot-Type) resonant tunneling, 207–208
Collector capacitance, 161
Collector current, 165
Collector saturation current, 91
Collisional broadening:
    hot-carrier degradation, 59
    incoherent resonant tunneling, 209, 210
    Monte Carlo methods, 49
Column-address line, 428
Common-emitter current gain, 157, 158
Common mode logic (CML) gate, 158
Communication network, 423
Complementary MOS design (CMOS):
    bipolar technology, 87–88
    chip architecture, 398
    GaAs digital ICs, 353
    MOSFET, 22–37
    VLSI technology, 350
Complexity:
    synthetic neural system, 423–424
    VLSI design, 349
Complex permittivity, 298
Compound semiconductor, 122. *See also* Submicron transistor(s)
Computational bandwidth, 426
Computational model:
    synthetic neural system, 414
    two-dimensional automata, 394–396
Computer analysis:
    hot-carrier degradation, 59
    MOSFET, 12, 14, 45–46
Computer architecture:
    balance considerations, 351–353
    classification of, 339–341
    interconnections, 269
    synthetic neural systems, 391, 425
    VLSI coherency, 386
    VLSI technology, 332
Computer vision, 355
Conductance, 294

Conductance matrix, 313
Conductivity, 298
Connection, 419
Connection machine, 401
Connection matrix, 427
Contact implant, 150
Contact resistance:
    bipolar scaling, 98
    MOSFET, 37
Contamination, 440, 463
Contrast enhancement, 472
Convergence time, 420
Coplanar line, 310–311
Coplanar waveguide, 312
$CoSi_2$, 455
Critically damped circuit, 285
Critically damped response, 293
Crosstalk:
    interconnections, 312–321
    RLC line circuits, 287
Crosstalk reduction, 321
Crowding effect, 91
Crystallization, 445–446
CsI, 487
Current gain, 91
Current-gain cutoff frequency, 181
Current-voltage characteristic, 131–146
Custom VLSI chip, 431
Cutoff, 163
Cutoff frequency, 144
CVD, 464

Data processing, 335, 337
DBHET, *see* Double base hot-electron transistor (DBHET)
DCFL MODFET, 130
dc stress, 62
DDD, *see* Doubly diffusioned drain
Deep-trench isolation, 106
Defect, 446, 447–449
Delay:
    crosstalk, 312
    interconnections, 269, 270, 271, 273
    RC line models, 277, 278, 280, 281, 282
    synthetic neural system, 426–427
    VLSI technology, 351
Delay component, 95
Dendrite, 392, 413
Density matrix, 394
Depletion (D) mode:
    high electron mobility transistor, 178, 188–189
    lateral surface superlattice, 365

Depletion region, 454

Depletion-type high electron mobility transistor, 190, 195, 196

Depth of focus, 477, 478–479

Determination of L and C, 304–305

Device structure, 12

Dielectric constant, 298

Dielectric regime, 295

Differential crosstalk, 314

Diffraction, 490, 494

Diffraction-limited optics, 476

Diffused junction, 454–455

Diffusion, 447, 449–451, 452, 453, 458

Diffusion capacitance, 94

Diffusion equation, 276

Diffusion of dopants, *see* Dopant diffusion

Diffusion length, 103

Diffusion sheet resistance, 37. *see also* Sheet resistance

Digital signal processor, 423

Dimensional stability, 492

Dislocation, 449, 452

Dispersion, 308, 310

Dissipation, *see also* Power dissipation
  interconnections, 269, 270, 272–273
  VLSI technology, 351

Distributed coupling analysis, 318–321

Distributed information processing, 378

Distributed RLC line circuit, 287–294

Distributed storage, 427

Distribution function, *see also* Electron distribution function
  Boltzmann equation, 47–48
  moment methods, 50
  Monte Carlo approach, 49
  MOSFET, 45

Distribution of line length, 274

Distribution of wire length, 273, 274, 275

Divide-by-two, 177

Divide-by-two/divide-by-four, 128

DNA, 391

Donor diffusion, 151

Dopant activation, 457–460

Dopant diffusion:
  current-voltage characteristics, 134
  rapid thermal processing, 446, 449

Dose, 25, 28, 29–30

Double base hot-electron transistor (DBHET), 165

Double diffused (DMOS) transistor, 74

Double injection field-effect transistor, 164

Double-poly self-aligned bipolar structure, 105

Doubly diffusioned drain, 64

Drift-diffusion approximation, 47

Driver crosstalk, 315. *See also* Nonlinear drive
  interconnection, 272, 273
  interconnection parameter calculation, 301
  pulse propagation, 309
  RLC line model, 292–293

DX center, 278

Dynamic temperature uniformity, 441

EBES, *see* Electron-beam exposure system (EBES)

EBMF2.5UHR, 485

ECL circuit, *see* Emitter-coupled-logic circuit

Eddy current, 306

Effective field, 44

Effective mass, 123

E-HEMT, 181, 184

Eigenvalue, 299

Einstein relation, 100–101

Ejected electron, 309

Electrical activation, 457

Electric modeling of interconnection, *see* Interconnection(s)

Electron-beam exposure system (EBES), 484, 502

Electron-beam lithography, 365

Electron distribution function, 46–51

Electron-hole plasma, 164

Electronic implementation, 415

Electron injection, 150

Electron lithography, 482–488

Electron mobility, 134. *See also* High electron mobility transistor (HEMT)

Electron saturation velocity, 135

Electron-temperature model, 56, 58–59

Electron velocity:
  current-voltage characteristic, 137
  submicron transistor, 123

Electro-optic effect, *see* Pockels effect

Electrostatic scaling, 15–17, 18, 19

Elements, 122

Elmore delay, 277, 278, 281, 289

Elmore response, 277–280

Emitter-coupled-logic circuit:
  bipolar scaling, 97
  bipolar transistor, 87
  delay components of, 95

Emitter current, 165

Emitter-on-top structure, 161

Emitter resistance, 104

Emitter stripe, 159

End-of-range dislocation, 449

Energy balance equation, 58

Energy band diagram, 54
Energy band engineering, 162–163
Energy density, 303
Energy distribution, 165
Energy relaxation time, 50–51
Enhancement (E) mode, 178, 188–190, 195, 196
Epitaxial multilayer structure, 464
Excimer laser, 478
Excitation, 413
Excitatory connection, 427
Expert computer system, 338
Extended defect, 449, 453
Extrinsic base, 105
Extrinsic transconductance, 139

Fabricational tolerance, 75
Fabricational variation, 9
Fabry–Perot-Type (coherent) resonant tunneling, 207–208
Fan-in/fan-out, 128
Fast crosstalk, 314
Fault tolerance, 385
Feedback loop, 416
FET, *see* Field-effect transistor
Field amplitude, 472
Field-effect transistor, *see also* Submicron transistor(s)
    high electron mobility transistor, 182
    resonant tunneling unipolar transistor, 263, 264
    scaling considerations, 126–130
    short-channel effects in, 146–152
    submicron transistor, 124
Field-effect transistor logic, 181
Finite difference approximation, 305
Finite difference method, 305
Finite elements, 305
Flat-band voltage, 441, 445
Fluctuation in carrier density, *see* Carrier density fluctuation
Forward crosstalk, 314
Forward direction, 416, 418
Foundries, 4
Fourier's law of heat conduction, 50
Fourier transform, 474
Free carrier, 437
Free-carrier contribution, 301
Frequency, 163
Frequency divider, 177, 181
Frequency multiplier, 222
Fringing capacitance, 128
Full-adder, 191
Full channel current, 137

Functionally partitioned circuit, 381
Fuzzy logic, 406, 408

GaAlAs, 361, 365
GaAs, 122, 130
    current-voltage characteristics, 131–146
    high electron mobility transistor, 176, 177, 179, 181, 182, 183, 185, 186, 188, 193, 196–197
    incoherent resonant tunneling, 209
    quantum well, 208
    resonant tunneling bipolar transistor, 229, 236, 237, 238, 239, 240
    resonant tunneling diode, 213, 214, 218–219, 220
    resonant tunneling unipolar transistor, 245, 247, 248, 256
    short-channel effects, 146–152
    superlattice, 361, 364–369, 371, 372, 381
    unipolar vertical device, 163
    VLSI technology, 353
GaAs/AlGaAs:
    high electron mobility transistor, 176, 177, 185, 186, 187, 189
    resonant tunneling unipolar transistor, 256
$GaAs/Al_xGa_{1-x}As$, 213
$Ga_xIn_{1-x}As$, 123
$Ga_{0.47}In_{0.53}As/Al_{0.48}In_{0.52}As$ resonant tunneling diode, 210–212
GaInAs/InP, 125
Gallium arsenide, 123
GaP, 123
Gate, 348
Gate array:
    high electron mobility transistor, 190
    Rent's rule, 380
    superlattice concept, 360, 361
Gate capacitance, 126
Gate current:
    hot-carrier degradation, 54–59, 60–61
    MOSFET, 76
    response time improvement design, 73
Gate dielectrics, 442–445
Gated quantum well resonant tunneling transistor, 259–264
Gate leakage current, 13υ
Gate length dependence, 150–152
Gate length modulation, 146
Gate material, 25, 31
Gate overlap, 65
Gate propagation delay, 159–160
Gate structure, 464
Gate-to-channel separation, 139
General-purpose processor, 422
Germanium, 122, 452

Glass reflow, 463
Graded heterojunction, 158
Graded heterostructure, 162
Grain boundary, 459
Grain boundary segregation, 463
Grain growth, 456–457
Grain size, 457
Green's function, 304
Ground plane:
  crosstalk, 313, 321
  interconnection parameters calculation, 306
  pulse propagation, 308, 309
  RLCG model, 295, 296

Half-adder, 191
Halo structure, 73
Hard limiter, 414
Harvard machine, 335
HBT, *see* Heterojunction bipolar transistor
  (HBT)
Heating, 273
HEMT, *see* High electron mobility transistor
  (HEMT)
HET, *see* Hot-electron transistor (HET)
Heterointerface:
  band discontinuity at, 145
  current-voltage characteristics, 134
  submicron transistor, 123
Heterojunction biopolar transistor (HBT),
  125, 126
  field-effect transistor, 130
  resonant tunneling bipolar transistor, 240
  vertical device structures, 154–162
Heterojunction superlattice, 360
Heterostructure field-effect transistor, 133
High electron mobility transistor (HEMT),
  176–199
  compared, 179–181
  current-voltage characteristics, 131–132
  generally, 176–177
  large scale integration circuit implementa-
    tion, 190–198
  lateral surface superlattice, 368
  logic circuits, 190–194
  material technology, 185–188
  memory circuits, 194–197
  performance advantages of, 177–184
  principle of, 177–179
  self-alignment device fabrication technol-
    ogy, 188–190
  submicron dimensional range performance
    of, 181–184
  superlattice, 361
  VLSI integration, 184–190

High-energy tail, 49, 57, 59
Higher-order mode, 299
High-frequency response, 371–374
High-performance MOS design, 19–37
High resolution electron microscopy, 457
High T_c superconductor, 125
HMOS, *see* High-performance MOS design
Hole mobility, 130
Hot carrier, 74
Hot-carrier degradation:
  Monte Carlo approach, 49
  MOSFET, 50–62
Hot-carrier reliability, 63–68
Hot electron:
  hot-carrier degradation, 59, 62
  MOSFET, 63
  VLSI, 379
Hot-electron degradation, 37
Hot-electron effect, 73
Hot-electron spectrometer, 165
Hot-electron transistor (HET), 125, 163
Hot hole, 56
Human brain, 413
Hydrodynamic model, *see* Moment methods
  (hydrodynamic models)

IBT, *see* Induced base transistor (IBT)
IC design, 11–12, 75
I^2L bipolar circuit, *see* MTL bipolar circuit
Image current, 309
Image processing, 355
Imaging system, 473
Implant, 19
Implant annealing, 151
Implantation, 151
Implant dose, 29
Incoherent (sequential) resonant tunneling,
  208–210
Indium:
  InAs, 122, 364
  InGa/AlGaAs, 125
  $In_xGa_{1-x}As$, 123
  InGaAs/InP, 159
  InP, 122, 123
  InSb, 122, 364
Induced base transistor (IBT), 125, 126, 163
Inductance:
  interconnection parameters calculation,
    303, 305
  RLC line model, 283, 290
Inductance matrix, 301, 302
Inductive noise, 317, 318
Inelastic mean path, 360, 364, 367
Information flow, 270

Information processing, 335, 337
Infrared pyrometer, 439–440
Inhibition, 413
Inhibitory connection, 427
Injection efficiency, 158
Input–output relationship, 418
In situ processing, 463–464
Integral base charge, 91
Integration level, 107
Intelligence processing, 335, 338
Interconnection(s), 269–322
 crosstalk, 312–321
 dissipation scenario, 272–273
 generally, 269–276
 parameter calculations, 297–306
 performance scenario, 270–272
 pulse propagation, 307–312
 RC line models, 276–283
 Rent's rule, 273–276, 380
 RLCG line models, 294–297
 RLC line models, 283–294
 synthetic neural system, 426
 VLSI technology, 349–351
Interconnection capacitance, 382
Interconnection metalization, 378
Interconnection parameters calculation,
 297–306
 beyond static approximation, 305–306
 ground planes, 306
 multiple line formulations, 301–304
 numerical determination of L and C
  in static approximation, 304–305
 single line formulation, 298–301
Interconnection strength, 396
Interfacial oxide, 104
Interline capacitance, 301
Intrinsic-base region, 106
Intrinsic-base sheet resistance, 91
Intrinsic drain conductance, 37
Inverse-narrow width effect, 379
Inversion layer, 361
Inverter, 428, 431
I/O, 273–274, 275, 425
Ion lithography, 494–499
ISL bipolar circuit, 109
Iterated tessalation network, 378

JEOL JBX5D2, 485
JEOL JBX6A3, 485
JFET, *see* Junction field-effect transistor
Josephson effect, 373
Junction capacitance, 20, 70, 72, 73
Junction depth, 21, 25, 70, 72
Junction field-effect transistor, 131

Junction formation, 452–455
Junction isolation, 106
Junction leakage, 70, 453
Junction profile, 39

Kirk effect, 70, 107
Knowledge processing, 335, 337

LaB_6, 482–483
Laplace's equation, 304, 305
Large-scale integration (LSI), 176, 177, 181,
 185, 187, 190–198
Laser, 125
Lateral *pnp* transistor, 98
Lateral surface superlattice, 360–375
 described, 361–364
 GaAs structures, 364–369
 transport theory, 369–374
Lattice constant, 123
Layered media, 302–303
LDD, *see* Lightly doped drain
Leakage control, 75
Leakage current:
 bipolar scaling, 99
 CMOS design, 22
 hot-carrier reliability, 67
 rapid thermal processing, 455
 synthetic neural system, 426
Learning curve, 1
Length distribution, 274
Light emitting diode, 125
Lightly doped drain (LDD), 64, 65, 67
Linear region, 148
Line capacitance, 294
Line parameter, 309, 312
Line resistance, 276, 283, 293. *See also RC*
 line models
Lithography, 153, 470–504
 electron lithography, 482–488
 generally, 470
 ion lithography, 494–499
 metrology, 499–503
 optical lithography, 470–482
 X-ray lithography, 488–494
Load, *see also* Nonlinear load
 crosstalk, 315, 318
 interconnection, 272, 273
 interconnection parameter calculation, 301
 pulse propagation, 309
 RLC line models, 292–293
Logic, 107–110
Logic circuit, 190–194
Long-channel device, 9, 23, 24–35, 36, 37,
 128–129

Long-channel subthreshold behavior, 18
Longitudinal field component, 297, 298–300, 303
Low-field mobility, 129, 141
LSI, *see* Large-scale integration
Lucky-electron model, 56, 57, 59
Lumped coupling analysis, 315–318
Lumped *RLC* circuit, 283–287

Masks, 491–492
Master-slice-type chip, 380
Matrix of weights, 416
Maximum concentration of two-dimensional gas, 137
Maximum device transconductance, 136
Maximum drain-to-source current, 136
Maximum field, 21
Maximum frequency, 291, 292
Maximum gain, 156–157
Maximum oscillation frequency, 163
Maximum transconductance, 139, 145
Maximum voltage swing, 137
Maxwell's equation, 298, 299
MBE, *see* Molecular-beam epitaxy (MBE)
MEBES electron-beam pattern generator, 482, 502
Memory:
  associative memory, 431–432
  bipolar circuits, 107, 110–113
  high electron mobility transistor, 194–197
  interconnections, 273
  VLSI, 335, 353
MESFET, *see* Metal semiconductor field-effect transistor (MESFET)
Metal organic chemical vapor deposition, 123, 125
Metal semiconductor field-effect transistor (MESFET), 126, 130
  current-voltage characteristics, 131–146
  gate length, 151
  high electron mobility transistor, 176
  lateral surface superlattice, 365
  resonant tunneling unipolar transistor, 256
Metrology, 499–503
Microcomputer, 423
Microprocessor, 379
Microprocessor chip, 381
Microstrip, 310
Miniband, 361, 364, 368, 369
Minigap, 361
Minimum device size, 122
Ministry of International Trade and Industry (MITI), 1–2
Minizone, 368

Mobility, 44, 45–46
MOCVD, *see* Metal organic chemical vapor deposition
Mode, 306
Modeling, 431
Mode-matching method, 306
Mode of propagation, *see* Propagation mode
Modulation doped field-effect transistor (MODFET), 125, 128, 130
  current-voltage characteristics, 131–146
  lateral surface superlattice, 364–367
  output conductance, 150
Modulation transfer function (MTF) 475
Molecular-beam epitaxy (MBE), 123, 125, 360
  heterojunction bipolar junction transistor, 158
  high electron mobility transistor, 181, 185, 190
  resonant tunneling diode, 213
  resonant tunneling unipolar transistor, 247
Moment equation, 49, 50
Moment methods (hydrodynamic models), 49–51
Momentum relaxation time, 50–51
Monolithic integration, 125
Monte Carlo methods, 47, 48–49, 51, 52, 371
MOS device, 377, 426
MOSFET, 9–76
  bipolar design, 89, 108
  bipolar scaling, 97
  chip architecture, 400
  device designs, 62–75
  environment overview, 10–14
  HMOS design, 19–37
  hot-carrier degradation, 51–62
  overview of, 9–10
  parasitic resistance of source and drain, 37–41
  scaling, 14, 15–19
  short-channel effects, 149
  transport, 42–51
MOS gate structure, 464
Mott–Gurney law, 148
MP2400, 480
MTF, 487–488
MTL bipolar circuit, 109
MTL memory, 115
Multidimensional superlattice, 361
Multilayer resist, 472
Multilevel distribution, 275
Multilevel interconnection, 271, 274
Multiple-level metal process, 107–108
Multiple pipeline structure, 341

Multiple-state memory, 232–234
Multiple valued logic, 223–225, 230
Multiplication, 421
Multiplier, 191, 193
Mutual capacitance, 315
Mutual inductance, 301, 315

Native oxide, 457
Natural language, 338
n-channel MOS design, 20–21, 400–401
NDR, see Negative differential resistance
   (NDR)
Nearest-neighbor coupling, 378
Negative differential conductivity, 361,
   366–368
Negative differential resistance (NDR):
   coherent resonant tunneling, 207
   incoherent resonant tunneling, 208
   origin of, 205–207
   resonant tunneling, 204
   resonant tunneling bipolar transistor, 228,
     243
   resonant tunneling diode, 218
   resonant tunneling unipolar transistor, 253,
     259, 260
NERFET device, 367
Neural network, see Synthetic neural system
   (SNS)
Neural state change, 420
Neuron, 391–392, 393, 410, 413
Neuroscience, 390–393
Neurotransmitter, 392
Newton's law, 50
Nitridation, 442
Nitrogen, 445
Nitroxide, 444
NMOS, see n-channel MOS design
Noise:
   crosstalk, 314, 315, 317–320
   interconnections, 270
   synthetic neural system, 425
   thermal noise, 377
Noise margin:
   electrostatic scaling, 17
   hot-carrier degradation, 51
   hot-carrier reliability, 63
   MOSFET, 40
   NMOS, 21
Nonconservative field, 309
Nonlinear drive, see also Driver crosstalk,
   320–321
   RC line model, 280–283
Nonlinearity, 414, 431

Nonlinear load, see also Load
   crosstalk, 320–321
   RC line model, 280–283
   RLC line models, 294
Nonrecurrent neural network, 416–418
Normal field dependence, 43–45
N^+/p junction, 452
N^+ polysilicon/silicide gate, 23
NTL bipolar circuit, 109–110
Nucleation, 269, 270, 273, 274
Numerical aperture, 474, 483

Object plane, 473
Off-carrier density, 25, 31
Off-current:
   CMOS design, 23, 25, 37
   MOSFET, 75
   response time improvement design, 73
One-dimensional quantum wire, 143
Optical delay line, 307
Optical lithography, 470–482
Optically activated semiconductor switch,
   307, 308
Optical pumping, 361
Optical pyrometer, 440
Optimization problem, 431
Optoelectronic measurement, 310
Optoelectronics, 125, 307
Orientation effect, 150
Output conductance, 146–150, 151
Output drain conductance, 144
Overdamped circuit, 284
Overlap capacitance, 20, 69, 70
Overshoot:
   current-voltage characteristics, 134
   field-effect transistor, 127, 130–138
   inductance, 283
   MOSFET, 51, 76
   submicron transistor, 124
   unipolar vertical device, 162
Oxide margin, 21
Oxide tapping, 73
Oxide thickness, 24, 72

P, 463
Pad driver, 426
Parabolic quantum well, 212–218
Parallel field dependence, 45
Parallelism, 423
Parallel processing, 388
Parasitic capacitance:
   CMOS design, 31, 35
   heterojunction bipolar junction transistor,
     159–160

Parasitic capacitance (*Continued*)
  hot-carrier reliability design, 65
  response time improvement design, 69, 70–72
Parasitic device, 63
Parasitic resistance, 37–41
Parity generator, 231, 255–258
Partial coherence, *see* Spatial partial coherence
Particle–particle correlation, 48
Partitioning, 269, 273
Passivation layer, 134, 150
PBT, *see* Permeable base transistor (PBT)
p-channel device, 130, 142, 143
PDBT, *see* Planar doped barrier transistor (PDBT)
Penumbral blur, 489, 498
Performance, 270–272
Perkin Elmer printer, 470–471
Permeability, 298
Permeable base transistor (PBT), 125, 126, 163
Permittivity, 298
Phosphosilicate glass (PSG), 463
Photolithography, *see* Optical lithography
Photoresist, 480
Piezoelectric charge, 150
Pinch-off velocity, 135
Pinch-off voltage, 151
Pipelined computer, 340–341
Planar doped barrier transistor (PDBT), 125, 163
Planarity, 345
Planar structure, 104
Plasma-discharge source, 493
Plasma-enhanced CVD, 189
PMMA, 488, 496, 497, 499
pn junction leakage, 73
pnp load cell, 112
Pockels effect, 309
Point-spread function, 488
Polar optical scattering, 157
Polycrystalline silicon anneal, 455–460
Polycrystalline silicon gate, 441
Poly-monosilicon interface, 104
Polysilicon, 442, 459–460
Polysilicon emitter, 105
Polysilicon transmitter, 104
Post-implantation anneal, 445–452
Power:
  interconnection parameter calculation, 300, 303
  pulse propagation, 312
  VLSI technology, 351

Power consumption, 269
Power-density requirement, 425
Power dissipation, *see also* Dissipation
  interconnections, 272–273
  synthetic neural system, 425
Poynting's vector, 300
p^+ polysilicon/silicide gate, 23
Princeton architecture, 335
Probability density, 209
Process centering, 12, 68
Process characterization, 13
Process design, 12
Process modeling, 75
Process sensitivity, 13
Process variation, 9, 63, 72
Programmable systolic array, 378
Projection printing, 487, 496
Propagation constant, 288, 298
Propagation delay, 128
Propagation mode, 298, 299, 300, 301, 302, 303
Propagation velocity:
  crosstalk, 321
  interconnection parameters calculation, 302, 303
  RLCG line model, 295
  RLC line model, 288
Proximity effect:
  crosstalk, 313
  interconnection parameters calculation, 305
  optical lithography, 477
Proximity printing:
  electron lithography, 482, 487
  ion lithography, 496
  X-ray lithography, 489–491
PSG, *see* Phosphosilicate glass (PSG)
Pulse propagation, 307–312
Punchthrough:
  CMOS design, 36, 37
  hot-carrier degradation, 51
  response time improvement design, 73
Punchthrough criterion, 35. *See also* Bulk punchthrough
Punchthrough voltage, 62
Pyrometer, 439–440

Q, 284–286, 293
Quantization, 420
Quantization of two-dimensional electron gas, 136
Quantized inversion layer, 360
Quantum box, 366
Quantum well:
  bipolar transistor, 239

gated quantum well resonant tunneling transistor, 259–264
incoherent resonant tunneling, 208
resonant tunneling, 204
resonant tunneling bipolar transistor, 238
resonant tunneling diode, 205, 207, 212–218, 219
resonant tunneling unipolar transistor, 245, 246
Quantum well superlattice, 360–361
Quantum wire transistor, 256–259
Quasi-Fermi level, 137
Quasi-static mode, 299

Radiation hardness, 125
Radiation losses, 436
RAM, 194–198, 431
Rapid thermal nitridation, 444–445
Rapid thermal oxidation, 442–444
Rapid thermal processing, 434–465
advantages of, 435
equipment for, 439–440
fundamentals of, 435–439
gate dielectrics, 442–445
generally, 434–435
glass reflow, 463
in situ processing, 463–464
junction formation, 452-455
polycrystalline silicon anneal, 455–460
post-implantation anneal, 445–452
silicide formation, 460–462
stress, 441
temperature measurement, 440–441
Rate of temperature rise, 436
Rayleigh unit of defocus, 476
RBT, 240
RC line models:
interconnections, 276–283
RLC line model and, 283–284, 287, 289–290
RC line response, 277–280
Recessed field oxide, 70
Recombination current, 157
Recurrent neural network, 418–419
Reflection, 292–293, 310
Regime, 295
Relaxation time, 51, 58, 369
Reliability, 385–386
Rent exponent, 273, 275, 276
Rent's rule:
information/interconnections, 384
interconnections, 273–276, 380
two-dimensional automata, 379–383
VLSI technology, 347

Repeater, 271, 273, 282–283
Residual damage, 452
Residual defect, 449
Resistance matrix, 313
Resistive connection, 427
Resistor-load cell, 112
Resolution:
electron lithography, 483, 487–488
optical lithography, 477, 478–479
Resonant tunneling, 367–368
Resonant tunneling bipolar transistor:
alternate designs of, 243
circuit applications of, 230–234
generally, 228–230
RT spectroscopy, 234–238
thermal injection with, 238–243
Resonant tunneling device, 204–264
diodes, 205–228. See also Resonant tunneling diode
generally, 204–205
resonant tunneling bipolar transistor, 228–243. See also Resonant tunneling bipolar transistor
resonant tunneling unipolar transistor, 243–264. See also Resonant tunneling unipolar transistor
Resonant tunneling diode, 205–228
integration of circuit applications, 218–228
parabolic quantum wells, 212–218
physics of, 205–210
room temperature operation, 210–212
Resonant tunneling gate field-effect transistor, 247–256
Resonant tunneling hot-electron transistor, 210, 245–247
Resonant tunneling unipolar transistor, 243–264
gated quantum well resonant tunneling transistor, 259–264
generally, 243–245
quantum wire transistor, 256–259
resonant tunneling gate field-effect transistor, 247–256
resonant tunneling hot-electron transistor, 245–247
Resonant tunneling spectroscopy, 234–238
Response time improvement design, 68–75
Reticle, 479
Retrograde drain, 68
Return current, 295, 306
Reverse direction, 418
RHET, see Resonant tunneling hot-electron transistor
Ringing, 283, 287, 292–293

Ring oscillator, 176, 191
Rise length, 318–320
RLCG line model, 294–297
RLC line model, 283–294
ROM, 399
Row-address line, 428
RTBT, *see* Resonant tunneling bipolar transistor
Rutherford backscattering (RBS), 449

Saturation regime:
  current-voltage characteristics, 136, 144
  short-channel effects, 146
  space charge current, 149
Saturation velocity, 91, 129
Scaling:
  bipolar, 96–98
  current-voltage characteristics, 140
  field-effect transistor, 126–130
  MOSFET, 14, 15–19, 40
  synthetic neural systems, 426
Scanning system, 482–487
Schottky barrier diode (SBD), 98
Schottky barrier height, 134, 136, 140
Secondary ion mass spectroscopy (SIMS), 452
Segregation, 459
Self-aligned contact, 39
Self-aligned device:
  bipolar transistor, 105
  high electron mobility transistor, 188–190
  output conductance, 147–148, 150
  rapid thermal processing, 455
  short-channel effect, 151
Sematech, 2
Semiconductor Research Corporation, 2
Semi-insulating substrate, 124–125, 146
Sensitivity, 72
Sequential operation, 422
Sequential (incoherent) resonant tunneling, 208–210
Serial implementation, 422
Serial-in parallel-out shift register, 428
Serial-parallel implementation, 422–423
Series resistance:
  CMOS design, 37
  hot-carrier reliability design, 64, 65, 67
  MOSFET, 40, 63, 76
  response time improvement design, 70, 72–73, 75
Shallow emitter, 103–104
Shallow junction, 20
Sheet resistance, 441, 446, 458, 462
Shockley–Read–Hall (SRH) recombination current, 101

Short-channel device, 145–146, 150
Short-channel effect(s):
  CMOS design, 35
  output conductance, 150
  response time improvement design, 72
  VLSI and, 379
Sidewall, 150, 152
Sigmoid, 414
Signal flow graph, 422
Signal processing, 354–355
Signal transfer, 425
Silicide, 455, 460–462
Silicide on lightly doped drain structure, 70
Silicon, 125
  boron and, 451
  rapid thermal processing, 434–465
Silicon bipolar transistor, *see* Bipolar transistor
Silicon nitride, 125
Silicon-on-insulator (SOI) technology, 72
Simple model, 14
Simulation, 76
Simulation-based studies, 421
SiO_2, 442, 444
$Si-SiO_2$, 441, 445
$Si-TiSi_2$, 455
Skin depth, 305
Skin depth region, 296
Skin effect, 303, 305
Skin-effect layer, 296, 297
Slow crosstalk, 314
Slow-wave regime, 295, 296, 297
Slow-wave structure, 306
Sodium, 445
Soft error, 113
Solid-phase epitaxy (SPE), 446
SOLID structure, *see* Silicide on lightly doped drain structure
Space charge injection, 146, 147
Space charge injection current, 148–149
Space charge limited current, 150
Spatial coherence, 474
Spatial incoherence, 475
Spatial partial coherence, 476
Special-purpose accelerator, 423
Spectral domain method, 306
Speed, 40–41
Spherical aberration, 483
Spike-notch structure, 157
Spreading resistance, 37
Square law, 135
Stability, 416, 418
Standby power, 22
Stark effect transistor (SET), 259

Stark ladder, 374
State transition, 420
Static RAM, 184, 185, 194, 196, 197
Statistical mechanics, 395
Steady transport, 369–371
Step response, 287
STL bipolar circuit, 109
Storage ring source, 493–494
Straggle, 496
Strain, 143
Stress-induced electric field, 150
Submicron transistor(s), 122–168. *See also*
    Field-effect transistor
  current-voltage characteristics, 131–146
  field-effect transistor scaling considera-
      tions, 126–130
  overview of, 122–126
  short–channel effects in, 146–152
  synthetic neural system, 425, 426
  vertical device structures, 153–165
  VLSI technology, 353
Substrate current, 39, 52–54, 60–61
Substrate doping, 20, 31–32
Substrate injection, 126
Subthreshold region, 427
Subthreshold scaling, 15, 17–19
Subthreshold slope, 51
Subthreshold turn-off, 17
Summation circuit, 427
Supercomputer, 413
Superlattice, *see also* Lateral surface super-
    lattice
  concept of, 360
  current-voltage characteristics, 133
  heterojunction bipolar junction transistor,
      162
  high electron mobility transistor, 176
  lateral surface superlattice, 363
Surface channel, 23, 62
Surface recombination velocity, 103
Switching event, 127
Symmetric CMOS design, 23–37
Symmetric technology, 27
Synapse, 392, 401, 413
Synaptic connection, 419
Synaptic strength, 418
Synchrotron, 493
Synthetic neural system (SNS), 413, 432
  applications, 431–432
  basic models, 414–416
  cellular array, 387, 388
  chip architecture, 398, 399, 402, 405–406
  implementation, 421–431
  network state evaluation, 419–420

nonrecurrent, 416–418
rationale for, 413–414
recurrent, 418–419
technological constraints, 425–426
two-dimensional automata, 378, 388–393,
    396
VLSI design, 354, 426–431
Systolic array, 348, 397

Tailored-barrier-height (TBH), 145
TEM, 454
Tensile stress, 150
Terminal conditions, 280–282
Termination, 301
T-gate, 150, 152
Thermal conductivity, 126, 425
Thermal field emitter (TFE), 482–483
Thermal noise, 377
Thermal resistance, 272
Thermal stress, 441
Thermionic emission, 58, 59, 103
Thermionic injection, 104
Thermocouple, 439–440
Thin base, 99–103
Three-dimensional wiring, 274
Threshold degradation, 60
Thresholding operation, 421
Threshold logic, 414
Threshold slope, 23–24
Threshold variation, 34–35, 37
Threshold voltage:
  CMOS design, 31
  current-voltage characteristics, 135, 136,
      140
  gate length dependence, 150–152
  hot-carrier degradation, 51
  NMOS design, 20
  output conductance, 149
  synthetic neural system, 426
  unipolar vertical device, 163
Threshold voltage shift, 132
Time-dependent-dielectric-breakdown, 51–52
Time-domain analysis, 305, 306
Time-reversal symmetry, 395
Titanium silicide, 460–461
Trade-off:
  CMOS design, 33
  crosstalk, 321
  interconnection, 270, 276
  MOSFET, 75, 76
  response time improvement design, 70, 72,
      74
  subthreshold scaling, 19
  synthetic neural system, 426

Transconductance:
  current-voltage characteristics, 133, 136,
    139, 140, 142, 144
  hot-carrier degradation, 51
  MOSFET, 40
  response time improvement design, 69
  synthetic neural system, 426
Transfer function:
  pulse propagation, 309
  RLC line circuits, 285–286, 288, 291, 293
  synthetic neural system, 415–416
Transient, 449
Transient-enhanced diffusion, 451
Transient temperature, 440
Transition matrix, 406
Transition probability, 395
Transit time, 159
Transmission electron microscopy, 444
Transmission line, 276
Transmission line equation, 287, 288, 299
Transport, 42–51
Transport theory, 369–374
Transverse field, 299
Traps, 149–150
Traveling salesman problem, 398, 431
Trilayer resist, 477
TTL bipolar circuit, 109
Tungsten lamp, 435–436, 439
Tungsten plate, 464
Tunneling, 58, 72, *See also* Resonant tunnel-
  ing
Tunneling emitter bipolar transistor (TEBT),
  161–162
Tunneling process, 374
Turing machine, 395
Two-dimensional automata, 377–410
  chip architecture, 397–409
  future prospects, 409–410
  generally, 377–379
  information/interconnections, 383–385
  massively interconnected systems, 387–397
  reliability, 385–386
  Rent's rule, 379–383
  VLSI and, 379, 386
Two-dimensional gas, 131
  current-voltage characteristics, 136, 137
  unipolar vertical device, 165
Two-dimensional modeling, 146–147
Two-particle distribution function, 48

Ultrahigh-speed HEMT LSI circuit, *see* High
  electron mobility transistor (HEMT)
Ultralarge scale integration (ULSI):
  cellular array, 388

chip architecture, 402
reliability, 385–386
Rent's rule, 381
synthetic neural system, 390
Ultratech stepper, 470
Underdamped circuit, 284
Unipolar vertical device structure, 162–165
Unpinned velocity saturation model, 136
UPMOS process, 71–72

Valence electron, 122
Van Hove singularity, 369
Variable-shaped beam, 485
Variation in threshold, *see* Threshold voltage;
  Threshold voltage shift
VBT, *see* Vertical ballistic transistor
Vector scan, 484
Velocity-field model, 42–46
Velocity-field relation, 45–46
Velocity overshoot, 73. *See also* Overshoot
Velocity profile, 135
Velocity of propagation, *see* Propagation
  velocity
Velocity saturation, 135
Vertical ballistic transistor (VBT), 125, 163
Vertical field-effect transistor (VFET),
  153–165
VHSIC chip, 379
VLSI technology, 332–356
  applications of, 353–355
  balance considerations, 351–353
  cellular array, 388
  chip architecture, 398, 399–400, 401
  coherency in, 386
  computer architecture classification,
    339–341
  CPU requirements, 341–351
  custom realization, 423
  GaAs digital Ics, 353
  generally, 332–334
  high electron mobility transistor, 176, 177,
    184–190
  historical perspective on, 334–339
  interconnections, 269
  massively interconnected systems, 395
  MOSFET, 10
  rationale for, 379
  RC line models, 277
  reliability, 385–386
  synthetic neural networks, 390–391,
    413–432
  two-dimensional automata, 377–410
Vision, *see* Computer vision

von Neumann architecture, 335
von Neumann bottleneck, 339–340

Wafer scale integration, 409
Warnier–Orr diagram, 10, 11, 14, 75
Wire capacitance, 98–99
Wire length distribution, *see* Distribution of
    wire length

Wiring track, 108

X-ray lithography, 488–494

Z-component, 300–303